Intermediate Algebra
Programmed

THOMAS J. McHALE
ALLAN A. CHRISTENSON
KEITH J. ROBERTS

ADDISON-WESLEY PUBLISHING COMPANY

Reading, Massachusetts • Menlo Park, California
Don Mills, Ontario • Wokingham, England • Amsterdam
Sydney • Singapore • Tokyo • Madrid • San Juan

Library of Congress Cataloging-in-Publication Data

McHale, Thomas J., 1931-
Intermediate algebra-programmed.

1. Algebra--Programmed instruction. I. Christenson, Allan A. II. Roberts, Keith J. III. Title.
QA154.2.M395 1986 512'.07'7 86-10918
ISBN 0-201-15880-9

Reprinted with corrections, May 1988

Reproduced by Addison-Wesley from camera-ready copy supplied by the author.

ISBN 0-201-15880-9
10 - CRS - 98 97 96 95 94

Preface

Intermediate Algebra is written for students who just completed a first course in algebra or who completed a first course in the past. Though it reviews all topics, it proceeds at a faster pace than Introductory Algebra and covers the usual topics of intermediate algebra. It uses a unique approach to deal with the learning difficulties experienced by many students. That is, the content is presented in an interactive format based on a learning task analysis, and the learning is supported by a set of diagnostic and criterion tests. This unique approach, which differs significantly from that of standard texts, has been used successfully with almost 500,000 students in similar courses.

LEARNING TASK ANALYSIS

A learning task analysis was done by the authors to determine how to present the content of the course. That is, each course objective was analyzed to identify the skills, concepts, and procedures needed by the students to master all of the problems covered by that objective. Then based on that analysis, the content of each objective is presented by a carefully planned sequence of worked-out examples proceeding from the simple to the complex. Worked-out examples are given for the full range of problems covered by each objective. In that sense, there are no "gaps" in the instruction from the student's point of view. Because a learning task analysis was used, this book has many more worked-out examples than comparable texts.

INTERACTIVE FORMAT

The format of this book is unique because it forces the students to interact with the content as it is presented. The significant features of the format are described below.

Frames. The content is presented in a step-by-step manner with each step called a frame. Each frame begins with some instruction, including one or more examples, and ends with some problem or problems that the students must do. Since answers for each frame are given to the right of the following frame, the students are given immediate feedback as they proceed from frame to frame. Because they are continually responding and getting feedback, the students can monitor their own learning process.

Assignment Self-Tests. Each chapter is divided into a number of assignments which can ordinarily be covered in one class period. A self-test is provided in the text at the end of each assignment. Students can either use the self-tests as assignment tests or save them for a chapter review after all assignments in the chapter are completed.

Supplementary Problems. Supplementary problems for each assignment are provided at the end of each chapter. They can be assigned as needed by instructors or simply used as further practice by students. Answers for all supplementary problems are given in the back of the book.

DIAGNOSTIC AND CRITERION TESTS

The text is accompanied by the book Tests For Intermediate Algebra which contains the various diagnostic and criterion tests described below. Answer keys are provided for all tests. The test book is provided only to instructors. Copies for student use must be made by some copying process.

Diagnostic Pre-Test. This test covers all chapters. It can be used either to simply get a measure of the entry skills of the students or as a basis for prescribing an individualized program.

Assignment Tests. Each chapter is divided into a number of assignments with a total of thirty-nine assignments in the complete text. After the students have completed each assignment and the assignment self-test (in the text), the assignment test (from the test book) can be administered and used as a basis for tutoring or assigning supplementary problems. The assignment tests are simply a diagnostic tool and need not be graded.

Chapter And Multi-Chapter Tests. After the appropriate assignments are completed, either a chapter test or a multi-chapter test can be administered. Ordinarily these tests should be graded. Three parallel forms are provided to facilitate the test administration, including the retesting of students who do not achieve a satisfactory score.

Comprehensive Test. The comprehensive test covers all chapters. Three parallel forms are provided. Since the comprehensive test is a parallel form of the diagmostic pre-test, the difference score can be used as a measure of each student's improvement in the course.

TEACHING MODES

This text can be used in various ways. Some possibilities are described below.

Lecture Class. The text can be used to reinforce lectures and to provide highly structured outside assignments related to the lectures.

Mini-Lecture Class. An instructor can give a brief lecture on difficult points in a completed assignment before the assignment test is administered. An instructor can also give a brief overview of each new assignment before it is begun by students.

No-Lecture Class. The text is ideally suited for a no-lecture class that is either paced or self-paced. Class time can then be used to administer tests and to tutor individual students when tutoring is needed.

Learning Laboratory. The text is also ideally suited for a learning laboratory where students proceed at their own pace. The instructor can manage the instruction by administering tests and tutoring when necessary.

ACKNOWLEDGMENTS

The authors wish to thank Jeffrey M. Pepper of the Addison-Wesley Publshing Company for many suggestions that improved this text. They also thank Arleen D'Amore who typed the camera-ready copy, Peggy McHale who prepared the drawings and made the corrections, and Gail W. Davis who did the final proofreading.

ASSIGNMENTS FOR INTERMEDIATE ALGEGBRA

Chapter 1:	#1	pp. 1-16
	#2	pp. 17-29
	#3	pp. 30-42
	#4	pp. 43-52
Chapter 2:	#5	pp. 58-72
	#6	pp. 73-88
	#7	pp. 89-100
	#8	pp. 101-113
Chapter 3:	#9	pp. 119-131
	#10	pp. 132-143
	#11	pp. 144-158
	#12	pp. 159-175
Chapter 4:	#13	pp. 180-191
	#14	pp. 192-206
	#15	pp. 207-218
	#16	pp. 219-229
Chapter 5:	#17	pp. 234-246
	#18	pp. 247-257
	#19	pp. 258-268
	#20	pp. 269-283
Chapter 6:	#21	pp. 288-302
	#22	pp. 303-315
	#23	pp. 316-329
Chapter 7:	#24	pp. 333-346
	#25	pp. 347-363
	#26	pp. 364-377
	#27	pp. 378-394
Chapter 8:	#28	pp. 402-414
	#29	pp. 415-428
	#30	pp. 429-440
Chapter 9:	#31	pp. 444-456
	#32	pp. 457-468
	#33	pp. 469-477
	#34	pp. 478-493
Chapter 10:	#35	pp. 500-514
	#36	pp. 515-528
	#37	pp. 529-542
Chapter 11:	#38	pp. 548-559
	#39	pp. 560-570

Contents

FOR STUDENTS

This textbook is written so that you can effectively learn on your own. Therefore, it is different than an ordinary math textbook. It is written in a programmed format. That is, the content is presented in a step-by-step manner with each step called a frame. Each frame begins with some instruction, including one or more examples, and ends with a problem or problems for you to do. Since answers for each frame are given to the right of the next frame, you will immediately know whether you are right or wrong.

Each page of the textbook has two columns. The left column contains the frames. The right column contains the answers. Remember that the answers are given to the right of the next frame. Some students find it helpful to cover the answers until they have done the problem or problems. After finishing a frame you should check your answers. If you are correct, go on to the next frame. If you are incorrect, begin by checking your work. If you can't get the correct answer, review the frame and some previous frames if necessary. If you still can't get the correct answer, make a note of the frame and page numbers and ask your instructor about it.

Each chapter is divided into a number of assignments. A self-test with answers is given after each assignment. Supplementary problems for each assignment are given at the end of each chapter. Answers for all supplementary problems are given in the back of the textbook.

Good luck!

Real Numbers

1

In this chapter, we will define real numbers and discuss the basic operations with real numbers. We will define powers with integral exponents and discuss the basic laws of exponents. We will discuss the proper order of operations for expressions involving more than one operation. We will also discuss algebraic expressions and the procedures used to simplify algebraic expressions.

1-1 SETS AND REAL NUMBERS

In this section, we will define sets and real numbers. We will also define the following subsets of the real numbers: natural numbers, whole numbers, integers, rational numbers, and irrational numbers.

1. A set is a collection of objects. The objects in a set are the elements or members of the set. To represent a set, we can list the elements and enclose them in braces { }. For example, the set of the first five letters of the alphabet is shown below.

{a,b,c,d,e }

The set of natural numbers or counting numbers is shown below. The three dots mean that the list goes on and on.

{1,2,3,4,5,6,7,...}

Write the set of the first four natural numbers below.

____

{1,2,3,4}

2. If all the elements of set A are also elements of set B, then set A is a <u>subset</u> of set B. For example, the set of the first five even natural numbers is shown below. It is a subset of the set of natural numbers. {2,4,6,8,10} Write each of the following subsets of the set of natural numbers. a) The set of the first four odd natural numbers. ________ b) The set of the natural numbers between 4 and 11. ________	
3. Instead of listing its elements, we can represent a set by using the variable <u>x</u> to state a rule. An example is shown. Notice how the rule is stated in words. {x|x is a natural number less than 6.} "The set of all x such that x is a natural number less than 6." We represented the same set by listing its elements below. {1,2,3,4,5} Represent each set below by listing its elements. a) {x|x is a natural number between 6 and 10.} ________ b) {x|x is a natural number greater than 4.} ________	a) {1,3,5,7} b) {5,6,7,8,9,10}
4. The rule below represents a set that contains no elements. {x|x is a natural number less than 1} A set with no elements is called the <u>empty</u> <u>set</u> or <u>null</u> <u>set</u>. Either { } or ∅ is used as the symbol for the empty set. Represent each set by listing its elements. a) {x|x is an odd natural number.} ________ b) {x|x is a negative natural number. } ________	a) {7,8,9} b) {5,6,7,8,9,...}
5. Any number that can be represented by a point on the number line is called a <u>real</u> <u>number</u>. For example, all of the numbers shown below are real numbers. $-3.4 \quad -\frac{2}{3} \quad .8 \quad \sqrt{5} \quad \frac{7}{2}$ −4 −3 −2 −1 0 1 2 3 4	a) {1,3,5,7,9,...} b) ∅ or { }

Continued on following page.

5. Continued

From the number line, you can see these facts:

1. Positive numbers go to the right of 0.
2. Negative numbers go to the left of 0.
3. The number 0 is neither positive nor negative.

Positive and negative numbers are called signed numbers.

For negative numbers, we always use the - sign.

"Negative 3" is always written -3.

For positive numbers, we can use the + sign or no sign, but we usually use no sign.

"Positive 5" is usually written 5 instead of +5.

a) Instead of writing +9 for "positive 9", we usually write ______.

b) Would it make sense to write either +0 or -0? ________

---

6. There are various subsets of the real numbers. Two subsets are:

The set of natural numbers or counting numbers.

{1,2,3,4,5,6,7,...}

The set of whole numbers, which includes the natural numbers and 0.

{0,1,2,3,4,5,6,7,...}

Which number is a whole number, but not a natural number? _____

Answers to frame 5: a) 9 b) No. The number 0 is not a signed number.

---

7. Another subset of the real numbers is the set of integers which is shown below. Integers include the whole numbers plus negative numbers like -1, -2, -3, and so on.

{...-3,-2,-1,0,1,2,3,...}

a) Is 4 both a whole number and an integer? ________

b) Is 0 both a whole number and an integer? ________

c) Is -5 both a whole number and an integer? ________

Answer to frame 6: 0

---

8. Another subset of the real numbers is the set of rational numbers which is shown below. A rational number is a number that can be expressed in the form $\frac{a}{b}$, where a and b are integers and b is not 0.

$$\left\{\frac{a}{b} \,\middle|\, \text{a and b are integers, and b is not 0}\right\}$$

Continued on following page.

Answers to frame 7: a) Yes b) Yes c) No. -5 is an integer, but not a whole number.

8. Continued

All of the following are rational numbers because each is a division of integers with a non-zero denominator.

$\frac{1}{2}$ $\frac{7}{6}$ $\frac{-4}{5}$ $\frac{9}{-2}$ $\frac{256}{87}$

Any negative fraction is a rational number because it can be written as a division of integers with a non-zero denominator. For example:

$-\frac{5}{6}$ can be written $\frac{-5}{6}$ or $\frac{5}{-6}$

$-\frac{13}{9}$ can be written _____ or _____

---

9. To convert a rational number to a decimal, we divide. Two examples are shown. Notice that we get either a terminating decimal (.625) or a nonterminating decimal with a repeating pattern (.3636...).

Answer: $\frac{-13}{9}$ or $\frac{13}{-9}$

```
      .625                    .3636...
5                  4
- = 8)5.000        -- = 11)4.0000
8     -4 8         11      -3 3
        20                   70
       -16                  -66
         40                   40
        -40                  -33
                               70
                              -66
                                4
```

For non-terminating decimals with a repeating pattern, the three dots are often replaced by a bar over the repeating part. For example:

Instead of .3636..., we write $.\overline{36}$.

In applied problems, we usually round a repeating pattern to a specific place.

Rounded to thousandths, $\frac{4}{11}$ = _______

---

10. A fraction is a proper fraction if its numerator is smaller than its denominator.

Answer: .364

$\frac{2}{3}$, $\frac{4}{11}$, and $\frac{1}{4}$ are proper fractions.

A fraction is an improper fraction if its numerator is equal to or larger than its denominator.

$\frac{7}{4}$, $\frac{15}{5}$, and $\frac{8}{8}$ are improper fractions.

Continued on following page.

10. Continued

When an improper fraction converts to a decimal, the decimal is greater than or equal to "1". An example is shown. Complete the other conversion. Round to hundredths.

$$\frac{9}{4} = 4\overline{)9.00} = 2.25 \qquad \begin{array}{r} -8 \\ \hline 10 \\ -8 \\ \hline 20 \\ -20 \\ \hline \end{array} \qquad\qquad \frac{7}{6} =$$

---

11. Any decimal is a rational number because it can be converted to a fraction. For example:

$2.4 = \frac{24}{10}$ (One decimal place in 2.4; one 0 in 10.)

$.61 = \frac{61}{100}$ (Two decimal places in .61; two 0's in 100.)

Convert each decimal to a fraction.

a) 13.7 = ______ b) 4.19 = ______ c) .854 = ______

Answer to 10: 1.17

---

12. Any mixed number is a rational number because it can be converted to an improper fraction. Two examples are shown. (Note: In algebra, we usually use improper fractions instead of mixed numbers.)

$$1\frac{2}{5} = 1 + \frac{2}{5} = \frac{5}{5} + \frac{2}{5} = \frac{7}{5} \qquad\qquad 3\frac{1}{2} = 3 + \frac{1}{2} = \frac{6}{2} + \frac{1}{2} = \frac{7}{2}$$

Convert each mixed number to an improper fraction.

a) $1\frac{3}{4}$ = ______ b) $2\frac{1}{3}$ = ______ c) $4\frac{1}{2}$ = ______

Answers to 11: a) $\frac{137}{10}$ b) $\frac{419}{100}$ c) $\frac{854}{1000}$

---

13. Any integer is a rational number because it can be written as a division of itself and "1". For example:

$8 = \frac{8}{1}$ $-3 = \frac{-3}{1}$ a) 140 = ______ b) -79 = ______

Answers to 12: a) $\frac{7}{4}$ b) $\frac{7}{3}$ c) $\frac{9}{2}$

Answers to 13: a) $\frac{140}{1}$ b) $\frac{-79}{1}$

14. Rational numbers are one major subset of the real numbers. Irrational numbers are the other major subset of the real numbers. The set of irrational numbers includes all real numbers that are not rational.

$$\{x|x \text{ is a real number that is not rational}\}$$

Any nonterminating decimal without a repeating pattern is an irrational number. For example, the number π is an irrational number.

$$\pi = 3.1415926535\ldots$$

State whether each number is rational or irrational.

a) 0.125 (terminating)

b) 3.424242... (repeating pattern)

c) 4.030030003... (pattern does not repeat)

---

15. When a number is a perfect square, its square root is rational because it is an integer. For example:

a) rational

b) rational

c) irrational

$$\sqrt{36} \text{ is rational, because } \sqrt{36} = 6$$

When a number is not a perfect square, its square root is irrational because it is a nonterminating, nonrepeating decimal.

$$\sqrt{15} \text{ is irrational, because } \sqrt{15} = 3.8729833\ldots$$

State whether each number is rational or irrational.

a) $\sqrt{3}$ b) $\sqrt{16}$ c) $\sqrt{49}$ d) $\sqrt{88}$

---

16. Some subsets of the real numbers are listed below.

a) irrational

b) rational

c) rational

d) irrational

Subsets Of The Real Numbers

Natural numbers or counting numbers	{1,2,3,4,5,6,7,...}
Whole numbers	{0,1,2,3,4,5,6,7,...}
Integers	{...-3,-2,-1,0,1,2,3,...}
Rational numbers	$\left\{\frac{a}{b} \middle| \begin{array}{l}\text{a and b are integers} \\ \text{and b is not 0}\end{array}\right\}$
Irrational numbers	{x|x is a real number that is not rational}

Continued on following page.

16. Continued

The diagram below shows that real numbers are either rational or irrational.

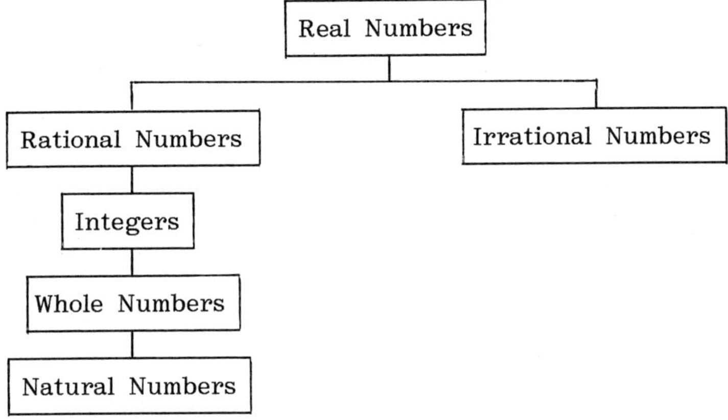

1-2 INEQUALITIES AND ABSOLUTE VALUE

In this section, we will define inequalities and absolute value. The graphs of sets containing inequalities are discussed.

17. The symbol = is used for equalities. The symbol ≠ is used for inequalities. The two symbols are defined below. Notice that the slash line in ≠ means "not".

= means "is equal to".
≠ means "is not equal to".

Therefore: 5 = 5 is read "5 is equal to 5".
4 ≠ 7 is read "4 is not equal to 7".

In an inequality, one number must be either greater than the other or less than the other. The two symbols > and < are used for "is greater than" and "is less than". That is:

> means "is greater than".
< means "is less than".

Therefore: 7 > 4 is read "7 is greater than 4".
4 < 7 is read "4 is less than 7".

Write either > or < in each blank.

a) 6___9 b) 10___4 c) 1___5

18. The definitions of "is greater than" and "is less than" in terms of the number line are given below.

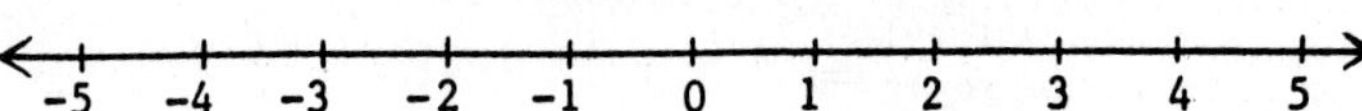

One number is greater than another if it is to the right of the other on the number line. That is:

$a > b$ if a is to the right of b.

Therefore: since 4 is to the right of -1, $4 > -1$.
since -2 is to the right of -5, $-2 > -5$.

One number is less than another if it is to the left of the other on the number line. That is:

$a < b$ if a is to the left of b.

Therefore: since -2 is to the left of 3, $-2 < 3$.
since -4 is to the left of 0, $-4 < 0$.

Write either $>$ or $<$ in each blank.

a) 7___-4 d) 0___-7

b) -5___0 e) -6___-2

c) -8___9 f) -4.5___-9.5

a) $6 < 9$

b) $10 > 4$

c) $1 < 5$

---

19. The symbols $\geq$ and $\leq$ are defined below.

$\geq$ means "is greater than or equal to".
$\leq$ means "is less than or equal to".

Inequalities containing $\geq$ or $\leq$ can be either true or false. For example:

$4 \geq 2$ is true, since $4 > 2$ is true.

$-3 \leq -3$ is true, since $-3 = -3$ is true.

$0 \leq -4$ is false, since both $0 < -4$ and $0 = -4$ are false.

Answer either true or false for these.

a) $2 \geq 2$ b) $3 \leq 0$ c) $-4 \leq 0$ d) $-5.2 \geq -4.6$

a) $7 > -4$

b) $-5 < 0$

c) $-8 < 9$

d) $0 > -7$

e) $-6 < -2$

f) $-4.5 > -9.5$

20. The symbols $\ngtr$ and $\nless$ are defined below. Notice again that the slash line means "not".

> $\ngtr$ means "is not greater than".
>
> $\nless$ means "is not less than".

Inequalities containing $\ngtr$ or $\nless$ can be either true or false. For example:

$5 \ngtr 7$ is true, since $5 > 7$ is false.

$-4 \nless -2$ is false, since $-4 < -2$ is true.

Answer either true or false for these.

a) $3 \ngtr -2$ b) $-1 \nless -6$ c) $0 \ngtr 5$ d) $-4.3 \nless 0$

a) True
b) False
c) True
d) False

---

21. In the sets $\{x|x < 3\}$ and $\{x|x \leq 3\}$, the rule is an inequality. We can graph each set on the number line.

The set $\{x|x < 3\}$ includes all real numbers less than 3. Its graph is the heavy line to the left of 3 below. The open circle at 3 means that 3 is not included.

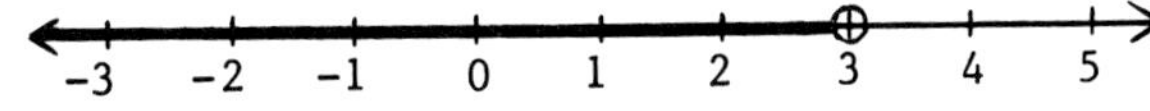

The set $\{x|x \leq 3\}$ includes 3 and all real numbers less than 3. Its graph is the heavy line to the left of 3 below. The closed circle at 3 means that 3 is included.

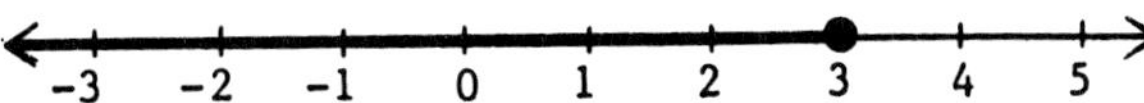

Graph each set below.

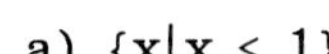

a) $\{x|x < 1\}$

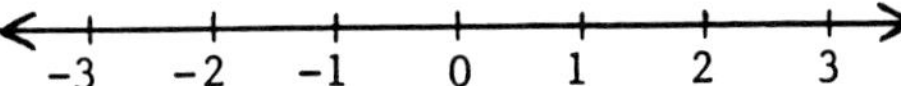

b) $\{x|x \leq -2\}$

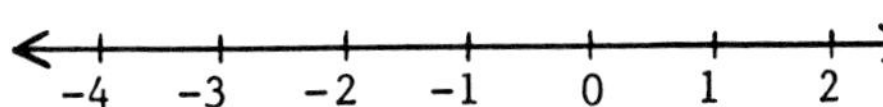

a) False
b) True
c) True
d) False

---

22. We graphed $\{x|x > -2\}$ and $\{x|x \geq -2\}$ below. The only difference is the open or closed circle at -2.

$\{x|x > -2\}$

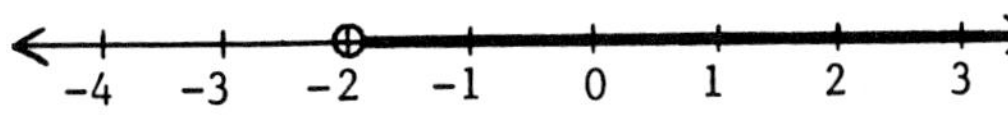

$\{x|x \geq -2\}$

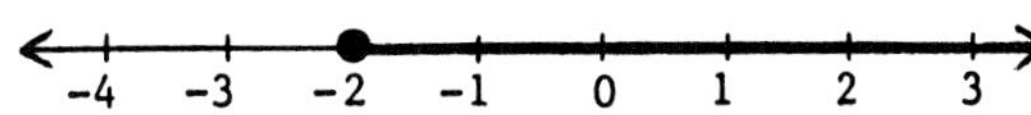

Continued on following page.

a) 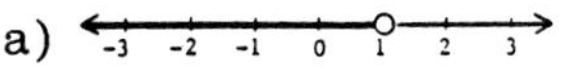

b)

22. Continued

Graph each set below.

a) $\{x|x > 0\}$

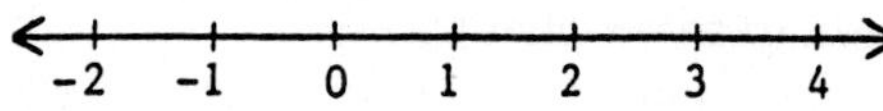

b) $\{x|x \geq -3\}$

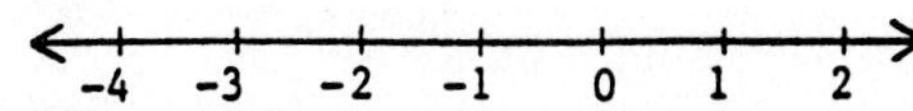

---

23. The inequality $-3 < x < 2$ is a combined inequality.

$-3 < x < 2$ is read "$\underline{x}$ is greater than -3 and less than 2".
or "$\underline{x}$ is between -3 and 2".

The set $\{x|-3 < x < 2\}$ includes all real numbers between -3 and 2. Its graph is shown below. The open circles at -3 and 2 mean that those two numbers are not included.

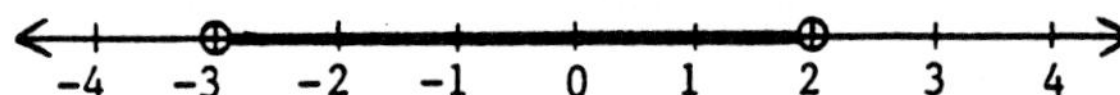

Graph $\{x|0 < x < 5\}$ below.

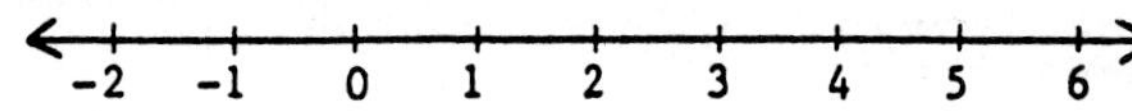

Answers to 22: a) b)

---

24. The inequality $-2 < x \leq 4$ is also a combined inequality.

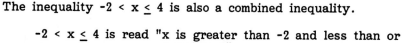

$-2 < x \leq 4$ is read "x is greater than -2 and less than or equal to 4".

or "x is between -2 and 4, including 4".

The set $\{x|-2 < x \leq 4\}$ is graphed below. The open circle at -2 means that -2 is not included. The closed circle at 4 means that 4 is included.

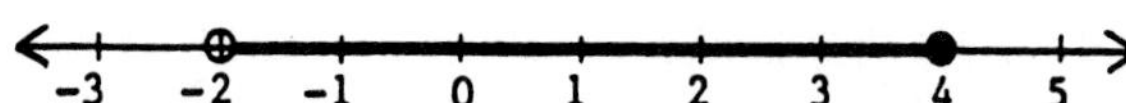

Graph each set below.

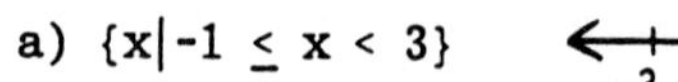

a) $\{x|-1 \leq x < 3\}$

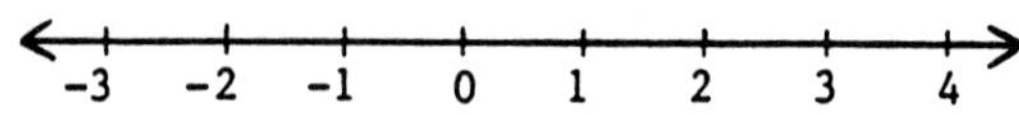

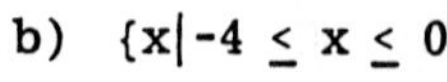

b) $\{x|-4 \leq x \leq 0\}$

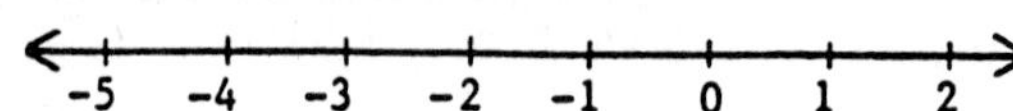

---

a) 

b)

25. The absolute value of a number is its distance from 0 on the number line, with no regard for direction. The symbol | | is used for absolute value.

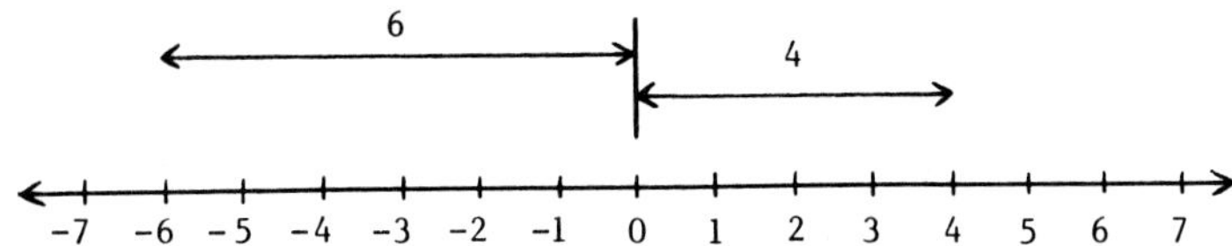

The absolute values of 4 and -6 are shown on the number line above.

Since 4 is 4 units from 0, $|4| = 4$

Since -6 is 6 units from 0, $|-6| = 6$

Write the absolute value of each number.

a) $|10| =$ ______ c) $|-1| =$ ______

b) $\left|\frac{3}{5}\right| =$ ______ d) $|-9.7| =$ ______

Answers: a) 10 c) 1 b) $\frac{3}{5}$ d) 9.7

26. Since both 3 and -3 are 3 units from 0, they have the same absolute value. That is:

$|3| = 3$ and $|-3| = 3$

The absolute value of 0 is 0. Complete these:

a) $|-12| =$ ______ c) $|0| =$ ______

b) $|12| =$ ______ d) $|-45| =$ ______

Answers: a) 12 c) 0 d) 12 d) 45

1-3 ADDITION AND SUBTRACTION

In this section, we will discuss the procedures for adding and subtracting real numbers. We will also define additive inverses or opposites.

---

27. The following two rules are used to add real numbers.

> **Rules For Adding Real Numbers**
>
> 1. When the numbers have the same sign:
>   a) Add their absolute values.
>   b) Give the sum the same sign as both numbers.
> 2. When the numbers have different signs:
>   a) Subtract their absolute values.
>   b) Give the sum the same sign as the number with the larger absolute value.

An example of the first rule is shown below.

-2 + (-4) = -6
- Sum of absolute values (2 + 4 = 6)
- Same sign as both numbers.

Two examples of the second rule are shown below.

5 + (-2) = +3
- Difference of absolute values (5 - 2 = 3)
- Sign of number with the larger absolute value (5)

-9 + 5 = -4
- Difference of absolute values (9 - 5 = 4)
- Sign of number with the larger absolute value (-9)

Use the rules for these:

a) -3 + (-7) = _____   c) 4 + (-9) = _____

b) -10 + (-2) = _____   d) -3 + 8 = _____

---

28. Complete: a) -40 + 10 = _____   c) -20 + (-70) = _____

b) 80 + (-55) = _____   d) -30 + 90 = _____

Answers (27): a) -10 c) -5 b) -12 d) 5

Answers (28): a) -30 c) -90 b) 25 d) 60

29. The same rules are used to add fractions. For example:

$$-\frac{2}{5} + \left(-\frac{4}{5}\right) = -\frac{6}{5}$$

$$-\frac{5}{6} + \frac{1}{3} = -\frac{5}{6} + \frac{2}{6} = -\frac{3}{6} = -\frac{1}{2}$$

Add. Reduce each sum to lowest terms.

a) $-\frac{1}{4} + \left(-\frac{1}{4}\right) =$ ______ c) $\frac{3}{8} + \left(-\frac{1}{4}\right) =$ ______

b) $-\frac{5}{2} + \left(-\frac{1}{2}\right) =$ ______ d) $-\frac{9}{10} + \frac{1}{2} =$ ______

---

30. The same rules are used to add decimals. For example:

-10.2 + (-20.5) = -30.7

-.7 + .55 = -.70 + .55 = -.15

Complete these:

a) -6.6 + (-9.9) = ______ c) -.18 + .129 = ______

b) 5.6 + (-3.1) = ______ d) 1 + (-.4067) = ______

Answers to 29: a) $-\frac{1}{2}$ c) $\frac{1}{8}$ b) -3 d) $-\frac{2}{5}$

---

31. To add three or more numbers, we can add "two at a time" from left to right as we have done below.

3 + (-5) + 6 + (-8) =

-2 + 6 + (-8) =

4 + (-8) = -4

By adding "two at a time", find each sum.

a) -5 + (-4) + 10 = b) 20 + (-10) + (-30) + 5 =

Answers to 30: a) -16.5 c) -.051 b) 2.5 d) .5933

---

32. Two numbers that are the same distance from 0 on the number line, but on opposite sides of 0, are called <u>additive</u> <u>inverses</u> or <u>opposites</u> of each other. For example:

-8 is the additive inverse of 8.

5 is the additive inverse of -5.

Continued on following page.

Answers to 31: a) 1 b) -15

32. Continued

The additive inverse of any number a is -a. That is:

The additive inverse of 7 is -7.
The additive inverse of -3 is -(-3) = 3.
The additive inverse of 0 is -0 = 0.

Write the additive inverse of each number.

a) -10 _____ b) 17 _____ c) $\frac{3}{4}$ _____ d) -6.5 _____

---

33. When two additive inverses are added, their sum is 0. That is:

4 + (-4) = 0 a) -1.75 + 1.75 = _____ b) $\frac{7}{8} + \left(-\frac{7}{8}\right) =$ _____

Answers to 32: a) 10 b) -17 c) $-\frac{3}{4}$ d) 6.5

---

34. If we subtract 2 from 5, we get 3. That is, 5 - 2 = 3. We also get 3 as an answer if we add -2 to 5. That is, 5 + (-2) = 3. Therefore:

5 - 2 = 5 + (-2)

Based on the example above, we can define subtraction for any real numbers a and b. Notice that subtracting b from a is the same as adding the additive inverse of b to a.

a - b = a + (-b)

Therefore, to perform a subtraction, we convert to an equivalent addition and then add. To convert to addition, we ADD THE ADDITIVE INVERSE OF THE SECOND NUMBER. For example:

Change - to +
Additive inverse of 7

4 - 7 = 4 + (-7) = -3

Convert to an equivalent addition and then add.

a) 4 - 10 = ____________________ = _____

b) -6 - 1 = ____________________ = _____

c) -1 - 3 = ____________________ = _____

Answers to 33: a) 0 b) 0

---

35. To perform the subtraction below, we also ADDED THE ADDITIVE INVERSE OF THE SECOND NUMBER.

Change - to +
Additive inverse of -3

6 - (-3) = 6 + 3 = 9

Continued on following page.

Answers to 34: a) 4 + (-10) = -6 b) -6 + (-1) = -7 c) -1 + (-3) = -4

35. Continued

Convert to an equivalent addition and then add.

a) $5 - (-2) =$ ______________ = ______

b) $-1 - (-6) =$ ______________ = ______

c) $-9 - (-3) =$ ______________ = ______

---

36. For each subtraction below, we converted to an equivalent addition and then added.

$$\frac{3}{8} - \frac{7}{8} = \frac{3}{8} + \left(-\frac{7}{8}\right) = -\frac{4}{8} = -\frac{1}{2}$$

$$-\frac{1}{4} - \frac{3}{2} = -\frac{1}{4} + \left(-\frac{3}{2}\right) = -\frac{1}{4} + \left(-\frac{6}{4}\right) = -\frac{7}{4}$$

Use the same method for these.

a) $-\frac{1}{5} - \frac{9}{5} =$ ______________________

b) $\frac{1}{6} - \frac{1}{2} =$ ______________________

Answers to 35: a) $5 + 2 = 7$ b) $-1 + 6 = 5$ c) $-9 + 3 = -6$

---

37. For each subtraction below, we converted to an equivalent addition and then added.

$$-\frac{5}{4} - \left(-\frac{1}{4}\right) = -\frac{5}{4} + \frac{1}{4} = -\frac{4}{4} = -1$$

$$\frac{1}{4} - \left(-\frac{7}{12}\right) = \frac{1}{4} + \frac{7}{12} = \frac{3}{12} + \frac{7}{12} = \frac{10}{12} = \frac{5}{6}$$

Use the same method for these.

a) $\frac{11}{16} - \left(-\frac{3}{16}\right) =$ ______________________

b) $-\frac{5}{12} - \left(-\frac{1}{3}\right) =$ ______________________

Answers to 36: a) -2 b) $-\frac{1}{3}$

---

38. Following the examples, complete the other subtractions.

$$1.4 - 8.5 = 1.4 + (-8.5) = -7.1$$

$$-.25 - (-.32) = -.25 + .32 = .07$$

a) $-3.7 - 9.9 =$ ______________________

b) $1 - (-.55) =$ ______________________

Answers to 37: a) $\frac{7}{8}$ b) $-\frac{1}{12}$

Answers to 38: a) -13.6 b) 1.55

SELF-TEST 1 (pages 1-16)

Given this set of numbers: $\{-9, -\sqrt{2}, -\frac{3}{4}, 0, \frac{1}{2}, 2, \sqrt{3}, \frac{5}{2}, 6.4\}$

1. List the integers. ________ 2. List the irrational numbers. ________

Represent each set by listing its elements.

3. $\{x|x$ is an even natural number less than $10\}$ ________________

4. $\{x|x$ is an integer between 5 and $9\}$ ________________

Write either > or < in each blank.

5. -3 ___ -7 6. -4 ___ 0

Answer "true" or "false" for these.

7. $10 \geq 12$ 8. $9 \nless 4$

Graph each set.

9. $\{x|x \geq -1\}$

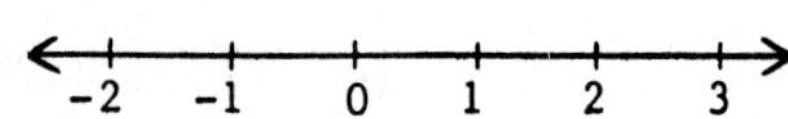

10. $\{x|-2 \leq x < 1\}$

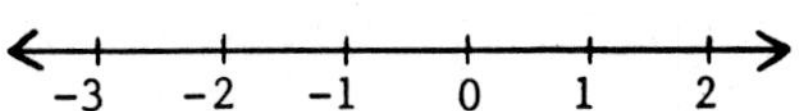

Find each absolute value.

11. $|15| =$ ______ 12. $|-2| =$ ______

Write the additive inverse of each number.

13. 9 ______ 14. -20 ______

Do each addition and subtraction.

15. $-10 + 2 =$ ______

16. $-6.5 + (-2.4) =$ ______

17. $8 + (-5) + 0 + (-1) =$ ______

18. $7 - (-3) =$ ______

19. $\frac{1}{2} - \frac{3}{4} =$ ______

20. $-5 - (-1) - 3 =$ ______

ANSWERS:
1. $-9, 0, 2$
2. $-\sqrt{2}, \sqrt{3}$
3. $\{2,4,6,8\}$
4. $\{6,7,8\}$
5. >
6. <
7. false
8. true
9. (graph)
10. (graph)
11. 15
12. 2
13. -9
14. 20
15. -8
16. -8.9
17. 2
18. 10
19. $-\frac{1}{4}$
20. -7

1-4 MULTIPLICATION AND DIVISION

In this section, we will discuss the procedures for multiplying and dividing real numbers. We will also define reciprocals or multiplicative inverses.

---

39. In a multiplication, the numbers multiplied are called factors; the answer is called the product. The following two rules are used to multiply real numbers.

> **Rules For Multiplying Real Numbers**
>
> 1. When the factors have the same sign, their product is positive.
> 2. When the factors have different signs, their product is negative.

Two examples of the first rule are shown below.

$5(4) = 20$ $\qquad$ $(-10)(-7) = 70$

Two examples of the second rule are shown below.

$3(-8) = -24$ $\qquad$ $(-9)(2) = -18$

Use the rules for these:

a) $8(-7) = $ ______ $\qquad$ c) $(-2)(40) = $ ______

b) $(-6)(-9) = $ ______ $\qquad$ d) $(-20)(-5) = $ ______

---

40. We used the same rules for the multiplications below.

$\left(-\frac{2}{3}\right)\left(-\frac{1}{5}\right) = \frac{2}{15}$ $\qquad$ $4\left(-\frac{1}{6}\right) = -\frac{4}{6} = -\frac{2}{3}$

$-8(-1.2) = 9.6$ $\qquad$ $(-.9)(7.5) = -6.75$

Complete these:

a) $\left(-\frac{5}{2}\right)\left(\frac{2}{7}\right) = $ ______ $\qquad$ c) $\left(-\frac{3}{5}\right)(-2) = $ ______

b) $(-7)(-.21) = $ ______ $\qquad$ d) $10(-67.5) = $ ______

Answers to 39: a) -56 b) 54 c) -80 d) 100

Answers to 40: a) $-\frac{5}{7}$ b) 1.47 c) $\frac{6}{5}$ d) -675

41. The rules for dividing real numbers are similar to the rules for multiplying real numbers.

> Rules For Dividing Real Numbers
>
> 1. When the numbers have the same sign, their quotient is positive.
> 2. When the numbers have different signs, their quotient is negative.

Two examples of the first rule are shown below.

$\frac{35}{7} = 5$ $\qquad\qquad$ $\frac{-12}{-4} = 3$

Two examples of the second rule are shown below.

$\frac{-32}{8} = -4$ $\qquad\qquad$ $\frac{14}{-2} = -7$

Complete these:

a) $\frac{-10}{5} =$ _____ $\qquad$ b) $\frac{-56}{-7} =$ _____ $\qquad$ c) $\frac{28}{-4} =$ _____

---

a) -2

b) 8

c) -7

42. When a division is done correctly, the product of the denominator and the quotient equals the numerator. For example:

$\frac{12}{3} = 4$, since $3(4) = 12$

The fact above can be used to justify the rules for dividing real numbers. That is:

$\frac{-20}{4} = -5$, since $4(-5) = -20$ $\qquad$ $\frac{-18}{-6} = 3$, since $-6(3) =$ _____

-18

43. Two numbers whose product is "1" are called reciprocals or multiplicative inverses of each other. For example:

Since $(3)\left(\frac{1}{3}\right) = 1$: $\qquad$ The reciprocal of 3 is $\frac{1}{3}$.

The reciprocal of $\frac{1}{3}$ is 3.

Since $\left(-\frac{2}{7}\right)\left(-\frac{7}{2}\right) = 1$: $\qquad$ The reciprocal of $-\frac{2}{7}$ is $-\frac{7}{2}$.

The reciprocal of $-\frac{7}{2}$ is $-\frac{2}{7}$.

Write the reciprocals of these:

a) 10 _____ $\qquad$ b) -4 _____ $\qquad$ c) $-\frac{1}{6}$ _____ $\qquad$ d) $\frac{5}{3}$ _____

---

a) $\frac{1}{10}$ $\qquad$ c) -6

b) $-\frac{1}{4}$ $\qquad$ d) $\frac{3}{5}$

44. There are two numbers that are their own reciprocals.

Since (1)(1) = 1, the reciprocal of "1" is "1".

Since (-1)(-1) = 1, the reciprocal of -1 is -1.

The number 0 has no reciprocal. To see that fact, answer these:

a) When one factor is 0, the product is always ______.

b) Can we get "1" as a product when one factor is 0? ______

c) Therefore, does 0 have a reciprocal? ______

---

45. Any division is the same as <u>multiplying the numerator by the reciprocal of the denominator</u>. Below, for example, we multiplied 12 by $\frac{1}{3}$, the reciprocal of 3.

$$\frac{12}{3} = 12\left(\frac{1}{3}\right) = 4$$

Complete these:

a) $\frac{-16}{8} = -16\left(\frac{1}{8}\right) =$ ______ b) $\frac{-35}{-5} = -35\left(-\frac{1}{5}\right) =$ ______

Answers to 44: a) 0 b) No c) No

---

46. To divide fractions, we <u>multiply the numerator by the reciprocal of the denominator</u>. Below, for example, we multiplied $\frac{3}{7}$ by $\frac{9}{4}$, the reciprocal of $\frac{4}{9}$.

$$\frac{\frac{3}{7}}{\frac{4}{9}} = \frac{3}{7}\left(\frac{9}{4}\right) = \frac{27}{28}$$

The same rules for signs apply to divisions of fractions. That is:

$$\frac{-\frac{2}{3}}{\frac{7}{6}} = \left(-\frac{2}{3}\right)\left(\frac{6}{7}\right) = -\frac{12}{21} = -\frac{4}{7}$$

Complete. Reduce each quotient to lowest terms.

a) $\frac{-\frac{1}{5}}{-\frac{3}{2}} =$ ______ b) $\frac{\frac{9}{8}}{-\frac{3}{4}} =$ ______

Answers to 45: a) -2 b) 7

---

47. In the division below, the denominator is -6, an integer.

$$\frac{\frac{3}{4}}{-6} = \frac{3}{4}\left(-\frac{1}{6}\right) = -\frac{3}{24} = -\frac{1}{8}$$

Continued on following page.

Answers to 46: a) $\frac{2}{15}$ b) $-\frac{3}{2}$

47. Continued

Complete. Reduce each quotient to lowest terms.

a) $\frac{2}{-\frac{3}{7}}$ = ______

b) $\frac{-\frac{2}{3}}{-8}$ = ______

Answers: a) $-\frac{14}{3}$ b) $\frac{1}{12}$

48. Two special types of division are given below.

1. When a non-zero number is divided by itself, the quotient is "1".

$\frac{9}{9} = 1$

$\frac{-\frac{7}{2}}{-\frac{7}{2}} = 1$

2. When a number is divided by "1", the quotient is identical to the number.

$\frac{-8}{1} = -8$

$\frac{\frac{3}{5}}{1} = \frac{3}{5}$

Complete these:

a) $\frac{-7}{-7}$ = ______ b) $\frac{4}{1}$ = ______ c) $\frac{\frac{5}{8}}{\frac{5}{8}}$ = ______ d) $\frac{-\frac{12}{11}}{1}$ = ______

Answers: a) 1 b) 4 c) 1 d) $-\frac{12}{11}$

1-5 OPERATIONS WITH ZERO

In this section, we will discuss operations involving the number 0.

49. When 0 is one number in an addition, the sum is identical to the other number. For example:

$8 + 0 = 8$

$0 + (-5) = -5$

Using the fact above, complete these:

a) $-7.6 + 0$ = ______

b) $0 + \frac{3}{4}$ = ______

Answers: a) -7.6 b) $\frac{3}{4}$

50. In each example below, a number is subtracted from 0.

$$0 - 7 = 0 + (-7) = -7$$

$$0 - (-3) = 0 + 3 = 3$$

In each example below, 0 is subtracted from a number. The answer is identical to the number.

$$4 - 0 = 4 \qquad -8 - 0 = -8$$

Following the examples, complete these:

a) $0 - \frac{2}{3} =$ ______ c) $-\frac{5}{4} - 0 =$ ______

b) $17.5 - 0 =$ ______ d) $0 - (-3.96) =$ ______

---

51. When any number is multiplied by 0, the product is 0. For example:

$$7(0) = 0 \qquad 0(-9) = 0$$

Using the fact above, complete these:

a) $\frac{1}{2}(0) =$ ______ b) $0(-6.4) =$ ______

Answers to 50: a) $-\frac{2}{3}$ c) $-\frac{5}{4}$ b) 17.5 d) 3.96

---

52. When a division is done correctly, the product of the denominator and the quotient equals the numerator. For example:

$$\frac{12}{3} = 4, \quad \text{since} \quad 3(4) = 12$$

When 0 is divided by any other number, the quotient is 0. We can use the relationship above to see that fact:

$$\frac{0}{5} = 0, \quad \text{since} \quad 5(0) = 0$$

$$\frac{0}{-9} = 0, \quad \text{since} \quad -9(0) = 0$$

Using the fact above, complete these:

a) $\frac{0}{-5.9} =$ ______ b) $\frac{0}{\frac{1}{4}} =$ ______

Answers to 51: a) 0 b) 0

---

53. If we divide 5 by 0, we must get a quotient such that 0 times the quotient equals 5. But 0 times any number is 0, not 5.

$$\frac{5}{0} = ?$$

Therefore, <u>division of any other number by 0 is IMPOSSIBLE or UNDEFINED</u>. Perform each possible division below.

a) $\frac{0}{1} =$ ______ b) $\frac{1}{0} =$ ______ c) $\frac{0}{-1.88} =$ ______ d) $\frac{-\frac{3}{4}}{0} =$ ______

Answers to 52: a) 0 b) 0

54. If we divide 0 by 0, we could use any number as the quotient since 0 times any number is 0. For example:

$$\frac{0}{0} = 7, \quad \text{since} \quad 0(7) = 0$$

$$\frac{0}{0} = -4, \quad \text{since} \quad 0(-4) = 0$$

Since we could use any number as the quotient, we say that division of 0 by 0 cannot be determined. Answer "0", "impossible", or "indeterminate" for these:

a) $\frac{12}{0}$ = ______ b) $\frac{0}{0}$ = ______ c) $\frac{0}{12}$ = ______ d) $\frac{-1}{0}$ = ______

a) 0

b) impossible

c) 0

d) impossible

---

a) impossible b) indeterminate c) 0 d) impossible

1-6 PROPERTIES OF ADDITION AND MULTIPLICATION

In this section, we will discuss the commutative, associative, and identity properties of addition and multiplication.

---

55. If we interchange the two numbers in an addition, we get the same sum. That is:

$$10 + (-15) = -15 + 10 \qquad \text{(Both equal -5.)}$$

The property above is called the commutative property of addition. The commutative property is stated for any numbers a and b below.

Commutative Property Of Addition
$a + b = b + a$

Using the commutative property, complete these:

a) $-8 + 7 = 7 + (\quad)$ b) $\frac{1}{2} + \left(-\frac{1}{4}\right) = (\quad) + \frac{1}{2}$

---

56. If we interchange the two numbers in a multiplication, we get the same product. That is:

$$3(-4) = (-4)(3) \qquad \text{(Both equal -12.)}$$

The property above is called the commutative property of multiplication. The commutative property is stated for any numbers a and b below.

Commutative Property Of Multiplication
$a \cdot b = b \cdot a$

Continued on following page.

a) $7 + (-8)$

b) $\left(-\frac{1}{4}\right) + \frac{1}{2}$

56. Continued

Using the commutative property, complete these:

a) (-5)(6) = ( )( ) b) $\left(-\frac{2}{3}\right)(-7) = (\quad)(\quad)$

---

57. Each addition below contains the same three numbers. The parentheses ( ) are grouping symbols which are used to show which two numbers are to be added first. We get the same sum both ways.

(5 + 3) + 6 → 8 + 6 = 14

5 + (3 + 6) → 5 + 9 = 14

The property above is called the associative property of addition. The associative property is stated for any numbers a, b, and c below.

Associative Property Of Addition
(a + b) + c = a + (b + c)

Using the associative property, complete these:

a) (-3 + 5) + 9 = -3 + (__ + 9)

b) [-6 + (-7)] + (-1) = -6 + [(-7) + (__)]

Answers to 56:

a) (6)(-5)

b) $(-7)\left(-\frac{2}{3}\right)$

---

58. For 4(-2)(5), we get the same product whether we multiply 4 and -2 or -2 and 5 first. That is:

4(-2)(5) → (-8)(5) = -40

4(-2)(5) → 4 (-10) = -40

The property above is called the associative property of multiplication. The associative property is stated for any numbers a, b, and c below.

Associative Property Of Multiplication
(a · b) · c = a · (b · c)

Using the associative property, complete these:

a) (-4 · 5) · 6 = -4 · (5 ·)

b) 3 · (-5 · -9) = (3 ·) · -9

Answers to 57:

a) -3 + (5 + 9)

b) -6 + [(-7) + (-1)]

Answers to 58:

a) -4 · (5 · 6)

b) (3 · -5) · -9

59. When 0 is one term in an addition, the sum is <u>identical</u> to the other term. For example:

$8 + 0 = 8$ $\qquad$ $0 + (-5) = -5$

The property above is called the <u>identity property of addition</u>. The number 0 is called the <u>identity element for addition</u>. The identity property is stated for any number <u>a</u> below.

Identity Property Of Addition
$a + 0 = a$ and $0 + a = a$

Using the identity property, complete these:

a) $6.5 + 0 =$ _____ $\qquad$ b) $0 + \left(-\frac{2}{3}\right) =$ _____

---

60. When "1" is one factor in a multiplication, the product is <u>identical</u> to the other factor. For example:

a) 6.5 $\quad$ b) $-\frac{2}{3}$

$1(5) = 5$ $\qquad$ $(-3)(1) = -3$

The property above is called the <u>identity property of multiplication</u>. The number "1" is called the <u>identity element for multiplication</u>. The identity property is stated for any number <u>a</u> below.

Identity Property Of Multiplication
$a \cdot 1 = a$ and $1 \cdot a = a$

Using the identity property, complete these:

a) $1(-3.44) =$ _____ $\qquad$ b) $\frac{5}{3}(1) =$ _____

---

61. The properties discussed in this section are summarized below.

a) -3.44 $\quad$ b) $\frac{5}{3}$

Properties Of Real Numbers

For any real numbers <u>a</u>, <u>b</u>, and <u>c</u>:

Commutative properties	$a + b = b + a$
	$ab = ba$
Associative properties	$(a + b) + c = a + (b + c)$
	$(ab)c = a(bc)$
Identity properties	$a + 0 = a$ and $0 + a = a$
	$a \cdot 1 = a$ and $1 \cdot a = a$

Continued on following page.

61. Continued

Using either "commutative", "associative", or "identity", identify the property.

a) $4(-7) = -7(4)$ ________

b) $0 + (-1) = -1$ ________

c) $(3 \cdot 4)8 = 3(4 \cdot 8)$ ________

d) $(-5 + 3) + 4 = -5 + (3 + 4)$ ________

e) $-3 + 2 = 2 + (-3)$ ________

f) $1(-7) = -7$ ________

---

a) commutative c) associative e) commutative
b) identity d) associative f) identity

1-7 INTEGRAL EXPONENTS

In this section, we will define powers whose exponents are integers.

---

62. Exponential form is a short way of writing a <u>multiplication of identical factors</u>. For example:

$$3 \cdot 3 \cdot 3 \cdot 3 = 3^4$$

In 3^4, 3 is called the <u>base</u>; 4 is called the <u>exponent</u>. The exponent 4 means that 3 appears as a factor 4 times. Two more examples are:

$$x \cdot x \cdot x = x^3$$

$$(-2)(-2)(-2)(-2)(-2)(-2) = (-2)^6$$

Write each multiplication in exponential form.

a) $y \cdot y =$ ____ b) $\left(-\frac{1}{3}\right)\left(-\frac{1}{3}\right)\left(-\frac{1}{3}\right)\left(-\frac{1}{3}\right)\left(-\frac{1}{3}\right) =$ ____

---

63. Any exponential expression is called a <u>power</u> of the base. The names of some powers are given below.

5^2 is called "5 <u>squared</u>" or "5 to the <u>second</u> power".

x^3 is called "<u>x</u> <u>cubed</u>" or "<u>x</u> to the <u>third</u> power".

$(-4)^6$ is called "-4 to the <u>sixth</u> power".

Write the exponential expression for these:

a) -8.5 cubed = ______ b) 10 to the fifth power = ______

a) y^2 b) $\left(-\frac{1}{3}\right)^5$

64.	We evaluated 5^3 below. The exponent 3 tells us to use 5 as a factor three times. $$5^3 = (5)(5)(5) = 125$$ Evaluate each power. a) 3^4 = ______ b) $\left(\frac{1}{2}\right)^5$ = ______	a) $(-8.5)^3$ b) 10^5
65.	When the base is a negative number, the value of a power is positive for even exponents and negative for odd exponents. For example: $$(-4)^4 = (-4)(-4)(-4)(-4) = 256$$ $$(-2)^5 = (-2)(-2)(-2)(-2)(-2) = -32$$ Evaluate each power. a) $\left(\frac{3}{4}\right)^2$ = ______ b) $(-4)^3$ = ______ c) $(-3)^4$ = ______	a) $(3)(3)(3)(3) = 81$ b) $\left(\frac{1}{2}\right)\left(\frac{1}{2}\right)\left(\frac{1}{2}\right)\left(\frac{1}{2}\right)\left(\frac{1}{2}\right) = \frac{1}{32}$
66.	When the base is negative, it is <u>always</u> written in parentheses. Therefore, don't confuse $(-5)^2$ with -5^2. $(-5)^2$ means $(-5)(-5) = 25$ -5^2 means $-(5)(5) = -25$ Following the examples, complete these: a) $(-2)^4$ = ______ b) -2^4 = ______	a) $\frac{9}{16}$ b) -64 c) 81
67.	Any positive power of "1" equals "1". Any positive power of 0 equals 0. Any positive power of -1 equals either "1" or "-1". For example: $1^3 = (1)(1)(1) = 1$ $(-1)^2 = (-1)(-1) = 1$ $0^4 = (0)(0)(0)(0) = 0$ $(-1)^3 = (-1)(-1)(-1) = -1$ Complete these: a) 1^7 = ____ b) 0^{10} = ____ c) $(-1)^4$ = ____ d) $(-1)^5$ = ____	a) 16 b) -16
68.	To fit the pattern below, 2^1 must be 2 and 5^1 must be 5. $2^3 = (2)(2)(2)$ $5^3 = (5)(5)(5)$ $2^2 = (2)(2)$ $5^2 = (5)(5)$ $2^1 = (2)$ or 2 $5^1 = (5)$ or 5 Continued on following page.	a) 1 b) 0 c) 1 d) -1

68. Continued.

Therefore, we agree to the following definition for any base a.

Definition: $a^1 = a$

Following the examples, complete these:

$10^1 = 10$ $\left(\frac{3}{4}\right)^1 = \frac{3}{4}$ a) $6.7^1 =$ _____   b) $(-9)^1 =$ _____

---

69. To fit the pattern below, 10^0 must be "1".

$$10^3 = 1000$$
$$10^2 = 100$$
$$10^1 = 10$$
$$10^0 = 1$$

Therefore, we agree to the following definition for any nonzero base a. We will show later why the base a cannot be 0.

Definition: $a^0 = 1 \quad (a \neq 0)$

Following the examples, complete these:

$75^0 = 1$ $\left(\frac{5}{2}\right)^0 = 1$ a) $x^0 =$ _____   b) $(-4.5)^0 =$ _____

Answers to 68: a) 6.7   b) -9

---

70. To fit the pattern below, 10^{-1} must be $\frac{1}{10}$ and 10^{-2} must be $\frac{1}{100}$ or $\frac{1}{10^2}$.

$$10^2 = 100$$
$$10^1 = 10$$
$$10^0 = 1$$
$$10^{-1} = \frac{1}{10}$$
$$10^{-2} = \frac{1}{100}$$

Therefore, we agree to the following definition, provided that the base a is not 0.

Definition: $a^{-n} = \frac{1}{a^n} \quad (a \neq 0)$

Following the examples, complete these:

$5^{-3} = \frac{1}{5^3}$ $x^{-6} = \frac{1}{x^6}$ a) $7^{-2} =$ _____   b) $(-3)^{-4} =$ _____

Answers to 69: a) 1   b) 1

71. Using the definition, we converted each power below to a fraction.

$7^{-2} = \frac{1}{7^2} = \frac{1}{49}$ $\qquad$ $(-2)^{-3} = \frac{1}{(-2)^3} = \frac{1}{-8} = -\frac{1}{8}$

Convert each power to a fraction.

a) 6^{-2} = ____ b) 4^{-3} = ____ c) $(-2)^{-5}$ = ____

a) $\frac{1}{7^2}$ b) $\frac{1}{(-3)^4}$

---

72. We converted each power below to a fraction.

$5^{-1} = \frac{1}{5^1} = \frac{1}{5}$ $\qquad$ $(-3)^{-1} = \frac{1}{(-3)^1} = \frac{1}{-3} = -\frac{1}{3}$

Convert each power to a fraction.

a) 6^{-1} = ____ b) 2^{-1} = ____ c) $(-9)^{-1}$ = ____

a) $\frac{1}{36}$

b) $\frac{1}{64}$

c) $-\frac{1}{32}$

---

73. The definitions for one, zero, and negative exponents are reviewed below.

Definitions

One exponent	$a^1 = a$	
Zero exponent	$a^0 = 1$	$(a \neq 0)$
Negative exponent	$a^{-n} = \frac{1}{a^n}$	$(a \neq 0)$

Evaluate each of these:

a) 5^1 = ____ b) 7^0 = ____ c) 4^{-1} = ____ d) $(-8)^0$ = ____

a) $\frac{1}{6}$

b) $\frac{1}{2}$

c) $-\frac{1}{9}$

---

a) 5

b) 1

c) $\frac{1}{4}$

d) 1

Self-Test 2 (pages 17-29)

Do each multiplication.

1. $3(-7) =$ ____	2. $(-6)(-9) =$ ____	3. $(-1)(6.5) =$ ____	4. $(0)(-4.9) =$ ____

Do each division.

5. $\frac{-24}{6} =$ ____	6. $\frac{0}{-3} =$ ____	7. $\frac{-5.5}{-5.5} =$ ____	8. $\frac{9}{0} =$ ____

Complete. Reduce each answer to lowest terms.

9. $6\left(-\frac{1}{8}\right) =$ ____	11. $\left(-\frac{3}{5}\right)\left(-\frac{2}{3}\right) =$ ____
10. $\frac{-\frac{1}{4}}{\frac{2}{3}} =$ ____	12. $\frac{-\frac{4}{5}}{-8} =$ ____

Using either "commutative", "associative", or "identity", identify the property.

13. $-5 + 0 = -5$ ____	15. $(2 \cdot 4)6 = 2(4 \cdot 6)$ ____
14. $\left(\frac{2}{3}\right)\left(\frac{1}{5}\right) = \left(\frac{1}{5}\right)\left(\frac{2}{3}\right)$ ____	16. $1\left(-\frac{6}{7}\right) = -\frac{6}{7}$ ____

Evaluate each power.

17. $(-4)^3 =$ ____	19. $17.5^1 =$ ____	21. $2^{-5} =$ ____
18. $\left(\frac{2}{3}\right)^2 =$ ____	20. $(-9)^0 =$ ____	22. $(-3)^{-1} =$ ____

ANSWERS:

1. -21
2. 54
3. -6.5
4. 0
5. -4
6. 0
7. 1
8. impossible
9. $-\frac{3}{4}$
10. $-\frac{3}{8}$
11. $\frac{2}{5}$
12. $\frac{1}{10}$
13. identity
14. commutative
15. associative
16. identity
17. -64
18. $\frac{4}{9}$
19. 17.5
20. 1
21. $\frac{1}{32}$
22. $-\frac{1}{3}$

1-8 LAWS OF EXPONENTS

In this section, we will define the laws of exponents for multiplication, division, and raising a power to a power.

---

74. We multiplied a^2 and a^3 below. Since a is a factor five times, the exponent of the product is 5.

$$a^2 \cdot a^3 = (a)(a) \cdot (a)(a)(a) = a^5$$

We can get the exponent of the product by adding the exponents of the factors. That is:

$$a^2 \cdot a^3 = a^{2+3} = a^5$$

Therefore, the law of exponents for multiplying powers with the same base is:

Multiplication Law

$$a^m \cdot a^n = a^{m+n}$$

The law above applies to positive, zero, and negative exponents. For example:

$$(-2)^5 \cdot (-2)^4 = (-2)^{5+4} = (-2)^9$$

$$5^{-3} \cdot 5^0 = 5^{-3+0} = 5^{-3}$$

$$x^{-1} \cdot x^{-6} = x^{-1+(-6)} = x^{-7}$$

Using the law, complete these:

a) $8^6 \cdot 8^7 =$ ______ c) $m^0 \cdot m^4 =$ ______

b) $(-1)^5 \cdot (-1)^{-3} =$ ______ d) $a^{-3} \cdot a^{-4} =$ ______

---

75. To use the law below, we substituted 3^1 for 3.

Answers to 74: a) 8^{13} b) $(-1)^2$ c) m^4 d) a^{-7}

$$3 \cdot 3^4 = 3^1 \cdot 3^4 = 3^{1+4} = 3^5$$

Use the same method for these:

a) $x \cdot x^7 =$ ______ b) $5^0 \cdot 5 =$ ______ c) $a \cdot a^{-3} =$ ______

---

76. The law applies only for powers with the same base. It does not apply to either multiplication below.

Answers to 75: a) x^8 b) 5^1 c) a^{-2}

$$2^3 \cdot 5^2 \qquad\qquad x^{-4} \cdot y^5$$

Use the law if it applies:

a) $x^4 \cdot x^4 =$ ____ b) $7^{-2} \cdot 8^4 =$ ____ c) $(-3)^{-5} \cdot (-3) =$ ____

77. The law also applies to multiplications with more than two factors. For example:

$$4^3 \cdot 4^{-2} \cdot 4^5 = 4^{3+(-2)+5} = 4^6$$

Complete these:

a) $7^4 \cdot 7^{-6} \cdot 7^{-3}$ = ______ b) $x^{-3} \cdot x \cdot x^5 \cdot x^{-1}$ = ______

Answers: a) x^8 b) does not apply c) $(-3)^{-4}$

---

78. Expressions like $5x$ and $3y^4$ stand for multiplications. That is:

$5x$ means $5 \cdot x$

$3y^4$ means $3 \cdot y^4$

To multiply below, we multiplied the numbers and the powers.

$$(5x)(2x) = 5 \cdot 2 \cdot x \cdot x = 10x^2$$

$$(4x^3)(-3x^4) = 4 \cdot (-3) \cdot x^3 \cdot x^4 = -12x^7$$

Following the examples, complete these:

a) $(3y)(6y)$ = ______ c) $(-2x^7)(5x^{-5})$ = ______

b) $(10y)(3y^{-6})$ = ______ d) $(x^{-1})(-4x^{-3})$ = ______

Answers: a) 7^{-5} b) x^2

---

79. Expressions like $5x^2y$ and $p^2q^{-5}r^4$ also stand for multiplications. That is:

$5x^2y$ means $5 \cdot x^2 \cdot y$

$p^2q^{-5}r^4$ means $p^2 \cdot q^{-5} \cdot r^4$

To multiply below, we multiplied the numbers and the powers.

$$(5x^2y)(4x^{-3}y^2) = 5 \cdot 4 \cdot x^2 \cdot x^{-3} \cdot y \cdot y^2 = 20x^{-1}y^3$$

Following the example, complete these:

a) $(2xy)(-3x^3y^{-2})$ = ________ b) $(p^2q^{-5}r^4)(pq^3r^{-2})$ = ________

Answers: a) $18y^2$ b) $30y^{-5}$ c) $-10x^2$ d) $-4x^{-4}$

---

80. We divided a^5 by a^2 below. Notice that $\frac{a \cdot a}{a \cdot a} = 1$.

$$\frac{a^5}{a^2} = \frac{a \cdot a \cdot a \cdot a \cdot a}{a \cdot a} = \left(\frac{a \cdot a}{a \cdot a}\right)(a \cdot a \cdot a) = 1(a \cdot a \cdot a) = a^3$$

We can get the exponent of the quotient by subtracting exponents. That is:

$$\frac{a^5}{a^2} = a^{5-2} = a^3$$

Continued on following page.

Answers: a) $-6x^4y^{-1}$ b) $p^3q^{-2}r^2$

80. Continued

Therefore, the law of exponents for dividing powers with the same base is:

Division Law

$$\frac{a^m}{a^n} = a^{m-n} \qquad (a \neq 0)$$

The law applies to positive, zero, and negative exponents. For example:

$$\frac{4^6}{4} = \frac{4^6}{4^1} = 4^{6-1} = 4^5 \qquad \frac{x^{-4}}{x^3} = x^{-4-3} = x^{-7}$$

Following the examples, complete these:

a) $\frac{2^9}{2^6}$ = ______ b) $\frac{x}{x^5}$ = ______ c) $\frac{y^{-7}}{y^2}$ = ______

---

81. Another division of powers is shown below. Notice how we subtracted the exponents.

$$\frac{x^{-2}}{x^{-5}} = x^{-2-(-5)} = x^{-2+5} = x^3$$

Following the example, complete these:

a) $\frac{10^4}{10^{-1}}$ = ______ b) $\frac{y^{-9}}{y^{-7}}$ = ______

Answers to 80: a) 2^3 b) x^{-4} c) y^{-9}

---

82. The law does not apply to $\frac{6^4}{3^2}$ because the powers have different bases. If it applies, use the law for these:

a) $\frac{3^8}{4^5}$ = ______ b) $\frac{m^{-2}}{m^{-3}}$ = ______ c) $\frac{p^{-4}}{q^{-1}}$ = ______

Answers to 81: a) 10^5 b) y^{-2}

---

83. To divide below, we divided the numbers and the powers.

$$\frac{15x^6y^{-1}}{5x^2y^{-3}} = \left(\frac{15}{5}\right)\left(\frac{x^6}{x^2}\right)\left(\frac{y^{-1}}{y^{-3}}\right) = 3x^4y^2$$

Following the example, do these:

a) $\frac{-10x}{2x^4}$ = ______ b) $\frac{-12a^{-3}b^9}{-3ab^{-1}}$ = ______

Answers to 82: a) does not apply b) $m^1 = m$ c) does not apply

---

Answers to 83: a) $-5x^{-3}$ b) $4a^{-4}b^{10}$

84. When a power is divided by itself, the exponent of the quotient is 0. Also, when any non-zero quantity is divided by itself, the quotient is "1". That is:

$$\frac{4^2}{4^2} = 4^{2-2} = 4^0 \qquad \frac{4^2}{4^2} = 1$$

The facts above confirm the definition we gave earlier for a^0.

Definition:	$a^0 = 1$	$(a \neq 0)$

As you can see from the definition, we do not define 0^0. We do not define it because $0^0 = \frac{0}{0}$ which is indeterminate. That is:

$$0^0 = 0^{1-1} = \frac{0^1}{0^1} = \frac{0}{0}$$

Answer "1" or "indeterminate" for these:

a) $x^0 =$ ______ b) $\frac{0^3}{0^3} =$ ______ c) $\frac{2^{-4}}{2^{-4}} =$ ______ d) $0^0 =$ ______

---

85. In the division below, we get a^0 which equals "1".

$$\frac{12a^{-3}b^4}{4a^{-3}b} = 3a^0b^3 = 3(1)b^3 = 3b^3$$

Following the example, complete these:

a) $\frac{8x^2y}{2x^2y^5} =$ ______ b) $\frac{24mp^{-2}}{4mp^{-7}} =$ ______

Answers to 84: a) 1 b) indeterminate c) 1 d) indeterminate

---

86. We raised x^4 to the third power below by converting to a multiplication.

$$(x^4)^3 = x^4 \cdot x^4 \cdot x^4 = x^{12}$$

We can get the exponent 12 by multiplying the 4 and 3. That is:

$$(x^4)^3 = x^{(4)(3)} = x^{12}$$

Therefore, the law of exponents for raising a power to a power is:

Power Law:	$(a^m)^n = a^{mn}$

The law above applies to both positive and negative exponents. For example:

$$(10^{-5})^2 = 10^{(-5)(2)} = 10^{-10} \qquad (y^{-4})^{-1} = y^{(-4)(-1)} = y^4$$

Use the law of exponents for these:

a) $(8^7)^2 =$ ______ b) $(a^{-3})^4 =$ ______ c) $(p^{-2})^{-5} =$ ______

Answers to 85: a) $4y^{-4}$ b) $6p^5$

87. We raised x^2y^5 to the third power below by converting to a multiplication.

$$(x^2y^5)^3 = (x^2y^5)(x^2y^5)(x^2y^5) = x^6y^{15}$$

However, it is simpler to multiply each exponent by 3. That is:

$$(x^2y^5)^3 = x^{(2)(3)}y^{(5)(3)} = x^6y^{15}$$

Based on the above example, we get the following law of exponents.

Power Law For Multiplications: $(a^mb^n)^p = a^{mp}b^{np}$

The law above applies to both positive and negative exponents. For example:

$$(p^3q^{-2})^4 = p^{(3)(4)}q^{(-2)(4)} = p^{12}q^{-8}$$

$$(c^{-4}d^2)^{-5} = c^{(-4)(-5)}d^{(2)(-5)} = c^{20}d^{-10}$$

Use the law to complete these:

a) $(b^{-1}t^{-3})^2$ = ________ b) $(x^{-2}y^9)^{-3}$ = ________

Answers: a) 8^{14} b) a^{-12} c) p^{10}

---

88. Notice how we substituted x^1 for $\underline{x}$ below.

$$(xy^{-2})^4 = (x^1y^{-2})^4 = x^4y^{-8}$$

Following the example, do these:

a) $(p^{-1}q)^8$ = ________ b) $(ab^4)^{-6}$ = ________

Answers: a) $b^{-2}t^{-6}$ b) x^6y^{-27}

---

89. Notice how we substituted 2^1 for 2 below.

$$(2x^4y^{-1})^3 = (2^1x^4y^{-1})^3 = 2^3x^{12}y^{-3} \quad \text{or} \quad 8x^{12}y^{-3}$$

Following the example, do these:

a) $(5a^{-3}b^4)^2$ = ________ b) $(3pq^{-3})^4$ = ________

Answers: a) $p^{-8}q^8$ b) $a^{-6}b^{-24}$

---

90. To raise the fraction below to a power, we expanded to a multiplication.

$$\left(\frac{x^{-2}}{y^4}\right)^3 = \left(\frac{x^{-2}}{y^4}\right)\left(\frac{x^{-2}}{y^4}\right)\left(\frac{x^{-2}}{y^4}\right) = \frac{x^{-6}}{y^{12}}$$

However, it is simpler to multiply each exponent by 3. That is:

$$\left(\frac{x^{-2}}{y^4}\right)^3 = \frac{x^{(-2)(3)}}{y^{(4)(3)}} = \frac{x^{-6}}{y^{12}}$$

Continued on following page.

Answers: a) $25a^{-6}b^8$ b) $81p^4q^{-12}$

90. Continued

Based on the example, we get the following law of exponents.

Power Law For Divisions:	$\left(\frac{a^m}{b^n}\right)^p = \frac{a^{mp}}{b^{np}}$	$(b \neq 0)$

Another example of the law is shown below.

$$\left(\frac{c}{d^{-4}}\right)^{-2} = \frac{c^{(1)(-2)}}{d^{(-4)(-2)}} = \frac{c^{-2}}{d^8}$$

Use the law to complete these:

a) $\left(\frac{p^4}{q}\right)^2 =$ _______ b) $\left(\frac{x^5}{y^{-3}}\right)^{-4} =$ _______

a) $\frac{p^8}{q^2}$ b) $\frac{x^{-20}}{y^{12}}$

91. Another example of the same law is shown below.

$$\left(\frac{2a^2b^{-3}}{c^4}\right)^3 = \frac{(2a^2b^{-3})^3}{(c^4)^3} = \frac{2^3a^6b^{-9}}{c^{12}} = \frac{8a^6b^{-9}}{c^{12}}$$

Following the example, complete this one.

$$\left(\frac{4x^{-4}y^3}{5z}\right)^2 =$$

$\frac{16x^{-8}y^6}{25z^2}$

92. The basic laws of exponents are summarized below.

Laws Of Exponents

Multiplication	$a^m \cdot a^n = a^{m+n}$	
Division	$\frac{a^m}{a^n} = a^{m-n}$	$(a \neq 0)$
Powers	$(a^m)^n = a^{mn}$	
	$(a^mb^n)^p = a^{mp}b^{np}$	
	$\left(\frac{a^m}{b^n}\right)^p = \frac{a^{mp}}{b^{np}}$	$(b \neq 0)$

Using the laws, complete these:

a) $(3x)(4x) =$ _______ c) $(xy)^3 =$ _______

b) $\frac{20y}{4y} =$ _______ d) $\left(\frac{ab}{c}\right)^{-1} =$ _______

a) $12x^2$ b) 5 c) x^3y^3 d) $\frac{a^{-1}b^{-1}}{c^{-1}}$

1-9 ORDER OF OPERATIONS

In this section, we will discuss the proper order of operations for evaluating expressions that contain more than one operation.

---

93. When evaluating an expression containing more than one operation, the following order of operations is used:

Order Of Operations

1. If the expression contains grouping symbols like parentheses ( ) or brackets [ ] or braces { }, do the operations within the grouping symbols first before evaluating the whole expression.

2. If the expression does not contain grouping symbols, use the following steps.

   a) Evaluate any power.

   b) Do all multiplications and divisions from left to right.

   c) Do all additions and subtractions from left to right.

   d) If the expression contains a fraction bar, do the operations above and below the bar before dividing.

The expression below does not contain grouping symbols.

$$10 - 2^3 + 3(4) - 8$$

To evaluate it, we evaluate the power 2^3, do the multiplication 3(4), and then add or subtract from left to right. We get:

$$10 - 2^3 + 3(4) - 8$$
$$10 - 8 + 3(4) - 8$$
$$10 - 8 + 12 - 8$$
$$2 + 12 - 8$$
$$14 - 8 = 6$$

Following the example, evaluate these:

a) $3^3 - 4(5) - 4$　　　b) $6(-1) + (-9) - 5(2)$

94. To evaluate the expression below, we evaluated the power before multiplying. Evaluate the other expression.

$$2(5)^2 - 30$$
$$2(25) - 30$$
$$50 - 30 = 20$$

$$5(-2)^3 + 40$$

a) 3 b) -25

---

95. The expression below contains the grouping (7 - 9).

$$(7 - 9) - 3 + \frac{12}{2} - 4$$

To evaluate it, we simplify (7 - 9), do the division $\frac{12}{2}$, and then add or subtract from left to right. We get:

$$(7 - 9) - 3 + \frac{12}{2} - 4$$
$$-2 - 3 + \frac{12}{2} - 4$$
$$-2 - 3 + 6 - 4$$
$$-5 + 6 - 4$$
$$1 - 4 = -3$$

Following the example, evaluate these:

a) $\frac{-10}{5} + 7 + (6 - 10)$ b) $4 - [3 - (-6)] + \frac{15}{-3}$

0

---

96. Below we simplified (3 + 4) first, then multiplied and subtracted.

$$5(3 + 4) - 10$$
$$5(7) - 10$$
$$35 - 10 = 25$$

Use the same method to evaluate these:

a) $5(-2) - 3(-2 + 4)$ b) $4[-2 - (-3)] + \frac{-40}{-5}$

a) 1 b) -10

---

97. Below we simplified (-1 - 4) and (-3 - 5) before multiplying and then subtracting. Evaluate the other expression.

$$(-1 - 4)(-3 - 5) - 10$$
$$(-5)(-8) - 10$$
$$40 - 10 = 30$$

$$(5 - 7)[3 - (-1)] + 2^3$$

a) -16 b) 12

---

98. To simplify the expression in the brackets below, we began by simplifying (4 - 7). Evaluate the other expression. 10 - [9 - (4 - 7)] 20 - [8 - 2(3 - 5)] 10 - [9 - (-3)] 10 - [9 + 3] 10 - 12 = -2	0
99. To simplify the expression below, we worked from the inside out. That is, first we simplified the parentheses, then the brackets, and then the braces. 10 - {8 - 3[7 - (6 + 3)]} 10 - {8 - 3[7 - 9]} 10 - {8 - 3(-2)} 10 - {8 - (-6)} 10 - {8 + 6} 10 - 14 = -4 Following the example, simplify this expression. 6{-4 + 2[9 - 5(8 - 7)]}	8
100. When an expression contains a fraction bar, *we do all operations above and below the bar before dividing*. For example: $\frac{4(3 + 5) - 2}{10 - 4} = \frac{4(8) - 2}{6} = \frac{32 - 2}{6} = \frac{30}{6} = 5$ Using the same method, evaluate this one: $\frac{5(-1 - 9)}{3(4) - 2} =$	24
101. Evaluate these: a) $\frac{2(-4) + (-8)(-5)}{-7 + 5 - 2} =$ b) $\frac{30 - 2(7 - 4)}{3(-1) + 7} =$	-5
	a) -8 b) 6

102. Following the example, evaluate the other expressions.

$$2\left(-\frac{1}{3} - 1\right) = 2\left(-\frac{1}{3} - \frac{3}{3}\right) = 2\left(-\frac{4}{3}\right) = -\frac{8}{3}$$

a) $3\left(\frac{5}{8}\right) - 1 =$

b) $2\left(\frac{1}{2}\right)^2 - \frac{1}{2} + 1 =$

---

a) $\frac{7}{8}$

b) 1

103. Following the example, evaluate the other expression.

$$\frac{4\left(-\frac{3}{5}\right) + 1}{5} = \frac{-\frac{12}{5} + \frac{5}{5}}{5} = \frac{-\frac{7}{5}}{5} = \left(-\frac{7}{5}\right)\left(\frac{1}{5}\right) = -\frac{7}{25}$$

$$\frac{2}{7\left(\frac{5}{4} - 1\right)} =$$

$\frac{8}{7}$

1-10 ALGEBRAIC EXPRESSIONS AND FORMULAS

In this section, we will do some evaluations with algebraic expressions and formulas.

---

104. An algebraic expression is a collection of numbers, letters, operation symbols, and grouping symbols. Some examples are:

$$\frac{2(x + 1)}{5} \qquad 5y^2 + 3y - 1 \qquad \frac{a + 3}{b - 2}$$

Each letter in an algebraic expression is a variable. A variable represents a number or set of numbers. To evaluate an algebraic expression, we must substitute some number for each variable. An example is shown. Do the other evaluation.

Evaluate $2x + 5$ when $x = -1$.

$$2x + 5 = 2(-1) + 5 = -2 + 5 = 3$$

Evaluate $\frac{3(y - 1)}{6}$ when $y = 9$.

$$\frac{3(y - 1)}{6} =$$

4

105. Following the example, do the other evaluation.

Evaluate $2x - 3y$ when $x = 10$ and $y = 2$.

$$2x - 3y = 2(10) - 3(2) = 20 - 6 = 14$$

Evaluate $\frac{x + 1}{y - 1}$ when $x = 9$ and $y = 3$.

$$\frac{x + 1}{y - 1} =$$

---

106. Evaluate each expression when $x = 10$ and $y = -2$.

Answer to 105: 5

a) $x - (2y + 5) =$ ______

b) $2x - (y - 6) =$ ______

---

107. Following the example, do the other evaluation.

Answers to 106: a) 9 b) 28

Evaluate $2xy - 15$ when $x = -8$ and $y = 5$.

$$2xy - 15 = 2(-8)(5) - 15 = -80 - 15 = -95$$

Evaluate $\frac{cd - 8}{10}$ when $c = 6$ and $d = -7$.

$$\frac{cd - 8}{10} =$$

---

108. We evaluated the first expression when $y = \frac{3}{8}$. Evaluate the second expression when $t = \frac{5}{4}$.

Answer to 107: -5

$$\frac{2(y + 1)}{5} = \frac{2\left(\frac{3}{8} + 1\right)}{5} = \frac{2\left(\frac{11}{8}\right)}{5} = \frac{\frac{11}{4}}{5} = \frac{11}{4}\left(\frac{1}{5}\right) = \frac{11}{20}$$

$$3t - (1 - t) =$$

---

109. Following the example, do the other evaluation.

Answer to 108: 4

Evaluate $3x^2 + 2xy$ when $x = -2$ and $y = 4$.

$$3x^2 + 2xy = 3(-2)^2 + 2(-2)(4) = 3(4) + 2(-8) = 12 + (-16) = -4$$

Evaluate $2y^2 - y + 1$ when $y = -\frac{1}{3}$.

$$2y^2 - y + 1 =$$

Answer to 109: $\frac{14}{9}$

110. The same method is used to do evaluations with formulas. An example is shown. Do the other evaluation.

In $C = \frac{5}{9}(F - 32)$, find C when $F = 50$.

$$C = \frac{5}{9}(F - 32) = \frac{5}{9}(50 - 32) = \frac{5}{9}(18) = 10$$

In $P = 2L + 2W$, find P when $L = 25$ and $W = 10$.

$P = 2L + 2W =$ ________________

---

111. Complete each evaluation.

P = 70

a) In $s = \frac{1}{2}gt^2$, find s when $g = 32$ and $t = 3$.

$s = \frac{1}{2}gt^2 =$ ________________

b) In $A = \frac{B}{B + 1}$, find A when $B = 99$.

$A = \frac{B}{B + 1} =$ ________________

---

112. Complete each evaluation.

a) $s = 144$

b) $A = .99$

a) In $m = \frac{y_2 - y_1}{x_2 - x_1}$, find m when $y_2 = 18$, $y_1 = 10$, $x_2 = 10$, and $x_1 = 6$.

$m = \frac{y_2 - y_1}{x_2 - x_1} =$ ________________

b) In $A = P(1 + rt)$, find A when $P = 1000$, $r = 0.12$ and $t = 2$.

$A = P(1 + rt) =$ ________________

---

a) $m = 2$

b) $A = 1240$

SELF-TEST 3 (pages 30-42)

Use the laws of exponents for these:

1. $(3x)(6x) =$ ________
2. $(5a^2b^{-1})(2a^{-3}b^4) =$ ________
3. $\frac{a^5b^2}{ab^7} =$ ________
4. $\frac{12xy^2}{4xy^{-1}} =$ ________
5. $(2c^{-4}d)^{-1} =$ ________
6. $\left(\frac{3p^{-1}q^4}{4t}\right)^2 =$ ________

Evaluate each expression.

7. $4(-3)^2 - 20 =$ ________
8. $10 - [-5 - (-10)] + \frac{-10}{2} =$ ________
9. $\frac{2(-3) - 3(1 - 5)}{4(-1) + 2} =$ ________
10. $20 - [6 - 2(12 - 7)] =$ ________
11. $3\{-5 + 2[10 - 6(2 - 7)]\} =$ ________
12. $\frac{5\left(-\frac{3}{4}\right) + 1}{4} =$ ________

Evaluate each expression when $x = 4$ and $y = -1$.

13. $3x - 2(y - 4) =$ ________
14. $\frac{2xy - 4}{3y} =$ ________

Evaluate each expression when $y = \frac{2}{3}$.

15. $4y - (1 - y) =$ ________
16. $y^2 - 2y + 1 =$ ________

17. Find A when $h = 8$, $b = 11$, and $c = 14$.

$A = \frac{1}{2}h(b + c) =$ ________

18. Find m when $y_2 = 20$, $y_1 = 40$, $x_2 = 14$, and $x_1 = 10$.

$m = \frac{y_2 - y_1}{x_2 - x_1} =$ ________

ANSWERS:

1. $18x^2$
2. $10a^{-1}b^3$
3. a^4b^{-5}
4. $3y^3$
5. $2^{-1}c^4d^{-1}$
6. $\frac{9p^{-2}q^8}{16t^2}$
7. 16
8. 0
9. -3
10. 24
11. 225
12. $-\frac{11}{16}$
13. 22
14. 4
15. $\frac{7}{3}$
16. $\frac{1}{9}$
17. A = 100
18. m = -5

1-11 THE DISTRIBUTIVE PRINCIPLE

In this section, we will discuss the distributive principle of multiplication over addition and subtraction.

---

113. In the multiplication $3(4 + 2)$, the second factor is an addition. To evaluate, we can either add before multiplying or multiply before adding.

$$3(4 + 2) \rightarrow 3(6) = 18$$

$$3(4 + 2) = 3(4) + 3(2) = 12 + 6 = 18$$

The "multiply before adding" method uses the DISTRIBUTIVE PRINCIPLE OF MULTIPLICATION OVER ADDITION. That principle is diagrammed below. Notice that we multiply both 4 and 2 by 3.

$$3(4 + 2) = 3(4) + 3(2) = 12 + 6$$

We used a, b, and c to state the same principle below. Notice that both b and c are multiplied by a.

> The Distributive Principle Of Multiplication Over Addition
>
> $a(b + c) = ab + ac$

Two examples of multiplying by the distributive principle are shown below.

$$4(x + 6) = 4(x) + 4(6) = 4x + 24$$

$$2(5 + 8y) = 2(5) + 2(8y) = 10 + 16y$$

Following the examples, complete these:

a) $6(3x + 7) =$ ______ + ______ = ______ + ______

b) $5(10 + y) =$ ______ + ______ = ______ + ______

---

114. There is also a distributive principle for subtraction. It is stated below.

> The Distributive Principle Of Multiplication Over Subtraction
>
> $a(b - c) = ab - ac$

Two examples of multiplying by the distributive principle over subtraction are shown below.

$$9(x - 3) = 9(x) - 9(3) = 9x - 27$$

$$4(6 - 2y) = 4(6) - 4(2y) = 24 - 8y$$

Continued on following page.

Answers to 113:

a) $6(3x) + 6(7) = 18x + 42$

b) $5(10) + 5(y) = 50 + 5y$

114. Continued

Following the examples, complete these:

a) $3(5x - 2) =$ ________ - ________ = ________ - ________

b) $5(4 - y) =$ ________ - ________ = ________ - ________

---

115. When using the distributive principle, we usually write the final product in one step. For example:

$$7(x + y) = 7x + 7y$$

$$4(7x - 3y) = 28x - 12y$$

Following the examples, complete these:

a) $2(5x + y) =$ ________ b) $8(2a - 6b) =$ ________

Answers to 114: a) $3(5x) - 3(2) = 15x - 6$ b) $5(4) - 5(y) = 20 - 5y$

---

116. Two more examples of the distributive principle are:

$$P(1 + rt) = P + Prt$$

$$ab(c - d) = abc - abd$$

Following the examples, complete these:

a) $2\pi r(h + 1) =$ ________ b) $2c(4p - q) =$ ________

Answers to 115: a) $10x + 2y$ b) $16a - 48b$

---

117. In an algebraic expression, the parts separated by a plus sign or a minus sign are called <u>terms</u>. The signs go with the terms. For example:

In $x + 2y - 3$, the terms are x, $2y$, and -3.

In $-3x - 4y + 5$, the terms are $-3x$, $-4y$, and 5.

The distributive principle also applies when the grouping contains three or more terms. For example:

$$2(3x + 5y + 4) = 6x + 10y + 8$$

$$5(2a - 7b - 9) = 10a - 35b - 45$$

Following the examples, complete these:

a) $3(2x + y - 3) =$ ________________________

b) $4a(c - 2d + h) =$ ________________________

Answers to 116: a) $2\pi rh + 2\pi r$ b) $8cp - 2cq$

Answers to 117: a) $6x + 3y - 9$ b) $4ac - 8ad + 4ah$

118. In each multiplication below, the first factor is negative. Notice how we wrote the final products to minimize the number of signs between terms. Both $-2x - 6$ and $-4t + 20$ have <u>only one sign between terms</u>.

$$-2(x + 3) = (-2)(x) + (-2)(3) = -2x + (-6) = -2x - 6$$

$$-4(t - 5) = (-4)(t) - (-4)(5) = -4t - (-20) = -4t + 20$$

Following the examples, do these:

a) $-3(y + 7) =$ ______________________

b) $-10(4 - 2d) =$ ______________________

c) $-2b(c - 3d + 5) =$ ______________________

a) $-3y - 21$ b) $-40 + 20d$ c) $-2bc + 6bd - 10b$

1-12 COMBINING LIKE TERMS

In this section, we will show how <u>like</u> terms can be combined by factoring by the distributive principle.

119. We can also multiply by the distributive principle when the addition or subtraction is the first factor. For example:

$$(2 + 3)x = 2x + 3x$$

$$(8 - 5)y = 8y - 5y$$

By reversing the process, we can break up each product into the original factors. Doing so is called <u>factoring by the distributive principle</u>. That is:

$2x + 3x$ can be factored to get $(2 + 3)x$

$8y - 5y$ can be factored to get ______________

120. Though the definition will be stated more precisely later, for now we will define <u>like terms</u> as terms with the same variable. For example:

$(8 - 5)y$

3x and 9x are <u>like</u> terms.

7y and -5y are <u>like</u> terms.

Like terms can be combined by factoring by the distributive principle.

$$3x + 9x = (3 + 9)x = 12x$$

$$7y + (-5y) = [7 + (-5)]y = \text{\_\_\_\_\_\_\_\_\_\_}$$

2y

121. In expressions like 3x or -5y, the numerical factor is called the <u>coefficient</u> of the variable. That is:

In 3x, 3 is the coefficient of $\underline{x}$.

In -5y, -5 is the coefficient of $\underline{y}$.

Using the distributive principle over addition to combine like terms is the same as adding their coefficients. That is:

$7x + 5x = 12x$, since $7 + 5 = 12$

$-8y + 3y = -5y$, since $-8 + 3 = -5$

By simply adding coefficients, combine the like terms below.

a) $9t + (-6t) =$ ________ b) $-5x + (-5x) =$ ________

---

122. To combine like terms below, we factored by the distributive principle over subtraction. As you can see, doing so is the same as simply subtracting coefficients.

$8x - 5x = (8 - 5)x = 3x$

$2y - 7y = (2 - 7)y = -5y$

By subtracting coefficients, combine the like terms below.

a) $10R - 2R =$ ________ b) $-2x - 5x =$ ________

Answers: a) 3t b) -10x

---

123. Any variable without a coefficient has a coefficient of "1". That is:

$x = 1x$ $y = 1y$

To combine like terms below, we wrote the coefficient "1" explicitly.

$7x + x = 7x + 1x = 8x$

$y - 3y = 1y - 3y = -2y$

Complete these:

a) $x + 3x =$ ________ c) $10d - d =$ ________

b) $m + m =$ ________ d) $V - 4V =$ ________

Answers: a) 8R b) -7x

---

124. Just as $x = 1x$ and $y = 1y$, $-x = -1x$ and $-y = -1y$. It is helpful to write the -1 coefficient explicitly in additions and subtractions like those below.

$-x + 3x = -1x + 3x = 2x$

$4y + (-y) = 4y + (-1y) = 3y$

$-t - 6t = -1t - 6t = -7t$

Continued on following page.

Answers: a) 4x b) 2m c) 9d d) -3V

124. Continued

Following the examples, complete these:

a) $-m + 10m =$ ________ c) $6d + (-d) =$ ________

b) $-R - 2R =$ ________ d) $-p - p =$ ________

---

125. Since 0x means "0 times x", $0x = 0$. Therefore, each sum below is 0.

$$5x + (-5x) = 0x = 0$$

$$x + (-x) = 1x + (-1x) = 0x = 0$$

Two quantities are called additive inverses if their sum is 0. Two like terms are additive inverses if their coefficients are additive inverses. That is:

5x and -5x are additive inverses, since $5 + (-5) = 0$

x and -x are additive inverses, since $1 + (-1) = 0$

Write the additive inverse of each term.

a) 8y _____ b) d _____ c) -12t _____ d) -V _____

Answers to 124: a) 9m c) 5d b) -3R d) -2p

---

126. We simplified the expressions below by combining like terms.

$$4x + 7y - 6x + 2y = 4x - 6x + 7y + 2y = -2x + 9y$$

$$10a - 5b - 6a + 2b = 10a - 6a - 5b + 2b = 4a - 3b$$

Simplify each expression:

a) $2y + 5x - 4y + x =$ ____________________

b) $7c - 9d - 6c - d =$ ____________________

Answers to 125: a) -8y c) 12t b) -d d) V

---

127. We simplified the expressions below by combining like terms.

$$7x - 9 - 3x + 7 = 4x - 2$$

$$9 - y - 4 - 2y = -3y + 5$$

Simplify each expression:

a) $12a - 7 + 5a - 1 =$ ____________________

b) $6x - y + 9 - 10x + 3 - 4y =$ ____________________

Answers to 126: a) -2y + 6x b) c - 10d

Answers to 127: a) 17a - 8 b) -4x - 5y + 12

1-13 REMOVING GROUPING SYMBOLS

In this section, we will discuss the procedures for removing grouping symbols to simplify expressions.

---

128. If a grouping is preceded by a + sign or no sign, we can simply drop the grouping symbols. For example:

$+(x - 4)$ can be written $x - 4$

$(3y + 1)$ can be written $3y + 1$

Using the fact above, we simplified the expressions below. Simplify the other expressions.

$$7 + (x - 4) = 7 + x - 4 = 3 + x$$

$$(3y + 1) - y = 3y + 1 - y = 2y + 1$$

a) $4t + (3t - 8) =$ ______________________

b) $(7 + 2d) - 5 - 6d =$ ______________________

Answers: a) $7t - 8$ b) $2 - 4d$

129. If a grouping is preceded by a - sign, we <u>cannot</u> simply drop the grouping symbols. To get rid of the grouping symbols, we can substitute -1 for the - sign and multiply by the distributive principle. That is:

$$-(3y + 2) = -1(3y + 2) = (-1)(3y) + (-1)(2) = -3y + (-2) = -3y - 2$$

$$-(6x - 4) = -1(6x - 4) = (-1)(6x) - (-1)(4) = -6x - (-4) = -6x + 4$$

As you can see, we can remove the grouping symbols above <u>by simply changing the sign of each term</u>. That is:

$$-(3y + 2) = -3y - 2$$

$$-(6x - 4) = -6x + 4$$

Remove the grouping symbols from these <u>by simply changing the sign of each term</u>.

a) $-(7x + 1) =$ ____________ c) $-(a + 2b) =$ ____________

b) $-(4y - 5) =$ ____________ d) $-(9x - y) =$ ____________

Answers: a) $-7x - 1$ b) $-4y + 5$ c) $-a - 2b$ d) $-9x + y$

130. To remove the grouping symbols below, we changed the sign of each term in (3y + 2). We then simplified.

$$5 - (3y + 2) = 5 - 3y - 2 = 3 - 3y$$

Following the example, remove the grouping symbols and simplify.

a) $2d - (4d + 7) =$ ______

b) $4x - 9 - (x + 1) =$ ______

c) $3a - (2a + 5b) =$ ______

---

131. To remove the grouping symbols below, we changed the sign of each term in (6x - 4). We then simplified.

$$7 - (6x - 4) = 7 - 6x + 4 = 11 - 6x$$

Following the example, remove the grouping symbols and simplify.

a) $5 - (x - 3) =$ ______

b) $6y - 7 - (5 - 3y) =$ ______

c) $4c - (2d - c) =$ ______

Answers to 130: a) $-2d - 7$ b) $3x - 10$ c) $a - 5b$

---

132. 3(x + 5) and 4(2y - 6) are instances of the distributive principle. To remove the grouping symbols, we multiply.

$$3(x + 5) = 3x + 15$$

$$4(2y - 6) = 8y - 24$$

Remove the grouping symbols in these.

a) $5(4d + 1) =$ ______ b) $7(p - q) =$ ______

Answers to 131: a) $8 - x$ b) $9y - 12$ c) $5c - 2d$

---

133. When an instance of the distributive principle is preceded by a + sign or no sign, we can remove the grouping symbols by multiplying. For example:

$$3(x + 4) + 7 = 3x + 12 + 7 = 3x + 19$$

$$10 + 6(2y - 3) = 10 + 12y - 18 = -8 + 12y$$

Following the examples, remove the grouping symbols and simplify.

a) $2(4d - 6) - 3d =$ ______

b) $15x + 4(1 - 3x) - 5 =$ ______

c) $a + 2(a + b) - b =$ ______

Answers to 132: a) $20d + 5$ b) $7p - 7q$

Answers to 133: a) $5d - 12$ b) $3x - 1$ c) $3a + b$

134. When an instance of the distributive principle is preceded by a - sign, we multiply first and then remove the grouping symbols in the usual way. <u>It is very helpful to draw brackets around the instance of the distributive principle before multiplying</u>. For example:

$10 - 2(x + 3)$	$= 10 - [2(x + 3)]$	Drawing brackets
	$= 10 - [2x + 6]$	Multiplying
	$= 10 - 2x - 6$	Removing the grouping symbols
	$= 4 - 2x$	Simplifying

Following the example, simplify these. <u>Begin by drawing brackets around the instance of the distributive principle</u>.

a) $5 - 3(d + 1)$ b) $x - 2(4x + y)$

---

a) $2 - 3d$

b) $-7x - 2y$

135. To simplify the expression below, we began by drawing brackets around $5(x - 1)$.

$$\begin{aligned} 6 - 5(x - 1) &= 6 - [5(x - 1)] \\ &= 6 - [5x - 5] \\ &= 6 - 5x + 5 \\ &= 11 - 5x \end{aligned}$$

Following the example, simplify these. <u>Draw brackets</u>.

a) $20 - 4(d - 2)$ b) $10a - 3(2a - b)$

---

a) $28 - 4d$

b) $4a + 3b$

136. $-2(x + 6)$ and $-5(3y - 4)$ are instances of the distributive principle. To remove the grouping symbols, we multiply.

$$-2(x + 6) = -2x - 12$$

$$-5(3y - 4) = -15y + 20$$

Remove the grouping symbols by multiplying.

a) $-3(7p + 1) =$ ________ b) $-4(2s - t) =$ ________

---

137. Following the example, simplify the other expressions.

a) $-21p - 3$
b) $-8s + 4t$

$$-3(t + 4) + 5 = -3t - 12 + 5 = -3t - 7$$

a) $-2(3y + 1) - 7 =$ ________________

b) $-5(c - d) - 2d =$ ________________

---

138. Simplify each expression.

a) $-6y - 9$
b) $-5c + 3d$

a) $2(x + 3) - 5(x - 2)$ b) $-3(2y - 5) - 4(y + 5)$

---

139. Notice the steps we used to simplify the expression below.

a) $-3x + 16$
b) $-10y - 5$

$$10x - [9 - 4(2x - 3)] = 10x - [9 - (8x - 12)]$$
$$= 10x - [9 - 8x + 12]$$
$$= 10x - [21 - 8x]$$
$$= 10x - 21 + 8x$$
$$= 18x - 21$$

Following the example, simplify these:

a) $30 - [7x - 2(5x + 4)]$ b) $5y - [6 - 3(2y - 1)]$

---

a) $3x + 38$ b) $11y - 9$

140. Notice the steps we used to simplify the expression below.

$$7y - \{3[2(y - 1) - 5(y + 1)] - 6\}$$
$$7y - \{3[2y - 2 - 5y - 5] - 6\}$$
$$7y - \{3[-3y - 7] - 6\}$$
$$7y - \{-9y - 21 - 6\}$$
$$7y - \{-9y - 27\}$$
$$7y + 9y + 27$$
$$16y + 27$$

Following the example, simplify this one.

$$9x - \{2[4(3x + 5) - 3(2x - 1)] - 30\}$$

$-3x - 16$

SELF-TEST 4 (pages 43-53)

Multiply.

1. $4(1 - 2d) =$
2. $-3(x + 5) =$
3. $2(5x - 7y - 3) =$
4. $-5(p + 4q - 1) =$

Simplify by combining like terms.

5. $4t - t =$
6. $-y + 6y =$
7. $3x - 5y + 4x + y =$
8. $b - 9 - 3b + 2 =$

Continued on following page.

SELF-TEST 4 (Continued)

Simplify each expression.

9. $7 - (3x + 2)$	10. $10y - (7y - 4)$	11. $2(5t - 3) - 7t$
12. $-(x - 1) + 7$	13. $12d - 3(d + 5)$	14. $20a - 5(2a - 3b)$
15. $7x - [6 - 5(3x - 1)]$	16. $6y - \{3[2(3y - 1) - 4(y + 2)] - 20\}$	

ANSWERS:

1. $4 - 8d$	5. $3t$	9. $5 - 3x$	13. $9d - 15$
2. $-3x - 15$	6. $5y$	10. $3y + 4$	14. $10a + 15b$
3. $10x - 14y - 6$	7. $7x - 4y$	11. $3t - 6$	15. $22x - 11$
4. $-5p - 20q + 5$	8. $-2b - 7$	12. $-x + 8$	16. 50

SUPPLEMENTARY PROBLEMS - CHAPTER 1

Note: Answers for all supplementary problems are in the back of the text.

Assignment 1

Given this set of numbers: $\{-4, -\sqrt{3}, -\frac{1}{2}, 0, \frac{2}{3}, 2, \sqrt{5}, 7, 8.9\}$

1. List the natural numbers. __________
2. List the whole numbers. __________

Given this set of numbers: $\{-6, -\sqrt{5}, -\frac{3}{2}, 0, \frac{5}{6}, 3, \sqrt{11}, 6.1, 13\}$

3. List the integers. __________
4. List the irrational numbers. __________

Answer "true" or "false" for these.

5. Every whole number is an integer.
6. Every square root is an irrational number.
7. Every whole number is a natural number.
8. Every integer is a rational number.

Write either $>$ or $<$ in each blank.

9. -5 ___ -7
10. 0 ___ -1
11. 3 ___ 0
12. -10 ___ -3

Answer "true" or "false" for these.

13. $5 \leq 9$
14. $-5 \geq -3$
15. $0 \not< -4$
16. $0 \not> -1$

Graph each set.

17. $\{x \mid x > 1\}$

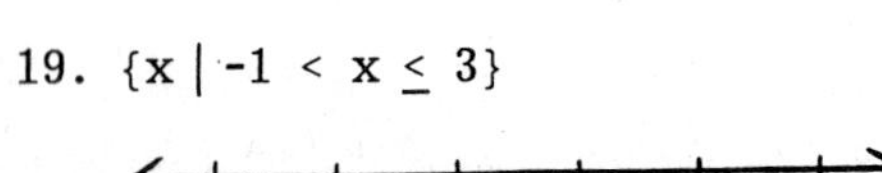

18. $\{x \mid x \leq 2\}$

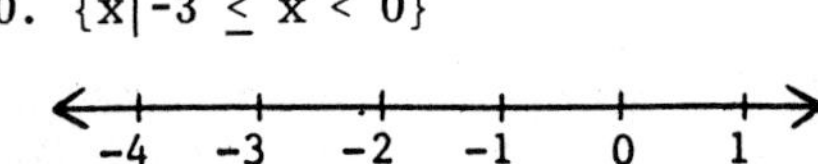

19. $\{x \mid -1 < x \leq 3\}$

-2 -1 0 1 2 3

20. $\{x \mid -3 \leq x < 0\}$

-4 -3 -2 -1 0 1

Find each absolute value.

21. $|7| =$ _____
22. $\left|-\frac{1}{2}\right| =$ _____
23. $|0| =$ _____
24. $|-15.7| =$ _____

Do each addition.

25. $-3 + (-1)$
26. $4 + (-7)$
27. $-2 + 8$
28. $0 + (-30)$
29. $\frac{1}{3} + \left(-\frac{2}{3}\right)$
30. $-\frac{1}{5} + \left(-\frac{3}{5}\right)$
31. $-\frac{1}{2} + \frac{3}{4}$
32. $\frac{3}{4} + \left(-\frac{7}{8}\right)$
33. $6.5 + (-6.5)$
34. $-9.9 + 2.2$
35. $.64 + (-.5)$
36. $-1 + .2033$
37. $2 + (-3) + 9 + (-5) + 0$
38. $-8 + 9 + (-7) + (-6) + 9 + (-1)$

Write the additive inverse of each number.

39. 8 _____
40. -5 _____
41. $\frac{3}{4}$ _____
42. -9.4 _____

Do each subtraction.

43. $5 - 13$ 44. $6 - (-4)$ 45. $-1 - 5$ 46. $-7 - (-8)$

47. $-9 - (-9)$ 48. $1.2 - 6.4$ 49. $-3.6 - 3.8$ 50. $-.33 - (-.77)$

51. $\frac{1}{6} - \frac{5}{6}$ 52. $-\frac{1}{2} - \frac{1}{4}$ 53. $-\frac{7}{8} - \left(-\frac{1}{2}\right)$ 54. $\frac{7}{12} - \left(-\frac{2}{3}\right)$

Assignment 2

Do each multiplication.

1. $8(-7)$ 2. $(-5)(4)$ 3. $(-9)(-3)$ 4. $0(-3)$

5. $(-1)(-10)$ 6. $1(-3.9)$ 7. $10(-.63)$ 8. $(-1.2)(-1.2)$

9. $\left(\frac{1}{2}\right)\left(-\frac{3}{4}\right)$ 10. $\left(-\frac{5}{6}\right)\left(-\frac{3}{5}\right)$ 11. $(-4)\left(\frac{5}{8}\right)$ 12. $\left(-\frac{2}{3}\right)(-9)$

13. $(-7)(-1)(4)$ 14. $(-3)(-5)(-2)$ 15. $(-9)(0)(-1)(-2)$ 16. $(1)(-6)(3)(-7)$

Do each division.

17. $\frac{42}{-7}$ 18. $\frac{-20}{4}$ 19. $\frac{-32}{-4}$ 20. $\frac{0}{-6}$

21. $\frac{59}{-1}$ 22. $\frac{-17.6}{1}$ 23. $\frac{4.4}{0}$ 24. $\frac{-8.8}{-1.1}$

25. $\frac{\frac{1}{8}}{-\frac{3}{4}}$ 26. $\frac{-\frac{45}{4}}{-\frac{15}{8}}$ 27. $\frac{-\frac{8}{7}}{2}$ 28. $\frac{-15}{-\frac{3}{5}}$

Using either "commutative", "associative", or "identity", identify the property.

29. $0 + (-3) = -3$

30. $(-9)(1) = -9$

31. $(-2 + 5) + 3 = -2 + (5 + 3)$

32. $5 + (-9) = -9 + 5$

33. $(2 \cdot 3) \cdot 5 = 2 \cdot (3 \cdot 5)$

34. $(-1)(-8) = (-8)(-1)$

35. $1(10) = 10$

36. $-1 + 2 = 2 + (-1)$

37. $12 + 0 = 12$

Evaluate each power.

38. 2^3 39. $(-4)^4$ 40. 1^8 41. $(-9)^1$ 42. 0^5

43. 7^0 44. 8^{-1} 45. $(-7)^{-2}$ 46. 6^{-3} 47. $(-2)^{-5}$

48. $(-.5)^3$ 49. $(1.5)^2$ 50. $\left(\frac{7}{8}\right)^0$ 51. $\left(-\frac{9}{5}\right)^1$ 52. $\left(\frac{1}{4}\right)^3$

Assignment 3

Multiply.

1. $4^5 \cdot 4^{-3}$ 2. $x^{-1} \cdot x^{-5}$ 3. $y \cdot y^{-4}$ 4. $b^5 \cdot b \cdot b^{-7} \cdot b^4$

5. $(3a)(4a)$ 6. $(-2x^{-1})(6x)$ 7. $(ab)(7a^5b^{-6})$ 8. $(cd^{-4}t^2)(c^3dt^{-5})$

Divide.

9. $\frac{3^6}{3}$ 10. $\frac{x^2}{x^6}$ 11. $\frac{y^{-1}}{y^{-7}}$ 12. $\frac{a^5b}{a^4b^{-1}}$

13. $\frac{16x}{2x}$ 14. $\frac{-5c^2d^3}{5cd}$ 15. $\frac{-20x^2y^{-3}}{-4x^5y^{-2}}$ 16. $\frac{32p^2t^5}{8p^2t}$

Raise to the indicated power.

17. $(4^2)^{-3}$ 18. $(x^{-3})^{-5}$ 19. $(b^{-2}c^3)^{-1}$ 20. $(xy^{-3})^4$

21. $(4p^{-4}q)^3$ 22. $\left(\frac{x^{-1}}{y^2}\right)^4$ 23. $\left(\frac{cd}{t}\right)^{-2}$ 24. $\left(\frac{2p^{-2}q^5}{3t^{-1}}\right)^2$

Evaluate each expression.

25. $4(-2) - (-1)(9)$ 26. $2(3)^2 + 5$ 27. $\frac{-18}{6} + 5 + (1 - 9)$

28. $7(-1) - 2[-3 - (-4)]$ 29. $20 - [10 - (5 - 10)]$ 30. $7\{-3 + 4[8 - 5(9 - 7)]\}$

31. $\frac{(-2 - 3)[5 - (-4)]}{3(-4) + 7}$ 32. $3\left(\frac{1}{3}\right)^2 + \frac{1}{3} + 1$ 33. $\frac{2\left(-\frac{7}{8}\right) + 1}{3}$

Evaluate each expression when $x = -1$ and $y = 3$.

34. $\frac{3x - 5}{2}$ 35. $\frac{x - 2y}{2x + y}$ 36. $\frac{xy - 7}{5}$

37. $y - (2x + 5)$ 38. $3x - 2(y - 7)$ 39. $2y^2 - 5xy - 1$

Evaluate each expression when $x = \frac{3}{4}$ and $y = -\frac{3}{2}$.

40. $2x^2 + x - 1$ 41. $6y^2 - 9$ 42. $\frac{4x - 9}{y}$

Do each formula evaluation.

43. In $A = LW$, find A when $L = 12$ and $W = 8$.

44. In $P = 2L + 2W$, find P when $L = 25$ and $W = 15$.

45. In $B = 180 - (A + C)$, find B when $A = 25$ and $C = 85$.

46. In $V = \frac{1}{3}BH$, find V when $B = 40$ and $h = 30$.

47. In $F = \frac{9}{5}C + 32$, find F when $C = 20$.

48. In $s = \frac{1}{2}gt^2$, find s when $g = 32$ and $t = 2$.

49. In $m = \frac{y_2 - y_1}{x_2 - x_1}$, find m when $y_2 = 10$, $y_1 = 2$, $x_2 = 2$ and $x_1 = 6$.

50. In $A = P(1 + rt)$, find A when $P = 1000$, $r = .09$, and $t = 2$.

Assignment 4

Do each multiplication.

1. $3(x + 7)$
2. $10(5y + 1)$
3. $7(x - 2)$
4. $5(1 - 4d)$
5. $4(5a - 6b)$
6. $bc(R - P)$
7. $-4(m + 3)$
8. $-1(7 - 6P)$
9. $3(2x - 4y + 3)$
10. $10(b - 2c - d)$
11. $-2a(p + 3q - 1)$

Simplify by combining like terms.

12. $5x + x$
13. $-y + 8y$
14. $-5m + (-2m)$
15. $9d - 4d$
16. $r - 7r$
17. $-t - t$
18. $4b + (-4b)$
19. $2 + 5p - 3$
20. $8a - 2b + a - b$
21. $2x - 7 - 5x + 1$
22. $-p + 2q + 3p - 7 - q + 9$

Simplify each expression.

23. $2x + (3x - 4)$
24. $(y + 9) - 4 + y$
25. $-(5t + 1) + 3t$
26. $-(10 - 3d) + 5$
27. $7 - (2b + 5)$
28. $4a - 3 - (a + 7)$
29. $5x - (x - 2y)$
30. $10 - (3t - 8) + 6t$
31. $5(x - 3) - x$
32. $20 + 5(3m + 2)$
33. $-4(y + 3) + 7y$
34. $-8(b - 2c) - 3b$
35. $15 - 3(2x + 7)$
36. $5a - 3(a + 2b)$
37. $9y - 4(y - 6)$
38. $p - 5(p - 3q)$
39. $3(x + 1) - 6(x - 1)$
40. $-2(3y - 4) - 4(y + 5)$
41. $12t - [11 - 3(t + 2)]$
42. $20 - [9y - 3(2y - 4)] + 4y$
43. $2x - \{3[4(x - 1) - 2(x + 1)] - 9\}$
44. $6p - \{2[5(2p + 3) - 4(3p - 1)] - 25\}$

2 Linear Equations and Inequalities

In this chapter, we will discuss the principles and processes that are used to solve linear equations and inequalities. We will also discuss formula evaluations and rearrangements, word problems involving equations and inequalities, and absolute value equations and inequalities.

2-1 SOLVING EQUATIONS

In this section, we will show how the addition axiom and multiplication axiom are used to solve equations.

---

1. Two examples of linear equations in one variable are shown below.

$$x + 4 = 9 \qquad\qquad 5y - 3 = 2y + 9$$

A linear equation is neither true nor false until we substitute a number for the variable. Then it becomes either true or false. For example:

If we substitute 3 for x, $x + 4 = 9$ is false.

If we substitute 5 for x, $x + 4 = 9$ is true.

The substitution that makes a linear equation true is called the solution or root of the equation. The solution or root of $x + 4 = 9$ is 5.

a) Is 9 the solution or root of $y + 10 = 19$? ________

b) Is 7 the solution or root of $3 + d = 9$? ________

---

a) Yes

b) No

2. The set of all numbers that are solutions of an equation is called the solution set of the equation. For example:

The solution set of $x + 4 = 9$ is $\{5\}$.

However, since a linear equation like $x + 4 = 9$ has only one solution, we do not usually write its solution in set notation. Instead of saying "the solution set is $\{5\}$", we simply say $x = 5$.

The solution set of $y + 2 = 10$ is $\{8\}$. We say: $y = 8$.

The solution set of $9 - t = 6$ is $\{3\}$. We say: ________

Answer: $t = 3$

---

3. To solve an equation means to find its solution or root. One principle used to solve equations is the addition axiom for equations. It is stated below.

> **The Addition Axiom For Equations**
>
> IF WE ADD THE SAME QUANTITY TO BOTH SIDES OF AN EQUATION THAT IS TRUE, THE NEW EQUATION IS ALSO TRUE. That is:
>
> If: $A = B$
>
> Then: $A + C = B + C$

To get the variable alone below, we added -3 to both sides. -3 is the inverse of 3. Solve the other equation by adding 5 to both sides. Check your solution.

$$x + 3 = 9$$
$$x + 3 + (-3) = 9 + (-3)$$
$$x + 0 = 6$$
$$x = 6$$

Check

$$x + 3 = 9$$
$$6 + 3 = 9$$
$$9 = 9$$

$$y - 5 = 0$$

Answer: $y = 5$

Check
$y - 5 = 0$
$5 - 5 = 0$
$0 = 0$

---

4. To solve the equation below, we added 8.1 to both sides. Solve the other equation by adding $-\frac{2}{5}$ to both sides.

$$2.4 = -8.1 + m$$
$$8.1 + 2.4 = 8.1 - 8.1 + m$$
$$10.5 = 0 + m$$
$$m = 10.5$$

$$y + \frac{2}{5} = -\frac{1}{5}$$

5. A second principle used to solve equations is <u>the multiplication axiom for equations</u>. It is stated below.

$y = -\frac{3}{5}$

<u>The Multiplication Axiom For Equations</u>

IF WE MULTIPLY BOTH SIDES OF AN EQUATION THAT IS TRUE BY THE SAME QUANTITY, THE NEW EQUATION IS ALSO TRUE. That is:

If: $B = C$

Then: $AB = AC$

To get the variable alone below, we multiplied both sides by $\frac{1}{5}$, the reciprocal of 5. Solve the other equation by multiplying both sides by $-\frac{1}{3}$. Check your solution.

$$5x = 20$$
$$\frac{1}{5}(5x) = \frac{1}{5}(20)$$
$$1 \quad x = 4$$
$$x = 4$$

<u>Check</u>

$$5x = 20$$
$$5(4) = 20$$
$$20 = 20$$

$$18 = -3y$$

6. We used the multiplication axiom to solve each equation below.

$y = -6$

<u>Check</u>
$18 = -3y$
$18 = -3(-6)$
$18 = 18$

$$1.2x = 4.8$$
$$\frac{1}{1.2}(1.2x) = \frac{1}{1.2}(4.8)$$
$$1 \quad x = \frac{4.8}{1.2}$$
$$x = 4$$

$$\frac{1}{2} = \frac{3}{5}y$$
$$\frac{5}{3}\left(\frac{1}{2}\right) = \frac{5}{3}\left(\frac{3}{5}y\right)$$
$$\frac{5}{6} = 1\ y$$
$$y = \frac{5}{6}$$

Following the examples, solve these:

a) $\frac{2}{3}m = -5$

b) $-7.5 = -1.5t$

7. We used the multiplication axiom to solve two equations below.

$6x = 24$	$35 = -5y$
$\frac{1}{6}(6x) = \frac{1}{6}(24)$	$-\frac{1}{5}(35) = -\frac{1}{5}(-5y)$
$1x = 4$	$-7 = 1\ y$
$x = 4$	$y = -7$

You can see that using the multiplication axiom is the same as dividing the number by the coefficient of the variable. Therefore, we can use the shorter method below.

$6x = 24$	$35 = -5y$
$x = \frac{24}{6} = 4$	$y = \frac{35}{-5} = -7$

Use the shorter method for these:

a) $4x = 4$ b) $-8 = 8R$ c) $-20 = -2t$

Answers: a) $m = -\frac{15}{2}$ b) $t = 5$

---

8. Notice at the left below that the solution is 0. Notice at the right below that we substituted -1y for -y.

$5x = 0$	$-y = 10$
$x = \frac{0}{5} = 0$	$-1y = 10$
	$y = \frac{10}{-1} = -10$

Solve these:

a) $-t = 7$ b) $0 = -9d$ c) $-2 = -p$

Answers: a) $x = 1$ b) $R = -1$ c) $t = 10$

---

9. When the solution is a fraction, the fraction is always reduced to lowest terms. For example:

$7y = -3$	$15 = 9m$
$y = \frac{-3}{7} = -\frac{3}{7}$	$m = \frac{15}{9} = \frac{5}{3}$

Solve. Reduce to lowest terms.

a) $30p = 45$ b) $-1 = 6y$ c) $-6 = -8m$

Answers: a) $t = -7$ b) $d = 0$ c) $p = 2$

10. To solve the equation below, we used the addition axiom to isolate 3x. Then we used the multiplication axiom to complete the solution.

	Check
$3x + 5 = 26$	$3x + 5 = 26$
$3x + 5 + (-5) = 26 + (-5)$	$3(7) + 5 = 26$
$3x + 0 = 21$	$21 + 5 = 26$
$3x = 21$	$26 = 26$
$x = \frac{21}{3} = 7$	

Using both axioms, solve each equation.

a) $3 - 6y = 7$ b) $0 = 7m - 28$

a) $p = \frac{3}{2}$

b) $y = -\frac{1}{6}$

c) $m = \frac{3}{4}$

---

11. To solve the equation below, we substituted -1x for -x. Solve the other equation.

$7 - x = 3$

$-7 + 7 - x = 3 + (-7)$

$-x = -4$

$-1x = -4$

$x = \frac{-4}{-1} = 4$

$5.49 = 2.18 - d$

a) $y = -\frac{2}{3}$

b) $m = 4$

---

12. To solve the equation below, we used the addition axiom to get both like terms on one side so that they could be combined. Then we used the multiplication axiom to complete the solution.

	Check
$7x = 2x + 20$	$7x = 2x + 20$
$-2x + 7x = -2x + 2x + 20$	$7(4) = 2(4) + 20$
$5x = 0 + 20$	$28 = 8 + 20$
$5x = 20$	$28 = 28$
$x = \frac{20}{5} = 4$	

Using both axioms, solve these:

a) $3d - 60 = 7d$ b) $4 - y = 4y$

$d = -3.31$

13. Following the example, solve the other equation.

$$2x - \frac{1}{3} = \frac{1}{6} \qquad\qquad \frac{1}{2}y + \frac{1}{4}y = 2$$

$$2x - \frac{1}{3} + \frac{1}{3} = \frac{1}{6} + \frac{1}{3}$$

$$2x = \frac{1}{6} + \frac{2}{6}$$

$$2x = \frac{1}{2}$$

$$\frac{1}{2}(2x) = \frac{1}{2}\left(\frac{1}{2}\right)$$

$$x = \frac{1}{4}$$

a) $d = -15$

b) $y = \frac{4}{5}$

---

14. To solve $7x + 9 = 4x + 21$, we must get both variable terms on one side and both numbers on the other side so they can be combined. To do so, we must move a variable term from one side and a number from the other side. We have a choice of two pairs:

1) moving 4x and 9

or 2) moving 7x and 21

With either choice, we must use the addition axiom twice. We solved below by moving 4x and 9.

$$7x + 9 = 4x + 21$$

$$-4x + 7x + 9 = -4x + 4x + 21$$

$$3x + 9 = 21$$

$$3x + 9 + (-9) = 21 + (-9)$$

$$3x = 12$$

$$x = \frac{12}{3} = 4$$

By moving either possible pair, solve these:

a) $8t + 1 = 4t - 9$ b) $15 - 2h = 8 - h$

$y = \frac{8}{3}$

15. Before using the addition axiom, we combine like terms if possible on both sides of an equation. An example is shown. Solve the other equation.

$$5x - 3x - 7 = 4 + 6x + 5 \qquad\qquad 9y - 13 - 3y = 30 - 9y - 8$$

$$2x - 7 = 6x + 9$$

$$-2x + 2x - 7 = -2x + 6x + 9$$

$$-7 = 4x + 9$$

$$-9 - 7 = 4x + 9 - 9$$

$$-16 = 4x$$

$$x = \frac{-16}{4} = -4$$

a) $t = -\frac{5}{2}$

b) $h = 7$

---

16. An equation can have no solution or an infinite number of solutions. The latter is called an identity. An example of each type is discussed below.

No Solution. If we add -2x to both sides below, we get 3 = 4 which is a false statement.

$$2x + 3 = 2x + 4$$

$$3 = 4 \qquad \text{(a false statement)}$$

Since we get a false statement no matter what number is substituted for x, the equation has no solution.

Identity. If we add -2x to both sides below, we get 5 = 5 which is a true statement.

$$2x + 5 = 2x + 5$$

$$5 = 5 \qquad \text{(a true statement)}$$

Since we get a true statement no matter what number is substituted for x, the equation has an infinite number of solutions.

$y = \frac{7}{3}$

2-2 EQUATIONS WITH GROUPING SYMBOLS

In this section, we will solve equations that contain grouping symbols.

---

17. To solve the equation below, we removed the grouping symbols, simplified the left side, and then used the axioms. Solve the other equation.

$$x + 3x + (x - 2) = 73$$
$$x + 3x + x - 2 = 73$$
$$5x - 2 = 73$$
$$5x - 2 + 2 = 73 + 2$$
$$5x = 75$$
$$x = 15$$

$$t + (t - 3) - 2 = 0$$

$t = \frac{5}{2}$

18. To solve the equation below, we simplified the left side and then used the axioms. Solve the other equation.

$$5x - (2x + 1) = 4$$
$$5x - 2x - 1 = 4$$
$$3x - 1 = 4$$
$$3x - 1 + 1 = 4 + 1$$
$$3x = 5$$
$$x = \frac{5}{3}$$

$$F = 11 - (6F + 7)$$

$F = \frac{4}{7}$

19. Following the example, solve the other equation.

$$6y - (2y - 3) = 1$$
$$6y - 2y + 3 = 1$$
$$4y + 3 = 1$$
$$4y + 3 + (-3) = 1 + (-3)$$
$$4y = -2$$
$$y = \frac{-2}{4}$$
$$y = -\frac{1}{2}$$

$$x = 10 - (5x - 1)$$

$x = \frac{11}{6}$

20. To solve the equation below, we began by multiplying by the distributive principle. Solve the other equation.

$$3(x + 2) = 21$$
$$3x + 6 = 21$$
$$3x + 6 + (-6) = 21 + (-6)$$
$$3x = 15$$
$$x = 5$$

$$3(4 - m) = m$$

---

m = 3

21. To solve the equation below, we multiplied by the distributive principle on both sides. Solve the other equation.

$$7(d + 2) = 2(3 + d)$$
$$7d + 14 = 6 + 2d$$
$$-2d + 7d + 14 = 6 + 2d - 2d$$
$$5d + 14 = 6$$
$$5d + 14 + (-14) = 6 + (-14)$$
$$5d = -8$$
$$d = -\frac{8}{5}$$

$$5(t + 3) = 3(t - 5)$$

---

t = -15

22. Following the example, solve the other equation.

$$3(m + 4) - (m + 8) = 12$$
$$3m + 12 - m - 8 = 12$$
$$2m + 4 = 12$$
$$2m + 4 - 4 = 12 - 4$$
$$2m = 8$$
$$m = 4$$

$$4t = 3(2t - 3) - (5t - 5)$$

$t = -\frac{4}{3}$

23. To solve the equation below, we began by simplifying the expression on the left side. Solve the other equation.

$$10 - 3(x + 4) = 16 \qquad\qquad 9y - 7(y + 2) = 4$$

$$10 - [3(x + 4)] = 16$$

$$10 - [3x + 12] = 16$$

$$10 - 3x - 12 = 16$$

$$-2 - 3x = 16$$

$$2 - 2 - 3x = 16 + 2$$

$$-3x = 18$$

$$x = -6$$

---

24. To solve the equation below, we began by simplifying the expression on the right side. Solve the other equation.

$y = 9$

$$35 = 47 - 6(4 - x) \qquad\qquad 20 = t - 3(5 - 2t)$$

$$35 = 47 - [6(4 - x)]$$

$$35 = 47 - [24 - 6x]$$

$$35 = 47 - 24 + 6x$$

$$35 = 23 + 6x$$

$$-23 + 35 = -23 + 23 + 6x$$

$$12 = 6x$$

$$x = 2$$

---

25. To solve the equation below, we simplified both sides and then used the axioms.

$t = 5$

$$2(a - 5) - 3(a + 2) = 7a - (a + 10)$$

$$2a - 10 - [3a + 6] = 7a - a - 10$$

$$2a - 10 - 3a - 6 = 6a - 10$$

$$-a - 16 = 6a - 10$$

$$a - a - 16 = a + 6a - 10$$

$$-16 = 7a - 10$$

$$10 - 16 = 7a - 10 + 10$$

$$-6 = 7a$$

$$a = -\frac{6}{7}$$

Continued on following page.

25. Continued

Using the same method, solve this equation.

$$12 - 2(x - 5) = 5(x + 1) - (x - 7)$$

$x = \frac{5}{3}$

2-3 EQUATIONS CONTAINING FRACTIONS

To solve an equation containing one or more fractions, we can begin by clearing the fraction or fractions. We will discuss the method in this section.

26. To solve the equation below, we began by clearing the fraction. To do so, we used the multiplication axiom, multiplying both sides by 2, the denominator of the fraction. Solve the other equation. Begin by multiplying both sides by 3.

	Check	
$\frac{3x}{2} = 12$	$\frac{3x}{2} = 12$	$6 = \frac{2m}{3}$
$\cancel{2}\left(\frac{3x}{\cancel{2}}\right) = 2(12)$	$\frac{3(8)}{2} = 12$	
$3x = 24$	$\frac{24}{2} = 12$	
$x = 8$	$12 = 12$	

$m = 9$

27. To solve the equation below, we began by multiplying both sides by 5. Solve the other equation.

$$\frac{x + 3}{5} = 2 \qquad\qquad 1 = \frac{2y + 7}{4}$$

$$\cancel{5}\left(\frac{x + 3}{\cancel{5}}\right) = 5(2)$$

$$x + 3 = 10$$

$$x + 3 + (-3) = 10 + (-3)$$

$$x = 7$$

$y = -\frac{3}{2}$

28. The equation below contains two terms on the left side. To clear the fraction, we multiplied both sides by 4. Notice that we multiplied by the distributive principle on the left side. Solve the other equation.

$$\frac{x}{4} + 2 = 3 \qquad\qquad 1 = \frac{5y}{2} - 3$$

$$4\left(\frac{x}{4} + 2\right) = 4(3)$$

$$\cancel{4}\left(\frac{x}{\cancel{4}}\right) + 4(2) = 12$$

$$x + 8 = 12$$

$$x = 4$$

$y = \frac{8}{5}$, from:

$2 = 5y - 6$

29. When an equation contains more than one fraction, we clear the fraction by multiplying both sides <u>by the lowest common denominator (LCD)</u> of the denominators. For example, we cleared the fractions below by multiplying both sides by 24. Solve the other equation.

$$\frac{3x}{8} = \frac{1}{6} \qquad\qquad \frac{3}{2} = \frac{7y}{10}$$

$$\overset{3}{\cancel{24}}\left(\frac{3x}{\cancel{8}}\right) = \overset{4}{\cancel{24}}\left(\frac{1}{\cancel{6}}\right)$$

$$9x = 4$$

$$x = \frac{4}{9}$$

$y = \frac{15}{7}$, from:

$15 = 7y$

30. Notice how we had to multiply by the distributive principle on the left side below. Solve the other equation.

$$\frac{x+4}{5} = \frac{x}{3} \qquad\qquad \frac{2y-1}{5} = \frac{1}{2}$$

$$\overset{3}{\cancel{15}}\left(\frac{x+4}{\cancel{5}}\right) = \overset{5}{\cancel{15}}\left(\frac{x}{\cancel{3}}\right)$$

$$3(x + 4) = 5(x)$$

$$3x + 12 = 5x$$

$$12 = 2x$$

$$x = 6$$

---

31. Notice below how we multiplied both terms on the left by 12. Solve the other equation.

$y = \frac{7}{4}$, from:

$4y - 2 = 5$

$$\frac{3m}{4} + \frac{2m}{3} = 1 \qquad\qquad \frac{5x}{2} - \frac{4x}{3} = 1$$

$$12\left(\frac{3m}{4} + \frac{2m}{3}\right) = 12(1)$$

$$\overset{3}{\cancel{12}}\left(\frac{3m}{\cancel{4}}\right) + \overset{4}{\cancel{12}}\left(\frac{2m}{\cancel{3}}\right) = 12$$

$$9m + 8m = 12$$

$$17m = 12$$

$$m = \frac{12}{17}$$

---

32. To clear the fraction below, we multiplied both sides by 15. Solve the other equation.

$x = \frac{6}{7}$, from:

$15x - 8x = 6$

$$\frac{y}{3} + \frac{y+1}{5} = 5 \qquad\qquad \frac{m-1}{3} + \frac{m}{4} = 2$$

$$15\left(\frac{y}{3} + \frac{y+1}{5}\right) = 15(5)$$

$$\overset{5}{\cancel{15}}\left(\frac{y}{\cancel{3}}\right) + \overset{3}{\cancel{15}}\left(\frac{y+1}{\cancel{5}}\right) = 75$$

$$5(y) + 3(y + 1) = 75$$

$$5y + 3y + 3 = 75$$

$$8y + 3 = 75$$

$$8y = 72$$

$$y = 9$$

33. To clear the fractions below, we multiplied both sides by 12, the LCD of all three denominators. Solve the other equation. $\frac{x}{2} - \frac{x}{3} = \frac{1}{4}$ $\qquad$ $\frac{y}{4} + \frac{y}{8} = \frac{1}{2}$ $12\left(\frac{x}{2} - \frac{x}{3}\right) = 12\left(\frac{1}{4}\right)$ $\overset{6}{\cancel{12}}\left(\frac{x}{\cancel{2}}\right) - \overset{4}{\cancel{12}}\left(\frac{x}{\cancel{3}}\right) = 3$ $6x - 4x = 3$ $2x = 3$ $x = \frac{3}{2}$	$m = 4$, from: $4m - 4 + 3m = 24$
34. To clear the fractions below, we multiplied both sides by 6. Solve the other equation. $\frac{m-2}{3} + \frac{m}{6} = \frac{1}{2}$ $\qquad$ $\frac{y-8}{5} + \frac{y}{3} = \frac{2}{5}$ $6\left(\frac{m-2}{3} + \frac{m}{6}\right) = 6\left(\frac{1}{2}\right)$ $\overset{2}{\cancel{6}}\left(\frac{m-2}{\cancel{3}}\right) + \cancel{6}\left(\frac{m}{\cancel{6}}\right) = \overset{3}{\cancel{6}}\left(\frac{1}{\cancel{2}}\right)$ $2m - 4 + m = 3$ $3m - 4 = 3$ $3m = 7$ $m = \frac{7}{3}$	$y = \frac{4}{3}$, from: $2y + y = 4$
35. To clear the fractions below, we multiplied both sides by 4. $\frac{b-4}{2} - \frac{b+1}{4} = \frac{2b-3}{4}$ $4\left(\frac{b-4}{2} - \frac{b+1}{4}\right) = \overset{1}{\cancel{4}}\left(\frac{2b-3}{\cancel{4}}\right)$ $\overset{2}{\cancel{4}}\left(\frac{b-4}{\cancel{2}}\right) - \overset{1}{\cancel{4}}\left(\frac{b+1}{\cancel{4}}\right) = 2b - 3$ $2b - 8 - (b + 1) = 2b - 3$ $2b - 8 - b - 1 = 2b - 3$ $b - 9 = 2b - 3$ $-9 = b - 3$ $b = -6$ Continued on following page.	$y = \frac{15}{4}$, from: $3y - 24 + 5y = 6$

35. Continued

Following the example, solve this equation.

$$\frac{y + 8}{3} - \frac{y - 1}{6} = \frac{3y - 5}{6}$$

$y = 11$, from: $2(y + 8) - (y - 1) = 3y - 5$

SELF-TEST 5 (pages 58-72)

Solve each equation. Report each solution in lowest terms.

1. $3x + 2 = 0$	2. $3.9 = 1.3 - y$	3. $w = 10 - 5w$
4. $6h - 13 = 22 - 9h$	5. $2w + 7 = 3(5 - 4w)$	6. $5 = 2E - (4E + 1)$
7. $5x - 2(3x - 1) = 1$	8. $\frac{d + 2}{3} = \frac{1}{2}$	9. $\frac{m - 1}{2} + \frac{5m}{6} = \frac{2}{3}$

ANSWERS:

1. $x = -\frac{2}{3}$	4. $h = \frac{7}{3}$	7. $x = 1$
2. $y = -2.6$	5. $w = \frac{4}{7}$	8. $d = -\frac{1}{2}$
3. $w = \frac{5}{3}$	6. $E = -3$	9. $m = \frac{7}{8}$

2-4 FORMULAS

In this section, we will discuss some formula evaluations that require solving an equation. We will also discuss some rearrangements of formulas.

---

36. To complete the evaluation below, we had to solve an equation after substituting. Complete the other evaluation.

In the formula below, find K when C = 50.

$$C = K - 273$$

$$50 = K - 273$$

$$K = 323$$

In the formula below, find R when E = 150 and I = 15.

$$E = IR$$

R = 10

37. In this frame and the following frames, we will give an example that requires equation-solving and ask you to do a similar evaluation.

In the formula below, find L when V = 40, W = 2, and H = 4.

$$V = LWH$$

$$40 = L(2)(4)$$

$$40 = 8L$$

$$L = \frac{40}{8} = 5$$

In the formula below, find b when A = 28 and h = 8.

$$A = \frac{1}{2}bh$$

---

b = 7

38. In the formula below, find r when e = 20, E = 50, and I = 2.

$$e = E - Ir$$

$$20 = 50 - 2r$$

$$-30 = -2r$$

$$r = \frac{-30}{-2} = 15$$

In the formula below, find L when P = 70 and W = 15.

$$P = 2L + 2W$$

L = 20

39. In the formula below, find F when C = 10.

$$C = \frac{5}{9}(F - 32)$$

$$10 = \frac{5}{9}(F - 32)$$

$$\frac{9}{5}(10) = \frac{9}{5}\left[\frac{5}{9}(F - 32)\right]$$

$$18 = F - 32$$

$$F = 50$$

In the formula below, find b when A = 400, h = 20, and a = 15.

$$A = \frac{1}{2}h(a + b)$$

---

b = 25

40. To complete the evaluation below, we had to solve the equation 120 = (5)(4)H.

Find H when V = 120, L = 5, and W = 4.

$$V = LWH$$

$$120 = (5)(4)H$$

$$120 = 20H$$

$$H = \frac{120}{20} = 6$$

However, we can rearrange the above formula to solve for H. That is:

Rearranging V = LWH, we get $H = \frac{V}{LW}$

Using the rearranged formula, we can do the same evaluation without solving an equation.

$$H = \frac{V}{LW} = \frac{120}{(5)(4)} = \frac{120}{20} = 6$$

One major purpose of formula rearrangement is to avoid the need for equation-solving in evaluations. A rearrangement is especially useful when the same evaluation has to be repeated a number of times.

---

41. Before rearranging formulas, we must extend some algebraic principles to literal expressions. For example:

Just as $\frac{1}{5}(3) = \frac{3}{5}$, $\frac{1}{WH}(V) = \frac{V}{WH}$

Complete these:

a) $\frac{1}{2a}(v^2) =$ ______ b) $\frac{1}{I^2}(P) =$ ______ c) $\frac{1}{Q + R}(D) =$ ______

---

a) $\frac{v^2}{2a}$ b) $\frac{P}{I^2}$ c) $\frac{D}{Q + R}$

42. When any expression is divided by itself, the quotient is "1". For example:

$\frac{I^2}{I^2} = 1$ a) $\frac{LW}{LW} =$ ______ b) $\frac{p + q}{p + q} =$ ______

---

43. Two quantities are a pair of reciprocals if their product is "1".

Since $\frac{1}{LW}(LW) = \frac{LW}{LW} = 1$, the reciprocal of LW is $\frac{1}{LW}$.

Since $\frac{1}{c + d}(c + d) = \frac{c + d}{c + d} = 1$, the reciprocal of c + d is $\frac{1}{c + d}$.

You can see that the reciprocal of a literal expression is "1" divided by the expression. That is:

The reciprocal of Pr is $\frac{1}{Pr}$.

The reciprocal of a + b is __________.

Answers to 42: a) 1 b) 1

---

44. In any term, the coefficient of a variable is <u>the other factor or factors</u>. For example:

In rt, the coefficient of <u>t</u> is <u>r</u>.

In LWH, the coefficient of W is LH.

In P(Q + R), the coefficient of P is (Q + R).

Following the examples, complete these:

a) In IR, the coefficient of I is __________.

b) In Prt, the coefficient of <u>t</u> is __________.

c) In h(a + b), the coefficient of <u>h</u> is __________.

Answer to 43: $\frac{1}{a + b}$

---

45. We used the multiplication axiom to solve for <u>t</u> below. That is, we multiplied both sides by $\frac{1}{r}$, <u>the reciprocal of the coefficient of t</u>. Solve for L in the other formula by multiplying both sides by $\frac{1}{WH}$.

$d = rt$

$\frac{1}{r}(d) = \frac{1}{r}(rt)$

$\frac{d}{r} = 1t$

$t = \frac{d}{r}$

V = LWH

Answers to 44: a) R b) Pr c) (a + b)

Answer to 45: $L = \frac{V}{WH}$

46. We used the multiplication axiom to solve for P below. Solve for b in the other formula.

$$D = P(Q + R)$$

$$a = b(c - d)$$

$$\left(\frac{1}{Q + R}\right)(D) = P(Q + R)\left(\frac{1}{Q + R}\right)$$

$$\frac{D}{Q + R} = P \cdot 1$$

$$P = \frac{D}{Q + R}$$

Answer: $b = \frac{a}{c - d}$

47. Two solutions from the preceding frames are shown below.

If $V = LWH$, $L = \frac{V}{WH}$

If $D = P(Q + R)$, $P = \frac{D}{Q + R}$

As you can see from the examples above, there is a shortcut for the multiplication axiom. That is, to solve for a variable, we can simply divide the other side of the formula by the coefficient of that variable. For example:

In $I = Prt$: The coefficient of r is Pt.

Therefore, $r = \frac{I}{Pt}$

In $2V = h(a + b)$: The coefficient of h is $(a + b)$.

Therefore, h = ________

Answer: $h = \frac{2V}{a + b}$

48. Use the shortcut of the multiplication axiom for these.

a) Solve for r^2.

$$A = \pi r^2$$

r^2 = ________

b) Solve for s.

$$v^2 = 2as$$

s = ________

Answer: a) $r^2 = \frac{A}{\pi}$ b) $s = \frac{v^2}{2a}$

49. To solve for h below, we began by multiplying both sides by 2 to get rid of the $\frac{1}{2}$. Then we proceeded in the usual way. Solve for B in the other formula.

$$A = \frac{1}{2}bh$$

$$V = \frac{1}{3}Bh$$

$$2(A) = 2\left(\frac{1}{2}bh\right)$$

$$2A = bh$$

$$h = \frac{2A}{b}$$

50. To solve for h below, we began by multiplying both sides by 2 to get rid of the $\frac{1}{2}$. Solve for v^2 in the other formula.

$$A = \frac{1}{2}h(a + b)$$

$$2(A) = 2\left[\frac{1}{2}h(a + b)\right]$$

$$2A = h(a + b)$$

$$h = \frac{2A}{a + b}$$

$$E = \frac{1}{2}(M + m)v^2$$

Answer: $B = \frac{3V}{h}$

---

51. Two terms are opposites if their sum is 0. Therefore:

Since Ir + (-Ir) = 0, Ir and -Ir are opposites.

To solve for E below, we used the addition axiom, adding Ir to both sides. Solve for A in the other formula.

$$e = E - Ir$$

$$e + Ir = E - Ir + Ir$$

$$e + Ir = E + 0$$

$$E = e + Ir$$

$$C = A - L$$

Answer: $v^2 = \frac{2E}{M + m}$

---

52. To solve for C below, we added -M to both sides. To solve for V below, we added -2D to both sides.

$$R = C + M \qquad\qquad F = V + 2D$$

$$R + (-M) = C + M + (-M) \qquad\qquad F + (-2D) = V + 2D + (-2D)$$

$$R + (-M) = C + 0 \qquad\qquad F + (-2D) = V + 0$$

$$C = R + (-M) \qquad\qquad V = F + (-2D)$$

Ordinarily the additions above are converted to subtractions. That is:

Instead of C = R + (-M), we write C = R - M

Instead of V = F + (-2D), we write ________________

Answer: A = C + L

---

53. Solve for the indicated variable. Write each solution as a subtraction.

a) Solve for E.

$$H = E + PV$$

b) Solve for b.

$$a + b + c = 180$$

Answer: V = F - 2D

54. We used both axioms to solve for P below. Solve for L in the other formula.

$$H = E + PV \qquad\qquad P = 2L + 2W$$

$$H + (-E) = (-E) + E + PV$$

$$H - E = PV$$

$$P = \frac{H - E}{V}$$

a) $E = H - PV$

b) $b = 180 - a - c$

---

55. To solve for F below, we began by multiplying both sides by $\frac{9}{5}$. Solve for S in the other formula.

$$C = \frac{5}{9}(F - 32) \qquad\qquad T = \frac{3}{2}(S - 40)$$

$$\frac{9}{5}(C) = \frac{9}{5}\left[\frac{5}{9}(F - 32)\right]$$

$$\frac{9C}{5} = F - 32$$

$$\frac{9C}{5} + 32 = F - 32 + 32$$

$$F = \frac{9C}{5} + 32$$

$L = \frac{P - 2W}{2}$

---

56. Notice how we multiplied by the distributive principle to solve for a below. Solve for d in the other formula.

$$A = \frac{1}{2}h(a + b) \qquad\qquad P = \frac{1}{3}R(c + d)$$

$$2A = h(a + b)$$

$$2A = ah + bh$$

$$2A + (-bh) = ah + bh + (-bh)$$

$$2A - bh = ah$$

$$a = \frac{2A - bh}{h}$$

$S = \frac{2T}{3} + 40$

$d = \frac{3P - cR}{R}$

2-5 WORD PROBLEMS

In this section, we will translate English phrases to algebraic expressions and English sentences to equations. Then we will solve word problems of various types.

---

57. The table on page 80 gives some translations of English phrases to algebraic expressions. The phrases involve addition, subtraction, multiplication, division, and powers. After studying the table, translate each phrase below to an algebraic expression. Use x as the variable.

    a) 7 more than 6 times a number __________

    b) 5 times the square of a number, minus 2 __________

    c) 10 divided by triple a number __________

    d) 50 minus the cube of a number __________

---

58. Translate each phrase to an algebraic expression. Use x and y as the variables.

    a) the product of two numbers, less 10 __________

    b) the difference of the squares of two numbers __________

    c) the sum of two numbers minus 5 __________

    d) the quotient of two numbers __________

Answers to 57:

a) $6x + 7$

b) $5x^2 - 2$

c) $\frac{10}{3x}$

d) $50 - x^3$

---

59. Sometimes we need parentheses to translate to algebraic expressions. For example:

    10 minus the sum of two numbers $10 - (x + y)$

    the product of 2 and the difference of a number and 5 $2(x - 5)$

    Using x for one number and x and y for two numbers, translate these.

    a) 7 times the sum of the square of a number and 3 __________

    b) 50 minus the difference of two numbers __________

    c) the product of 10 and the sum of two numbers __________

Answers to 58:

a) $xy - 10$

b) $x^2 - y^2$

c) $x + y - 5$

d) $\frac{x}{y}$

Answers to 59:

a) $7(x^2 + 3)$

b) $50 - (x - y)$

c) $10(x + y)$

Translating English Phrases To Algebraic Expressions

English Phrase	Algebraic Expression
Addition	
the sum of a number and 3	$x + 3$
9 more than a number	$x + 9$
a number plus 7	$x + 7$
15 added to a number	$x + 15$
a number increased by 10	$x + 10$
the sum of two numbers	$x + y$
Subtraction	
5 less than a number	$x - 5$
20 minus a number	$20 - x$
a number decreased by 6	$x - 6$
a number reduced by 12	$x - 12$
a number subtracted from 50	$50 - x$
the difference between 18 and a number	$18 - x$
the difference between two numbers	$x - y$
Multiplication	
14 times a number	$14x$
the product of 4 and a number	$4x$
a number multiplied by 8	$8x$
twice a number or double a number	$2x$
three times a number or triple a number	$3x$
$\frac{3}{4}$ of a number	$\frac{3}{4}x$
the product of two numbers	xy
Division	
a number divided by 5	$\frac{x}{5}$
the quotient of 7 and some number	$\frac{7}{x}$
the ratio of two numbers or the quotient of two numbers	$\frac{x}{y}$
Powers	
the square of a number	x^2
the cube of a number	x^3

60. We translated the English sentence below to an equation.

Four times a number, minus 10 equals 74. $4x - 10 = 74$

Translate each sentence to an equation. Use x as the variable.

a) 100 minus three times a number is 48. ________

b) If the difference between a number and 5 is multiplied by 3, the result is 60. ________

c) 48 divided by twice a number equals 3. ________

d) If a number is divided by the sum of that number and 2, the result is 5. ________

---

a) $100 - 3x = 48$

b) $3(x - 5) = 60$

c) $\frac{48}{2x} = 3$

d) $\frac{x}{x + 2} = 5$

---

61. The following steps are useful for solving word problems.

> Four Steps For Solving Word Problems
>
> 1. Represent the unknown or unknowns with a letter or algebraic expression.
> 2. Translate the problem to an equation.
> 3. Solve the equation.
> 4. Check the solution in the original words of the problem.

The four steps are used to solve the problem below.

Problem: If we subtract 5 from double a number, we get 11. Find the number.

Step 1: Represent the unknown with a letter.

Let x equal the unknown number.

Step 2: Translate the problem to an equation.

$$2x - 5 = 11$$

Step 3: Solve the equation.

$$2x - 5 = 11$$
$$2x = 16$$
$$x = 8$$

Step 4: Check the solution in the original words of the problem.

Double 8 is 16. If we subtract 5 from 16, we get 11. The solution checks.

Continued on following page.

61. Continued

Using the same steps, solve these:

a) If 5 is subtracted from 3 times a number, the result is the number plus 11. Find the number.

b) If the sum of a number and 10 is divided by 3, the result is 8. Find the number.

---

62. Three consecutive integers are integers that are next to each other, like 5, 6, and 7. The problem below involves three consecutive integers.

Problem: The sum of three consecutive integers is 42. What are the integers?

If the smallest is x, the middle number is x + 1 and the largest is x + 2. Therefore, we get the following translation.

$$\begin{array}{ccccccc} \text{smallest} & + & \text{middle number} & + & \text{largest} & = & 42 \\ \downarrow & \downarrow & \downarrow & \downarrow & \downarrow & \downarrow & \downarrow \\ x & + & (x+1) & + & (x+2) & = & 42 \end{array}$$

Solving the equation, we get:

$$x + (x + 1) + (x + 2) = 42$$
$$3x + 3 = 42$$
$$3x = 39$$
$$x = 13$$

Check. The answers are 13, 14, and 15. They are consecutive integers and their sum is 42. The solution checks.

Three consecutive odd integers are integers like 7, 9, and 11. If the smallest is x, the middle one is x + 2 and the largest is x + 4. Use that fact for these:

a) The sum of three consecutive odd integers is 111. What are the integers?

b) If 11 is added to the largest of three consecutive odd integers, the answer equals the sum of the smaller two. Find the integers.

a) 8, from:

$$3x - 5 = x + 11$$

b) 14, from:

$$\frac{x + 10}{3} = 8$$

63. The problem below involves dividing a whole into two parts.

Problem: A wire 50 centimeters long is cut into two parts. The larger part is 10 centimeters longer than the smaller part. How long is each part?

We drew a diagram for the problem below. In the diagram, we let x equal the length of the smaller part. Therefore, the length of the larger part is x + 10.

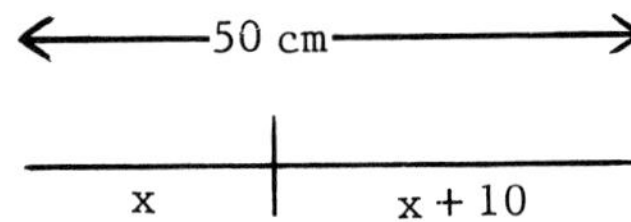

We translated to an equation and solved the equation below.

$$\begin{array}{l} \text{smaller} + \text{larger} = 50 \\ \downarrow \quad \downarrow \quad \downarrow \quad \downarrow \quad \downarrow \\ x + (x + 10) = 50 \\ 2x + 10 = 50 \\ 2x = 40 \\ x = 20 \quad (\text{and } x + 10 = 30) \end{array}$$

Check: The smaller part is 20 cm and the larger part is 30 cm. 30 is 10 longer than 20 and 30 + 20 equals 50. Therefore, the solution checks.

Using the same steps, solve these:

a) A rope 56 feet long is cut into two parts. One part is three times as long as the other. How long is each part?

b) There were 27 more females than males in a graduation class. If the total number of graduates was 207, find the number of males and females.

a) 35, 37, and 39

b) 13, 15, and 17

---

64. The following problem involves a percent.

$$\begin{array}{l} 108 \text{ is } 12\% \text{ of what number?} \\ \downarrow \quad \downarrow \quad \downarrow \quad \downarrow \qquad \downarrow \\ 108 = 12\% \cdot x \end{array}$$

$$108 = .12x$$

$$x = \frac{108}{.12} = 900$$

Check. $12\% \cdot 900 = .12(900) = 108$

Continued on following page.

a) 14 feet and 42 feet

b) 90 males and 117 females

64. Continued

Use the same method for these.

a) 14 is 8% of what number? b) 63 is what percent of 84?

---

65. The following problem also involves a percent.

a) 175 b) 75%

If the annual interest on a $500 loan is $60, what is the interest rate?

Rewording: $60 is what percent of $500?

Equation: $60 = x \cdot 500$

Solution: $x = \frac{60}{500} = 12\%$

Use the same method for these:

a) How much money must be invested at 5% annual interest to earn $500 interest annually?

b) A woman with an annual salary of $24,000 gets a 7% raise. How large is the raise? What is her new salary?

---

a) $10,000

b) Raise = $1,680

New salary = $25,680

66. To solve the problem below. we used the formula P = 2L + 2W. We let W = x and L = x + 20. Use the same method for the other problem. Draw an appropriate figure and label the sides.

The perimeter of a rectangle is 160 cm. The length is 20 cm greater than the width. Find the dimensions.

The perimeter of a rectangle is 236 ft. The width is 22 ft less than the length. Find the dimensions.

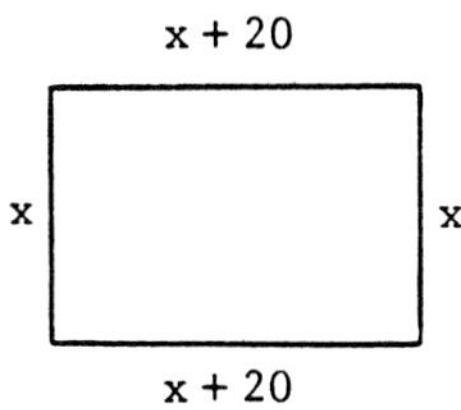

$$P = 2L + 2W$$

$$160 = 2(x + 20) + 2(x)$$

$$160 = 2x + 40 + 2x$$

$$160 = 4x + 40$$

$$120 = 4x$$

$$x = 30$$

Therefore, the width is 30 cm and the length is 30 + 20 = 50 cm.

---

67. We used the fact that the sum of the three angles of a triangle is 180° to solve the problem below. Solve the other problem.

L = 70 ft; W = 48 ft

The second angle of a triangle is three times as large as the first. The third angle is 30° larger than the first. How large are the angles?

The second angle of a triangle is twice as large as the first. The third angle is 20° less than the first. How large are the angles?

1st angle = x

2nd angle = 3x

3rd angle = x + 30

$$x + 3x + (x + 30) = 180$$

$$5x + 30 = 180$$

$$5x = 150$$

$$x = 30$$

Therefore, we get these sizes for the angles:

1st angle = 30°

2nd angle = 90°

3rd angle = 60°

68. The following problem involves a mixture.

Problem: How many liters of a 20% acid solution must be mixed with 6 liters of a 70% acid solution to get a 50% acid solution?

Letting x equal the number of liters of the 20% solution, we can organize the data in the following table. To get the numbers in the right column, we multiplied the strengths and the number of liters.

Strength	Liters of solution	Liters of pure acid
70%	6	$.70(6) = 4.2$
20%	x	$.20x$
50%	6 + x	$.50(6 + x)$

Since the number of liters of pure acid in the 50% solution is the sum of the number of liters of pure acid in the 70% and 20% solution, we can set up and solve the following equation.

$$4.2 + .20x = .50(6 + x)$$

$$4.2 + .20x = 3 + .50x$$

$$1.2 = .3x$$

$$x = \frac{1.2}{.3} = 4$$

Therefore, 4 liters of the 20% solution must be used.

Using the same method, solve this mixture problem.

How many pounds of nuts worth $3 per pound must be mixed with 20 pounds of nuts worth $6 per pound to get a mixture worth $4 per pound?

1st angle = 50°

2nd angle = 100°

3rd angle = 30°

40 pounds of the $3 nuts

69. We use the formula $d = rt$ for the problem below. In that formula, d = distance traveled, r = rate (or speed), and t = time traveled.

Problem: One car leaves a motel traveling 42 mph. A second car leaves the same motel one hour later traveling 56 mph on the same highway. How long does it take the second car to catch up to the first car?

We diagrammed the problem below. The two distances are equal. Since we want to find the time traveled by the second car, we used t for the time of the second car.

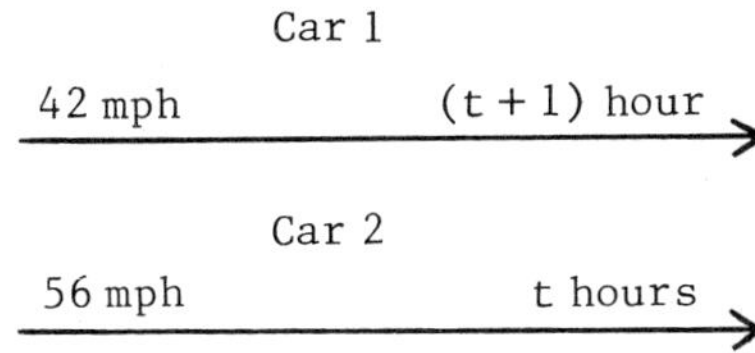

We organized the data in the chart below. We used $d = rt$ to get $42(t + 1)$ and $56t$.

	Distance	Rate	Time
Car 1	$d = 42(t + 1)$	42	$t + 1$
Car 2	$d = 56t$	56	t

From the diagram, we can see that the two distances are equal. Therefore, we can set up and solve the following equation.

$$42(t + 1) = 56t$$

$$42t + 42 = 56t$$

$$42 = 14t$$

$$t = 3 \text{ hours (and } t + 1 = 4 \text{ hours)}$$

Therefore, it takes the second car 3 hours to catch up to the first car (which travels 4 hours). This checks since $4(42) = 168$ and $3(56) = 168$.

Solve this one. Use a diagram and a chart.

Problem: A small plane leaves an airport and flies due north at 180 km/h. A jet leaves the same airport two hours later and flies due north at 900 km/h. How long will it take the jet to catch up to the small plane?

$t = \frac{1}{2}$ hour

SELF-TEST 6 (pages 73-88)

1. In $V = \frac{1}{3}BH$, find B when V = 20 and H = 12.

2. In P = 2L + 2W, find W when P = 84 and L = 24.

3. Solve for b.

$$A = \frac{1}{2}bh$$

4. Solve for C.

$$P = C + M$$

5. Solve for b.

$$A = \frac{1}{2}h(a + b)$$

6. The sum of three consecutive integers is 162. What are the three integers?

7. On an algebra test, the highest grade is 54 points more than the lowest grade. The sum of the two grades is 138. Find the lowest grade.

8. How much money must be invested at 12% annual interest to earn $600 interest annually?

9. The second angle of a triangle is four times as large as the first. The third angle is 30° larger than the first. How large are the three angles?

10. How many liters of a 10% alcohol solution must be mixed with 20 liters of a 60% solution to get a 30% solution?

11. A plane leaves an airport traveling 160 mph. A second plane leaves the same airport two hours later traveling 400 mph in the same direction. How long will it take the second plane to catch up to the first plane?

ANSWERS:

1. B = 5
2. W = 18
3. $b = \frac{2A}{h}$
4. C = P - M
5. $b = \frac{2A - ah}{h}$
6. 53, 54, and 55
7. 42 points
8. $5,000
9. 25°, 100°, and 55°
10. 30 liters
11. $1\frac{1}{3}$ hours

2-6 SOLVING INEQUALITIES

In this section, we will show how the addition axiom and multiplication axiom are used to solve inequalities.

---

70. Statements containing $<$, $>$, $\leq$, or $\geq$ are called inequalities. Two examples of linear inequalities in one variable are shown below.

$x + 3 < 7$ $\qquad$ $2y - 1 \geq 5$

A linear inequality is neither true nor false until we substitute a number for the variable. Then it becomes either true or false. For example:

If we substitute 1, 2, or 3 for x, $x + 3 < 7$ is true.

If we substitute 4, 5, or 6 for x, $x + 3 < 7$ is false.

The substitutions that make a linear inequality true are called solutions of the inequality.

a) Are 2, 3, or 4 solutions of $y - 1 > 5$? __________

b) Are 5, 6, or 7 solutions of $y + 2 \leq 9$? __________

---

71. To solve an inequality means to find its solutions. One principle used to solve inequalities is the addition axiom for inequalities. It is stated below.

a) No
b) Yes

> The Addition Axiom For Inequalities
>
> IF WE ADD THE SAME QUANTITY TO BOTH SIDES OF AN INEQUALITY THAT IS TRUE, THE NEW INEQUALITY IS ALSO TRUE. That is:
>
> If: $A > B$
>
> Then: $A + C > B + C$
>
> Note: The addition axiom also applies to $<$, $\geq$, and $\leq$.

We used the addition axiom to solve $x + 2 > 5$ below. We added -2 to both sides.

$$x + 2 > 5$$

$$x + 2 - 2 > 5 - 2$$

$$x > 3$$

The solution of $x + 2 > 5$ is $x > 3$. That is, any number greater than 3 satisfies the inequality. We can express the solution as a solution set. We get:

The solution set of $x + 2 > 5$ is $\{x|x > 3\}$.

Continued on following page.

71. Continued

We can also graph the solution as we have done below. The open circle at 3 means that 3 is not included.

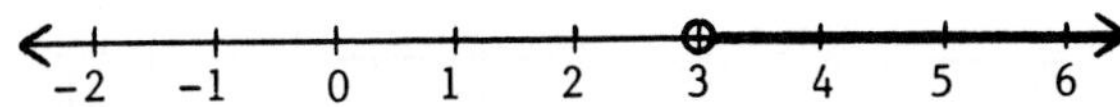

Solve the inequality at the left below and then answer the questions at the right below.

$y + 4 < 6$

a) The solution is ________.

b) The solution set is ________.

c) The graph of the solution is:

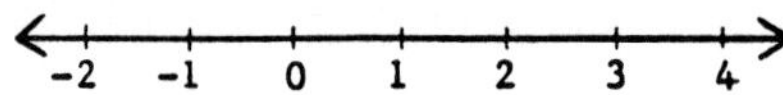

---

a) $y < 2$

b) $\{y|y < 2\}$

c) 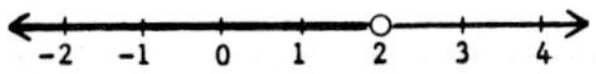 (graph: open circle at 2, shaded to the left)

72. We solved $x - 1 \leq 0$. Then we stated the solution and solution set and graphed the solution. On the graph, the closed circle at "1" means that "1" is included. Solve the other inequality and then answer the questions.

$x - 1 \leq 0$ $\qquad$ $y - 3 \geq -5$

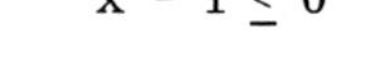

$x - 1 + 1 \leq 0 + 1$

$x \leq 1$

The solution is $x \leq 1$.

The solution set is $\{x|x \leq 1\}$.

The graph of the solution is:

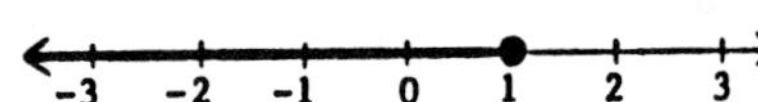

a) The solution is ____________.

b) The solution set is ____________.

c) The graph of the solution is:

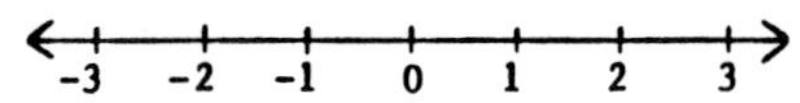

---

a) $y \geq -2$

b) $\{y|y \geq -2\}$

c) 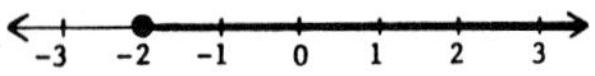 (graph: closed circle at −2, shaded to the right)

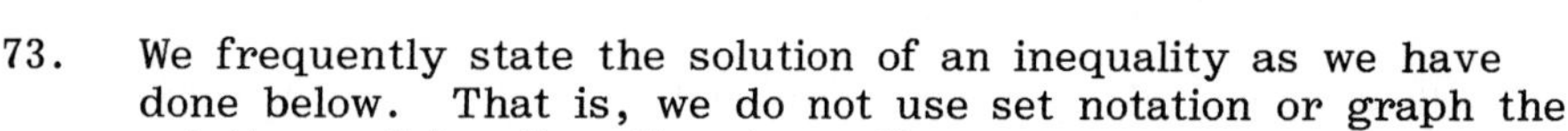

73. We frequently state the solution of an inequality as we have done below. That is, we do not use set notation or graph the solution. Solve the other inequality.

$y + 1.4 \leq 6.7$ $\qquad$ $x - 2.6 \geq 3.3$

$y + 1.4 - 1.4 \leq 6.7 - 1.4$

$y \leq 5.3$

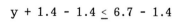

$x \geq 5.9$

74. Following the example, solve the other inequality.

$$x - \frac{1}{4} < \frac{1}{2}$$

$$y + \frac{2}{3} \geq \frac{1}{6}$$

$$x - \frac{1}{4} + \frac{1}{4} < \frac{1}{2} + \frac{1}{4}$$

$$x < \frac{3}{4}$$

$y \geq -\frac{1}{2}$

75. We solved two inequalities below.

$$5 > x + 3 \qquad\qquad -7 \leq y - 2$$

$$5 - 3 > x + 3 - 3 \qquad\qquad -7 + 2 \leq y - 2 + 2$$

$$2 > x \qquad\qquad -5 \leq y$$

It is easier to read the solutions when the variable is on the left. Therefore:

Instead of $2 > x$, we can write $x < 2$.

Instead of $-5 \leq y$, we can write $y \geq -5$.

Solve these. Write each solution with the variable on the left side.

a) $0 < x - 1.5$

b) $\frac{1}{2} \geq y + \frac{1}{3}$

---

a) $x > 1.5$

b) $y \leq \frac{1}{6}$

76. The inequality $3 < 5$ is true. Let's multiply both sides by 2 and -4 and see whether we get true inequalities.

$$3 < 5 \qquad\qquad 3 < 5$$

$$2(3) < 2(5) \qquad\qquad -4(3) < -4(5)$$

$$6 < 10 \quad \text{True} \qquad\qquad -12 < -20 \quad \text{False}$$

Though $6 < 10$ is true, $-12 < -20$ is false. To get a true inequality when multiplying by a negative number like -4, we have to reverse the direction of the inequality symbol. That is:

$$3 < 5$$

$$-4(3) > -4(5)$$

↑ Reversed < to >

$$-12 > -20$$

Is $-12 > -20$ a true inequality? ________

Yes

77. A second principle used to solve inequalities is the multiplication axiom for inequalities. It is stated below.

The Multiplication Axiom For Inequalities

IF WE MULTIPLY BOTH SIDES OF AN INEQUALITY THAT IS TRUE BY A POSITIVE NUMBER, THE NEW INEQUALITY IS ALSO TRUE. That is:

If : $B > C$ and $A > 0$

Then: $AB > AC$

IF WE MULTIPLY BOTH SIDES OF AN INEQUALITY THAT IS TRUE BY A NEGATIVE NUMBER AND REVERSE THE INEQUALITY SYMBOL, THE NEW INEQUALITY IS ALSO TRUE. That is:

If : $B > C$ and $A < 0$

Then: $AB < AC$

Note: The multiplication axiom also applied to $<$, $\geq$, and $\leq$.

To solve the inequality below, we multiplied both sides by the positive number $\frac{1}{4}$.

$$4x > 12$$

$$\frac{1}{4}(4x) > \frac{1}{4}(12)$$

$$x > 3$$

The solution of $4x > 12$ is $x > 3$. The solution set is $\{x|x > 3\}$.

Use the same method to solve these.

a) $5x < -20$

b) $3y \geq 18$

---

78. To solve the inequality below, we multiplied both sides by the negative number $-\frac{1}{2}$. Notice that we reversed the inequality symbol.

$$-2x \leq 18$$

$$-\frac{1}{2}(-2x) \geq -\frac{1}{2}(18)$$

$$x \geq -9$$

Use the same method to solve these. Be sure to reverse the inequality symbol.

a) $-7x > -14$

b) $-6y \leq 48$

Answers to 77:

a) $x < -4$

b) $y \geq 6$

79. Solve these. Reverse the inequality symbol when multiplying by a negative number. a) $3x < 2$ b) $-5y \geq 0$ c) $-7t \leq -8$	a) $x < 2$ b) $y \geq -8$
80. Following the example, solve the other inequality. $3x < \frac{2}{5}$ $\frac{1}{3}(3x) < \frac{1}{3}\left(\frac{2}{5}\right)$ $x < \frac{2}{15}$ $2y \geq -\frac{1}{2}$	a) $x < \frac{2}{3}$ b) $y \leq 0$ c) $t \geq \frac{8}{7}$
81. Following the example, solve the other inequality. $-\frac{1}{2}x > 3$ $-2\left(-\frac{1}{2}x\right) < -2(3)$ $x < -6$ $-\frac{3}{5}y \leq -1$	$y \geq -\frac{1}{4}$
82. To solve for <u>x</u> below, we multiplied both sides by -1. Solve the other inequalities. $-x < 6$ $-1(-x) > -1(6)$ $x > -6$ a) $-y > -3$ b) $-t \leq \frac{3}{5}$	$y \geq \frac{5}{3}$
83. After solving the inequality below, we wrote the solution with the variable on the left. Write the other solution with the variable on the left side. $4 \geq 3x$ $\frac{1}{3}(4) \geq \frac{1}{3}(3x)$ $\frac{4}{3} \geq x$ $x \leq \frac{4}{3}$ $-6 < -7y$	a) $y < 3$ b) $t \geq -\frac{3}{5}$
	$y < \frac{6}{7}$

84. We used both axioms to solve the inequality below. Solve the other inequality.

$$3x - 2 < 5$$
$$3x - 2 + 2 < 5 + 2$$
$$3x < 7$$
$$\frac{1}{3}(3x) < \frac{1}{3}(7)$$
$$x < \frac{7}{3}$$

$$2x + 6 \geq 7$$

---

85. We used both axioms below. Solve the other inequality.

$x \geq \frac{1}{2}$

$$4 - 3x > 5$$
$$-4 + 4 - 3x > 5 - 4$$
$$-3x > 1$$
$$-\frac{1}{3}(-3x) < -\frac{1}{3}(1)$$
$$x < -\frac{1}{3}$$

$$2 - 5y \leq 6$$

---

86. Notice that we multiplied both sides by -1 below. Solve the other inequality.

$y \geq -\frac{4}{5}$

$$3 - x \geq 9$$
$$-3 + 3 - x \geq 9 - 3$$
$$-x \geq 6$$
$$-1(-x) \leq -1(6)$$
$$x \leq -6$$

$$7 - y < 4$$

---

87. To solve the inequality below, we used the addition axiom twice to get -7x on the left side and -9 on the right side. Solve the other inequality.

$y > 3$

$$8 - 3x < 4x - 1$$
$$-8 + 8 - 3x < 4x - 1 - 8$$
$$-3x < 4x - 9$$
$$-4x - 3x < -4x + 4x - 9$$
$$-7x < -9$$
$$-\frac{1}{7}(-7x) > -\frac{1}{7}(-9)$$
$$x > \frac{9}{7}$$

$$2y - 5 \geq 7y + 5$$

$y \leq -2$

88. In solving the inequality below, we got the variable on the right side. We rewrote the solution to get the variable on the left side. Solve the other inequality.

$$3 - 5x > x - 5$$
$$3 - 5x + 5x > 5x + x - 5$$
$$3 > 6x - 5$$
$$5 + 3 > 6x - 5 + 5$$
$$8 > 6x$$
$$\frac{1}{6}(8) > \frac{1}{6}(6x)$$
$$\frac{4}{3} > x$$
$$x < \frac{4}{3}$$

$$9 - 6x \leq 7 - 4x$$

---

89. To solve the inequality below, we began by simplifying the expression on the left side. Solve the other inequality.

$x \geq 1$

$$3(x + 4) - (x - 5) < 25$$
$$3x + 12 - x + 5 < 25$$
$$2x + 17 < 25$$
$$2x + 17 - 17 < 25 - 17$$
$$2x < 8$$
$$x < 4$$

$$4t \geq 2(3t - 5) - (5t + 8)$$

---

90. Solve the following inequality by simplifying both sides and then using the axiom.

$t \geq -6$

$$3(y - 1) - 2(y + 4) > 10y - (y - 9)$$

$y < -\frac{5}{2}$

2-7 INEQUALITIES IN WORD PROBLEMS

In this section, we will solve some word problems involving inequalities.

---

91. We translated each phrase below to an inequality.

All numbers greater than -1.	$x > -1$
5 and all numbers less than 5.	$x \leq 5$
All positive numbers.	$x > 0$

Using x as the variable, translate these to inequalities.

a) All numbers less than 7. ________

b) -1 and all numbers greater than -1. ________

c) All negative numbers. ________

---

92. We translated each sentence below to an inequality.

The sum of a number and 3 is less than 10.	$x + 3 < 10$
Twice a number, minus 1, is greater than or equal to -4.	$2x - 1 \geq -4$

Using x as the variable, translate each sentence to an inequality.

a) The difference between a number and 4 is greater than 8. ________

b) Four times a number is less than or equal to 20. ________

c) 10 is greater than or equal to three times a number, plus 5. ________

Answers to 91:

a) $x < 7$

b) $x \geq -1$

c) $x < 0$

---

93. We used an inequality to solve the problem below. Solve the other problem.

If 4 is subtracted from twice a number, the result is less than 8. Find the numbers satisfying this condition.

$$2x - 4 < 8$$
$$2x < 12$$
$$x < 6$$

Any number less than 6 satisfies this condition.

The sum of a number and 10 is equal to or greater than 25. Find the numbers satisfying this condition.

Answers to 92:

a) $x - 4 > 8$

b) $4x \leq 20$

c) $10 \geq 3x + 5$

94. Following the example, solve the other problem.

The length of a rectangle is 20 meters. The area of the rectangle can be no more than 240 square meters. Find the longest possible width of the rectangle.

$$LW \le 240$$

$$20W \le 240$$

$$W \le 12$$

The longest possible width is 12 meters.

The width of a rectangle is 8 ft. The area of the rectangle must be at least 160 ft^2. Find the shortest possible length of the rectangle.

$x \ge 15$

15 or any number greater than 15 satisfies this condition.

---

95. Following the example, solve the other problem.

Three tests are given in an algebra course. To get a B, a student needs a total of 240 points. If a student gets scores of 82 and 75 on the first two tests, what score can he get on the third test and get a B?

$$T_1 + T_2 + T_3 \ge 240$$

$$82 + 75 + T_3 \ge 240$$

$$157 + T_3 \ge 240$$

$$T_3 \ge 83$$

A score of 83 or more on the third test will get him a B.

Four tests are given in a business course. To get an A, a student needs a total of 360 points. If a student gets scores of 85, 93, and 89 on the first three tests, what score can she get on the fourth test to get an A?

$L \ge 20$

The shortest possible length is 20 ft.

$T_4 \ge 93$

A score of 93 or more on the fourth test will get her an A.

96. To solve the problem below, we let x equal the number of miles. Notice how we multiplied both sides of $1.00 < .02x$ by 100 to get rid of the decimals. Solve the other problem.

Company A rents compact cars for \$22.95 plus 20¢ per mile. Company B rents compact cars for \$21.95 plus 22¢ per mile. For what mileages would the cost of renting from Company A be cheaper?

$$22.95 + .20x < 21.95 + .22x$$
$$1.00 < .02x$$
$$100 < 2x$$
$$50 < x$$
$$x > 50$$

If you drive more than 50 miles, it is cheaper to rent from Company A.

Company A rents intermediate-size cars at \$26.95 plus 20¢ per mile. Company B rents intermediate-size cars at \$24.95 plus 22¢ per mile. For what mileages would the cost of renting from Company A be cheaper?

$x > 100$

If you drive more than 100 miles, it is cheaper to rent from Company A.

---

97. To solve the problem below, we set up an inequality stating that the income from Plan A is greater than the income from Plan B. Letting x equal gross sales, the salesman receives a commission on x in Plan A and on x - 8000 in Plan B. Solve the other problem.

A salesman can be paid in one of two ways:

Plan A: A salary of \$1000 per month, plus a commission of 4% of gross sales.

Plan B: A salary of \$1250 per month, plus a commission of 5% of gross sales over \$8000.

For what gross sales is Plan A better than Plan B, assuming that gross sales are always more than \$8000?

$$1000 + 4\%(x) > 1250 + 5\%(x-8000)$$
$$1000 + .04x > 1250 + .05(x-8000)$$
$$1000 + .04x > 1250 + .05x - 400$$
$$1000 + .04x > 850 + .05x$$
$$150 > .01x$$
$$15000 > x$$
$$x < 15000$$

Plan A is better for gross sales under \$15,000.

A salesman can be paid in one of two ways:

Plan A: A salary of \$1200 per month, plus a commission of 3% of gross sales.

Plan B: A salary of \$1300 per month, plus a commission of 5% of gross sales over \$10,000.

For what gross sales is Plan B better than Plan A, assuming that gross sales are always more than \$10,000?

98. To solve the problem below, we let x equal the amount invested at 7% and 20,000 - x equal the amount invested at 9%. Solve the other problem.

You are going to invest $20,000, part at 7% and part at 9%. What is the most that can be invested at 7% in order to make at least $1500 interest per year?

$7\%(x)+9\%(20000-x) \geq 1500$

$.07x+.09(20000-x) \geq 1500$

$.07x + 1800 - .09x \geq 1500$

$-.02x \geq -300$

$-2x \geq -30{,}000$

$x \leq 15{,}000$

$15,000 is the most that could be invested at 7%.

You are going to invest $15,000, part at 8% and part at 10%. What is the most that can be invested at 8% in order to make at least $1300 interest per year?

$x > 20{,}000$

Plan B is better for gross sales over $20,000.

$x \leq 10{,}000$

$10,000 is the most that could be invested at 8%.

SELF-TEST 7 (pages 89-100)

Solve each inequality and graph the solution.

1. $x + 2 < 1$

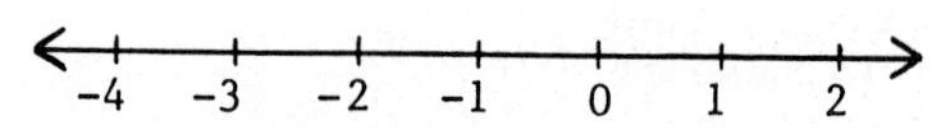

2. $x - 2 \geq 0$

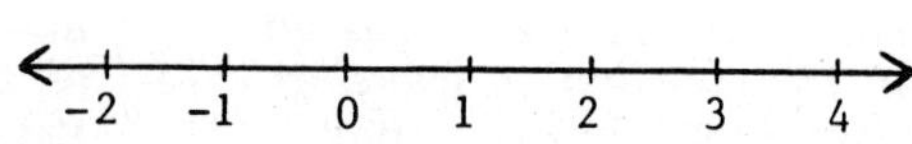

Solve each inequality.

3. $y - \frac{1}{6} \leq \frac{1}{2}$

4. $0 > x + 2.6$

5. $-4y \geq 12$

6. $2x < \frac{1}{4}$

7. $-1 \geq 5t$

8. $5 - x > 9$

9. $3y - 4 \leq y + 1$

10. $2(x - 5) - 3(x - 4) > 4x - (2x - 7)$

Solve each problem.

11. The length of a rectangle is 15 ft. The area of the rectangle can be no more than 135 ft^2. Find the longest possible width of the rectangle.

12. You are going to invest $14,000, part at 6% and part at 9%. What is the most that can be invested at 6% in order to make at least $1200 interest per year?

ANSWERS:

1. $x < -1$

2. $y \geq 2$

3. $y \leq \frac{2}{3}$

4. $x < -2.6$

5. $y \leq -3$

6. $x < \frac{1}{8}$

7. $t \leq -\frac{1}{5}$

8. $x < -4$

9. $y \leq \frac{5}{2}$

10. $x < -\frac{5}{3}$

11. $W \leq 9$ ft

12. $2000 is the most that could be invested at 6%

2-8 COMBINED INEQUALITIES

A combined inequality is made up of two or more inequalities. We will discuss inequalities of that type in this section.

---

99. The inequality $-1 < x < 3$ is a <u>combined</u> <u>inequality</u>. It is a shorthand way of writing the two inequalities $-1 < x$ <u>and</u> $x < 3$. Any solution of $-1 < x < 3$ must satisfy both of the simpler inequalities.

We can find the solution of $-1 < x < 3$ by graphing. To do so, we begin by graphing the two simpler inequalities.

$-1 < x$

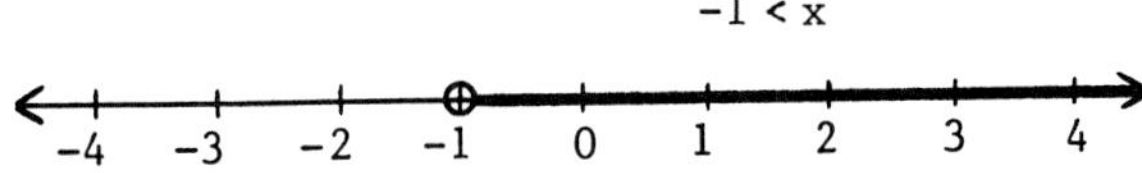

$x < 3$

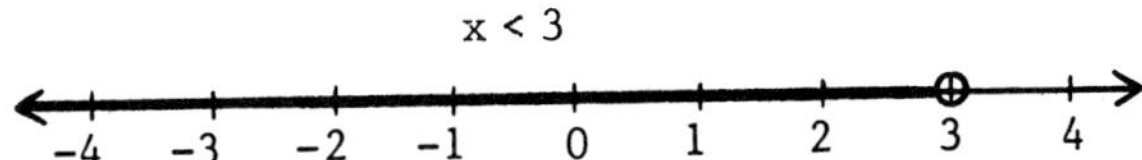

The solution of $-1 < x < 3$ includes the points common to the two graphs above. It is shown below. Finding the points common to the first two graphs is called finding their <u>intersection</u>.

$-1 < x < 3$

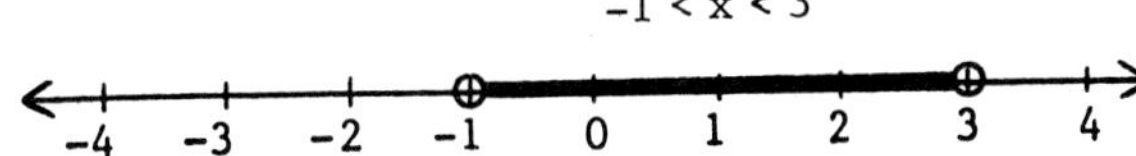

The solution set is $\{x \mid -1 < x < 3\}$.

---

100. Since the combined inequality $-3 < x \le 2$ means "$\underline{x}$ lies between -3 and 2", we graphed it directly below. Notice that the point 2 is included in the graph.

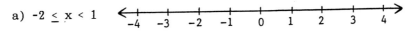

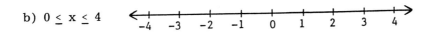

$-3 < x \le 2$

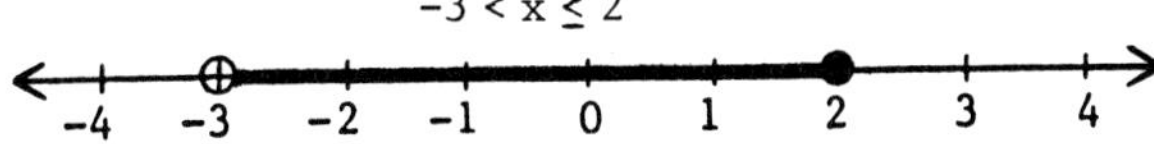

Graph $-2 \le x < 1$ and $0 \le x \le 4$ below.

a) $-2 \le x < 1$ -4 -3 -2 -1 0 1 2 3 4

b) $0 \le x \le 4$ -4 -3 -2 -1 0 1 2 3 4

---

a) -4 -3 -2 -1 0 1 2 3 4

b) -4 -3 -2 -1 0 1 2 3 4

101. The inequality $-2 < x + 3 < 6$ is a shorthand way of writing $-2 < x + 3$ and $x + 3 < 6$. We can solve it in either of two ways:

1. Solve each simpler inequality and then translate back.

$$-2 < x + 3 \quad \text{and} \quad x + 3 < 6$$

$$-2 - 3 < x + 3 - 3 \quad \text{and} \quad x + 3 - 3 < 6 - 3$$

$$-5 < x \quad \text{and} \quad x < 3$$

$$-5 < x < 3$$

2. Add -3 to the three expressions in the combined inequality.

$$-2 < x + 3 < 6$$

$$-2 - 3 < x + 3 - 3 < 6 - 3$$

$$-5 < x < 3$$

Using the second method, solve these:

a) $-1 < x + 2 < 7$ b) $-5 < y + 4 \leq 9$

---

a) $-3 < x < 5$

b) $-9 < y \leq 5$

102. To solve the inequality below, we added 5 to all three expressions. Solve the other inequality.

$$-2 < x - 5 < 1$$

$$-2 + 5 < x - 5 + 5 < 1 + 5$$

$$3 < x < 6$$

$$0 < y - 6 < 7$$

$6 < y < 13$

103. To solve the inequality below, we multiplied all three expressions by $\frac{1}{2}$. Solve the other inequality.

$$-4 \leq 2x < 3$$

$$\frac{1}{2}(-4) \leq \frac{1}{2}(2x) < \frac{1}{2}(3)$$

$$-2 \leq x < \frac{3}{2}$$

$$2 < 3y \leq 9$$

$\frac{2}{3} < y \leq 3$

104. We used both axioms to solve the inequality below. Solve the other inequality.

$$-6 < 4x + 1 < -3 \qquad\qquad -11 < 5y + 4 < -2$$

$$-6 - 1 < 4x + 1 - 1 < -3 - 1$$

$$-7 < 4x < -4$$

$$\frac{1}{4}(-7) < \frac{1}{4}(4x) < \frac{1}{4}(-4)$$

$$-\frac{7}{4} < x < -1$$

$-3 < y < -\frac{6}{5}$

105. Following the example, solve the other inequality.

$$2 < 3x - 1 < 8 \qquad\qquad 0 < 2t - 1 \le 7$$

$$2 + 1 < 3x - 1 + 1 < 8 + 1$$

$$3 < 3x < 9$$

$$\frac{1}{3}(3) < \frac{1}{3}(3x) < \frac{1}{3}(9)$$

$$1 < x < 3$$

$\frac{1}{2} < t \le 4$

106. To solve the inequality below, we multiplied all three expressions by 4 to clear the fraction. Then we continued in the usual way. Solve the other inequality.

$$-8 \le \frac{3x + 1}{4} \le 3 \qquad\qquad -1 < \frac{5y - 3}{6} < 7$$

$$4(-8) \le \not{4}\left(\frac{3x + 1}{\not{4}}\right) \le 4(3)$$

$$-32 \le 3x + 1 \le 12$$

$$-32 - 1 \le 3x + 1 - 1 \le 12 - 1$$

$$-33 \le 3x \le 11$$

$$\frac{1}{3}(-33) \le \frac{1}{3}(3x) \le \frac{1}{3}(11)$$

$$-11 \le x \le \frac{11}{3}$$

$-\frac{3}{5} < y < 9$

107. A second type of combined inequality contains the word "or". An example is given.

$$x < -2 \quad \underline{\text{or}} \quad x \geq 1$$

We can find the solution by graphing. To do so, we begin by graphing the two simpler inequalities.

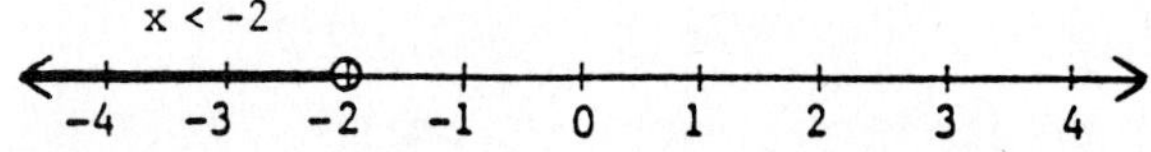

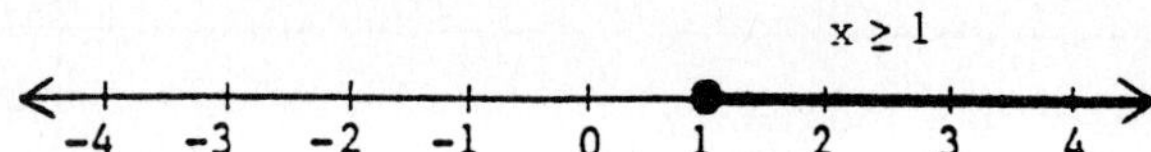

The solution of $x < -2$ or $x \geq 1$ includes all numbers satisfying $x < -2$ and all numbers satisfying $x \geq 1$. It is shown below. Finding the combination of the two graphs is called finding their union.

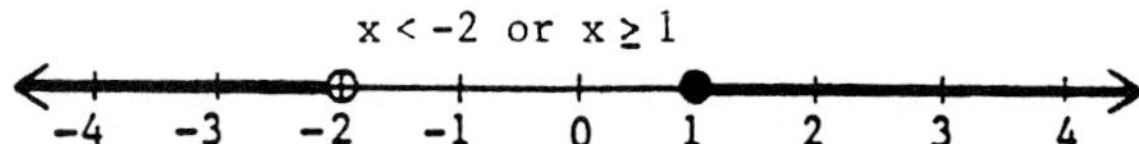

Note: There is no shorthand way of writing "$x < -2$ or $x \geq 1$". When the word "or" occurs, you must keep it.

Following the example, graph the solution of $x \leq -1$ or $x > 3$ below.

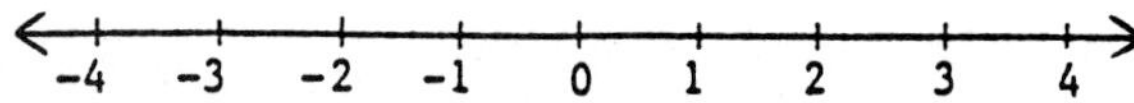

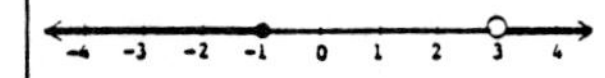

---

108. To solve $x + 3 < 2$ or $5x > 10$, we solved each inequality separately below.

$$x + 3 < 2 \quad \text{or} \quad 5x > 10$$

$$x + 3 - 3 < 2 - 3 \quad \text{or} \quad \frac{1}{5}(5x) > \frac{1}{5}(10)$$

$$x < -1 \quad \text{or} \quad x > 2$$

We graphed the solution below.

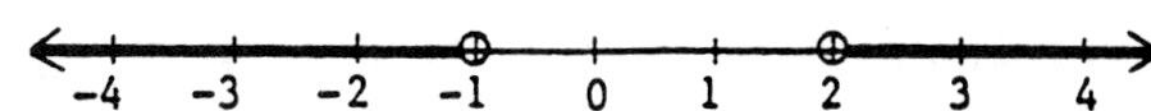

Solve and graph $2x + 7 < 1$ or $x - 1 > 0$

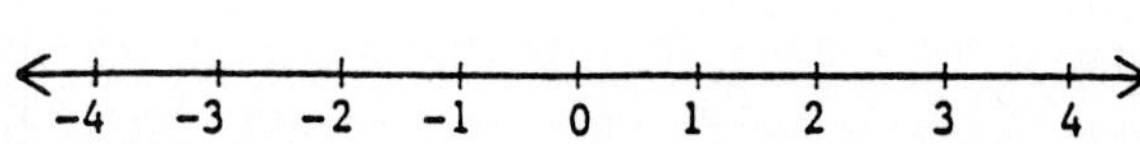

$x < -3$ or $x > 1$

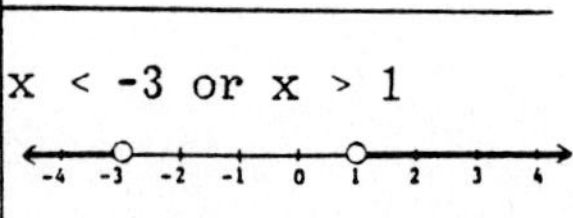

2-9 ABSOLUTE VALUE EQUATIONS

In this section, we will solve some equations in which the variable is part of an absolute value expression.

---

109. Both 3 and -3 have the same absolute value because they are the same distance from 0 on the number line. That is:

$$|3| = 3 \qquad \text{and} \qquad |-3| = 3$$

The equation $|x| = 3$ says: "Find all numbers whose absolute value is 3". The two numbers are 3 and -3. That is:

If $|x| = 3$, then $x = 3$ or $x = -3$.

The solution set of $|x| = 3$ is $\{3,-3\}$. We graphed the two solutions below.

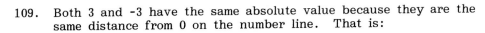

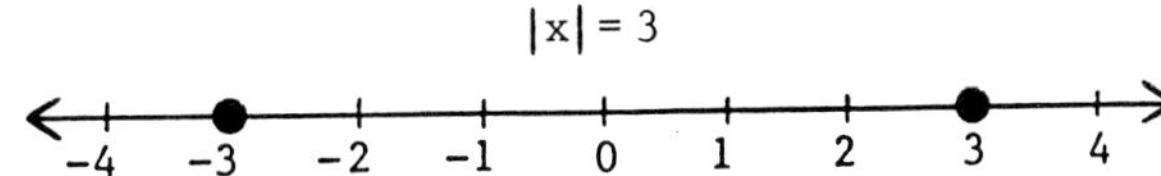

Write the solution set for these:

a) $|x| = 7$ b) $|y| = \frac{2}{3}$

---

110. The two equations $|x| = -5$ and $|x| = 0$ are discussed below.

$|x| = -5$ This equation has no solution because the absolute value of a number is never negative.

$|x| = 0$ Since only 0 has an absolute value of 0, the only solution of the equation is 0.

Solve these:

a) $|x| = -1$ b) $|x| = 1$ c) $|x| = 0$

Answers to 109: a) $\{7,-7\}$ b) $\left\{\frac{2}{3},-\frac{2}{3}\right\}$

---

111. The following principle can be used to solve absolute value equations.

> Principle For Absolute Value Equations
>
> If $|X| = b$, then $X = b$ or $X = -b$

We can apply the principle above to equations like the one below.

If $|x + 3| = 8$, then $x + 3 = 8$ or $x + 3 = -8$.

Continued on following page.

Answers to 110: a) no solution b) 1 and -1 c) 0

111. Continued

We can solve the original equation by solving the two new equations.

$$x + 3 = 8 \qquad\qquad x + 3 = -8$$
$$x = 5 \qquad\qquad x = -11$$

The solutions for $|x + 3| = 8$ are 5 and -11. The solution set is {5,-11}. We checked 5 below. Check -11 in the space provided.

$$|x + 3| = 8 \qquad\qquad |x + 3| = 8$$
$$|5 + 3| = 8$$
$$|8| = 8$$
$$8 = 8$$

---

112. Using the method in the last frame, solve these.

a) $|x + 4| = 9$ b) $|y - 3| = 1$

$$|-11 + 3| = 8$$
$$|-8| = 8$$
$$8 = 8$$

---

113. We used the same method to solve the equation below.

$$|3x - 5| = 7$$

$$3x - 5 = 7 \quad \text{or} \quad 3x - 5 = -7$$
$$3x = 12 \qquad\qquad 3x = -2$$
$$x = 4 \qquad\qquad x = -\frac{2}{3}$$

The solutions for the equation above are 4 and $-\frac{2}{3}$. Solve these:

a) $|2x + 9| = 15$ b) $\left|\frac{1}{2}y + 3\right| = 4$

a) 5 and -13

b) 4 and 2

---

a) 3 and -12

b) 2 and -14

114. We solved the equation below.

$$\left|\frac{5x - 6}{3}\right| = 8$$

$$\frac{5x - 6}{3} = 8 \quad \text{or} \quad \frac{5x - 6}{3} = -8$$

$$5x - 6 = 24 \qquad 5x - 6 = -24$$

$$5x = 30 \qquad 5x = -18$$

$$x = 6 \qquad x = -\frac{18}{5}$$

The solutions for the equation above are 6 and $-\frac{18}{5}$. Solve this one.

$$\left|\frac{2y + 1}{5}\right| = 3$$

---

7 and -8

2-10 ABSOLUTE VALUE INEQUALITIES

In this section, we will solve some inequalities in which the variable is part of an absolute value expression.

---

115. The inequality $|x| < 3$ says: "Find all numbers whose absolute value is less than 3". The solution includes all numbers from -3 to 3. In other words, the solution includes all numbers $\underline{x}$ such that $-3 < x < 3$. That is:

If $|x| < 3$, then $-3 < x < 3$.

The solution set of $|x| < 3$ is $\{x|-3 < x < 3\}$. We graphed the solution below.

$|x| < 3$

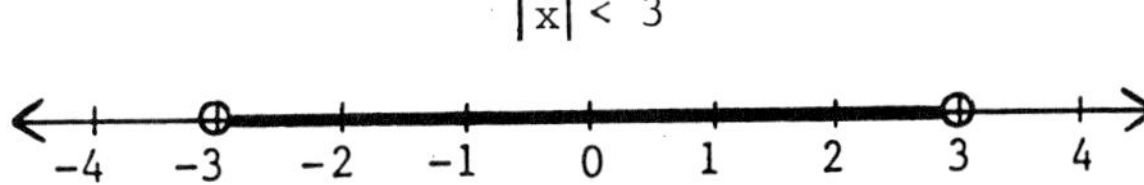

Solve each inequality and graph the solution.

a) $|x| < 2$

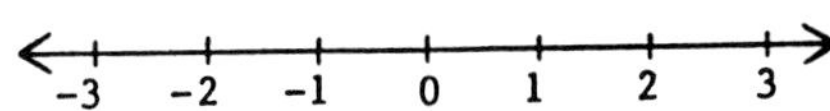

b) $|x| < \frac{1}{2}$

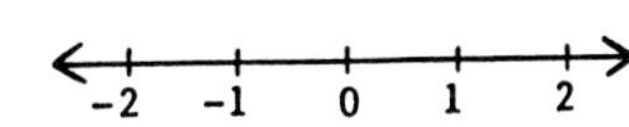

116. The inequality $|x| \leq 4$ is satisfied by all numbers between -4 and 4, and also by -4 and 4. Therefore:

For $|x| \leq 4$: the solution is $-4 \leq x \leq 4$.

the solution set is $\{x | -4 \leq x \leq 4\}$.

The graph of the solution is shown below.

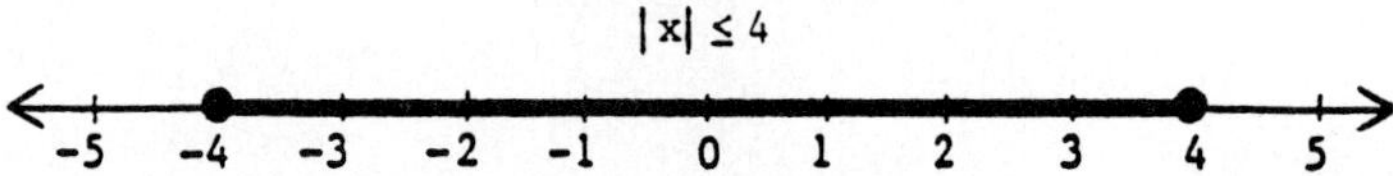

Solve each inequality and graph the solution.

a) $|x| \leq 1$

b) $|x| \leq 3.5$

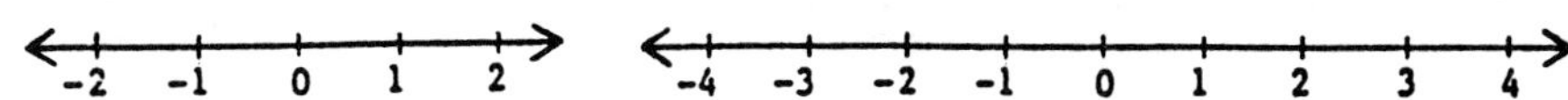

a) $-2 < x < 2$

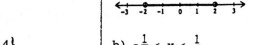

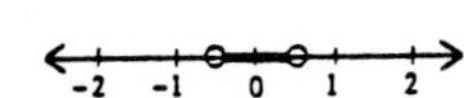

b) $-\frac{1}{2} < x < \frac{1}{2}$

-2 -1 0 1 2

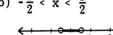

117. The following principle can be used to solve absolute value inequalities of the form $|X| < b$.

<u>Principle</u> <u>For</u> <u>Absolute</u> <u>Value</u> <u>Inequalities</u> <u>With</u> <

If $|X| < b$, then $-b < X < b$

<u>Note</u>: This principle also applies to $\leq$.

We can apply the principle above to inequalities like the one below.

If $|x + 2| < 5$, then $-5 < x + 2 < 5$

To solve the inequality, we solve $-5 < x + 2 < 5$. We get:

$$-5 < x + 2 < 5$$

$$-5 - 2 < x + 2 - 2 < 5 - 2$$

$$-7 < x < 3$$

Solve each inequality.

a) $|x + 4| < 10$

b) $|y - 3| \leq 8$

a) $-1 \leq x \leq 1$

-2 -1 0 1 2

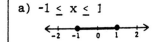

b) $-3.5 \leq x \leq 3.5$

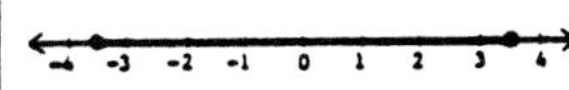

---

a) $-14 < x < 6$

b) $-5 \leq y \leq 11$

118. We solved one inequality below. Solve the other inequality.

$$|2x - 5| \leq 7 \qquad\qquad |3y + 2| < 4$$

$$-7 \leq 2x - 5 \leq 7$$

$$-7 + 5 \leq 2x - 5 + 5 \leq 7 + 5$$

$$-2 \leq 2x \leq 12$$

$$\frac{1}{2}(-2) \leq \frac{1}{2}(2x) \leq \frac{1}{2}(12)$$

$$-1 \leq x \leq 6$$

$-2 < y < \frac{2}{3}$

119. We solved one inequality. Solve the other inequality.

$$\left|\frac{2x + 1}{3}\right| < 1 \qquad\qquad \left|\frac{3y - 1}{2}\right| \leq 4$$

$$-1 < \frac{2x + 1}{3} < 1$$

$$3(-1) < \cancel{3}\left(\frac{2x + 1}{\cancel{3}}\right) < 3(1)$$

$$-3 < 2x + 1 < 3$$

$$-4 < 2x < 2$$

$$-2 < x < 1$$

$-\frac{7}{3} \leq y \leq 3$

120. The inequality $|x| > 2$ says: "Find all numbers whose absolute value is greater than 2". The solution includes all numbers less than -2 and all numbers greater than 2. In other words, the solution includes all numbers $\underline{x}$ such that $x < -2$ or $x > 2$. That is:

If $|x| > 2$, then $x < -2$ or $x > 2$.

The solution set is $\{x | x < -2 \text{ or } x > 2\}$. We graphed the solution below.

$|x| > 2$

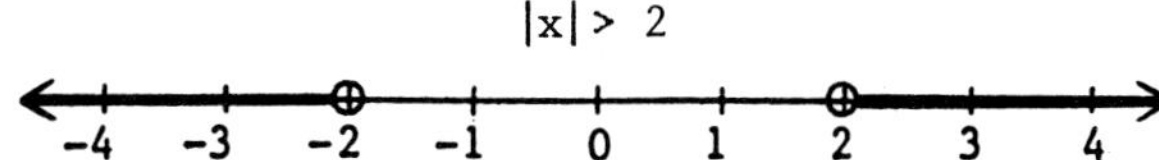

Solve each inequality and graph the solution.

a) $|x| > 1$

-2 -1 0 1 2

b) $|x| > \frac{1}{2}$

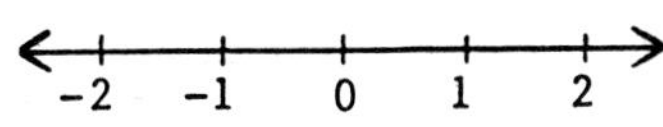

---

a) $x < -1$ or $x > 1$

-2 -1 0 1 2

b) $x < -\frac{1}{2}$ or $x > \frac{1}{2}$

-2 -1 0 1 2

121. The inequality $|x| \geq 3$ is satisfied by all numbers less than -3 and all numbers greater than 3. It is also satisfied by -3 and 3. Therefore:

For $|x| \geq 3$: the solution is $x \leq -3$ or $x \geq 3$.
the solution set is $\{x \mid x \leq -3 \text{ or } x \geq 3\}$.

The graph of the solution is shown below.

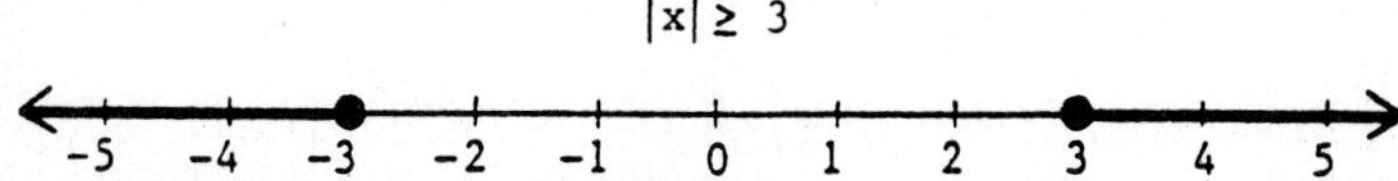

Solve the inequality below and graph the solution.

$|y| \geq 4$

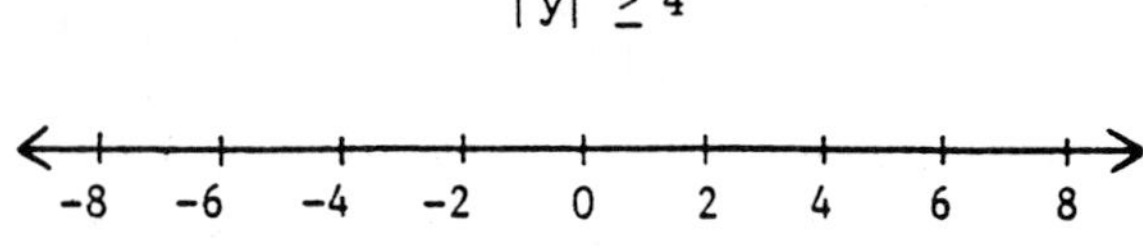

$y \leq -4$ or $y \geq 4$

---

122. The following principle can be used to solve absolute value inequalities of the form $|X| > b$.

> <u>Principle For Absolute Value Inequalities With</u> >
>
> If $|X| > b$, then $X < -b$ or $X > b$
>
> <u>Note</u>: This principle also applies to $\geq$.

We can apply the principle above to inequalities like the one below.

If $|x + 4| > 3$, then $x + 4 < -3$ or $x + 4 > 3$.

To solve the inequality, we solve both inequalities.

$$x + 4 < -3 \qquad \text{or} \qquad x + 4 > 3$$
$$x < -7 \qquad\qquad\qquad x > -1$$

The solution of $|x + 4| > 3$ is $x < -7$ or $x > -1$. The solution is graphed below.

$|x + 4| > 3$

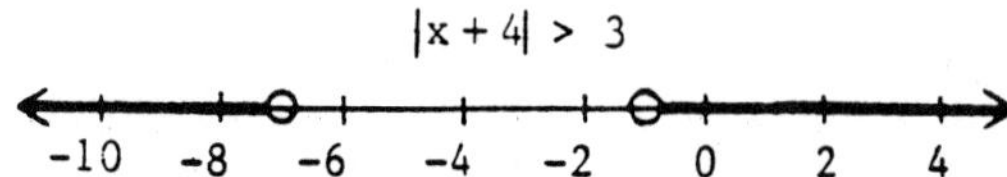

Solve each inequality.

a) $|x - 2| > 5$ b) $|y + 7| \geq 9$

123. We solved $|3x + 1| \geq 5$ below.

$$3x + 1 \leq -5 \quad \text{or} \quad 3x + 1 \geq 5$$
$$3x \leq -6 \qquad\qquad 3x \geq 4$$
$$x \leq -2 \qquad\qquad x \geq \frac{4}{3}$$

The solution is $x \leq -2$ or $x \geq \frac{4}{3}$.

Solve $|2y - 5| > 7$.

a) $x < -3$ or $x > 7$

b) $y \leq -16$ or $y \geq 2$

---

124. We solved $\left|\frac{3x - 2}{5}\right| > 1$ below.

$$\frac{3x - 2}{5} < -1 \quad \text{or} \quad \frac{3x - 2}{5} > 1$$
$$3x - 2 < -5 \qquad\qquad 3x - 2 > 5$$
$$3x < -3 \qquad\qquad 3x > 7$$
$$x < -1 \qquad\qquad x > \frac{7}{3}$$

The solution is $x < -1$ or $x > \frac{7}{3}$.

Solve $\left|\frac{2y - 7}{5}\right| \geq 3$.

$y < -1$ or $y > 6$

---

125. The three principles used to solve absolute value equations and inequalities are summarized below.

<u>Principles For Absolute Value Equations And Inequalities</u>

1. If $|X| = b$, then $X = b$ or $X = -b$
2. If $|X| < b$, then $-b < X < b$
3. If $|X| > b$, then $X < -b$ or $X > b$

<u>Note</u>: Principle 2 applies to $\leq$; principle 3 applies to $\geq$.

Continued on following page.

$y \leq -4$ or $y \geq 11$

125. Continued

Solve each equation and inequality.

a) $|x| < 10$ b) $|x| = 10$ c) $|x| \geq 10$

126. Solve.

a) $|x + 3| = 12$ b) $|y - 5| \leq 10$

a) $-10 < x < 10$

b) 10 or -10

c) $x \leq -10$ or $x \geq 10$

127. Solve.

a) $|2m - 9| > 1$ b) $|5d + 10| = 20$

a) $x = 9$ or -15

b) $-5 \leq y \leq 15$

a) $m < 4$ or $m > 5$

b) $d = 2$ or -6

SELF-TEST 8 (pages 101-113)

Graph each combined inequality.

1. $-3 \le x < 1$

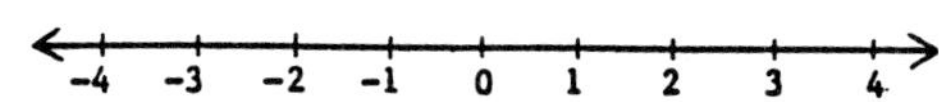

2. $x < -1$ or $x \ge 2$

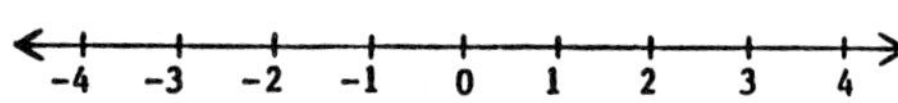

Solve each inequality.

3. $-3 < 2x + 1 \le 7$

4. $4x \le 8$ or $x - 3 > 2$

Solve each equation.

5. $|y + 3| = 8$

6. $\left|\frac{2y - 7}{5}\right| = 1$

Solve and graph each inequality.

7. $|x| \le 2.5$

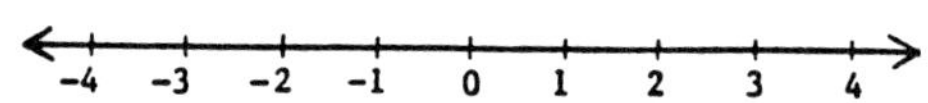

8. $|x| > 1.5$

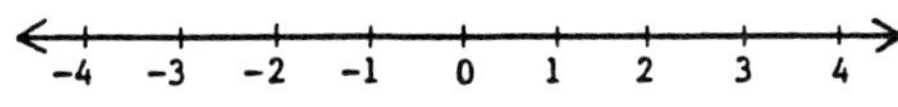

Solve each inequality.

9. $|2x + 3| \le 11$

10. $|y - 5| > 8$

ANSWERS:

1. 

2. 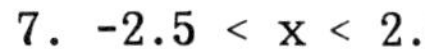

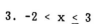

3. $-2 < x \le 3$

4. $x \le 2$ or $x > 5$

5. $y = 5$ and -11

6. $y = 6$ and 1

7. $-2.5 \le x \le 2.5$

8. $x < -1.5$ or $x > 1.5$

9. $-7 \le x \le 4$

10. $y < -3$ or $y > 13$

SUPPLEMENTARY PROBLEMS - CHAPTER 2

Assignment 5

Solve each equation. Report each solution in lowest terms.

1. $x + 9 = 5$
2. $29 = y - 12$
3. $7.5 + d = 0$
4. $t - \frac{1}{3} = \frac{1}{2}$
5. $9p = 4$
6. $-3h = 0$
7. $-36.6 = 12.2y$
8. $\frac{3}{4}p = 1$
9. $3x + 5 = 17$
10. $3 = 9 + 4H$
11. $7 - 12s = 4$
12. $4x + 7 = 3$
13. $2 - 11p = 2$
14. $9 = 5 - t$
15. $4.5 - m = 0$
16. $.7A - 1.4 = 5.6$
17. $2x - \frac{1}{4} = \frac{1}{8}$
18. $6r - 5 = 2r$
19. $11x = 5x - 4$
20. $a = 1 - a$
21. $1.9t - 2.4 = 1.1t$
22. $5N + 1 = 2N + 2$
23. $E + 1 = 3 - E$
24. $2 + 3y = 10 - y$
25. $15 - 6w = 25 - 2w$
26. $5 - k = 1 - 5k$

Solve each equation. Report each answer in lowest terms.

27. $y + 2y + (3y - 1) = 17$
28. $18 - (2s + 7) = 12$
29. $x = 3 - (5x + 2)$
30. $2r - (7r - 3) = 1$
31. $4t - (1 - t) = 0$
32. $7y = 5(4 + y)$
33. $3(V + 1) = 2(V - 1)$
34. $7 + 4(3d - 1) = 0$
35. $25 - 2(4x + 5) = 45$
36. $3 = 9 - 2(1 - 4E)$
37. $4 - 5(2r - 7) = 3r$
38. $8y - 3(2y + 7) = 9$
39. $3(b - 2) - 2(b + 5) = 9b - (b + 12)$
40. $10 - 3(x - 4) = 4(x + 3) - (2x - 9)$

Solve each equation. Report each answer in lowest terms.

41. $\frac{3y}{4} = 1$
42. $\frac{m - 6}{5} = 2$
43. $1 = \frac{7p + 5}{4}$
44. $\frac{6y}{7} + 4 = 5$
45. $\frac{4}{3} = \frac{5d}{6}$
46. $\frac{3y - 1}{7} = \frac{1}{2}$
47. $\frac{t}{4} = \frac{t + 5}{8}$
48. $\frac{x - 1}{2} + \frac{x}{5} = 1$
49. $\frac{3y}{2} - \frac{2y}{3} = 2$
50. $\frac{x}{3} - \frac{x}{4} = \frac{1}{2}$
51. $\frac{m - 8}{8} + \frac{m}{4} = \frac{7}{8}$
52. $\frac{d + 2}{5} - \frac{d - 3}{10} = \frac{4d - 1}{10}$

Assignment 6

1. In $A = LW$, find W when $A = 120$ and $L = 15$.
2. In $C = K - 273$, find K when $C = 50$.
3. In $V = LWH$, find W when $V = 960$, $L = 12$, and $H = 10$.
4. In $A = \frac{1}{2}bh$, find h when $A = 120$ and $b = 24$.
5. In $e = E - Ir$, find I when $e = 50$, $E = 100$ and $r = 5$.

6. In $P = 2L + 2W$, find L when $P = 200$ and $W = 38$.

7. In $C = \frac{5}{9}(F - 32)$, find F when $C = 30$.

8. In $T = \frac{3}{2}(S - 40)$, find S when $T = 60$.

Complete each rearrangement.

9. Solve for s.

$P = 4s$

10. Solve for L.

$A = LW$

11. Solve for r.

$C = 2\pi r$

12. Solve for H.

$V = LWH$

13. Solve for h.

$V = \pi r^2 h$

14. Solve for r.

$I = Prt$

15. Solve for p.

$m = p(q + r)$

16. Solve for D.

$2C = D(a - b)$

17. Solve for b.

$A = \frac{1}{2}bh$

18. Solve for h.

$V = \frac{1}{3}Bh$

19. Solve for T.

$S = \frac{1}{2}T(V + W)$

20. Solve for t^2.

$A = \frac{1}{2}(B + b)t^2$

21. Solve for K.

$D = K - L$

22. Solve for a^2.

$c^2 = a^2 + b^2$

23. Solve for a.

$a + b + c = 180$

24. Solve for V.

$H = E + PV$

25. Solve for W.

$P = 2L + 2W$

26. Solve for N.

$M = \frac{3}{4}(N - 23)$

27. Solve for t.

$R = \frac{1}{2}p(s + t)$

28. Solve for c.

$P = \frac{1}{3}R(c + d)$

Solve each problem.

29. If the difference between a number and 12 is multiplied by 5, we get 20. Find the number.

30. If we subtract 10 from double a number, the result is the number plus 25. Find the number.

31. The sum of two consecutive even integers is 234. Find the numbers.

32. If 100 is added to the middle of three consecutive integers, the result is 21 more than the sum of the largest and three times the smallest. Find the integers.

33. A wire 125 centimeters long is divided into two parts. The smaller part is 35 centimeters shorter than the longer part. How long is each part?

34. There were three times as many children as adults at a circus performance. If the total number attending was 3,196, how many adults attended?

35. What is 55% of 400? 36. 75 is what percent of 125? 37. 56 is 8% of what?

38. How much sales tax is paid for a $9,470 car if the sales tax rate is 5%?

39. If a student got 19 problems correct on a 25-problem test, what was his percent grade?

40. How much money must be invested at 12% interest to earn $1,500 interest annually?

41. The perimeter of a rectangle is 240 meters. The width is 20 meters less than the length. Find the dimensions.

40. The perimeter of a rectangle is 210 feet. The length is 25 feet greater than the width. Find the dimensions.

43. The second angle of a triangle is three times as large as the first. The third angle is 60° larger than the first. How large are the angles?

44. The second angle of a triangle is twice as large as the first. The third angle is 12° less than the sum of the other two. How large are the angles?

45. How many pounds of an alloy containing 8% chromium must be mixed with 200 pounds of an alloy containing 5% chromium to get an alloy containing 6% chromium?

46. How many liters of a 30% acid solution must be mixed with 20 liters of a 60% acid solution to get a 40% acid solution?

47. A small plane leaves an airport and flies due south at 240 km/h. A jet leaves the same airport two hours later and flies due south at 960 km/h. How long will it take the jet to catch up to the small plane?

48. One car leaves a motel traveling 44 mph. A second car leaves the same motel one hour later traveling 55 mph on the same freeway. How long does it take the second car to catch up to the first car?

Assignment 7

Solve each inequality and graph each solution.

1. $x + 3 > 2$

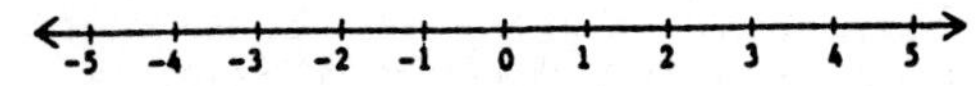

2. $x + 4 < 4$

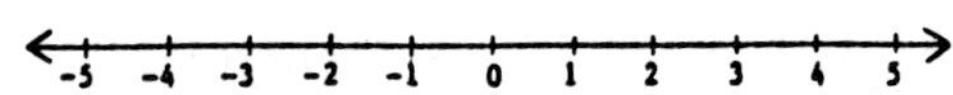

3. $x - 6 \geq -4$

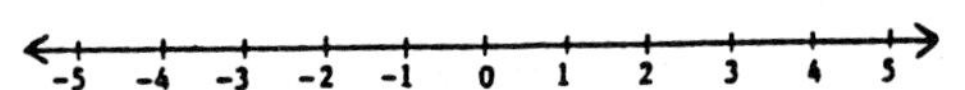

4. $x - 1 \leq 0$

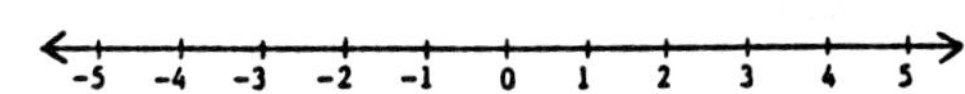

Solve each inequality.

5. $y - 1.2 \geq 2.4$
6. $x - \frac{1}{3} < \frac{1}{2}$
7. $\frac{1}{4} \leq y + \frac{1}{2}$
8. $-7 > x - 5$
9. $4x \geq 20$
10. $-2y < 10$
11. $5x \leq 0$
12. $-3y > -2$
13. $3x < -\frac{1}{4}$
14. $-y > \frac{2}{5}$
15. $4 \geq \frac{2}{3}x$
16. $-5 < -4y$
17. $2x - 3 < -2$
18. $4 - 3y \geq 1$
19. $10 - x > 6$
20. $3x - 2 \leq x - 5$
21. $7 - 4x < 5x - 1$
22. $5 - 3y \geq y - 7$
23. $9t \geq 3(4t - 5) - 2(t - 3)$
24. $2(b + 5) - 5(2b - 2) < 7b - (b + 8)$

Translate to an inequality. Use $\underline{x}$ as the variable.

25. -2 and all numbers less than -2.
26. All positive numbers.
27. All numbers less than 4.
28. -3 and all numbers greater than -3.
29. The sum of a number and 5 is less than 7.
30. The difference between a number and 4 is greater than -1.
31. Three times a number is greater than or equal to -6.
32. 5 is less than or equal to twice a number, minus 3.

Solve each problem.

33. The difference between a number and 10 is equal to or greater than 50. Find the numbers satisfying this condition.

34. Double a number, plus 5, is less than 11. Find the numbers satisfying this condition.

35. The width of a rectangle is 12 cm. The area of the rectangle must be at least 180 cm^2. Find the shortest possible length of the rectangle.

36. The length of a rectangle is 20 meters. The area of the rectangle can be no more than 300 m^2. Find the longest possible width of the rectangle.

37. Four tests are given in a math course. A total of 280 points is needed to get a C. If a student gets scores of 72, 65, and 68 on the first three tests, what score can she get on the fourth test to get a C.

38. Three tests are given in a history course. A total of 240 points is needed to get a B. If a student gets scores of 83 and 81 on the first two tests, what score can he get on the third test to get a B?

39. Company A rents compact cars for $21.95 plus 20¢ per mile. Company B rents compact cars for $20.95 plus 21¢ per mile. For what mileages would the cost of renting from Company A be cheaper?

40. Company A rents intermediate-size cars for $26.95 plus 21¢ per mile. Company B rents intermediate-size cars for $24.95 plus 23¢ per mile. For what mileages would the cost of renting from Company B be cheaper.

41. A salesman can be paid in one of two ways:

    Plan A: A salary of $1000 per month, plus a commission of 4% of gross sales.

    Plan B: A salary of $1200 per month, plus a commission of 5% of gross sales over $8000.

    Assuming that gross sales are always more than $8000, for what gross sales is Plan A better than Plan B?

42. A saleswoman can be paid in two ways:

    Plan A: A salary of $1100 per month, plus a commission of 3% of gross sales.

    Plan B: A salary of $1300 per month, plus a commission of 5% of gross sales over $10,000.

    For what gross sales is Plan B better than Plan A, assuming that gross sales are always more than $10,000?

43. You are going to invest $20,000, part at 8% and part at 10%. What is the most that can be invested at 8% in order to make at least $1800 interest per year?

44. You are going to invest $30,000, part at 6% and part at 9%. What is the most that can be invested at 6% in order to make at least $2100 interest per year?

Assignment 8

Graph each inequality.

1. $-2 < x < 3$

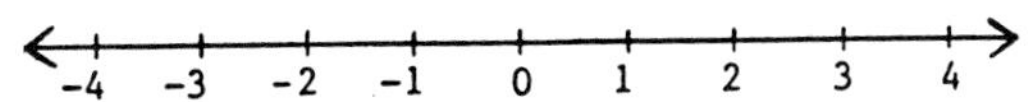

2. $-3 < x \leq 0$

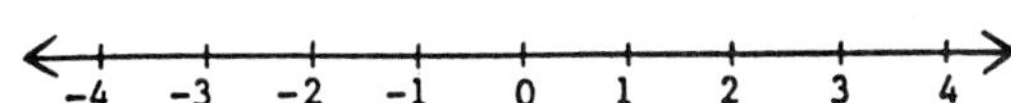

3. $-4 \leq x < -1$

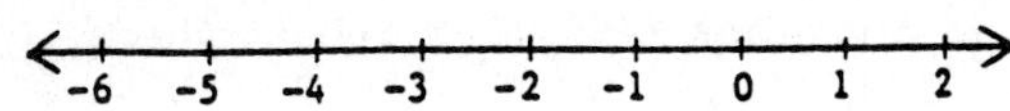

4. $2 \leq x \leq 5$

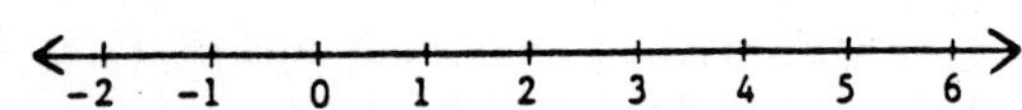

Solve each inequality.

5. $-3 < x + 1 < 5$

6. $1 \leq x + 4 < 9$

7. $-5 < x - 4 \leq 0$

8. $3 \leq 4x \leq 12$

9. $-6 < 3x < 1$

10. $0 < 5x - 3 \leq 8$

11. $-9 \leq 2x + 1 \leq 1$

12. $-3 < \frac{x - 3}{6} < 5$

Graph each inequality.

13. $x < -1$ or $x \geq 3$

14. $x \leq 0$ or $x > 2$

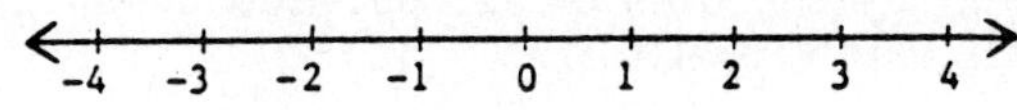

Solve each inequality.

15. $x + 5 < 1$ or $3x > 15$

16. $5x < -10$ or $x - 3 \geq 1$

17. $2x + 9 \leq 11$ or $3x - 15 > 0$

Solve each equation.

18. $|x| = 5$

19. $|x| = 0$

20. $|x| = \frac{1}{2}$

21. $|x| = -3$

22. $|x + 5| = 6$

23. $|y - 9| = 3$

24. $|x + 1| = -4$

25. $|4x - 1| = 3$

26. $|2x + 7| = 13$

27. $\left|\frac{1}{2}m - 1\right| = 2$

28. $\left|\frac{2}{3}x - 4\right| = 8$

29. $\left|\frac{4x - 1}{5}\right| = 3$

Solve and graph each inequality.

30. $|x| < 6$

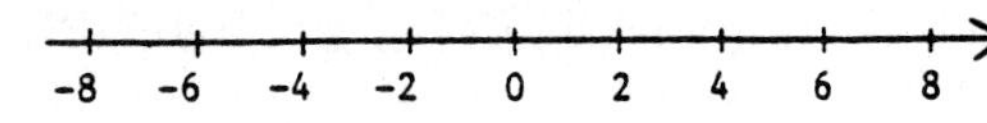

31. $|x| \leq 5$

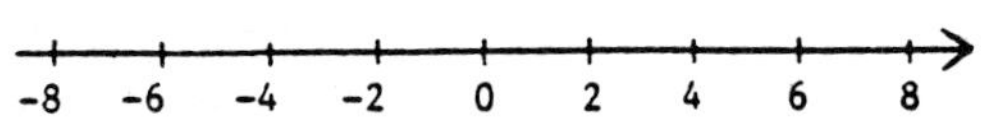

32. $|x| > 2.5$

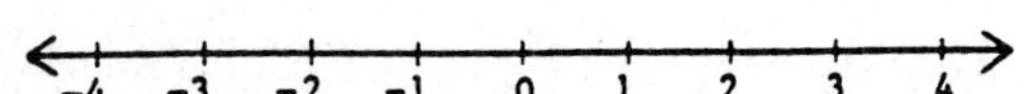

33. $|x| \geq 6$

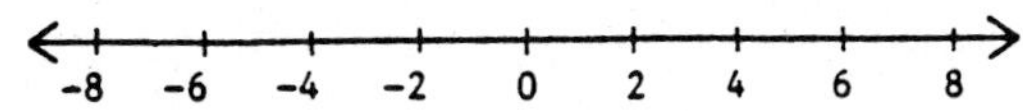

Solve each inequality.

34. $|x + 7| < 10$

35. $|y - 5| \leq 1$

36. $|3x + 2| < 5$

37. $\left|\frac{2x - 5}{3}\right| \leq 5$

38. $|a + 3| > 5$

39. $|b - 7| \geq 2$

40. $|5x + 3| > 7$

41. $\left|\frac{3x - 2}{5}\right| \geq 1$

Polynomials

3

In this chapter, we will define polynomials and show the procedures for adding, subtracting, multiplying, and dividing polynomials. Methods of factoring polynomials are discussed. Some equations are solved by the factoring method, and some word problems involving equations of that type are included.

3-1 POLYNOMIALS

In this section, we will define polynomials and some terms related to polynomials.

1. A polynomial in one variable is an expression containing only terms of the form ax^n, where the numerical coefficient a can be any real number and the exponent n is either a positive integer or 0. Some examples are shown.

$$-3x \qquad y^2 + 9y \qquad 2a^2 - 5a + 6 \qquad 4t^3 + \frac{1}{2}t^2 - t - 7$$

Terms like y^2, $-5a$, and $-t$ can be written explicitly in the form ax^n by writing the coefficient, the exponent, or both. That is:

$$y^2 = 1y^2 \qquad -5a = -5a^1 \qquad -t = -1t^1$$

Terms like 6 and -7 can be written in the form ax^n by using the zero power of the variable. For example:

6 can be written $6a^0$, since $a^0 = 1$.

-7 can be written $-7t^0$, since $t^0 =$ ______.

1

2. A <u>polynomial in more than one variable</u> is an expression containing only terms of the form ax^ny^m or $ax^ny^mz^p$, where <u>a</u> can be any number and the exponents are positive integers or 0. Some examples are shown. $x^4y^2 - 5x^2y^3 + 2$ $5pq$ $-a^5b^2c - 9$ Terms like x^4y^2, $5pq$, and $-a^5b^2c$ can be written explicitly in the form ax^ny^m or $ax^ny^mz^p$ by writing the coefficient or exponents. That is: $x^4y^2 = 1x^4y^2$ $5pq = 5p^1q^1$ $-a^5b^2c = -1a^5b^2c^1$ Terms like 2 and -9 can be written in the form ax^ny^m or $ax^ny^mz^p$ by using powers with 0 exponents. That is: $2 = 2x^0y^0$, since x^0 and $y^0 = 1$ $-9 = -9a^0b^0c^0$, since a^0, b^0, and c^0 = ______.	
3. Polynomials with one term are called <u>monomials</u>. Polynomials with two terms are called <u>binomials</u>. Polynomials with three terms are called <u>trinomials</u>. Some examples are shown. <u>Monomials</u>: $-5x$, x, $\frac{1}{3}xy$, $-t^2$, $7bc^3d^4$ <u>Binomials</u>: $2x + 9$, $x^2y^3 - 3xy$, $a^4b^2c + abc^3$ <u>Trinomials</u>: $2x^2 + 8x - 7$, $x^4y^2 - 5x^2y^3 + 1$ State whether each expression is a monomial, binomial, or trinomial. a) 8 ____________ d) $5m^6 - 6m^5$ ____________ b) $x^2y - 2.5$ ____________ e) $-\frac{1}{2}pq^3$ ____________ c) $2y^2 + y + 7$ ____________ f) $m^4t - m^2t^2 - 6$ ____________	1
4. There are no special names for polynomials with more than three terms. $2x^3y - x^2y^2 + 4x - 6$ is called a polynomial with <u>four</u> terms. $x^4 - 5x^3 + 2x^2 - x + 9$ is called a polynomial with ______ terms.	a) monomial b) binomial c) trinomial d) binomial e) monomial f) trinomial
5. The <u>degree of a term</u> in a polynomial is the sum of the exponents of the variables. Some examples are shown. <u>Term</u> — <u>Degree</u> $7x^2$ — 2 $4x^3y$ — 4 ($4x^3y = 4x^3y^1$, and $3 + 1 = 4$) 9 — 0 ($9 = 9x^0$) a^4bc^3 — 8 ($a^4bc^3 = a^4b^1c^3$, and $4 + 1 + 3 = 8$) Write the degree of each term. a) $2y^3$ _____ b) 10 _____ c) $4d^5p^2$ _____ d) pqr^8 _____	five

6.	The <u>degree of a polynomial</u> is the degree of the term with the highest degree. The degree of $5x^3 - 2x^2 + 4$ is 3. (From $5x^3$) The degree of $4x^2y + 3xy^3 - 7xy$ is 4. (From $3xy^3$) Write the degree of each polynomial. a) $7a^5 - a^3 + 4a$ ____ c) $a^4b^3 - 3a^3b^2 + 4ab$ ____ b) $-2y^7 + 4y^6 - 5y^3 + 2y$ ____ d) $x^5yz + x^4y^3z^2 - xyz^3$ ____	a) 3 b) 0 c) 7 d) 10
7.	Polynomials in one variable are usually written <u>in descending order</u>. That is, they are written so that the exponents of the powers decrease from left to right. For example: $3x^4 + 5x^3 + 6x^2 + 2x + 9$ Write each polynomial in descending order. a) $8b + b^2 + 1 + 4b^3$ = ____________ b) $5y^3 + y - 6y^4 - 2y^2$ = ____________	a) 5 c) 7 b) 7 d) 9
8.	Polynomials in more than one variable are usually written <u>in descending order</u> for one of the variables. Two examples are shown. $x^4y - 2x^2y^3 + 1$ (Written in descending order for <u>x</u>) $ab^2c^5 + a^4bc - a^2$ (Written in descending order for <u>b</u>) Write the polynomial below in descending order for <u>m</u>. $m^2t^5 - 3 - 5mt^4 + m^3t$ = ____________	a) $4b^3 + b^2 + 8b + 1$ b) $-6y^4+5y^3-2y^2+y$
9.	In the polynomial below, there is no term with x^3. We say that the x^3-term is <u>missing</u>. $3x^4 + x^2 - 5x - 1$ When there is a missing term, we can write it with a 0 coefficient or leave a space. For example: $3x^4 + 0x^3 + x^2 - 5x - 1$ $3x^4 + \qquad x^2 - 5x - 1$ However, we usually do not write a 0 coefficient or leave a space. That is: Instead of $y^2 + 0y + 5$, we write $y^2 + 5$. Instead of $x^3y - \qquad xy^2 - 1$, we write ____________.	$m^3t+m^2t^5-5mt^4-3$
		$x^3y - xy^2 - 1$

3-2 COMBINING LIKE TERMS

In this section, we will define like terms and show how they can be combined by addition or subtraction.

---

10. Two monomials are like terms if they contain the same variable or variables to the same power. Otherwise they are unlike terms. Some examples are shown.

Like Terms	Unlike Terms
$7y^3$ and $-4y^3$	$3x^2$ and $5x$
$3x^2y^3$ and $5x^2y^3$	$2p^4q^3$ and $5p^2q^3$
$2ab^4c^5$ and $-7ab^4c^5$	$-4c^2dt^5$ and $9c^2t^5$

Which of the following pairs are like terms? ______________

a) $5m^4$ and $2m^4$ d) $-3m^4t^6$ and $4m^4t^6$

b) $-3p^5$ and $4p^2$ e) $6abc^3$ and $-9abc^3$

c) $8xy^2$ and $2x^2y$ f) $5pq^27^7$ and $7pr^7$

---

11. When the terms in an addition or subtraction are like terms, we can factor by the distributive principle to combine them. Doing so is the same as adding or subtracting their numerical coefficients. That is:

(a), (d), and (e)

$$-ab^4 + 7ab^4 = (-1 + 7)ab^4 = 6ab^4$$

$$4cx^7 - cx^7 = (4 - 1)cx^7 = 3cx^7$$

Combine like terms.

a) $4t^2 + (-7t^2) =$ ________ c) $7y^5 - y^5 =$ ____________

b) $x^3y + x^3y =$ ____________ d) $mp^3t^2 - 9mp^3t^2 =$ __________

---

12. In the polynomial below, there are two pairs of like terms: ($4x^2y$ and x^2y) and ($2x$ and $-6x$). We simplified by combining like terms.

a) $-3t^2$

b) $2x^3y$

c) $6y^5$

d) $-8mp^3t^2$

$$4x^2y + 2x + x^2y - 6x = 5x^2y - 4x$$

Simplify these polynomials.

a) $5t^4 + 3at^2 - 2t^4 - at^2 =$

b) $6a^5b^4 - ab^2 + a^5b^4 - ab^2 =$

---

a) $3t^4 + 2at^2$

b) $7a^5b^4 - 2ab^2$

13.	When combining like terms to simplify a polynomial, we do not ordinarily write "1" and -1 coefficients or 0-terms. For example: $5x^3y^2 - 6x^2y^3 - 6x^3y^2 + 7x^2y^3 = -x^3y^2 + x^2y^3$ $3x^2 + xy - 3x^2 - 5xy = 0x^2 - 4xy = -4xy$ Simplify these polynomials. a) $2t^2 + bt - t^2 - 2bt =$ b) $p^2q + pq^3 + p^2q - pq^3 =$	
14.	There are three pairs of like terms in the polynomial below. Notice how we simplified it. $3x^2 - 5x + 1 + 2x^2 + 5x - 7 = 5x^2 + 0x - 6 = 5x^2 - 6$ Simplify each polynomial. a) $y^2 + 2y - 9 + y^2 - y - 3 =$ b) $at^4 - a^2t^2 + 3 - at^4 + 4a^2t^2 - 3 =$	a) $t^2 - bt$ b) $2p^2q$
15.	We simplified each polynomial below and wrote the terms in descending order for <u>x</u>. $5x^3 - x + 2x^2 + 8 + x^2 - 3x = 5x^3 + 3x^2 - 4x + 8$ $3xy^2 - 9 + 4x^2y - xy^2 - 1 = 4x^2y + 2xy^2 - 10$ Simplify each polynomial. Write the terms in descending order for <u>b</u>. a) $3b^5 + 4b + b^4 - b^5 - 1 - b^4 =$ b) $bc - 5b^3c^2 + 7b^2 + 2bc - b^3c^2 =$	a) $2y^2 + y - 12$ b) $3a^2t^2$
		a) $2b^5 + 4b - 1$ b) $-6b^3c^2+7b^2+3bc$

3-3 ADDITION AND SUBTRACTION OF POLYNOMIALS

In this section, we will discuss the procedures for adding and subtracting polynomials both horizontally and vertically. The additive inverses of polynomials are defined.

---

16. To add polynomials, we simply combine like terms. An example is shown.

Add $x^4 - 2x^2 + 1$ and $x^4 + 3x^2 - 9$.

$$(x^4 - 2x^2 + 1) + (x^4 + 3x^2 - 9)$$
$$= x^4 - 2x^2 + 1 + x^4 + 3x^2 - 9$$
$$= 2x^4 + x^2 - 8$$

Do each addition.

a) $(4ax^2 - bx - 1) + (ax^2 + bx - 3) =$

b) $(by^3 - 2cy^2 - y) + (2by^3 - cy^2) + (2cy^2 + y) =$

---

17. Add. Write each sum in descending order for x.

a) $(x^3 - 2) + (3x^4 - x^2 - 1) + (x^4 + 5) =$

b) $(x^2y^2 - 3x) + (x^3y + x^2y^2) + (x^3y + 3x) =$

Answers to 16:
a) $5ax^2 - 4$
b) $3by^3 - cy^2$

---

18. To add polynomials vertically, we line up like terms in columns and then find the sums of the column. An example is shown. Complete the other addition.

Add $3x^2 - 5x + 7$ and $-6x^2 + 3x - 1$.

$$\begin{array}{r} 3x^2 - 5x + 7 \\ \underline{-6x^2 + 3x - 1} \\ -3x^2 - 2x + 6 \end{array}$$

Add $7ay^4 + 3by^2 - 9$ and $2ay^4 - 8by^2 + 12$.

Answers to 17:
a) $4x^4 + x^3 - x^2 + 2$
b) $2x^3y + 2x^2y^2$

Answer to 18:

$$\begin{array}{r} 7ay^4 + 3by^2 - 9 \\ \underline{2ay^4 - 8by^2 + 12} \\ 9ay^4 - 5by^2 + 3 \end{array}$$

19. Notice how we left spaces for the missing terms below. Complete the other addition.

Add $2x^2 + 5x - 1$, $3x^2 + 5$, and $x^3 - 4x - 7$.

Add $ax^4 - 6$, $ax^4 + 3bx^2 - 1$, and $-bx^2 + 6$.

$$\begin{array}{rrrr} & 2x^2 & + 5x & - 1 \\ & 3x^2 & & + 5 \\ x^3 & & - 4x & - 7 \\ \hline x^3 & + 5x^2 & + x & - 3 \end{array}$$

Answer:

$$\begin{array}{rrr} ax^4 & & - 6 \\ ax^4 & + 3bx^2 & - 1 \\ & - bx^2 & + 6 \\ \hline 2ax^4 & + 2bx^2 & - 1 \end{array}$$

20. If the sum of two polynomials is 0, they are called <u>additive inverses</u> of each other. For example:

Since $(y^2 - 2y - 7) + (-y^2 + 2y + 7) = 0y^2 + 0y + 0 = 0$:

the additive inverse of $y^2 - 2y - 7$ is $-y^2 + 2y + 7$.

the additive inverse of $-y^2 + 2y + 7$ is $y^2 - 2y - 7$.

As you can see, we can get the additive inverse of a polynomial by changing the sign of each term. Write the additive inverse of each polynomial.

a) $3b^5 - 6b^4$ ________________

b) $5xy^2 - 1$ ________________

c) $-3ad^4 + 8bd^2 - d$ ________________

d) $xy^2 - 3x^2y - 2x^3 + 5$ ________________

Answers:

a) $-3b^5 + 6b^4$

b) $-5xy^2 + 1$

c) $3ad^4 - 8bd^2 + d$

d) $-xy^2+3x^2y+2x^3-5$

21. To subtract polynomials, we add the inverse of the second polynomial to the first polynomial. An example is shown.

Subtract $2x^2 - 3x + 1$ from $5x^2 + 4x - 7$

Inverse of $(2x^2 - 3x + 1)$

Changed - to +

$$(5x^2 + 4x - 7) - (2x^2 - 3x + 1) = (5x^2 + 4x - 7) + (-2x^2 + 3x - 1)$$

$$= 3x^2 + 7x - 8$$

Do these subtractions.

a) $(y^4 + 3y^2 - 5) - (y^2 - 3) =$

b) $(p^2q - 2pq^2 + 4q^3) - (p^2q + pq^2 - 3q^3) =$

Answers:

a) $y^4 + 2y^2 - 2$

b) $-3pq^2 + 7q^3$

22. Subtract. Write each difference in descending order for x.

a) $(2x^2 - 5x) - (x^4 + 3x) =$

b) $(4x^3y - 2xy + 3) - (x^2y^2 + xy - 9) =$

---

a) $-x^4 + 2x^2 - 8x$

b) $4x^3y - x^2y^2 - 3xy + 12$

23. To subtract polynomials vertically, we line up like terms in columns and then add the inverse of the bottom polynomial. An example is shown. Complete the other subtraction.

Subtract $2x^2 - 3x + 5$ from $7x^2 - x - 3$.

$$\begin{array}{r} 7x^2 - x - 3 \\ \underline{(-)\ 2x^2 - 3x + 5} \end{array} \quad \text{becomes} \quad \begin{array}{r} 7x^2 - x - 3 \\ \underline{-2x^2 + 3x - 5} \\ 5x^2 + 2x - 8 \end{array}$$

Subtract $t^4 - t^2 - 1$ from $t^4 + t^2 + 1$

$$\begin{array}{r} t^4 + t^2 + 1 \\ \underline{(-)\ t^4 - t^2 - 1} \end{array} \quad \text{becomes}$$

$$\begin{array}{r} t^4 + t^2 + 1 \\ \underline{-t^4 + t^2 + 1} \\ 2t^2 + 2 \end{array}$$

24. Following the example, complete the other subtraction.

Subtract $ay^2 + 3$ from $4ay^2 - by + 1$.

$$\begin{array}{r} 4ay^2 - by + 1 \\ \underline{(-)\ ay^2 \qquad + 3} \end{array} \quad \text{becomes} \quad \begin{array}{r} 4ay^2 - by + 1 \\ \underline{-\ ay^2 \qquad - 3} \\ 3ay^2 - by - 2 \end{array}$$

Subtract $2x^2y - 4x - 3$ from $x^3y^2 - 5x + 6$.

$$\begin{array}{r} x^3y^2 \qquad - 5x + 6 \\ \underline{(-)\qquad 2x^2y - 4x - 3} \end{array} \quad \text{becomes}$$

$$\begin{array}{r} x^3y^2 \qquad\qquad - 5x + 6 \\ \underline{-\ 2x^2y + 4x + 3} \\ x^3y^2 - 2x^2y - x + 9 \end{array}$$

3-4 MULTIPLICATION OF POLYNOMIALS

In this section, we will discuss multiplying by monomials, binomials, and trinomials. Multiplications by binomials and trinomials are shown both horizontally and vertically.

---

25. To multiply monomials, we multiply the numerical coefficient and then use the law of exponents to multipy the variables. For example:

$$(-2x^4y^2)(3xy^5) = (-2)(3)(x^4)(x)(y^2)(y^5) = -6x^5y^7$$

Do these multiplications.

a) $(-y)(5y^2)$ = ________ b) $(-6x^2y^2z)(-5x^3yz^7)$ = ________

---

26. We multiplied variables wherever possible in this multiplication.

	a) $-5y^3$
	b) $30x^5y^3z^8$

$$(apq^3)(pq^2r^5) = (a)(p)(p)(q^3)(q^2)(r^5) = ap^2q^5r^5$$

Do these multiplications.

a) $(9x^3y^4)(7bx)$ = ________________

b) $(-b^2c)(5ab^2)(c^4d)$ = ________________

---

27. To square a monomial, we multiply the monomial by itself. For example:

	a) $63bx^4y^4$
	b) $-5ab^4c^5d$

$$(4x^3y^4)^2 = (4x^3y^4)(4x^3y^4) = 16x^6y^8$$

As you can see, squaring a monomial is the same as <u>squaring the numerical coefficient and doubling each exponent</u>. Using the shorter method, complete these:

a) $(5d)^2$ = ________ c) $(3x^2y)^2$ = ________

b) $(-6v^3)^2$ = ________ d) $(c^3d^6f^9)^2$ = ________

---

28. To multiply a binomial by a monomial, we used the distributive principle. Each term in the binomial is multiplied by the monomial. For example:

	a) $25d^2$
	b) $36v^6$
	c) $9x^4y^2$
	d) $c^6d^{12}f^{18}$

$$-3(5x^2 + 2x) = -3(5x^2) + (-3)(2x) = -15x^2 - 6x$$

$$5t^2(t^3 - 4) = 5t^2(t^3) - 5t^2(4) = 5t^5 - 20t^2$$

Do each multiplication.

a) $-2x(5x^2 - 6)$ = ________________

b) $5a^3b^2(3a^4b^2 + a^2b)$ = ________________

---

	a) $-10x^3 + 12x$
	b) $15a^7b^4 + 5a^5b^3$

29. To multiply $2x^4 - x^2 + 7$ by $5x$ below, we multiplied each term in the trinomial by $5x$.

$$5x(2x^4 - x^2 + 7) = 5x(2x^4) - 5x(x^2) + 5x(7)$$
$$= 10x^5 - 5x^3 + 35x$$

Complete these by multiplying each term in the trinomial by the monomial.

a) $2t(t^4 - 5t^2 + 8) =$ ______________________

b) $pq(3p^2q - pq^5 - 4) =$ ______________________

---

30. The distributive principle is also used to multiply a binomial by a binomial. An example is shown. The arrows show that $(x^2 - 3)$ is multiplied by each term ($2x^2$ and 5) in the first binomial.

a) $2t^5 - 10t^3 + 16t$

b) $3p^3q^2 - p^2q^6 - 4pq$

$$(2x^2 + 5)(x^2 - 3) = 2x^2(x^2 - 3) + 5(x^2 - 3)$$

On the right side above, both $2x^2(x^2 - 3)$ and $5(x^2 - 3)$ are multiplications of a binomial by a monomial. We completed those multiplications and simplified below.

$$(2x^2 + 5)(x^2 - 3) = 2x^2(x^2 - 3) + 5(x^2 - 3)$$
$$= 2x^4 - 6x^2 + 5x^2 - 15$$
$$= 2x^4 - x^2 - 15$$

Do this multiplication.

$(xy + 5)(xy - 7) =$

---

31. We used the same method to multiply two binomials below. Notice that we multiplied $y + 5$ by both $2x$ and -3.

$x^2y^2 - 2xy - 35$

$$(2x - 3)(y + 5) = 2x(y + 5) - 3(y + 5)$$
$$= 2xy + 10x - 3y - 15$$

Do this multiplication.

$(a^2 - b)(a - 2b^2) =$

$a^3 - 2a^2b^2 - ab + 2b^3$

32. We used the same method to multiply a trinomial by a binomial below. Notice that we multiplied $x^2 - 4x + 5$ by both x and -2.

$$\begin{aligned}(x - 2)(x^2 - 4x + 5) &= x(x^2 - 4x + 5) - 2(x^2 - 4x + 5)\\ &= x^3 - 4x^2 + 5x - 2x^2 + 8x - 10\\ &= x^3 - 6x^2 + 13x - 10\end{aligned}$$

Do this multiplication.

$(x + y)(x^2 - xy + y^2) =$

Answer: $x^3 + y^3$

33. We can multiply polynomials vertically. As an example, we multiplied $3x - 4$ and $x + 2$ below. Notice that we multiplied the top row $(3x - 4)$ by both 2 and x and then added.

$$\begin{array}{rrrl} & 3x & -\ 4 & \\ & x & +\ 2 & \\ \hline & 6x & -\ 8 & \text{Multiplying } 3x - 4 \text{ by } 2.\\ 3x^2 & -\ 4x & & \text{Multiplying } 3x - 4 \text{ by } x.\\ \hline 3x^2 & +\ 2x & -\ 8 & \text{Adding}\end{array}$$

Use the same steps for these.

a) Multiply $x + 3$ and $x + 5$.

$$\begin{array}{r} x + 3\\ \underline{x + 5}\end{array}$$

b) Multiply $2y - 1$ and $3y + 4$.

$$\begin{array}{r} 2y - 1\\ \underline{3y + 4}\end{array}$$

Answers:

a)
$$\begin{array}{rrr} & x & +\ 3\\ & x & +\ 5\\ \hline & 5x & +\ 15\\ x^2 & +\ 3x & \\ \hline x^2 & +\ 8x & +\ 15\end{array}$$

b)
$$\begin{array}{rrr} & 2y & -\ 1\\ & 3y & +\ 4\\ \hline & 8y & -\ 4\\ 6y^2 & -\ 3y & \\ \hline 6y^2 & +\ 5y & -\ 4\end{array}$$

34. To multiply $3y^2 - 4y + 1$ and $y - 2$ below, we multiplied the top row by -2 and y and then added. Do the other multiplication.

$$\begin{array}{rrrr} & 3y^2 & -\ 4y & +\ 1\\ & & y & -\ 2\\ \hline & -\ 6y^2 & +\ 8y & -\ 2\\ 3y^3 & -\ 4y^2 & +\ y & \\ \hline 3y^3 & -\ 10y^2 & +\ 9y & -\ 2\end{array}$$

$$\begin{array}{r} m^2 - 3m - 2\\ \underline{4m - 1}\end{array}$$

Answer: $4m^3 - 13m^2 - 5m + 2$

35. To multiply $x^2 + 4x - 3$ and $2x^2 - x + 3$ below, we multiplied the top row by 3, -x, and $2x^2$ and then added. Do the other multiplication.

$$\begin{array}{r} x^2 + 4x - 3 \\ \underline{2x^2 - x + 3} \\ 3x^2 + 12x - 9 \\ -x^3 - 4x^2 + 3x \qquad \\ \underline{2x^4 + 8x^3 - 6x^2 \qquad\qquad} \\ 2x^4 + 7x^3 - 7x^2 + 15x - 9 \end{array}$$

$$\begin{array}{r} 2x^2 - xy + y^2 \\ \underline{x^2 - xy + 2y^2} \end{array}$$

$2x^4 - 3x^3y + 6x^2y^2 - 3xy^3 + 2y^4$

36. Sometimes we have to leave space for a missing term so that like terms are lined up in columns. Below, for example, we left space for a y-term in $4y^2 - 12$ and space for a y^2-term in $2y^3 - 6y$. Do the other multiplication.

$$\begin{array}{r} y^2 - 3 \\ \underline{2y + 4} \\ 4y^2 \qquad\quad - 12 \\ \underline{2y^3 \qquad\quad - 6y \qquad} \\ 2y^3 + 4y^2 - 6y - 12 \end{array}$$

$$\begin{array}{r} 3x^3 - 5x + 4 \\ \underline{x^2 + 1} \end{array}$$

$3x^5 - 2x^3 + 4x^2 - 5x + 4$

SELF-TEST 9 (pages 119-131)

State whether each expression is a monomial, binomial, or trinomial.

1. $x^2y - 1$ __________

2. $3t^2 + t - 5$ __________

State the degree of each polynomial.

3. $5y^4 - 2y^2 + 1$ __________

4. $a^3bc^2 + a^2b^4c - abc^3$ __________

Simplify by combining like terms.

5. $3x^2 + ax - x^2 - 2ax =$ __________

6. $by^5 - b^2y^3 + 3 - by^5 + 4b^2y^3 - 7 =$ __________

Add. Write each sum in descending order for x.

7. $(3x + 1) + (5x^2 - x - 4) + (x^2 + 4)$

8. Add $ax^2 - 1$, $bx^4 - 2ax^2 + 1$, and $ax^2 - cx$

Subtract. Write each difference in descending order for y.

9. $(2y^3 - 3y^2 - y) - (y^3 - 3y^2 + 5)$

10. Subtract $cy^2 - 2by$ from $ay^4 + cy^2 - 1$

Multiply or square.

11. $(-3xy^4)(5x^3y)$

12. $(-2ab^4)^2$

13. $x^2y(2x^3y - xy^3)$

14. $(x - 1)(x^2 + 5x - 3)$

15. $\begin{array}{r} 3y^2 - y + 4 \\ \underline{2y^2 + 1} \end{array}$

ANSWERS:

1. binomial
2. trinomial
3. four
4. seven
5. $2x^2 - ax$
6. $3b^2y^3 - 4$
7. $6x^2 + 2x + 1$
8. $bx^4 - cx$
9. $y^3 - y - 5$
10. $ay^4 + 2by - 1$
11. $-15x^4y^5$
12. $4a^2b^8$
13. $2x^5y^2 - x^3y^4$
14. $x^3 + 4x^2 - 8x + 3$
15. $6y^4 - 2y^3 + 11y^2 - y + 4$

3-5 SPECIAL MULTIPLICATIONS

In this section, we will discuss the FOIL method for multiplying two binomials, the pattern for multiplying the sum and difference of two terms, and the patterns for squaring binomials. All three help us to multiply faster.

---

37. The first two steps used to multiply (2x + 3) and (5x + 4) by the distributive-principle method are shown below.

$$(2x + 3)(5x + 4) = 2x(5x + 4) + 3(5x + 4)$$

$$= 10x^2 + 8x + 15x + 12$$

In the second step above, there are four terms in the product. We can skip the first step and use the FOIL method to write those four terms directly as we have done below.

$$(2x + 3)(5x + 4) = \underset{F}{10x^2} + \underset{O}{8x} + \underset{I}{15x} + \underset{L}{12}$$

(F and O arrows from 2x to 5x and 4; I and L arrows from 3 to 5x and 4.)

Note: 1) To get F, we multiplied the first terms:

$$(2x)(5x) = 10x^2$$

2) To get O, we multiplied the outside terms:

$$(2x)(4) = 8x$$

3) To get I, we multiplied the inside terms:

$$(3)(5x) = 15x$$

4) To get L, we multiplied the last terms:

$$(3)(4) = 12$$

After writing the four terms, we simplify the product by combining like terms.

$$(2x + 3)(5x + 4) = 10x^2 + 8x + 15x + 12$$

$$= 10x^2 + 23x + 12$$

Using the FOIL method, write the four-term product below and then simplify.

(a + 2b)(2a + b) = _______ + _______ + _______ + _______

= ______________________________

$2a^2 + ab + 4ab + 2b^2$

$2a^2 + 5ab + 2b^2$

38. We used the FOIL method below and then combined like terms. Do the other multiplication.

$$(2b + c)(b - 3c) = 2b^2 - 6bc + bc - 3c^2$$
$$= 2b^2 - 5bc - 3c^2$$

$(x + 2)(6x - 7) =$ ____________________

$=$ ____________________

---

39. Following the example, complete the other multiplication.

Answer to 38: $6x^2 - 7x + 12x - 14$; $6x^2 + 5x - 14$

$$(3d - 4)(4d + 3) = 12d^2 + 9d - 16d - 12$$
$$= 12d^2 - 7d - 12$$

$(p - 3q)(2p + q) =$ ____________________

$=$ ____________________

---

40. Following the example, do the multiplication.

Answer to 39: $2p^2 + pq - 6pq - 3q^2$; $2p^2 - 5pq - 3q^2$

$$(3xy - 1)(2xy - 4) = 6x^2y^2 - 12xy - 2xy + 4$$
$$= 6x^2y^2 - 14xy + 4$$

$(2x^2 - y^2)(x^2 - 4y^2) =$ ____________________

$=$ ____________________

---

41. When there are no like terms, we are unable to simplify the product. For example:

Answer to 40: $2x^4 - 8x^2y^2 - x^2y^2 + 4y^4$; $2x^4 - 9x^2y^2 + 4y^4$

$$(2a - b)(5c + 3d) = 10ac + 6ad - 5bc - 3bd$$

Do these:

a) $(c + d)(p - q) =$ ____________________

b) $(2x - 4xy)(y^4 - 5x^2) =$ ____________________

---

42. Below we multiplied the sum and difference of the same two terms. Since $-AB + AB = 0$, the product simplifies to the binomial $A^2 - B^2$.

Answer to 41: a) $cp - cq + dp - dq$; b) $2xy^4 - 10x^3 - 4xy^5 + 20x^3y$

$$(A + B)(A - B) = A^2 - AB + AB - B^2 = A^2 - B^2$$

Therefore, we get the following pattern. Memorize it.

> $$(A + B)(A - B) = A^2 - B^2$$
>
> The product of the sum and difference of two terms is the square of the first term minus the square of the second term.

Continued on following page.

42. Continued

We used this pattern for the multiplication below.

$$(x + 4)(x - 4) = (x)^2 - (4)^2 = x^2 - 16$$

Using the same pattern, complete these:

a) $(t + 9)(t - 9) =$ ________ b) $(10 + d)(10 - d) =$ ________

---

43. We used the same pattern for this multiplication.

$$(2p + q)(2p - q) = (2p)^2 - (q)^2 = 4p^2 - q^2$$

Do these:

a) $(4x + 1)(4x - 1) =$ ____________

b) $(3a + 5b)(3a - 5b) =$ ____________

Answers to 42: a) $t^2 - 81$ b) $100 - d^2$

---

44. Another example of the same pattern is shown below.

$$(x^2 + 2y^3)(x^2 - 2y^3) = (x^2)^2 - (2y^3)^2 = x^4 - 4y^6$$

Do these:

a) $(a^4 + b^5)(a^4 - b^5) =$ ____________

b) $(2m^3 + 6t^7)(2m^3 - 6t^7) =$ ____________

Answers to 43: a) $16x^2 - 1$ b) $9a^2 - 25b^2$

---

45. We squared the sum and difference of the same two terms below. To do so, we multiplied each binomial by itself.

$$(A + B)^2 = (A + B)(A + B)$$
$$= A^2 + AB + AB + B^2$$
$$= A^2 + 2AB + B^2$$

$$(A - B)^2 = (A - B)(A - B)$$
$$= A^2 - AB - AB + B^2$$
$$= A^2 - 2AB + B^2$$

Therefore, we get the following patterns. Memorize them.

> $$(A + B)^2 = A^2 + 2AB + B^2$$
> $$(A - B)^2 = A^2 - 2AB + B^2$$
> The square of a binomial is the square of the first term, plus or minus double the product of the two terms, plus the square of the second term.

We used the pattern to square x + 4 below.

$$(x + 4)^2 = x^2 + 2(x)(4) + 4^2 = x^2 + 8x + 16$$

Continued on following page.

Answers to 44: a) $a^8 - b^{10}$ b) $4m^6 - 36t^{14}$

45. Continued

Using the same pattern, square these. Be sure to double the product of the two terms to get the middle term of the trinomial.

a) $(y + 7)^2$ = ______________________

b) $(m + 1)^2$ = ______________________

---

46. We used the pattern to square $3x - 5$ below.

$$(3x - 5)^2 = (3x)^2 - 2(3x)(5) + 5^2 = 9x^2 - 30x + 25$$

Use the same pattern for these. Be sure to subtract the middle term of the trinomial.

a) $(6y - 2)^2$ = ______________________

b) $(9m - 1)^2$ = ______________________

Answers to 45: a) $y^2 + 14y + 49$ b) $m^2 + 2m + 1$

---

47. We used the same pattern below.

$$(6 + 4d)^2 = 6^2 + 2(6)(4d) + (4d)^2 = 36 + 48d + 16d^2$$

Complete these:

a) $(12 - t)^2$ = ______________________

b) $(1 + 4m)^2$ = ______________________

Answers to 46: a) $36y^2 - 24y + 4$ b) $81m^2 - 18m + 1$

---

48. The same pattern was used below.

$$(2p + 3q)^2 = (2p)^2 + 2(2p)(3q) + (3q)^2 = 4p^2 + 12pq + 9q^2$$

Complete these:

a) $(c + d)^2$ = ______________________

b) $(4x - 3y)^2$ = ______________________

Answers to 47: a) $144 - 24t + t^2$ b) $1 + 8m + 16m^2$

---

49. Another example of the same pattern is shown.

$$(3x^2 - xy)^2 = (3x^2)^2 - 2(3x^2)(xy) + (xy)^2 = 9x^4 - 6x^3y + x^2y^2$$

Complete these:

a) $(4p^2 - q^3)^2$ = ______________________

b) $(2a^2b + 3b)^2$ = ______________________

Answers to 48: a) $c^2 + 2cd + d^2$ b) $16x^2 - 24xy + 9y^2$

Answers to 49: a) $16p^4 - 8p^2q^3 + q^6$ b) $4a^4b^2 + 12a^2b^2 + 9b^2$

3-6 DIVIDING BY MONOMIALS

In this section, we will discuss the procedure for dividing a polynomial by a monomial.

---

50. To divide monomials, we divide the numerical coefficients and then use the law of exponents to divide the variables. For example,

$$\frac{-8a^3b^6}{2ab^5} = \left(\frac{-8}{2}\right)\left(\frac{a^3}{a}\right)\left(\frac{b^6}{b^5}\right) = -4a^2b$$

Use the same method for these:

a) $\frac{12b^2t^7}{3bt^3}$ = ________ b) $\frac{-6m^9}{-6m}$ = ________

---

51. Two more examples are shown below. Notice how we wrote -1 explicitly in the first denominator. Notice that b^0 can be eliminated in the second quotient because $b^0 = 1$.

Answers to 50: a) $4bt^4$ b) m^8

$$\frac{7a^4t^5}{-a^3t^2} = \frac{7a^4t^5}{-1a^3t^2} = -7at^3$$

$$\frac{12b^3x^4}{6b^3x} = 2b^0x^3 = 2(1)(x^3) = 2x^3$$

Complete these:

a) $\frac{5x^3}{x}$ = ________ b) $\frac{2a^2b}{a^2b}$ = ________

---

52. We saw the following definition for powers with a negative exponent.

Answers to 51: a) $5x^2$ b) 2

$$x^{-1} = \frac{1}{x} \qquad y^{-2} = \frac{1}{y^2} \qquad b^{-3} = \frac{1}{b^3}$$

By substituting $\frac{1}{x}$ for x^{-1}, we can write the expression below with only positive exponents.

$$5x^{-1}y^2 = 5\left(\frac{1}{x}\right)y^2 = \frac{5y^2}{x}$$

Write each expression with only positive exponents.

a) xy^{-1} = ________ b) $3a^{-1}b^{-2}$ = ________

---

53. We get a negative exponent in the division below. Notice how we substituted to write the quotient with only positive exponents.

Answers to 52: a) $\frac{x}{y}$ b) $\frac{3}{ab^2}$

$$\frac{15xy^3}{5x^3y} = 3x^{-2}y^2 = 3\left(\frac{1}{x^2}\right)y^2 = \frac{3y^2}{x^2}$$

Divide. Write each quotient with only positive exponents.

a) $\frac{12m^2}{6m^3}$ = ________ b) $\frac{8xy^2z^3}{2x^2y^4z^2}$ = ________

54. Notice how we wrote x^0 and a^0 explicitly in each division below.

$$\frac{4}{4x} = \frac{4x^0}{4x} = x^{-1} = \frac{1}{x}$$

$$\frac{14bc}{7abc} = \frac{14a^0bc}{7abc} = 2a^{-1} = 2\left(\frac{1}{a}\right) = \frac{2}{a}$$

Divide. Write each quotient with only positive exponents.

a) $\frac{24}{8m} =$ __________ b) $\frac{12x}{3x^2y} =$ __________

Answers: a) $\frac{2}{m}$ b) $\frac{4z}{xy^2}$

---

55. In each division below, we got a fractional coefficient in the quotient. Notice how we can write the quotient in an alternate form.

$$\frac{5x^2}{2x} = \frac{5}{2}x \text{ or } \frac{5x}{2}$$

$$\frac{x^3y^2}{3xy} = \frac{1x^3y^2}{3xy} = \frac{1}{3}x^2y \text{ or } \frac{x^2y}{3}$$

Do these divisions.

a) $\frac{4m^7}{3m^2} =$ __________ b) $\frac{2a^2b^2}{4ab^2} =$ __________

Answers: a) $\frac{3}{m}$ b) $\frac{4}{xy}$

---

56. The quotient below is a numerical fraction.

$$\frac{x}{4x} = \frac{1x}{4x} = \frac{1}{4}x^0 = \frac{1}{4}$$

Do these divisions.

a) $\frac{y^2}{5y^2} =$ __________ b) $\frac{2x}{3x} =$ __________

Answers: a) $\frac{4m^5}{3}$ b) $\frac{a}{2}$

---

57. Notice how we wrote the quotient below with a positive exponent.

$$\frac{5x}{2x^3} = \frac{5}{2}x^{-2} = \frac{5}{2}\left(\frac{1}{x^2}\right) = \frac{5}{2x^2}$$

Write each quotient with only positive exponents.

a) $\frac{y^2}{3y^4} =$ __________ b) $\frac{5pq}{6p^2q^2} =$ __________

Answers: a) $\frac{1}{5}$ b) $\frac{2}{3}$

---

58. To divide below, we divided each term in the trinomial by the number.

$$\frac{8x^2 + 10x + 6}{2} = \frac{8x^2}{2} + \frac{10x}{2} + \frac{6}{2} = 4x^2 + 5x + 3$$

$$\frac{9ay^2 - 6y - 12}{3} = \frac{9ay^2}{3} - \frac{6y}{3} - \frac{12}{3} = \text{\_\_\_\_\_\_\_\_\_\_}$$

Answers: a) $\frac{1}{3y^2}$ b) $\frac{5}{6pq}$

$3ay^2 - 2y - 4$

59. Following the example, complete the other division.

$$\frac{5x^2 - 10x + 15}{10} = \frac{5x^2}{10} - \frac{10x}{10} + \frac{15}{10} = \frac{x^2}{2} - x + \frac{3}{2}$$

$$\frac{8p^4 + 4p^2 - 20}{8} = \_\_\_\_\_\_\_\_\_\_$$

---

60. Following the example, complete the other division.

Answer to 59: $p^4 + \frac{p^2}{2} - \frac{5}{2}$

$$\frac{2y^3 + 6y^2 + 4y}{2y} = \frac{2y^3}{2y} + \frac{6y^2}{2y} + \frac{4y}{2y} = y^2 + 3y + 2$$

$$\frac{x^5y^2 - x^3y^3 + x^2y}{x^2y} = \_\_\_\_\_\_\_\_\_\_$$

---

61. Following the example, complete the other division.

Answer to 60: $x^3y - xy^2 + 1$

$$\frac{5x^3 - 3x^2 + 6x}{3x^2} = \frac{5x^3}{3x^2} - \frac{3x^2}{3x^2} + \frac{6x}{3x^2}$$

$$= \frac{5}{3}x - 1 + 2x^{-1}$$

$$= \frac{5x}{3} - 1 + \frac{2}{x}$$

$$\frac{3a^3b^2 - 6a^2b + 2a^2b^2}{2a^2b^2} =$$

---

62. Following the example, complete the other division.

Answer to 61: $\frac{3a}{2} - \frac{3}{b} + 1$

$$\frac{10x^3y + 8x^2y^2 - 12xy^2}{4x^3y} = \frac{10x^3y}{4x^3y} + \frac{8x^2y^2}{4x^3y} - \frac{12xy^2}{4x^3y}$$

$$= \frac{5}{2} + 2x^{-1}y - 3x^{-2}y$$

$$= \frac{5}{2} + \frac{2y}{x} - \frac{3y}{x^2}$$

$$\frac{18a^2b^2c - 12ab^2c^2 + 24abc^2}{6a^2b^3c} =$$

63. Following the example, do the other division.

$$\frac{9x^3 + 6x^2 - 5x}{3x^3} = \frac{9x^3}{3x^3} + \frac{6x^2}{3x^3} - \frac{5x}{3x^3}$$

$$= 3 + 2x^{-1} - \frac{5}{3}x^{-2}$$

$$= 3 + \frac{2}{x} - \frac{5}{3x^2}$$

$$\frac{8p^3q^3 + 10p^2q^3 - 6pq^2}{6p^3q^3} =$$

Answer: $\frac{3}{b} - \frac{2c}{ab} + \frac{4c}{ab^2}$

---

64. Notice how we wrote a "1" coefficient and an x^0 below. Do the other division.

$$\frac{10x^2 + x + 5}{5x} = \frac{10x^2}{5x} + \frac{1x}{5x} + \frac{5x^0}{5x}$$

$$= 2x + \frac{1}{5} + x^{-1}$$

$$= 2x + \frac{1}{5} + \frac{1}{x}$$

$$\frac{3ab + 4ac + bc}{2abc} =$$

Answer: $\frac{4}{3} + \frac{5}{3p} - \frac{1}{p^2q}$

Answer: $\frac{3}{2c} + \frac{2}{b} + \frac{1}{2a}$

3-7 DIVIDING BY OTHER POLYNOMIALS

To divide polynomials by a binomial or trinomial, we use a procedure similar to long division. We will discuss that procedure in this section.

---

65. The two basic steps needed to divide $x^2 + 6x + 8$ by $x + 2$ are shown below.

1) Dividing the first term of the dividend by the first term of the divisor.

$$\begin{array}{r l} x & \leftarrow \text{Dividing } x^2 \text{ by } x\text{: } \frac{x^2}{x} = x \\ x + 2\overline{)x^2 + 6x + 8} & \\ \underline{x^2 + 2x} & \leftarrow \text{Multiplying } x + 2 \text{ by } \underline{x} \\ 4x & \leftarrow \text{Subtracting } x^2 + 2x \text{ from } x^2 + 6x \end{array}$$

2) Bringing down the next term 8 of the dividend and dividing the first term of 4x + 8 by the first term of the divisor.

$$\begin{array}{r l} x + 4 & \leftarrow \text{Dividing } 4x \text{ by } x\text{: } \frac{4x}{x} = 4 \\ x + 2\overline{)x^2 + 6x + 8} & \\ \underline{x^2 + 2x} & \\ 4x + 8 & \\ \underline{4x + 8} & \leftarrow \text{Multiplying } x + 2 \text{ by } 4 \\ 0 & \leftarrow \text{Subtracting } 4x + 8 \text{ from } 4x + 8 \end{array}$$

Therefore, $x^2 + 6x + 8$ divided by $x + 2$ equals $x + 4$ with a 0 remainder. To check the division, multiply the divisor and quotient below.

$(x + 2)(x + 4) =$ ______________________

Answer: $x^2 + 6x + 8$

66. Using the same steps, we completed one division below. Notice how subtracting -2x from -5x is the same as adding +2x and -5x. Complete the other division.

$$\begin{array}{r} x - 3 \\ x - 2\overline{)x^2 - 5x + 6} \\ \underline{x^2 - 2x} \\ -3x + 6 \\ \underline{-3x + 6} \\ 0 \end{array} \qquad y - 6\overline{)y^2 - 11y + 30}$$

Answer:

$$\begin{array}{r} y - 5 \\ y - 6\overline{)y^2 - 11y + 30} \\ \underline{y^2 - 6y} \\ -5y + 30 \\ \underline{-5y + 30} \\ 0 \end{array}$$

67. We completed one division below. Notice how subtracting -8x from -5x is the same as adding +8x and -5x. Complete the other division.

$$\begin{array}{r} 2x + 1 \\ 3x - 4\overline{)6x^2 - 5x - 4} \\ \underline{6x^2 - 8x} \\ 3x - 4 \\ \underline{3x - 4} \\ 0 \end{array}$$

$$4x - 1\overline{)8x^2 + 2x - 1}$$

Answer to 67:

$$\begin{array}{r} 2x + 1 \\ 4x - 1\overline{)8x^2 + 2x - 1} \\ \underline{8x^2 - 2x} \\ 4x - 1 \\ \underline{4x - 1} \\ 0 \end{array}$$

68. There is a remainder of 44 in the division below. Notice how we wrote the remainder as the fraction $\frac{44}{m - 5}$ when writing the quotient. Complete the other division.

$$\begin{array}{r} m + 8 \\ m - 5\overline{)m^2 + 3m + 4} \\ \underline{m^2 - 5m} \\ 8m + 4 \\ \underline{8m - 40} \\ 44 \end{array}$$

$$3m + 4\overline{)6m^2 - m + 13}$$

The quotient is:

$$m + 8 + \frac{44}{m - 5}$$

The quotient is:

____

Answer to 68:

$$\begin{array}{r} 2m - 3 \\ 3m + 4\overline{)6m^2 - m + 13} \\ \underline{6m^2 + 8m} \\ -9m + 13 \\ \underline{-9m - 12} \\ 25 \end{array}$$

$$2m - 3 + \frac{25}{3m + 4}$$

69. There are two points to remember when dividing polynomials:

1) Arrange both the divisor and dividend in descending order.
2) Write missing terms with 0 coefficients.

In the division below, we wrote $0x^2$ and $0x$ for the two missing terms. Do the other division.

$(64x^3 + 27) \div (4x + 3)$

$(8y^3 - 125) \div (2y - 5)$

$$\begin{array}{r} 16x^2 - 12x + 9 \\ 4x + 3\overline{)64x^3 + 0x^2 + 0x + 27} \\ \underline{64x^3 + 48x^2} \\ -48x^2 + 0x \\ \underline{-48x^2 - 36x} \\ 36x + 27 \\ \underline{36x + 27} \\ 0 \end{array}$$

Answer to 69:

$$\begin{array}{r} 4y^2 + 10y + 25 \\ 2y - 5\overline{)8y^3 + 0y^2 + 0y - 125} \\ \underline{8y^3 - 20y^2} \\ 20y^2 + 0y \\ \underline{20y^2 - 50y} \\ 50y - 125 \\ \underline{50y - 125} \\ 0 \end{array}$$

70. Notice these two points about the division below.

1) We wrote 0x for the missing term.
2) Since the degree of $6x + 7$ is less than the degree of $x^2 - 3$, $6x + 7$ is the remainder. Therefore, we can stop dividing.

Using the same steps, do the other division.

$(2x^3 + 5x^2 - 8) \div (x^2 - 3)$ $(y^3 - 7y^2 + 5) \div (y^2 - 1)$

$$\begin{array}{r}
2x \quad + 5 \\
x^2 - 3 \overline{)\,2x^3 + 5x^2 + 0x - 8} \\
\underline{2x^3 - 6x} \\
5x^2 + 6x - 8 \\
\underline{5x^2 - 15} \\
6x + 7
\end{array}$$

The quotient is: The quotient is:

$$2x + 5 + \frac{6x + 7}{x^2 - 3}$$ ____________________

$y - 7 + \frac{y - 2}{y^2 - 1}$

71. In the division below, the divisor is a trinomial. We proceeded in the usual way.

$$(6x^3 - 19x^2 + 17x - 11) \div (2x^2 - 5x + 1)$$

$$\begin{array}{r}
3x \quad - 2 \\
2x^2 - 5x + 1 \overline{)\,6x^3 - 19x^2 + 17x - 11} \\
\underline{6x^3 - 15x^2 + 3x} \\
-4x^2 + 14x - 11 \\
\underline{-4x^2 + 10x - 2} \\
4x - 9
\end{array}$$

The quotient is: $3x - 2 + \frac{4x - 9}{2x^2 - 5x + 1}$

Use the same steps for this division.

$$(4y^3 - 11y^2 + 6y + 5) \div (y^2 - 3y + 2)$$

The quotient is: ____________________

$4y + 1 + \frac{y + 3}{y^2 - 3y + 2}$

SELF-TEST 10 (pages 132-143)

Multiply or square.

1. $(x + 3y)(2x - y)$

2. $(2x^2 - y)(xy^3 - 5y^2)$

3. $(3m^3 + 4t^4)(3m^3 - 4t^4)$

4. $(2a^2 - b^5)^2$

Divide.

5. $\dfrac{4x^5y^7}{x^4y^3}$

6. $\dfrac{2a^2b^2}{3a^3b^4}$

7. $\dfrac{pq^2}{5p^2q^2}$

8. $\dfrac{4x^3 + 2x^2 - 10x}{4x}$

9. $\dfrac{2ab^2 + 3ab^2c - ac}{3ab^2c}$

10. $(27x^3 - 8) \div (3x - 2)$

11. $(3y^3 + y^2 + 1) \div (y^2 - 2)$

ANSWERS:

1. $2x^2 + 5xy - 3y^2$
2. $2x^3y^3 - 10x^2y^2 - xy^4 + 5y^3$
3. $9m^6 - 16t^8$
4. $4a^4 - 4a^2b^5 + b^{10}$
5. $4xy^4$
6. $\dfrac{2}{3ab^2}$
7. $\dfrac{1}{5p}$
8. $x^2 + \frac{1}{2}x - \frac{5}{2}$
9. $\dfrac{2}{3c} + 1 - \dfrac{1}{3b^2}$
10. $9x^2 + 6x + 4$
11. $3y + 1 + \dfrac{6y + 3}{y^2 - 2}$

3-8 FACTORING

Factoring is the reverse of multiplication. That is, to factor a polynomial, we write it as a multiplication of two or more simpler polynomials. In this section, we will discuss factoring common monomial factors out of polynomials and factoring by grouping.

---

72. We factored out the largest common numerical factor below. As a check, we can multiply 6 and $x + 2$. The result should be $6x + 12$.

$$6x + 12 = 6(x + 2)$$

When factoring, we always factor out the largest common factor. Factor these. Check by multiplying.

a) $8a + 24 =$ ______ c) $12t^2 + 18 =$ ______

b) $9b^2 - 45 =$ ______ d) $16x - 12y =$ ______

---

73. Notice how we got a "1" in each binomial factor below.

$$5x + 5 = 5(x + 1)$$

$$6 - 12y = 6(1 - 2y)$$

Factor these:

a) $8a - 4 =$ ______ b) $7 + 21d =$ ______

Answers to 72: a) $8(a + 3)$ b) $9(b^2 - 5)$ c) $6(2t^2 + 3)$ d) $4(4x - 3y)$

---

74. We factored out a common numerical factor from the trinomial below.

$$12t^2 - 18t - 6 = 6(2t^2 - 3t - 1)$$

Factor each trinomial.

a) $16m^2 + 4m - 8 =$ ______

b) $9a - 12b + 3c =$ ______

Answers to 73: a) $4(2a - 1)$ b) $7(1 + 3d)$

---

75. When factoring a common power out of a polynomial, we always factor out the largest possible power. The largest possible power is the one with the smallest exponent. For example, we factored out y^2 below.

$$2y^6 - 7y^4 + y^2 = y^2(2y^4 - 7y^2 + 1)$$

Factor out the largest possible power.

a) $4t^2 - 7t =$ ______

b) $a^8 + 4a^6 - a^4 =$ ______

Answers to 74: a) $4(4m^2 + m - 2)$ b) $3(3a - 4b + c)$

Answers to 75: a) $t(4t - 7)$ b) $a^4(a^4 + 4a^2 - 1)$

76. In the example below, we factored out the largest possible numerical factor and power.

$$10t^5 - 30t^2 = 10t^2(t^3 - 3)$$

Factor out the largest possible numerical factor and power.

a) $8m^9 + 4m^5 =$ ____________________

b) $6x^3 - 12x^2 - 9x =$ ____________________

---

77. To check factorings, we can multiply. For example:

$$16x^3 - 20x^2 = 4x^2(4x - 5) \text{ is correct,}$$

$$\text{since } 4x^2(4x - 5) = 16x^3 - 20x^2.$$

Check each factoring. State whether it is "correct" or "incorrect".

a) $6y^5 + 4y^2 = 2y^2(3y^3 + 2y)$ ____________

b) $18m^3 - 12m^2 - 24m = 6m(3m^2 - 2m - 4)$ ____________

Answers to 76:

a) $4m^5(2m^4 + 1)$

b) $3x(2x^2 - 4x - 3)$

---

78. In the example below, we factored out the largest possible power of each variable.

$$3x^2y^2 - xy^3 = xy^2(3x - y)$$

Factor these and check your results.

a) $c^4d + 7c^2d =$ ____________

b) $2p^5q^3 - 5p^4q - p^3q^2 =$ ____________

Answers to 77:

a) Incorrect, since: $2y^2(3y^3 + 2y) = 6y^5 + 4y^3$

b) Correct

---

79. In the example below, we factored out the largest possible numerical factor and power.

$$12x^2y - 20x^3y = 4x^2y(3 - 5x)$$

Factor these and check your results.

a) $6p^4q^4 + 3p^2q^2 =$ ____________

b) $8c^5d - 4c^4d^2 + 4c^3d^3 =$ ____________

Answers to 78:

a) $c^2d(c^2 + 7)$

b) $p^3q(2p^2q^2 - 5p - q)$

---

80. We multiplied by the distributive principle below. To do so, we multiplied $(x + 3)$ by both $\underline{x}$ and 2.

$$(x + 2)(x + 3) = x(x + 3) + 2(x + 3)$$

By reversing the process, we can use the distributive principle to factor out $(x + 3)$ which is a <u>common</u> <u>binomial</u> <u>factor</u>.

$$x(x + 3) + 2(x + 3) = (x + 2)(x + 3)$$

Continued on following page.

Answers to 79:

a) $3p^2q^2(2p^2q^2 + 1)$

b) $4c^3d(2c^2 - cd + d^2)$

80. Continued Factor out the common binomial factor. a) $y(y - 1) + 5(y - 1) =$ ________________ b) $2t(4t + 7) - 1(4t + 7) =$ ________________	
81. We cannot factor out a common monomial factor from $x^2 + 5x + 3x + 15$. However, we can factor that expression by a process called <u>factoring by grouping</u>. That is, we can factor $x^2 + 5x$ and $3x + 15$ and then factor out a common <u>binomial</u> factor. We get: $x^2 + 5x + 3x + 15 = (x^2 + 5x) + (3x + 15)$ $= x(x + 5) + 3(x + 5)$ $= (x + 3)(x + 5)$ Use the same method to factor the expression below. $2y^2 - 16y + 5y - 40 = (2y^2 - 16y) + (5y - 40)$ = ________________ = ________________	a) $(y + 5)(y - 1)$ b) $(2t - 1)(4t + 7)$
82. Following the example, factor the other expression. $5m^2 - 15mt + 2mt - 6t^2 = (5m^2 - 15mt) + (2mt - 6t^2)$ $= 5m(m - 3t) + 2t(m - 3t)$ $= (5m + 2t)(m - 3t)$ $8k^2 + 6kq + 12kq + 9q^2 = (8k^2 + 6kq) + (12kq + 9q^2)$ = ________________ = ________________	$2y(y - 8) + 5(y - 8)$ $(2y + 5)(y - 8)$
83. Following the example, factor the other expression. $x^3 + x^2 + 4x + 4 = (x^3 + x^2) + (4x + 4)$ $= x^2(x + 1) + 4(x + 1)$ $= (x^2 + 4)(x + 1)$ $y^3 - y^2 + 7y - 7 = (y^3 - y^2) + (7y - 7)$ = ________________ = ________________	$2k(4k+3q)+3q(4k+3q)$ $(2k + 3q)(4k + 3q)$
	$y^2(y - 1) + 7(y - 1)$ $(y^2 + 7)(y - 1)$

84. Notice how we factored -4 out of (-4x - 12) below. Following the example, factor the other expression.

$$x^2 + 3x - 4x - 12 = (x^2 + 3x) + (-4x - 12)$$
$$= x(x + 3) - 4(x + 3)$$
$$= (x - 4)(x + 3)$$

$$y^2 + 5y - 2y - 10 = (y^2 + 5y) + (-2y - 10)$$
$$= \underline{\hspace{6cm}}$$
$$= \underline{\hspace{6cm}}$$

---

85. Notice how we factored -3 out of (-3t + 12) below. Following the example, factor the other expression.

$$t^2 - 4t - 3t + 12 = (t^2 - 4t) + (-3t + 12)$$
$$= t(t - 4) - 3(t - 4)$$
$$= (t - 3)(t - 4)$$

$$m^2 - 7m - 2m + 14 = (m^2 - 7m) + (-2m + 14)$$
$$= \underline{\hspace{6cm}}$$
$$= \underline{\hspace{6cm}}$$

Answers to 84:

$y(y + 5) - 2(y + 5)$

$(y - 2)(y + 5)$

---

86. Factor these:

a) $a^3 + a^2 - 2a - 2 =$

b) $9x^3 - 3x^2y^2 - 3xy + y^3 =$

Answers to 85:

$m(m - 7) - 2(m - 7)$

$(m - 2)(m - 7)$

---

87. Factor these:

a) $ap - aq + bp - bq =$

b) $cx + 2dx - cy - 2dy =$

Answers to 86:

a) $(a^2 - 2)(a + 1)$

b) $(3x^2 - y)(3x - y^2)$

Answers to 87:

a) $(a + b)(p - q)$

b) $(x - y)(c + 2d)$

3-9 FACTORING TRINOMIALS OF THE FORM: $x^2 + bx + c$

In this section, we will discuss the procedure for factoring trinomials when the coefficient of the x^2-term is "1".

---

88. When multiplying binomials, some products can be simplified to a trinomial. Three examples are shown below.

#1 $(x + 2)(x + 3) = x^2 + 5x + 6$

#2 $(x - 3)(x - 4) = x^2 - 7x + 12$

#3 $(x + 2)(x - 5) = x^2 - 3x - 10$

Notice that each product is a trinomial of the form $x^2 + bx + c$. In each case, b is the sum and c is the product of the numbers in the factors. That is:

For #1, $b = 2 + 3 = 5$ and $c = (2)(3) = 6$

For #2, $b = (-3) + (-4) = -7$ and $c = (-3)(-4) = 12$

For #3, $b = 2 + (-5) = -3$ and $c = (2)(-5) = -10$

Using these facts, we can write the product for $(x + 5)(x - 7)$.

a) Since $5 + (-7) = -2$, b = ________

b) Since $(5)(-7) = -35$, c = ________

c) Therefore, $(x + 5)(x - 7) =$ ____________

---

a) -2

b) -35

c) $x^2 - 2x - 35$

89. To factor a trinomial of the form $x^2 + bx + c$, we can use the same facts about b and c. Three examples are discussed.

Example 1: Factoring $x^2 + 6x + 8$

Since the product of the numbers in the binomials must be 8 and their sum must be 6, both numbers must be positive. The possible pairs of positive factors for 8 are (1 and 8) and (2 and 4). The pair whose sum is 6 is (2 and 4). Therefore:

$$x^2 + 6x + 8 = (x + 2)(x + 4)$$

Example 2: Factoring $x^2 - 7x + 10$

Since the product of the numbers in the binomials must be 10 and their sum must be -7, both numbers must be negative. The possible pairs of negative factors for 10 are (-1 and -10) and (-2 and -5). The pair whose sum is -7 is (-2 and -5). Therefore:

$$x^2 - 7x + 10 = (x - 2)(x - 5)$$

Continued on following page.

89. Continued

Example 3: Factoring $x^2 + 5x - 6$

Since the product of the numbers in the binomials must be -6, one number must be positive and the other negative. The possible pairs of factors for -6 are (-1 and 6), (-6 and 1), (-2 and 3), and (-3 and 2). The pair whose sum is 5 is (-1 and 6). Therefore:

$$x^2 + 5x - 6 = (x - 1)(x + 6)$$

Using the same method, factor these:

a) $y^2 + 15y + 36$ = ______________

b) $m^2 - 10m + 16$ = ______________

c) $p^2 - 5p - 14$ = ______________

---

90. To check the factoring of a trinomial, multiply the two binomial factors to see whether you obtain the original trinomial. For example:

$$y^2 + 7y - 18 = (y - 2)(y + 9) \text{ is correct, since:}$$

$$(y - 2)(y + 9) = y^2 + 7y - 18$$

Check each factoring. State whether it is "correct" or "incorrect".

a) $x^2 - 8x + 12 = (x - 2)(x - 6)$ ______________

b) $t^2 - 5t - 14 = (t - 2)(t + 7)$ ______________

Answers to 89:

a) $(y + 3)(y + 12)$

b) $(m - 2)(m - 8)$

c) $(p + 2)(p - 7)$

---

91. Remember that c is the key to factoring a trinomial. That is:

1) If c is positive, the numbers in the binomials have the same sign. b tells us whether both are positive or negative.

   If b is positive, both are positive.

   If b is negative, both are negative.

2) If c is negative, the numbers in the binomials have different signs.

Factor each trinomial below and check your results.

a) $y^2 - 3y - 18$ = ______________

b) $t^2 - 9t + 8$ = ______________

Answers to 90:

a) Correct

b) Incorrect, since:

$(t - 2)(t + 7) =$

$t^2 + 5t - 14$

Answers to 91:

a) $(y + 3)(y - 6)$

b) $(t - 1)(t - 8)$

92. Some trinomials cannot be factored into binomials in which the numbers are integers. For example, the trinomial below is not factorable since the possible pairs of numbers are (1 and 6) and (2 and 3), and neither pair has 4 as its sum.

$$x^2 + 4x + 6 \text{ is not factorable}$$

Factor if possible.

a) $x^2 - 11x + 16 =$ ________________

b) $m^2 + m - 2 =$ ________________

---

93. We factored a trinomial containing two variables below. Notice that the same general method is used.

$$a^2 - 3ab - 28b^2 = (a + 4b)(a - 7b)$$

Factor these and check your results.

a) $p^2 - 7pq + 6q^2 =$ ________________

b) $m^2 + 6mt - 16t^2 =$ ________________

Answers to 92:

a) Not possible

b) $(m - 1)(m + 2)$

---

94. We factored another trinomial containing two variables below.

$$a^2b^2 + 4ab - 32 = (ab - 4)(ab + 8)$$

Factor these and check your results.

a) $p^2q^2 + 11pq + 10 =$ ________________

b) $m^2v^2 - 5mv - 14 =$ ________________

Answers to 93:

a) $(p - q)(p - 6q)$

b) $(m - 2t)(m + 8t)$

---

95. We factored two trinomials containing higher powers below.

$$x^4 + 11x^2 + 30 = (x^2 + 5)(x^2 + 6)$$
$$p^6q^6 - 3p^3q^3 - 4 = (p^3q^3 + 1)(p^3q^3 - 4)$$

Factor these and check your results.

a) $y^6 - 10y^3 + 16 =$ ________________

b) $a^4b^4 + a^2b^2 - 6 =$ ________________

Answers to 94:

a) $(pq + 1)(pq + 10)$

b) $(mv + 2)(mv - 7)$

---

96. We always begin by factoring out a common monomial factor if possible. For example, we factored out an <u>x</u> below. After doing so, we could factor $x^2 - 8x + 15$.

$$x^3 - 8x^2 + 15x = x(x^2 - 8x + 15)$$
$$= x(x - 3)(x - 5)$$

Using the same method, factor these:

a) $y^3 - y^2 - 12y =$ ________________

b) $t^3 + 14t^2 + 48t =$ ________________

Answers to 95:

a) $(y^3 - 2)(y^3 - 8)$

b) $(a^2b^2+3)(a^2b^2-2)$

97. Sometimes a more complicated expression can be factored by substituting a monomial for a binomial. An example is discussed below.	a) $y(y + 3)(y - 4)$
	b) $t(t + 6)(t + 8)$

Factor $(x + 3)^2 - 7(x + 3) + 10$.

Substituting y for x + 3, we get $y^2 - 7y + 10$. We can factor that trinomial. We get:

$$y^2 - 7y + 10 = (y - 2)(y - 5)$$

Now we can substitute x + 3 for y and get:

$$(x + 3)^2 - 7(x + 3) + 10 = [(x + 3) - 2][(x + 3) - 5]$$
$$= (x + 1)(x - 2)$$

Using the same method, factor each expression below.

a) $(b - 5)^2 - 2(b - 5) - 8 =$

b) $(c + d)^2 + 7(c + d) + 12 =$

---

a) $(b - 9)(b - 3)$ b) $(c + d + 3)(c + d + 4)$

3-10 FACTORING TRINOMIALS OF THE FORM: $ax^2 + bx + c$

In this section, we will discuss the procedure for factoring trinomials in which the coefficient of the x^2-term is a number other than "1".

---

98. We multiplied 4x + 3 and x + 2 below. Notice that $4x^2$ is the product of the first terms (4x and x) and 6 is the product of the numbers (3 and 2).

$$(4x + 3)(x + 2) = 4x^2 + 8x + 3x + 6 = 4x^2 + 11x + 6$$

The facts above are used to factor the trinomial below.

$$3x^2 + 14x + 8$$

1) Since the product of the first terms must be $3x^2$, the only possible pair of first terms is (3x and x).
2) Since the product of the numbers must be 8, the possible pairs of numbers are (1 and 8) and (2 and 4).
3) Therefore, the possible pairs of binomial factors are:

    A: $(3x + 1)(x + 8)$

    B: $(3x + 8)(x + 1)$

    C: $(3x + 2)(x + 4)$

    D: $(3x + 4)(x + 2)$

4) Only one pair of binomials is correct. It is the pair that produces 14x as the middle term of the trinomial product.

Continued on following page.

98. Continued

a) Which pair has 14x as the middle term of its product?

Pair ______

b) Therefore: $3x^2 + 14x + 8 = (\qquad)(\qquad)$

---

99. Two more examples of factoring are given below.

Example 1: Factoring $2y^2 - 13y + 15$

The only possible pair of first terms is (2y and y). The possible pairs of numbers are (-1 and -15) and (-3 and -5). The possible pairs of binomial factors are:

A: $(2y - 1)(y - 15)$ C: $(2y - 3)(y - 5)$

B: $(2y - 15)(y - 1)$ D: $(2y - 5)(y - 3)$

Only pair C has -13y as the middle term of its product. Therefore:

$$2y^2 - 13y + 15 = (2y - 3)(y - 5)$$

Example 2: Factoring $6x^2 - 7x - 5$

The possible pairs of first terms are (x and 6x) and (2x and 3x). The possible pairs of numbers are (1 and -5) and (-1 and 5). The possible pairs of binomial factors are:

A: $(x + 1)(6x - 5)$ E: $(2x + 1)(3x - 5)$

B: $(x - 5)(6x + 1)$ F: $(2x - 5)(3x + 1)$

C: $(x - 1)(6x + 5)$ G: $(2x - 1)(3x + 5)$

D: $(x + 5)(6x - 1)$ H: $(2x + 5)(3x - 1)$

a) Which pair has -7x as the middle term of its product?

Pair ______

b) Therefore: $6x^2 - 7x - 5 = (\qquad)(\qquad)$

Answers (frame 98):

a) Pair C

b) $(3x + 2)(x + 4)$

---

100. When the coefficient of the first term is a number other than "1", factoring is a process of trial and error. Each possible pair of factors has to be checked by multiplying them. That is:

$2y^2 - y - 1 = (2y + 1)(y - 1)$ is correct, since:

$$(2x + 1)(y - 1) = 2y^2 - y - 1$$

Not every trinomial is factorable. For example, the trinomial below is not factorable because the only possible pairs of factors are $(2m + 3)(m + 1)$ and $(2m + 1)(m + 3)$, and neither pair has 6m as the middle term of its product.

$$2m^2 + 6m + 3 \text{ is not factorable.}$$

Continued on following page.

Answers (frame 99):

a) Pair E

b) $(2x + 1)(3x - 5)$

100. Continued

Factor if possible.

a) $5x^2 + 17x + 6 =$ ______

b) $2t^2 - 5t - 7 =$ ______

c) $6x^2 - 15x + 7 =$ ______

d) $15y^2 + 4y - 4 =$ ______

---

101. When the number is the first term, we can use the same general method. For example:

$$5 + 13t - 6t^2 = (5 - 2t)(1 + 3t)$$

Factor these:

a) $8 - 2y - y^2 =$ ______

b) $1 + 3x - 10x^2 =$ ______

a) $(5x + 2)(x + 3)$

b) $(2t - 7)(t + 1)$

c) Not possible

d) $(5y - 2)(3y + 2)$

---

102. We factored a trinomial containing two variables below. Notice that the same general method is used.

$$8t^2 - 2tv - 15v^2 = (4t + 5v)(2t - 3v)$$

Factor these:

a) $3a^2 - 5ab + 2b^2 =$ ______

b) $4p^2 + 8pq - 5q^2 =$ ______

a) $(4 + y)(2 - y)$

b) $(1 + 5x)(1 - 2x)$

---

103. We factored another trinomial containing two variables below.

$$5x^2y^2 - 12xy + 7 = (5xy - 7)(xy - 1)$$

Factor these:

a) $7p^2q^2 + 9pq + 2 =$ ______

b) $6a^2b^2 - 7ab - 3 =$ ______

a) $(3a - 2b)(a - b)$

b) $(2p + 5q)(2p - q)$

---

104. We factored two trinomials containing higher powers below.

$$3x^4 + 8x^2 + 5 = (3x^2 + 5)(x^2 + 1)$$

$$8a^6b^6 - 10a^3b^3 - 3 = (2a^3b^3 - 3)(4a^3b^3 + 1)$$

Factor these:

a) $2y^6 - 11y^3 + 5 =$ ______

b) $6p^4q^4 + 7p^2q^2 - 10 =$ ______

Answers to 103:

a) $(7pq + 2)(pq + 1)$

b) $(3ab + 1)(2ab - 3)$

105. Always look for a common factor before factoring a trinomial. For example, we were able to factor out a 3 below. Then we were able to factor $2x^2 - 11x + 5$.

$$6x^2 - 33x + 15 = 3(2x^2 - 11x + 5)$$
$$= 3(2x - 1)(x - 5)$$

Use the same method to factor these:

a) $2y^2 + 28y + 80$ = ________________

b) $12a^2 - 16ab - 28b^2$ = ________________

a) $(2y^3 - 1)(y^3 - 5)$

b) $(6p^2q^2-5)(p^2q^2+2)$

---

106. We began by factoring out 4x in the example below.

$$12x^3 - 20x^2 - 8x = 4x(3x^2 - 5x - 2)$$
$$= 4x(3x + 1)(x - 2)$$

Factor these:

a) $10y^3 - 55y^2 + 60y$ = ________________

b) $a^3b^3 + 5a^2b^2 - 14ab$ = ________________

a) $2(y + 4)(y + 10)$

b) $4(3a - 7b)(a + b)$

---

107. We can factor the expression below by making a substitution.

Factor $10(x - 5)^2 - 9(x - 5) - 9$.

Substituting <u>y</u> for $x - 5$, we get a trinomial that we can factor.

$$10y^2 - 9y - 9 = (5y + 3)(2y - 3)$$

Then substituting $x - 5$ for <u>y</u>, we get:

$$10(x - 5)^2 - 9(x - 5) - 9 = [5(x - 5) + 3][2(x - 5) - 3]$$
$$= (5x - 25 + 3)(2x - 10 - 3)$$
$$= (5x - 22)(2x - 13)$$

Factor these:

a) $2(m + 7)^2 - 5(m + 7) + 3$ =

b) $4(a + b)^2 + 7(a + b) + 3$ =

a) $5y(2y - 3)(y - 4)$

b) $ab(ab-2)(ab+7)$

---

a) $(2m + 11)(m + 6)$

b) $(4a + 4b + 3)(a + b + 1)$

3-11 FACTORING THE DIFFERENCE OF TWO SQUARES

In this section, we will discuss the pattern used to factor the difference of two squares.

---

108. When a monomial is squared, the result is called the square of the monomial. For example:

Since $(3x)^2 = 9x^2$, $9x^2$ is called the square of 3x.

Some monomials and their squares are shown in the table below.

Monomial	Square
5	25
x	x^2
$4y^2$	$16y^4$
$7ab^3$	$49a^2b^6$

As you can see from the table, when a power is squared, the exponent of its square is always an even number. To find the square root of a power that is a square, we divide its exponent by 2. Two examples are shown. (Note: In this section, we will assume that all variables represent positive numbers.)

$\sqrt{x^2} = x^{\frac{2}{2}} = x^1 = x$ $\sqrt{y^{10}} = y^{\frac{10}{2}} = y^5$

Complete these.

a) $\sqrt{t^2}$ = ____ b) $\sqrt{d^6}$ = ____ c) $\sqrt{x^{20}}$ = ____

---

109. We found the square roots of two squares below.

$\sqrt{25x^2} = 5x$ $\sqrt{81a^{10}b^6} = 9a^5b^3$

Complete these.

a) $\sqrt{9y^4}$ = ____ b) $\sqrt{x^6y^2}$ = ____ c) $\sqrt{64a^8b^{10}}$ = ____

Answers to 108: a) t b) d^3 c) x^{10}

---

110. As we saw earlier, the product of the sum and difference of two terms is the difference of the squares of the two terms. That is:

$$(A + B)(A - B) = A^2 - B^2$$

By reversing the pattern, we can factor the difference of two squares. We get:

$$\boxed{A^2 - B^2 = (A + B)(A - B)}$$

We used the pattern above to factor each difference below.

$$x^2 - 16 = (x + 4)(x - 4)$$

$$9y^2 - 4 = (3y + 2)(3y - 2)$$

Continued on following page.

Answers to 109: a) $3y^2$ b) x^3y c) $8a^4b^5$

110. Continued

Factor these:

a) $p^2 - 25 =$ ______ c) $36x^2 - 81 =$ ______

b) $t^2 - 100 =$ ______ d) $64h^2 - 1 =$ ______

---

111. We used the same pattern to factor the difference below.

$$9x^6 - 25y^2 = (3x^3 + 5y)(3x^3 - 5y)$$

Factor these:

a) $4d^8 - 81 =$ ______

b) $100p^2 - 49q^4 =$ ______

Answers to 110:
a) $(p + 5)(p - 5)$
b) $(t + 10)(t - 10)$
c) $(6x + 9)(6x - 9)$
d) $(8h + 1)(8h - 1)$

---

112. We used the same pattern to factor the difference below.

$$9a^2b^4 - 4 = (3ab^2 + 2)(3ab^2 - 2)$$

Factor these:

a) $x^2y^2 - 36 =$ ______

b) $1 - 16p^6q^8 =$ ______

Answers to 111:
a) $(2d^4 + 9)(2d^4 - 9)$
b) $(10p+7q^2)(10p-7q^2)$

---

113. After factoring below, we were able to use the same pattern to factor $x^2 - 4$.

$$x^4 - 16 = (x^2 + 4)(x^2 - 4)$$
$$= (x^2 + 4)(x + 2)(x - 2)$$

Factor this one:

$16y^4 - 81 =$

Answers to 112:
a) $(xy + 6)(xy - 6)$
b) $(1+4p^3q^4)(1-4p^3q^4)$

---

114. After factoring out a common monomial factor, we were able to use the pattern below.

$$7a^4b^2 - 7 = 7(a^4b^2 - 1)$$
$$= 7(a^2b + 1)(a^2b - 1)$$

$$x^4y - y = y(x^4 - 1)$$
$$= y(x^2 + 1)(x^2 - 1)$$
$$= y(x^2 + 1)(x + 1)(x - 1)$$

Answer to 113: $(4y^2+9)(2y+3)(2y-3)$

Continued on following page.

114. Continued

Factor these:

a) $3a^4 - 3b^4 =$

b) $36c^2d - 49d =$

---

115. The pattern we have been using applies only to the difference of two squares. <u>It does not apply to a sum of two squares like $9y^2 + 25$</u>. Factor these if the pattern applies.

a) $m^2 - 1 =$ __________

c) $64a^2 + 9 =$ __________

b) $v^{10} + 25 =$ __________

d) $a^8 - 64 =$ __________

Answers to 114:

a) $3(a^2+b^2)(a+b)(a-b)$

b) $d(6c + 7)(6c - 7)$

---

116. We can use the same pattern when the squares are not monomials. For example:

$$(x + 3)^2 - 16 = [(x + 3) + 4][(x + 3) - 4] = (x + 7)(x - 1)$$

Factor these:

a) $100 - (y - 7)^2 =$

b) $(x + y)^2 - (x - y)^2 =$

Answers to 115:

a) $(m + 1)(m - 1)$

b) Does not apply

c) Does not apply

d) $(a^4 + 8)(a^4 - 8)$

Answers to 116:

a) $(3 + y)(17 - y)$

b) $4xy$

SELF-TEST 11 (pages 144-158)

Factor out the largest common factor.

1. $3t^2 - t =$ ______
2. $5x^3y^2 + 10xy^3 =$ ______
3. $b^7 - 5b^5 - 5b^3 =$ ______
4. $6p^4q - 3p^3q^2 + 3p^2q^3 =$ ______

Factor by grouping.

5. $3y^3 - 3y^2 + 5y - 5 =$ ______
6. $4a^4 - 2a^2b^2 - 2a^2b + b^3 =$ ______

Factor.

7. $x^2y^2 + 2xy - 15 =$ ______
8. $a^4 + 7a^2 + 12 =$ ______
9. $6y^2 - 11y - 7 =$ ______
10. $8p^4q^4 - 11p^2q^2 + 3 =$ ______
11. $6x^2 - 27x - 15 =$ ______
12. $(y + 4)^2 - 5(y + 4) + 6 =$ ______

Factor.

13. $25y^2 - 1 =$ ______
14. $a^2x^4 - 9y^2 =$ ______
15. $4ab^2 - 49a =$ ______
16. $(t + 3)^2 - 81 =$ ______

ANSWERS:

1. $t(3t - 1)$
2. $5xy^2(x^2 + 2y)$
3. $b^3(b^4 - 5b^2 - 5)$
4. $3p^2q(2p^2 - pq + q^2)$
5. $(3y^2 + 5)(y - 1)$
6. $(2a^2 - b)(2a^2 - b^2)$
7. $(xy + 5)(xy - 3)$
8. $(a^2 + 3)(a^2 + 4)$
9. $(3y - 7)(2y + 1)$
10. $(8p^2q^2 - 3)(p^2q^2 - 1)$
11. $3(2x + 1)(x - 5)$
12. $(y + 2)(y + 1)$
13. $(5y + 1)(5y - 1)$
14. $(ax^2 + 3y)(ax^2 - 3y)$
15. $a(2b + 7)(2b - 7)$
16. $(t + 12)(t - 6)$

3-12 FACTORING PERFECT SQUARE TRINOMIALS

In this section, we will discuss the methods used to identify and factor perfect square trinomials.

---

117. Earlier we saw the following patterns for squaring binomials. The expressions on the right are called perfect square trinomials.

$$(A + B)^2 = A^2 + 2AB + B^2$$

$$(A - B)^2 = A^2 - 2AB + B^2$$

Notice these facts about the perfect square trinomials.

1. A^2 and B^2 are squares.
2. Though 2AB may be either positive or negative, A^2 and B^2 are always positive.
3. 2AB is double the product of the square roots of A^2 and B^2. That is:

$$2AB = 2\sqrt{A^2} \cdot \sqrt{B^2} = 2(A)(B)$$

Some trinomials are not perfect square trinomials because of facts 1 and 2 above. For example:

In $2x^2 + 10x + 25$, $2x^2$ is not a square.

In $y^2 - 6y - 9$, -9 is negative.

Which of the following cannot be perfect square trinomials for the two reasons above? ______________

a) $y^2 + 4y - 4$

b) $x^2 - 8x + 16$

c) $9a^2 + 12ab + 4b^2$

d) $4c^2 - 12cd + 7d^2$

Both (a) and (d)

118. Fact 3 in the last frame says that the middle term must be double the product of the square roots of the first and last terms. We tested that fact in each example below.

$x^2 + 14x + 49$ is a perfect square trinomial, since:

$$2 \cdot \sqrt{x^2} \cdot \sqrt{49} = 2(x)(7) = 14x$$

$25a^2 + 10ab + 4b^2$ is not a perfect square trinomial, since:

$$2\sqrt{25a^2} \cdot \sqrt{4b^2} = 2(5a)(2b) = 20ab$$

Use the above test to decide whether each trinomial below is a perfect square trinomial.

a) $x^2 + 2x + 1$ ________

b) $c^2 - 7cd + d^2$ ________

c) $100t^2 - 20t + 1$ ________

d) $4p^2 + 10pq + 9q^2$ ________

119. The patterns for factoring perfect square trinomials are shown below.

$$A^2 + 2AB + B^2 = (A + B)^2$$
$$A^2 - 2AB + B^2 = (A - B)^2$$

We used the patterns above to factor the trinomials below.

$$x^2 + 6x + 9 = (x + 3)^2$$
$$25p^2 - 20pq + 4q^2 = (5p - 2q)^2$$

Use the pattern to factor these:

a) $m^2 + 16m + 64 =$ ________ c) $25d^2 + 10dh + h^2 =$ ________

b) $F^2 - 14F + 49 =$ ________ d) $9x^2 - 12xy + 4y^2 =$ ________

a) Yes, since: $2\sqrt{x^2} \cdot \sqrt{1} = 2x$

b) No, since: $2\sqrt{c^2} \cdot \sqrt{d^2} = 2cd$

c) Yes, since: $2\sqrt{100t^2} \cdot \sqrt{1} = 20t$

d) No, since: $2\sqrt{4p^2} \cdot \sqrt{9q^2} = 12pq$

---

120. Each trinomial below is also a perfect square.

$$x^6 + 2x^3 + 1 = (x^3 + 1)^2$$
$$p^2q^2 - 10pq + 25 = (pq - 5)^2$$

Factor these:

a) $y^8 - 6y^4 + 9 =$ ________ b) $4a^2b^2 + 4ab + 1 =$ ________

a) $(m + 8)^2$

b) $(F - 7)^2$

c) $(5d + h)^2$

d) $(3x - 2y)^2$

---

121. After factoring out 2x below, we get a perfect square trinomial which can be factored.

$$2x^3 - 8x^2 + 8x = 2x(x^2 - 4x + 4) = 2x(x - 2)^2$$

Factor these:

a) $3y^2 + 30y + 75 =$ ____________________

b) $4a^3 - 32a^2 + 64a =$ ____________________

a) $(y^4 - 3)^2$

b) $(2ab + 1)^2$

---

a) $3(y + 5)^2$

b) $4a(a - 4)^2$

3-13 FACTORING THE SUM AND DIFFERENCE OF TWO CUBES

In this section, we will discuss the patterns used to factor the sum and difference of two cubes.

---

122. When a number is raised to the third power, the result is called the cube of the number. For example:

Since $(4)^3 = 64$, 64 is called the cube of 4.

The cubes of the numbers from 1 to 10 are given below.

Number	Cube	Number	Cube
1	1	6	216
2	8	7	343
3	27	8	512
4	64	9	729
5	125	10	1000

A cube root of a number N is a number whose cube is N. That is:

Since $5^3 = 125$, $\sqrt[3]{125} = 5$

Use the table above to find these cube roots.

a) $\sqrt[3]{8}$ = _____ b) $\sqrt[3]{64}$ = _____ c) $\sqrt[3]{216}$ = _____

---

123. When any monomial is raised to the third power, the result is called the cube of the monomial. For example:

Since $(3x)^3 = 27x^3$, $27x^3$ is called the cube of 3x.

Some monomials and their cubes are shown in the table below.

Monomial	Cube
x	x^3
$2y^2$	$8y^6$
$5ab^3$	$125a^3b^9$

As you can see from the table, when a power is cubed, the exponent of its cube is a multiple of 3. To find the cube root of such a power, we divide the exponent by 3. That is:

$$\sqrt[3]{m^3} = m^{\frac{3}{3}} = m^1 = m \qquad \sqrt[3]{b^9} = b^{\frac{9}{3}} = b^3$$

Complete these.

a) $\sqrt[3]{t^3}$ = _____ b) $\sqrt[3]{d^6}$ = _____ c) $\sqrt[3]{p^{15}}$ = _____

Answers to 122: a) 2 b) 4 c) 6

Answers to 123: a) t b) d^2 c) p^5

124. We found the cube roots of two cubes below.

$$\sqrt[3]{8x^3} = 2x \qquad \sqrt[3]{64y^6} = 4y^2$$

Complete these.

a) $\sqrt[3]{27t^3}$ = ______ b) $\sqrt[3]{125m^6}$ = ______ c) $\sqrt[3]{1000q^{12}}$ = ______

Answers: a) $3t$ b) $5m^2$ c) $10q^4$

125. We did two multiplications below. Notice that each product is either the sum or difference of two cubes.

$$\begin{aligned}(a + b)(a^2 - ab + b^2) &= a(a^2 - ab + b^2) + b(a^2 - ab + b^2)\\ &= a^3 - a^2b + ab^2 + a^2b - ab^2 + b^3\\ &= a^3 + b^3\end{aligned}$$

$$\begin{aligned}(a - b)(a^2 + ab + b^2) &= a(a^2 + ab + b^2) - b(a^2 + ab + b^2)\\ &= a^3 + a^2b + ab^2 - a^2b - ab^2 - b^3\\ &= a^3 - b^3\end{aligned}$$

By reversing the multiplications above, we get the following patterns for factoring the sum or difference of two cubes.

$$A^3 + B^3 = (A + B)(A^2 - AB + B^2)$$
$$A^3 - B^3 = (A - B)(A^2 + AB + B^2)$$

We used the pattern to factor $x^3 + 125$ below.

$$x^3 + 125 = x^3 + 5^3 = (x + 5)(x^2 - 5x + 25)$$

Note: In the trinomial: 1) x^2 is the square of x.
2) $5x$ is the product of 5 and x.
3) 25 is the square of 5.

Using the pattern, factor these:

a) $y^3 + 8$ = ______________________

b) $t^3 - 64$ = ______________________

Answers: a) $(y+2)(y^2-2y+4)$ b) $(t-4)(t^2+4t+16)$

126. We used the pattern to factor $64d^3 - t^3$ below.

$$64d^3 - t^3 = (4d)^3 - t^3 = (4d - t)(16d^2 + 4dt + t^2)$$

Note: In the trinomial: 1) $16d^2$ is the square of $4d$.
2) $4dt$ is the product of $4d$ and t.
3) t^2 is the square of t.

Factor these:

a) $m^3 - 27p^3$ = ______________________

b) $125d^3 + 1$ = ______________________

127.	We used the pattern to factor $x^6 + 8y^3$ below. $x^6 + 8y^3 = (x^2)^3 + (2y)^3 = (x^2 + 2y)(x^4 - 2x^2y + 4y^2)$ <u>Note</u>: In the trinomial: 1) x^4 is the square of x^2. 2) $2x^2y$ is the product of x^2 and $2y$. 3) $4y^2$ is the square of $2y$. Factor these: a) $a^3 + b^6$ = __________ b) $64d^6 - 27t^3$ = __________	a) $(m - 3p)$ $(m^2 + 3mp + 9p^2)$ b) $(5d + 1)$ $(25d^2 - 5d + 1)$
128.	After factoring out a <u>y</u> below, we were able to factor $8x^3 + y^3$. $8x^3y + y^4 = y(8x^3 + y^3) = y(2x + y)(4x^2 - 2xy + y^2)$ Factor these: a) $5x^3 - 5$ = __________ b) $3a^3 + 81b^3$ = __________	a) $(a + b^2)$ $(a^2 - ab^2 + b^4)$ b) $(4d^2 - 3t)$ $(16d^4+12d^2t+9t^2)$
129.	We used the same pattern to factor $x^3 + (x + 5)^3$ below. $x^3 + (x + 5)^3 = [x + (x + 5)][x^2 - x(x + 5) + (x + 5)^2]$ $= (2x + 5)(x^2 - x^2 - 5x + x^2 + 10x + 25)$ $= (2x + 5)(x^2 + 5x + 25)$ Factor this expression: $(p - 3)^3 - 8 =$	a) $5(x-1)(x^2+x+1)$ b) $3(a + 3b)$ $(a^2-3ab+9b^2)$
		$(p - 5)(p^2 - 4p + 7)$

3-14 A GENERAL STRATEGY FOR FACTORING

In this section, we will discuss a general strategy for factoring polynomials.

---

130. Here is a general strategy for factoring polynomials.

> A STRATEGY FOR FACTORING POLYNOMIALS
>
> 1. Factor out any common factor if there is one.
> 2. If the polynomial is a binomial, try to factor it as the difference of two squares, or the sum or difference of two cubes. Remember: do not try to factor a sum of two squares.
> 3. If the polynomial is a trinomial, check to see whether it is a perfect square trinomial. If it is not, factor it by trial and error.
> 4. If the polynomial contains four or more terms, try to factor it by grouping.
> 5. Always factor completely. That is, whenever a factor with more than one term can be factored, factor it.

When factoring a binomial, we always begin by looking for a common factor. Then to factor completely, we factor the remaining binomial if possible. For example:

$$3x^2 - 12 = 3(x^2 - 4) = 3(x + 2)(x - 2)$$

$$4y^3 + 8y = 4y(y^2 + 2)$$

$$5m^3 + 40 = 5(m^3 + 8) = 5(m + 2)(m^2 - 2m + 4)$$

Factor completely.

a) $7a^2 - 7b^2$ = ______________

b) $3pq^2 + 3pt^2$ = ______________

c) $10x^3 - 10y^6$ = ______________

---

131. When factoring a trinomial, we always begin by looking for a common factor. Then to factor completely, we factor the remaining trinomial if possible. For example:

$$x^3 + 5x^2 + 6x = x(x^2 + 5x + 6) = x(x + 2)(x + 3)$$

$$4y^2 - 6y - 8 = 2(2y^2 - 3y - 4)$$

$$bx^2 + 2bxy + by^2 = b(x^2 + 2xy + y^2) = b(x + y)^2$$

Continued on following page.

a) $7(a + b)(a - b)$

b) $3p(q^2 + t^2)$

c) $10(x - y^2)(x^2 + xy^2 + y^4)$

131. Continued

Factor completely.

a) $5b^2 - 10b + 15$ = ______

b) $ad^2 - 4ad - 5a$ = ______

c) $2x^2 - 12x + 18$ = ______

Answers: a) $5(b^2 - 2b + 3)$ b) $a(d - 5)(d + 1)$ c) $2(x - 3)^2$

132. After factoring $(1 - 16t^4)$ below, we were able to factor $(1 - 4t^2)$.

$$1 - 16t^4 = (1 + 4t^2)(1 - 4t^2) = (1 + 4t^2)(1 + 2t)(1 - 2t)$$

Factor completely.

$x^4 - y^4$ = ______

Answer: $(x^2 + y^2)(x+y)(x-y)$

133. After factoring the trinomial below, we were able to factor $(x^2 - 1)$.

$$x^4 - 3x^2 + 2 = (x^2 - 2)(x^2 - 1) = (x^2 - 2)(x + 1)(x - 1)$$

Factor completely.

$y^4 - 5y^2 + 4$ = ______

Answer: $(y+2)(y-2)(y+1)(y-1)$

134. When a polynomial contains four or more terms, we try to factor by grouping. For example:

$$ax + 3a + 5x + 15 = a(x + 3) + 5(x + 3)$$
$$= (a + 5)(x + 3)$$

Factor this one:

$cx + dx + cy + dy$ = ______

Answer: $(x + y)(c + d)$

135. Notice how we factored completely below.

$$x^3 - xy^2 - x^2y + y^3 = x(x^2 - y^2) - y(x^2 - y^2)$$
$$= (x - y)(x^2 - y^2)$$
$$= (x - y)(x + y)(x - y)$$
or
$$(x - y)^2(x + y)$$

Factor completely.

$a^3 - ab^2 + a^2b - b^3$

Answer: $(a + b)(a + b)(a - b)$ or $(a + b)^2(a - b)$

136. Factor completely.

a) $4bp^2 - 9bq^2 =$ ______

b) $2xy^3 - 250x =$ ______

c) $3x^2 + 12x - 36 =$ ______

d) $x^2 - 7x + 16 =$ ______

---

a) $b(2p + 3q)(2p - 3q)$

b) $2x(y - 5)(y^2 + 5y + 25)$

c) $3(x + 6)(x - 2)$

d) Not factorable

---

137. Factor completely.

a) $9a^2t + 12abt + 4b^2t =$ ______

b) $81y^4 - 1 =$ ______

c) $a^2 + 5a - 4a - 20 =$ ______

d) $x^4 - xy^3 + x^3y - y^4 =$ ______

---

a) $t(3a + 2b)^2$

b) $(9y^2 + 1)(3y + 1)(3y - 1)$

c) $(a - 4)(a + 5)$

d) $(x+y)(x-y)(x^2+xy+y^2)$

3-15 SOLVING EQUATIONS BY FACTORING

Some equations can be solved by factoring and then using the principle of zero products. We will discuss solutions of that type in this section.

---

138. The principle of zero products says this: If the product of two factors is 0, then at least one of the factors must be 0.

<u>Principle Of Zero Products</u>
If $ab = 0$, either $a = 0$ or $b = 0$

We can use the principle of zero products to solve this equation.

$$(x - 4)(x + 5) = 0$$

Since the product is 0, the equation is true when either $x - 4$ or $x + 5$ is 0. Therefore, we can find the two solutions by solving the two equations below.

$$x - 4 = 0 \qquad x + 5 = 0$$

$$x = 4 \qquad x = -5$$

Continued on following page.

138. Continued

We checked 4 as one solution of the equation below. Check -5 as the second solution.

$$(x - 4)(x + 5) = 0 \qquad\qquad (x - 4)(x + 5) = 0$$

$$(4 - 4)(4 + 5) = 0$$

$$0(9) = 0$$

$$0 = 0$$

$$(-5 - 4)(-5 + 5) = 0$$

$$-9(0) = 0$$

$$0 = 0$$

139. To solve the equation below, we factored the left side and then used the principle of zero products. That is, we set each factor equal to 0.

$$x^2 + 5x - 14 = 0$$

$$(x - 2)(x + 7) = 0$$

$x - 2 = 0$	$x + 7 = 0$
$x = 2$	$x = -7$

The two solutions are 2 and -7. The solution set is {2,-7}. We checked 2 as a solution of the original equation below. Check -7 as a solution.

$$x^2 + 5x - 14 = 0 \qquad\qquad x^2 + 5x - 14 = 0$$

$$(2)^2 + 5(2) - 14 = 0$$

$$4 + 10 - 14 = 0$$

$$14 - 14 = 0$$

$$0 = 0$$

$$(-7)^2 + 5(-7) - 14 = 0$$

$$49 + (-35) - 14 = 0$$

$$14 - 14 = 0$$

$$0 = 0$$

140. We used the factoring method to solve one equation below. The solution set is $\left\{\frac{2}{5}, -1\right\}$. Solve the other equation.

$$5d^2 + 3d - 2 = 0 \qquad\qquad 2y^2 - 7y + 6 = 0$$

$$(5d - 2)(d + 1) = 0$$

$5d - 2 = 0$	$d + 1 = 0$
$5d = 2$	$d = -1$
$d = \frac{2}{5}$	

$y = \frac{3}{2}$ and 2, from:

$$(2y - 3)(y - 2) = 0$$

141. Following the example, solve the other equation.

$$9x^2 - 16 = 0$$
$$(3x + 4)(3x - 4) = 0$$

$3x + 4 = 0$	$3x - 4 = 0$
$3x = -4$	$3x = 4$
$x = -\frac{4}{3}$	$x = \frac{4}{3}$

$$t^2 - 49 = 0$$

Answer: $t = -7$ and 7, from: $(t + 7)(t - 7) = 0$

142. We factored to solve the equation. Notice that one factor is <u>x</u> and one root is 0. Solve the other equation.

$$x^2 + 8x = 0$$
$$x(x + 8) = 0$$

$x = 0$	$x + 8 = 0$
	$x = -8$

$$t^2 - t = 0$$

Answer: $t = 0$ and 1

143. We used the factoring method to solve one equation below. Solve the other equation.

$$2x^2 - x = 0$$
$$x(2x - 1) = 0$$

$x = 0$	$2x - 1 = 0$
	$2x = 1$
	$x = \frac{1}{2}$

$$7y^2 + y = 0$$

Answer: $y = 0$ and $-\frac{1}{7}$

144. Following the example, solve the other equation.

$$8x^2 + 12x = 0$$
$$4x(2x + 3) = 0$$

$4x = 0$	$2x + 3 = 0$
$x = 0$	$2x = -3$
	$x = -\frac{3}{2}$

$$12y^2 - 6y = 0$$

Answer: $y = 0$ and $\frac{1}{2}$

145. To use the factoring method, all of the terms must be on one side with 0 on the other side. Therefore, we cannot use the factoring method with the equation below as it stands.

$$m^2 = 7m - 8$$

However, we can put it in the form needed for the factoring method by adding -7m and 8 to both sides. We get:

$$m^2 - 7m + 8 = 7m - 7m - 8 + 8$$

$$m^2 - 7m + 8 = 0$$

Using the same steps, put each of these in the form needed for the factoring method.

a) $7x^2 = 3x + 4$ b) $2R^2 = 8R$

---

a) $7x^2 - 3x - 4 = 0$

b) $2R^2 - 8R = 0$

146. To solve the equation below, we began by putting it in the form needed for the factoring method. Solve the other equation.

$$y^2 = y + 6$$

$$y^2 - y - 6 = 0$$

$$(y + 2)(y - 3) = 0$$

$y + 2 = 0$ | $y - 3 = 0$

$y = -2$ | $y = 3$

$$3m^2 + 2 = 7m$$

$m = \frac{1}{3}$ and 2,

from:

$3m^2 - 7m + 2 = 0$

147. To solve the equation below, we began by putting it in the form needed for the factoring method. Solve the other equation.

$$t^2 = 2t$$

$$t^2 - 2t = 0$$

$$t(t - 2) = 0$$

$t = 0$ | $t - 2 = 0$

| $t = 2$

$$4d^2 = 1$$

$d = -\frac{1}{2}$ and $\frac{1}{2}$,

from:

$4d^2 - 1 = 0$

148. To solve the equation below, we began by putting it in the form needed for the factoring method. Solve the other equation.

$$3 + p(2p - 5) = 15$$
$$3 + 2p^2 - 5p = 15$$
$$2p^2 - 5p - 12 = 0$$
$$(2p + 3)(p - 4) = 0$$

$2p + 3 = 0$	$p - 4 = 0$
$2p = -3$	$p = 4$
$p = -\frac{3}{2}$	

$$4y(y - 1) = 3$$

$y = -\frac{1}{2}$ and $\frac{3}{2}$, from:

$4y^2 - 4y - 3 = 0$

149. To solve the equation below, we began by factoring out the largest common numerical factor. Use the same method to solve the other equation.

$$3x^2 + 12x + 9 = 0$$
$$3(x^2 + 4x + 3) = 0$$
$$\frac{1}{3}(3)(x^2 + 4x + 3) = \frac{1}{3}(0)$$
$$x^2 + 4x + 3 = 0$$
$$(x + 1)(x + 3) = 0$$

$x + 1 = 0$	$x + 3 = 0$
$x = -1$	$x = -3$

$$30t^2 - 80t + 50 = 0$$

$t = \frac{5}{3}$ and 1, from:

$3t^2 - 8t + 5 = 0$

3-16 WORD PROBLEMS

In this section, we will solve some word problems involving equations that are solved by the factoring method.

---

150. We translated the English sentence below to an equation.

The square of a number	minus	three times the number	is	18.
x^2	$-$	$3x$	$=$	18

Following the example, translate each of these to an equation. Use x for the variable.

a) 10 more than the square of a number is seven times that number.

b) Two more than a number times one less than that number is 4.

---

151. To solve the problem below, we set up an equation and solved it by the factoring method. Solve the other problem.

Six less than the square of a number is five times that number. Find the number.

$$x^2 - 6 = 5x$$

$$x^2 - 5x - 6 = 0$$

$$(x - 6)(x + 1) = 0$$

$x - 6 = 0$ | $x + 1 = 0$

$x = 6$ | $x = -1$

The two solutions are 6 and -1.

If you subtract a number from three times its square, you get 2. Find the number.

Answers to 150: a) $x^2 + 10 = 7x$ b) $(x + 2)(x - 1) = 4$

---

152. Following the example, solve the other problem.

One more than a number times one less than that number is 15. Find the number.

$$(x + 1)(x - 1) = 15$$

$$x^2 - 1 = 15$$

$$x^2 - 16 = 0$$

$$(x + 4)(x - 4) = 0$$

$x + 4 = 0$ | $x - 4 = 0$

$x = -4$ | $x = 4$

The two solutions are -4 and 4.

Two more than a number times two less than that number is 45. Find the number.

Answer to 151: $-\frac{2}{3}$ and 1, from: $3x^2 - x = 2$

153. To solve the problem below, we let $\underline{x}$ equal the smaller integer and $\underline{x + 1}$ equal the larger integer. Solve the other problem.

The product of two consecutive integers is 56. Find the integers.

$$x(x + 1) = 56$$
$$x^2 + x = 56$$
$$x^2 + x - 56 = 0$$
$$(x - 7)(x + 8) = 0$$

$x - 7 = 0$	$x + 8 = 0$
$x = 7$	$x = -8$

If $x = 7$, $x + 1 = 8$. The two consecutive integers are 7 and 8.

If $x = -8$, $x + 1 = -7$. The two consecutive integers are -8 and -7.

The product of two consecutive integers is 110. Find the integers.

Answer: -7 and 7, from: $(x + 2)(x - 2) = 45$

---

154. To solve the problem below, we let $\underline{x}$ equal the smaller integer and $\underline{x + 2}$ equal the larger integer. Solve the other problem.

The product of two consecutive even integers is 48. Find the integers.

$$x(x + 2) = 48$$
$$x^2 + 2x = 48$$
$$x^2 + 2x - 48 = 0$$
$$(x - 6)(x + 8) = 0$$

$x - 6 = 0$	$x + 8 = 0$
$x = 6$	$x = -8$

If $x = 6$, $x + 2 = 8$. The two consecutive even integers are 6 and 8.

If $x = -8$, $x + 2 = -6$. The two consecutive even integers are -8 and -6.

The product of two consecutive odd integers is 99. Find the integers.

Answer: 10 and 11 or -11 and -10

Answer: 9 and 11 or -11 and -9

155. To solve the problem below, we let x equal the width and x + 5 equal the length. Solve the other problem.

The length of a rectangle is 5 ft more than the width. The area of the rectangle is 84 ft^2. Find the length and width.

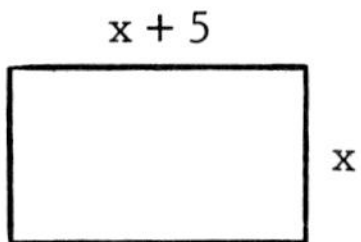

$$A = LW$$
$$84 = (x + 5)(x)$$
$$84 = x^2 + 5x$$
$$0 = x^2 + 5x - 84$$
$$0 = (x - 7)(x + 12)$$

$x - 7 = 0$	$x + 12 = 0$
$x = 7$	$x = -12$

Though the solutions of the equation are 7 and -12, the width of a rectangle cannot have a negative value.

Therefore, W = 7ft and L = 12 ft.

The width of a rectangle is 7m less than the length. The area of the rectangle is 120m^2. Find the length and width.

L = 15m, W = 8m

156. Following the example, solve the other problem.

The height of a triangle is 10 cm more than the base. If the area of the triangle is 48 cm^2, find the height and base.

b + 10

b

The height of a triangle is 6m less than the base. If the area of the triangle is 36m^2, find the height and base.

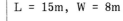

Continued on following page.

156. Continued

$A = \frac{1}{2}bh$

$48 = \frac{1}{2}(b)(b + 10)$

$48 = \frac{1}{2}(b^2 + 10b)$

$96 = b^2 + 10b$

$0 = b^2 + 10b - 96$

$0 = (b - 6)(b + 16)$

$b - 6 = 0$	$b + 16 = 0$
$b = 6$	$b = -16$

Though the solutions of the equation are 6 and -16, the base of a triangle cannot have a negative length.

Therefore, the base is 6 cm and the height is 16 cm.

---

157. If an object is dropped, the distance d it falls in t seconds (disregarding air resistance) is given by the formula below.

$$d = 16t^2$$

We used the formula to solve one problem below. Solve the other problem.

How long would it take an object to fall 64 ft?

$d = 16t^2$

$64 = 16t^2$

$0 = 16t^2 - 64$

$0 = 16(t^2 - 4)$

$0 = t^2 - 4$

$0 = (t + 2)(t - 2)$

$t + 2 = 0$	$t - 2 = 0$
$t = -2$	$t = 2$

Though the solutions of the equations are -2 and 2, a negative time interval does not make sense.

Therefore, it takes 2 seconds to drop 64 ft.

How long would it take an object to fall 400 ft?

Answers:

$b = 12m$, $h = 6m$

$t = 5$ seconds

SELF-TEST 12 (pages 159-175)

(pages 159-175)

Factor.

1. $9x^2 + 30xy + 25y^2$

2. $y^4 + 2y^2 + 1$

3. $27x^3 + 8y^3$

4. $a^6 - b^3$

Factor completely.

5. $ab^2 - 2ab - 35a$

6. $cx^2 - cy^2 + dx^2 - dy^2$

7. $2x^2 + 12x + 18$

8. $1 - 16y^4$

Solve by factoring.

9. $4d^2 - 5d = 0$

10. $k^2 - 7k - 18 = 0$

11. $x(x + 2) = 48$

12. $60t^2 + 10t = 20$

13. The product of two consecutive even integers is 80. Find the integers.

14. The length of a rectangle is 3m more than the width. The area of the rectangle is $88m^2$. Find the length and width.

ANSWERS:

1. $(3x + 5y)^2$
2. $(y^2 + 1)^2$
3. $(3x + 2y)(9x^2 - 6xy + 4y^2)$
4. $(a^2 - b)(a^4 + a^2b + b^2)$
5. $a(b + 5)(b - 7)$
6. $(c + d)(x + y)(x - y)$
7. $2(x + 3)^2$
8. $(1 + 4y^2)(1 + 2y)(1 - 2y)$
9. $d = 0$ and $\frac{5}{4}$
10. $k = 9$ and -2
11. $x = 6$ and -8
12. $t = -\frac{2}{3}$ and $\frac{1}{2}$
13. 8 and 10 or -10 and -8
14. $L = 11m$, $W = 8m$

SUPPLEMENTARY PROBLEMS - CHAPTER 3

Assignment 9

State whether each polynomial is a monomial, binomial, or trinomial.

1. $x^2 - 7$ 2. $4a^2b$ 3. $y^3 - y + 2$ 4. $at^4 - 4bt$

State the degree of each term.

5. $-3x$ 6. $4cd$ 7. $5y^4$ 8. $-2x^3y^2$ 9. ab^4c^5

State the degree of each polynomial.

10. $y^2 - 3y + 4$ 11. $x^2y^3 + x^5y^2 - 3xy$ 12. $x^3yz^4 - x^2y^3z^4 - xy^5z$

Simplify by combining like terms.

13. $x^2 + 7x - 3x^2 - x$ 14. $3ay^4 - 2by^2 - 2ay^4 + by^2$ 15. $4t^3 - 5t + 1 + t^3 + 5t - 6$

Simplify and write in descending order for x.

16. $5x + 3x^2 - 6x - 1$ 17. $xy - 3x^3y^2 + 5y^2 + 3xy$ 18. $2ax - 5 + 3bx^2 - ax - 1$

Add. Write each sum in descending order for x.

19. $(x^2y + 2) + (x^2y - 1)$ 20. $(x^3 - 2x - 3) + (2x^3 + x + 3)$

21. $(x^4 - x^2 + 1) + (3x^4 + 2x^2 - 3)$ 22. $(x^2 + x - 5) + (2x + 3) + (4x^3 - 3x)$

23. $(ax^2 + b) + (rx - 2b) + (kx^4 - rx)$

Subtract. Write each difference in descending order for y.

24. $(3y^2 + 2) - (y^2 - 2)$ 25. $(y^3 - 4y + 3) - (2y^3 - 7y + 1)$

26. $(2y + 3) - (3y - y^2)$ 27. $(2xy^4 - 1) - (xy^4 + 2)$

28. $(3by^2 + dy + 2) - (by^2 + 2)$ 29. $(ky^3 - ty + w) - (ky^3 - 2ty - w)$

Multiply or square.

30. $-t(5t)$ 31. $(-9x^2)(-2x)$ 32. $(3rt^2)(-4r^3t)$ 33. $(bx^2y)(axy^5)$

34. $(-dk)(5d^2k^3)(-2dk^2)$ 35. $(6p^3s)(-3a^2s)(ap)$ 36. $(-5x^3)^2$ 37. $(4ab^4c^6)^2$

Multiply.

38. $y(3y + 7)$ 39. $-5d(d + 1)$ 40. $3x^2(4x^3 - 2)$

41. $my(3y^3 - 2m)$ 42. $xy(x^2 + xy + y^2)$ 43. $5h^2s(3hs^2 - h^2s - 2hs)$

Multiply.

44. $(2x^2 + 3)(x^2 - 2)$ 45. $(ab - 6)(ab + 5)$ 46. $(x - y)(x^2 + xy + y^2)$

47. $\begin{array}{r} d - 1 \\ \underline{2d + 5} \end{array}$ 48. $\begin{array}{r} x^2 - 5x + 6 \\ \underline{x + 2} \end{array}$ 49. $\begin{array}{r} a^2 + ab + b^2 \\ \underline{a^2 - ab + 2b^2} \end{array}$ 50. $\begin{array}{r} y^2 - 2y + 3 \\ \underline{y^2 - 1} \end{array}$

Assignment 10

Multiply.

1. $(3x + 2)(2x + 5)$
2. $(p + 3q)(2p + q)$
3. $(y + 4)(5y - 1)$
4. $(2a + b)(a - 4b)$
5. $(5x - 3)(3x + 4)$
6. $(x - 2y)(4x + y)$
7. $(ab - 1)(ab + 7)$
8. $(4a - 3)(3a - 4)$
9. $(5pq - 2)(4pq - 1)$
10. $(x^2 - 2y^2)(3x^2 - y^2)$
11. $(3a - b)(p - 2q)$
12. $(3d - 2ab)(b^2 - 4a^2)$

Multiply.

13. $(x + 5)(x - 5)$
14. $(9 + 2y)(9 - 2y)$
15. $(7a + 1)(7a - 1)$
16. $(3a + b)(3a - b)$
17. $(4p + 3q)(4p - 3q)$
18. $(xy + 3)(xy - 3)$
19. $(x^2 + y^3)(x^2 - y^3)$
20. $(4b^4+5d^5)(4b^4-5d^5)$

Square.

21. $(x + 9)^2$
22. $(5y - 6)^2$
23. $(9 + 2t)^2$
24. $(10 - m)^2$
25. $(3x + 4y)^2$
26. $(ab - c)^2$
27. $(2t^2 - v^3)^2$
28. $(x^2y + 4x)^2$

Divide.

29. $\dfrac{6x^5y^4}{2xy^3}$
30. $\dfrac{-5p^2q^3}{p^2q}$
31. $\dfrac{12ab^2c^3}{4a^2bc^5}$
32. $\dfrac{18}{3t}$
33. $\dfrac{7c}{7c^3d}$
34. $\dfrac{2x^5y^4}{5xy^4}$
35. $\dfrac{8m^2}{6m^2}$
36. $\dfrac{ab^2}{3a^3b}$

Divide.

37. $\dfrac{6x^2 + 9x + 3}{3}$
38. $\dfrac{8y^4 - 6y^2 - 2}{6}$
39. $\dfrac{a^3b^4 - a^2b^2 + ab^5}{ab^2}$
40. $\dfrac{4x^3y + 6x^2y^2 - 12xy^3}{6x^3y}$
41. $\dfrac{6p^3q^3 - 8p^2q^2 + 10pq^4}{4p^3q^3}$
42. $\dfrac{6y^2 + y - 3}{3y}$

Divide.

43. $(x^2 + 14x + 48) \div (x + 8)$
44. $(a^2 - 10a + 24) \div (a - 4)$
45. $(6y^2 + 7y - 20) \div (2y + 5)$
46. $(20t^2 - 23t + 6) \div (4t - 3)$
47. $(m^2 + 5m + 7) \div (m + 2)$
48. $(8x^3 - 27) \div (2x - 3)$
49. $(2y^3 + 3y^2 - 4) \div (y^2 - 3)$
50. $(5b^3 - 9b^2 + 8b - 2) \div (b^2 - 2b + 1)$

Assignment 11

Factor out the largest common factor.

1. $5x + 35$
2. $4 - 8y$
3. $6t^2 - t$
4. $3m^4 + 9m^2$
5. $12d^3 - 8d^7$
6. $x^2y^2 + 7x^3y$
7. $8a^5b - a^3b$
8. $6x + 8y + 4z$
9. $y^7 + 3y^5 + y^3$
10. $10t^4 + 15t^3 - 30t^2$
11. $3p^4q^2 - p^3q^3 + 5p^2q^4$
12. $3a^4b^4 + 6a^3b^3 - 3a^2b^2$

Factor by grouping.

13. $4t^2 + 8t + 5t + 10$
14. $12p^2 - 8pq + 15pq - 10q^2$
15. $x^2 + 5x - 2x - 10$
16. $d^2 - 3d - 4d + 12$
17. $8a^3 - 4a^2b^2 + 2ab - b^3$
18. $px + 3qx - py - 3qy$

Factor.

19. $t^2 - 4t + 3$
20. $w^2 - w - 12$
21. $x^2 + 4x - 21$
22. $p^2 + 8pq + 15q^2$
23. $c^2d^2 - 6cd + 5$
24. $x^2y^2 + 5xy - 14$
25. $t^8 - 8t^4 + 15$
26. $p^6q^6 - 2p^3q^3 - 24$
27. $x^3 + 12x^2 + 35x$
28. $y^4 - y^3 - 30y^2$
29. $(y + 3)^2 - 2(y + 3) - 15$
30. $(a - b)^2 - 5(a - b) + 4$

Factor.

31. $6x^2 - 11x + 4$
32. $4y^2 + 4y - 3$
33. $1 - 4t - 12t^2$
34. $2b^2 + 7bt + 3t^2$
35. $3t^2w^2 - 7tw - 6$
36. $8a^2b^2 + 6ab - 9$
37. $6x^8 - 11x^4 + 4$
38. $10p^6q^6 + 19p^3q^3 + 6$
39. $3y^2 + 30y + 63$
40. $10x^3 - 15x^2 - 10x$
41. $c^3d^3 + 6c^2d^2 - 16cd$
42. $6(y + 4)^2 - 11(y + 4) + 5$

Factor.

43. $49x^2 - 1$
44. $9a^2 - 16b^2$
45. $25m^6 - 64t^4$
46. $4 - 9p^2q^8$
47. $y^4 - 81$
48. $2t^2 - 18w^2$
49. $64ax^2 - 49a$
50. $(b + 2)^2 - 25$

Assignment 12

Factor.

1. $x^2 + 8x + 16$
2. $y^2 - 16y + 64$
3. $9w^2 + 12w + 4$
4. $4a^2 - 20ab + 25b^2$
5. $16d^2h^2 - 8dh + 1$
6. $t^4 + 2t^2 + 1$
7. $49m^{10} - 42m^5 + 9$
8. $5x^2 + 20x + 20$
9. $3y^3 - 6y^2 + 3y$

Factor.

10. $x^3 + 1$
11. $y^3 - 27$
12. $8b^3 + 125d^3$
13. $64p^3 - t^3$
14. $x^6 - 27y^3$
15. $a^3b + 8b^4$
16. $2m^3 - 128$
17. $(y + 5)^3 + y^3$

Factor.

18. $5x^2 - 5y^2$
19. $2a^3 - 16b^6$
20. $3c^2 - 15c + 12$
21. $p^4 - q^4$
22. $y^4 - 6y^2 + 8$
23. $x^2 + 5x + 10x + 50$
24. $16ap^2 - a$
25. $16bt^3 - 54b$
26. $y^2 + 3y - 5y - 15$
27. $16x^4 - 1$
28. $3ax^2 - 24axy + 48ay^2$
29. $x^5 - x^2y^3 + x^3y^2 - y^5$

Solve by factoring.

30. $x^2 - 5x + 6 = 0$
31. $3y^2 - 2y - 1 = 0$
32. $p^2 - 9 = 0$
33. $25t^2 - 16 = 0$
34. $x^2 + 7x = 0$
35. $3y^2 - 2y = 0$
36. $6a^2 = a + 12$
37. $5b^2 = 3b$
38. $49h^2 = 1$
39. $20x^2 + 10x = 60$
40. $x(x + 2) = 80$
41. $(x + 3)(x - 3) = 16$

Solve each problem.

42. The square of a number minus three times that number equals 10. Find the number.
43. If a number is subtracted from double its square, the result is 15. Find the number.
44. Three more than a number times three less than that number is 27. Find the number.
45. The product of two consecutive integers is 72. Find the integers.
46. The product of two consecutive odd integers is 63. Find the integers.

47. The product of two consecutive even integers is 120. Find the integers.

48. The length of a rectangle is 6 ft more than the width. The area of the rectangle is 40 ft^2. Find the length and width.

49. The height of a triangle is 3m less than the base. If the area of the triangle is $35m^2$, find the height and base.

50. If an object is dropped, the distance $\underline{d}$ it falls in $\underline{t}$ seconds is given by the formula: $d = 16t^2$. How long would it take an object to fall 144 ft?

4 Rational Expressions

In this chapter, we will define rational expressions and discuss the basic operations with them. We will show a method for solving fractional equations and rearranging fractional formulas. Various types of word problems involving fractional equations are included.

4-1 EQUIVALENT RATIONAL EXPRESSIONS

In this section, we will define rational expressions and discuss the procedures for obtaining equivalent rational expressions. The procedure for reducing rational expressions to lowest terms is emphasized.

---

1. A quotient of two polynomials is called a rational expression. Some examples of rational expressions are:

$$\frac{3x}{5} \qquad \frac{x + y}{x - y} \qquad \frac{t^2 + 5t + 6}{t^2 - 9}$$

A rational expression indicates a division. That is:

$\frac{3x}{5}$ means $3x \div 5$ $\qquad$ $\frac{x + y}{x - y}$ means ________ $\div$ ________

$(x + y) \div (x - y)$

2. Some substitutions in rational expressions do not make sense because they make the denominator equal 0, and division by 0 is not possible. Some examples are given below.

$\frac{10}{x}$ — x cannot be 0.

$\frac{y + 5}{x - 3}$ — x cannot be 3.

$\frac{x^2 + 4}{x^2 - 4}$ — x cannot be 2 or -2.

Note: In this text, we will avoid substitutions in rational expressions that do not make sense. That is, we will assume that a variable does not take on a value that leads to a division by 0.

---

3. To multiply two rational expressions, we multiply their numerators and multiply their denominators. For example:

$$\left(\frac{3}{x + 2}\right)\left(\frac{x - 1}{x - 5}\right) = \frac{3(x - 1)}{(x + 2)(x - 5)}$$

$$\left(\frac{y + 9}{7}\right)\left(\frac{y^2}{y - 3}\right) = \_\_\_\_\_\_\_\_$$

$\frac{(y + 9)y^2}{7(y - 3)}$

4. Any rational expression with the same numerator and denominator equals "1". For example:

$$\frac{3x}{3x} = 1 \qquad \frac{a - b}{a - b} = 1 \qquad \frac{2(y^2 + 1)}{2(y^2 + 1)} = 1$$

Sometimes it is necessary to convert a rational expression to an equivalent form. To do so, we multiply it by an expression that equals "1". For example:

$$\frac{x + 5}{x - 1} = \left(\frac{x + 5}{x - 1}\right)\left(\frac{3x}{3x}\right) = \frac{(x + 5)(3x)}{(x - 1)(3x)}$$

$$\frac{a^2 + b}{a^2 - b} = \left(\frac{a^2 + b}{a^2 - b}\right)\left(\frac{a - b}{a - b}\right) = \_\_\_\_\_\_\_\_$$

$\frac{(a^2 + b)(a - b)}{(a^2 - b)(a - b)}$

5. To reduce a rational expression to lowest terms, we reverse the procedure in the last frame. That is, we factor out an expression that equals "1". For example:

$$\frac{3ab}{2ab} = \left(\frac{3}{2}\right)\left(\frac{ab}{ab}\right) = \left(\frac{3}{2}\right)(1) = \frac{3}{2}$$

$$\frac{2x}{8x} = \left(\frac{2x}{2x}\right)\left(\frac{1}{4}\right) = (1)\left(\frac{1}{4}\right) = \_\_\_\_\_\_\_\_$$

$\frac{1}{4}$

6. We reduced each rational expression below to lowest terms.

$$\frac{4t^2}{t} = \frac{4t^2}{1t} = \left(\frac{4t}{1}\right)\left(\frac{t}{t}\right) = (4t)(1) = 4t$$

$$\frac{x}{7x^2} = \frac{1x}{7x^2} = \left(\frac{1}{7x}\right)\left(\frac{x}{x}\right) = \left(\frac{1}{7x}\right)(1) = \frac{1}{7x}$$

Reduce these to lowest terms.

a) $\frac{y}{10y} =$ b) $\frac{3x^2}{9x} =$

---

7. Following the example, reduce the other expression to lowest terms.

$$\frac{4x^2y^2}{2x^4y} = \left(\frac{2x^2y}{2x^2y}\right)\left(\frac{2y}{x^2}\right) = (1)\left(\frac{2y}{x^2}\right) = \frac{2y}{x^2}$$

$$\frac{8a^2b}{12ab^2} =$$

a) $\frac{1}{10}$ b) $\frac{x}{3}$

---

8. Following the example, reduce the other expression to lowest terms.

$$\frac{6b+9}{3} = \frac{3(2b+3)}{3} = \left(\frac{3}{3}\right)(2b+3) = (1)(2b+3) = 2b+3$$

$$\frac{20y}{10y-5} =$$

$\frac{2a}{3b}$

---

9. Following the example, reduce the other expression to lowest terms.

$$\frac{8y^2-6y}{2y^2+4y} = \frac{2y(4y-3)}{2y(y+2)} = \left(\frac{2y}{2y}\right)\left(\frac{4y-3}{y+2}\right) = (1)\left(\frac{4y-3}{y+2}\right) = \frac{4y-3}{y+2}$$

$$\frac{3x^2+6x}{9x^2-3x} =$$

$\frac{4y}{2y-1}$

---

10. Following the example, reduce the other expression to lowest terms.

$$\frac{y^2-9}{2y^2+6y} = \frac{(y+3)(y-3)}{2y(y+3)} = \left(\frac{y+3}{y+3}\right)\left(\frac{y-3}{2y}\right) = (1)\left(\frac{y-3}{2y}\right) = \frac{y-3}{2y}$$

$$\frac{5x^3-10x^2}{x^2-4} =$$

$\frac{x+2}{3x-1}$

---

11. Following the example, reduce the other expression to lowest terms.

$$\frac{x^2+x-6}{x^2-3x+2} = \frac{(x+3)(x-2)}{(x-1)(x-2)} = \left(\frac{x+3}{x-1}\right)\left(\frac{x-2}{x-2}\right) = \left(\frac{x+3}{x-1}\right)(1) = \frac{x+3}{x-1}$$

$$\frac{2x^2-7x-4}{x^2-16} =$$

$\frac{5x^2}{x+2}$

$\frac{2x+1}{x+4}$

12. Following the example, reduce the other expression to lowest terms.

$$\frac{x^3 - y^3}{x^2 - y^2} = \frac{(x - y)(x^2 + xy + y^2)}{(x + y)(x - y)}$$

$$= \left(\frac{x - y}{x - y}\right)\left(\frac{x^2 + xy + y^2}{x + y}\right)$$

$$= (1)\left(\frac{x^2 + xy + y^2}{x + y}\right)$$

$$= \frac{x^2 + xy + y^2}{x + y}$$

$$\frac{a^3 + b^3}{a + b} =$$

$a^2 - ab + b^2$

13. In $\frac{x - 3}{3 - x}$, the numerator and denominator are additive inverses. To reduce to lowest terms, we multiply both by -1. Performing the multiplication in the denominator, we get: $(-1)(3 - x) = -3 + x = x - 3$.

$$\frac{x - 3}{3 - x} = \frac{(-1)(x - 3)}{(-1)(3 - x)} = \frac{(-1)(x - 3)}{x - 3} = (-1)\left(\frac{x - 3}{x - 3}\right) = -1$$

Reduce this expression to lowest terms.

$$\frac{a - b}{b - a} =$$

-1

14. Notice how we multiplied both terms by -1 below, and then performed $(-1)(3 - y)$ in the denominator. Reduce the other expression to lowest terms.

$$\frac{y^2 - 9}{3 - y} = \frac{(-1)(y + 3)(y - 3)}{(-1)(3 - y)} = \frac{(-1)(y + 3)(y - 3)}{y - 3} = (-1)(y + 3)\left(\frac{y - 3}{y - 3}\right) = -(y + 3)$$

$$\frac{p^2 - q^2}{q - p} =$$

$-(p + q)$

4-2 CANCELLING TO REDUCE TO LOWEST TERMS

Instead of reducing rational expressions to lowest terms by factoring out an expression that equals "1", we can use a shorter method called cancelling. We will discuss cancelling in this section.

---

15. When a rational expression contains the same factor in both its numerator and denominator, we can reduce it to lowest terms by cancelling the common factor. For example:

$$\frac{4x}{7x} = \frac{4\cancel{x}}{7\cancel{x}} = \frac{4}{7} \qquad \frac{at^2}{bt^2} = \frac{a\cancel{t^2}}{b\cancel{t^2}} = \frac{a}{b}$$

Before cancelling below, we substituted 1x for x and 1pq for pq.

$$\frac{9x}{x} = \frac{9\cancel{x}}{1\cancel{x}} = 9 \qquad \frac{pq}{3pq} = \frac{1\cancel{pq}}{3\cancel{pq}} = \frac{1}{3}$$

Reduce these to lowest erms.

a) $\frac{5y}{2y}$ = ________ b) $\frac{R}{bR}$ = ________

---

16. To reduce the expression below to lowest terms, we cancelled the x's and divided both the 6 and 8 by 2. Reduce the other expression to lowest terms.

Answers to 15: a) $\frac{5}{2}$ b) $\frac{1}{b}$

$$\frac{6x}{8x} = \frac{\overset{3}{\cancel{6}}\cancel{x}}{\underset{4}{\cancel{8}}\cancel{x}} = \frac{3}{4} \qquad \frac{12ab}{8ab} = \_\_\_\_\_\_$$

---

17. To reduce the expressions below to lowest terms, we cancelled the x and one of the x's in x^2. We also divided both the 8 and 4 by 2.

Answer to 16: $\frac{3}{2}$

$$\frac{3x^2}{5x} = \frac{3x^{\cancel{2}\,1}}{5\cancel{x}} = \frac{3x}{5} \qquad \frac{8t}{4t^2} = \frac{\overset{2}{\cancel{8}}\cancel{t}}{\underset{1}{\cancel{4}}t^{\cancel{2}\,1}} = \frac{2}{t}$$

Reduce these to lowest terms.

a) $\frac{7m}{4m^2}$ = ________ b) $\frac{10y^2}{6y}$ = ________

---

18. To reduce the expression below to lowest terms, we cancelled the x and one of the x's in x^2 and we cancelled the y^2 and two of the y's in y^5.

Answers to 17: a) $\frac{7}{4m}$ b) $\frac{5y}{3}$

$$\frac{2xy^5}{3x^2y^2} = \frac{2\cancel{x}y^{\cancel{5}\,3}}{3x^{\cancel{2}\,1}\cancel{y^2}} = \frac{2y^3}{3x}$$

Reduce these to lowest terms.

a) $\frac{5a^4b^2}{7ab^3}$ = ________ b) $\frac{6pq^5}{4p^3q^3}$ = ________

Answers to 18: a) $\frac{5a^3}{7b}$ b) $\frac{3q^2}{2p^2}$

19. After factoring each expression below, we cancelled to reduce to lowest terms.

$$\frac{3x + 6}{4x + 8} = \frac{3\cancel{(x + 2)}}{4\cancel{(x + 2)}} = \frac{3}{4}$$

$$\frac{5y^2 - 5y}{5y^2 + 5y} = \frac{\cancel{5y}(y - 1)}{\cancel{5y}(y + 1)} = \frac{y - 1}{y + 1}$$

Reduce these to lowest terms.

a) $\dfrac{6a + 3}{10a + 5} =$ ________ b) $\dfrac{3t^2 - 6t}{18t^2 + 12t} =$ ________

---

20. Following the example, reduce the other expression to lowest terms.

a) $\frac{3}{5}$

b) $\frac{t - 2}{2(3t + 2)}$

$$\frac{6x + 2}{4} = \frac{\overset{1}{\cancel{2}}(3x + 1)}{\underset{2}{\cancel{4}}} = \frac{3x + 1}{2}$$

$$\frac{6y}{12y - 3} =$$

---

21. Following the example, reduce the other expression to lowest terms.

$\frac{2y}{4y - 1}$

$$\frac{y^2 - 7y + 10}{y^2 - 2y - 15} = \frac{(y - 2)\cancel{(y - 5)}}{(y + 3)\cancel{(y - 5)}} = \frac{y - 2}{y + 3}$$

$$\frac{t^2 - 9}{t^2 + t - 12} =$$

---

22. Notice how we used -1 below to get identical factors. Reduce the other expression to lowest terms.

$\frac{t + 3}{t + 4}$

$$\frac{y - 7}{7 - y} = \frac{(-1)(y - 7)}{(-1)(7 - y)} = \frac{(-1)\cancel{(y - 7)}}{\cancel{y - 7}} = -1$$

$$\frac{x - y}{y - x} =$$

---

23. Following the example, reduce the other expression to lowest terms.

-1

$$\frac{x^2 - 4}{2 - x} = \frac{(-1)(x + 2)(x - 2)}{(-1)(2 - x)} = \frac{(-1)(x + 2)\cancel{(x - 2)}}{\cancel{x - 2}} = -(x + 2)$$

$$\frac{a^2 - b^2}{b - a} =$$

$-(a + b)$

24. When reducing fractions to lowest terms, we can only cancel common factors of the numerator and denominator. <u>A common error is to cancel common terms of the numerator and denominator.</u> For example:

$$\frac{\not{x} + 3}{\not{x}} = 3 \qquad \text{(Common error)}$$

$$\frac{x + \not{5}}{y + \not{5}} = \frac{x}{y} \qquad \text{(Common error)}$$

Reduce to lowest terms if possible.

a) $\frac{2x + 2}{2} =$ c) $\frac{x - 5}{y - 5} =$

b) $\frac{x + 2}{2} =$ d) $\frac{5x - 5}{5y - 5} =$

a) $x + 1$ b) cannot be reduced c) cannot be reduced d) $\frac{x - 1}{y - 1}$

4-3 MULTIPLICATION

In multiplications of rational expressions, the product should always be reduced to lowest terms. To avoid products that are not in lowest terms, we can cancel before multiplying. We will discuss the method in this section.

25. When multiplying rational expressions, the product should always be reduced to lowest terms. For example:

$$\left(\frac{2x}{5}\right)\left(\frac{5}{3x}\right) = \frac{\overset{2}{\not{10}\not{x}}}{\underset{3}{\not{15}\not{x}}} = \frac{2}{3}$$

To avoid having to reduce the product above to lowest terms, we ordinarily cancel before multiplying. That is:

$$\left(\frac{2\not{x}}{\not{5}}\right)\left(\frac{\not{5}}{3\not{x}}\right) = \frac{2}{3}$$

Do these. <u>Cancel before multiplying.</u>

a) $\left(\frac{6}{7y}\right)\left(\frac{4y}{3}\right) =$ ________ b) $\left(\frac{8m}{5}\right)\left(\frac{1}{4m}\right) =$ ________

26. We cancelled before multiplying in the example below.

$$\left(\frac{\overset{1}{\not{2}}x^{\overset{1}{\not{2}}}}{3}\right)\left(\frac{5}{\underset{3}{\not{6}}\not{x}}\right) = \frac{5x}{9}$$

Do these by cancelling before multiplying.

a) $\left(\frac{1}{ax^2}\right)\left(\frac{7ax}{8}\right) =$ b) $\left(\frac{3y^2}{4x}\right)\left(\frac{8x^2}{9y^4}\right)$

a) $\frac{8}{7}$, from $\left(\frac{\overset{2}{\not{6}}}{7\not{y}}\right)\left(\frac{4\not{y}}{\not{3}}\right)$

b) $\frac{2}{5}$, from $\left(\frac{\overset{2}{\not{8}}\not{m}}{5}\right)\left(\frac{1}{\not{4}\not{m}}\right)$

27. We cancelled after factoring below. Do the other multiplication.

$$\left(\frac{x+1}{5x^2-10x}\right)\left(\frac{5x}{x+1}\right) = \left[\frac{\cancel{x+1}}{\cancel{5x}(x-2)}\right]\left(\frac{\cancel{5x}}{\cancel{x+1}}\right) = \frac{1}{x-2}$$

$$\left(\frac{y+4}{3y}\right)\left(\frac{3y^2-3y}{y-1}\right) =$$

a) $\frac{7}{8x}$ b) $\frac{2x}{3y^2}$

---

28. We cancelled after factoring below. Do the other multiplication.

$$\left(\frac{x^2+5x+4}{x^2-1}\right)\left(\frac{x-1}{x-3}\right) = \left[\frac{\cancel{(x+1)}(x+4)}{\cancel{(x+1)}\cancel{(x-1)}}\right]\left(\frac{\cancel{x-1}}{x-3}\right) = \frac{x+4}{x-3}$$

$$\left(\frac{y-5}{y+2}\right)\left(\frac{y^2-4}{y^2-6y+5}\right) =$$

$y + 4$

---

29. In the multiplication below, $(x + 3)^2$ means $(x + 3)(x + 3)$. Notice how we cancelled one of those factors. Do the other multiplication.

$$\left[\frac{x^2-6x-7}{(x+3)^2}\right]\left(\frac{x+3}{x-7}\right) = \left[\frac{(x+1)\cancel{(x-7)}}{(x+3)^{\cancel{2}\,1}}\right]\left(\frac{\cancel{x+3}}{\cancel{x-7}}\right) = \frac{x+1}{x+3}$$

$$\left(\frac{y+2}{y-1}\right)\left[\frac{(y-1)^2}{y^2+7y+10}\right] =$$

$\frac{y-2}{y-1}$

---

30. Notice how we cancelled after factoring below. Do the other multiplication.

$$\left(\frac{4x^2}{x^2-1}\right)\left(\frac{x^2+4x-5}{2x^2+10x}\right) = \left[\frac{\overset{2}{\cancel{4}}\overset{1}{\cancel{x^2}}}{(x+1)\cancel{(x-1)}}\right]\left[\frac{\cancel{(x+5)}\cancel{(x-1)}}{\cancel{2x}\cancel{(x+5)}}\right] = \frac{2x}{x+1}$$

$$\left(\frac{3y^2-9y}{y^2+y-12}\right)\left(\frac{y^2-16}{6y^2}\right) =$$

$\frac{y-1}{y+5}$

---

31. Following the example, do the other multiplication.

$$\left(\frac{x^3-8}{x^2-4}\right)\left(\frac{x^2+4x+4}{x^2+2x+4}\right) = \left[\frac{\cancel{(x-2)}\cancel{(x^2+2x+4)}}{\cancel{(x+2)}\cancel{(x-2)}}\right]\left[\frac{\cancel{(x+2)}(x+2)}{\cancel{x^2+2x+4}}\right] = x+2$$

$$\left(\frac{a^2-b^2}{a^3-b^3}\right)\left(\frac{a^2+ab+b^2}{a^2+2ab+b^2}\right) =$$

$\frac{y-4}{2y}$

$\frac{1}{a+b}$

4-4 DIVISION

In this section, we will discuss reciprocals of rational expressions and then use that concept to perform divisions involving rational expressions.

---

32. Two quantities are reciprocals if their product is "1". For example:

Since $(y + 3)\left(\frac{1}{y + 3}\right) = \frac{y + 3}{y + 3} = 1$: the reciprocal of $y + 3$ is $\frac{1}{y + 3}$

the reciprocal of $\frac{1}{y + 3}$ is $y + 3$

Write the reciprocal of each quantity.

a) $3x^2y$ ______ b) $\frac{1}{m^3}$ ______ c) $x - y$ ______ d) $\frac{1}{3b^2 + 4}$ ______

Answers: a) $\frac{1}{3x^2y}$ c) $\frac{1}{x - y}$ b) m^3 d) $3b^2 + 4$

33. Two rational expressions are reciprocals if their product is "1". For example:

Since $\left(\frac{x + 3}{x - 1}\right)\left(\frac{x - 1}{x + 3}\right) = 1$: the reciprocal of $\frac{x + 3}{x - 1}$ is $\frac{x - 1}{x + 3}$

the reciprocal of $\frac{x - 1}{x + 3}$ is $\frac{x + 3}{x - 1}$

Write the reciprocal of each expression.

a) $\frac{2x^2}{3y^3}$ ______ b) $\frac{m + 2}{m - 5}$ ______ c) $\frac{y + 1}{y^2 - 4y + 5}$ ______

Answers: a) $\frac{3y^3}{2x^2}$ b) $\frac{m - 5}{m + 2}$ c) $\frac{y^2 - 4y + 5}{y + 1}$

34. Any division involving a rational expression is written as a <u>complex fraction</u> in algebra. For example:

$$\frac{x}{3} \div \frac{x}{5} \text{ is written } \frac{\frac{x}{3}}{\frac{x}{5}}$$

To perform divisions like those above, <u>we multiply the numerator by the reciprocal of the denominator</u>. That is:

Numerator unchanged

$$\frac{\frac{x}{3}}{\frac{x}{5}} = \left(\frac{\cancel{x}}{3}\right)\left(\frac{5}{\cancel{x}}\right) = \frac{5}{3}$$

Reciprocal of denominator

Following the example, do these:

a) $\frac{\frac{3x}{7}}{\frac{5x^2}{14}} = (\quad)(\quad) =$ ______ b) $\frac{\frac{a^3c}{b^4}}{\frac{a^2c}{b^2}} = (\quad)(\quad) =$ ______

35. To perform the division below, we also multiplied the numerator by the reciprocal of the denominator. Do the other division.

$$\frac{\frac{x+1}{x-3}}{\frac{x+1}{x+2}} = \left(\frac{\cancel{x+1}}{x-3}\right)\left(\frac{x+2}{\cancel{x+1}}\right) = \frac{x+2}{x-3}$$

$$\frac{\frac{c}{c+d}}{\frac{c}{c-d}}$$

a) $\left(\frac{3\cancel{x}}{\cancel{7}}\right)\left(\frac{\overset{2}{\cancel{14}}}{5x^{\cancel{2}\,1}}\right) = \frac{6}{5x}$

b) $\left(\frac{a^{\cancel{2}\,1}\cancel{c}}{b^{\cancel{4}\,2}}\right)\left(\frac{\cancel{b^2}}{\cancel{a^2}\cancel{c}}\right) = \frac{a}{b^2}$

36. Notice how we factored to complete the division below. Do the other division.

$$\frac{\frac{3x+9}{x-1}}{\frac{x+3}{x^2-1}} = \left(\frac{3x+9}{x-1}\right)\left(\frac{x^2-1}{x+3}\right)$$

$$= \left[\frac{3\cancel{(x+3)}}{\cancel{x-1}}\right]\left[\frac{(x+1)\cancel{(x-1)}}{\cancel{x+3}}\right] = 3(x+1)$$

$$\frac{\frac{y^2-16}{6y+24}}{\frac{y-4}{8}} =$$

$\frac{c-d}{c+d}$

37. Following the example, do the other division.

$$\frac{\frac{x^2+4x}{x^2+5x+4}}{\frac{x^2-3x}{x^2+4x+3}} = \left(\frac{x^2+4x}{x^2+5x+4}\right)\left(\frac{x^2+4x+3}{x^2-3x}\right)$$

$$= \left[\frac{\cancel{x}\cancel{(x+4)}}{\cancel{(x+1)}\cancel{(x+4)}}\right]\left[\frac{(x+3)\cancel{(x+1)}}{\cancel{x}(x-3)}\right] = \frac{x+3}{x-3}$$

$$\frac{\frac{y^2-7y+10}{y^2-2y-15}}{\frac{y^2-4y+4}{y^2+6y+9}} =$$

$\frac{4}{3}$

$\frac{y+3}{y-2}$

38. Following the example, do the other division.

$$\frac{\frac{y^2 - 1}{y^3 - 1}}{\frac{y + 1}{y^2 - 2y + 1}} = \left[\frac{(y + 1)(y - 1)}{(y - 1)(y^2 + y + 1)}\right]\left[\frac{(y - 1)(y - 1)}{y + 1}\right]$$

$$= \frac{(y - 1)(y - 1)}{y^2 + y + 1} \quad \text{or} \quad \frac{y^2 - 2y + 1}{y^2 + y + 1}$$

$$\frac{\frac{x^3 - 64}{x^2 - 16}}{\frac{x^2 + 4x + 16}{x^2 + 8x + 16}} =$$

39. In the division below, the numerator of the complex fraction is $3(x + 1)$. Do the other division.

$x + 4$

$$\frac{3(x + 1)}{\frac{x + 1}{x}} = 3(x + 1)\left(\frac{x}{x + 1}\right) = 3x$$

$$\frac{a^2 - b^2}{\frac{a + b}{b}} =$$

40. In the division below, the denominator of the complex fraction is $y + 3$. Do the other division.

$b(a - b)$

$$\frac{\frac{(y + 3)^2}{5}}{y + 3} = \left[\frac{(y + 3)^{\not 2\,1}}{5}\right]\left(\frac{1}{y + 3}\right) = \frac{y + 3}{5}$$

$$\frac{\frac{x^2 - 1}{x}}{x + 1} =$$

$\frac{x - 1}{x}$

SELF-TEST 13 (pages 180-191)

Reduce to lowest terms.

1. $\dfrac{6x^4y}{8x^2y^2} =$

2. $\dfrac{6a^2 - 3a}{3a^2 + 12a} =$

3. $\dfrac{2m^2 - 5m - 3}{m^2 - 9} =$

Multiply. Report each product in lowest terms.

4. $\left(\dfrac{4x}{9y}\right)\left(\dfrac{15y^2}{8x^2}\right) =$

5. $\left(\dfrac{x^2 - 4x}{2x - 6}\right)\left(\dfrac{x - 3}{3x}\right) =$

6. $\left(\dfrac{x^2 - 9}{x^3 - 27}\right)\left(\dfrac{x^2 + 3x + 9}{x^2 + 6x + 9}\right) =$

Divide. Report each quotient in lowest terms.

7. $\dfrac{\dfrac{m - 5}{8}}{\dfrac{m}{4}}$

8. $\dfrac{\dfrac{a^2 - b^2}{a}}{a + b}$

9. $\dfrac{\dfrac{x^2 + 9x + 18}{x^2 + 5x - 6}}{\dfrac{x^2 - 3x - 10}{x^2 - 6x + 5}}$

ANSWERS:

1. $\dfrac{3x^2}{4y}$
2. $\dfrac{2a - 1}{a + 4}$
3. $\dfrac{2m + 1}{m + 3}$
4. $\dfrac{5y}{6x}$
5. $\dfrac{x - 4}{6}$
6. $\dfrac{1}{x + 3}$
7. $\dfrac{m - 5}{2m}$
8. $\dfrac{a - b}{a}$
9. $\dfrac{x + 3}{x + 2}$

4-5 ADDITION AND SUBTRACTION WITH LIKE DENOMINATORS

In this section, we will discuss the procedure for adding and subtracting rational expressions with like or common denominators.

---

41. To add or subtract rational expressions with like denominators, we add or subtract their numerators and keep the same denominator. For example:

$$\frac{2x}{5} + \frac{1}{5} = \frac{2x + 1}{5} \qquad \frac{y}{y + 2} - \frac{3}{y + 2} = \frac{y - 3}{y + 2}$$

Complete these:

a) $\frac{a}{c} - \frac{b}{c} =$ ________ b) $\frac{5t}{t - 1} + \frac{1}{t - 1} =$ ________

Answers to 41: a) $\frac{a - b}{c}$ b) $\frac{5t + 1}{t - 1}$

42. Notice how we combined like terms to simplify the numerator below.

$$\frac{2x}{3} + \frac{5x}{3} = \frac{2x + 5x}{3} = \frac{7x}{3}$$

Complete these:

a) $\frac{4}{y} - \frac{1}{y} =$ ________ b) $\frac{m}{m + 4} + \frac{2m}{m + 4} =$ ________

Answers to 42: a) $\frac{3}{y}$ b) $\frac{3m}{m + 4}$

43. Notice how we reduced the sum below to lowest terms.

$$\frac{3x}{10} + \frac{x}{10} = \frac{\overset{2}{\cancel{4}x}}{\underset{5}{\cancel{10}}} = \frac{2x}{5}$$

Complete. Reduce to lowest terms if possible.

a) $\frac{7}{2t} + \frac{3}{2t} =$ ________ b) $\frac{11y}{8} - \frac{5y}{8} =$ ________

Answers to 43: a) $\frac{5}{t}$ b) $\frac{3y}{4}$

44. Notice how we combined like terms to simplify the numerator below and then reduced to lowest terms. Do the other addition.

$$\frac{x^2 + 6x}{x^2 - 3x} + \frac{x^2 - x}{x^2 - 3x} = \frac{x^2 + 6x + x^2 - x}{x^2 - 3x} = \frac{2x^2 + 5x}{x^2 - 3x} = \frac{\cancel{x}(2x + 5)}{\cancel{x}(x - 3)} = \frac{2x + 5}{x - 3}$$

$$\frac{y^2 - 3y}{y^2 + y} + \frac{y^2 - 4y}{y^2 + y} =$$

Answer to 44: $\frac{2y - 7}{y + 1}$

45. Notice how we wrote parentheses around x - 4 to make sure we subtracted that whole numerator. Complete the other subtraction.

$$\frac{3x - 1}{x} - \frac{x - 4}{x} = \frac{3x - 1 - (x - 4)}{x} = \frac{3x - 1 - x + 4}{x} = \frac{2x + 3}{x}$$

$$\frac{4y + 7}{y + 3} - \frac{3y - 1}{y + 3} =$$

46. Following the example, reduce the other answer to lowest terms.

$$\frac{5x}{8} - \frac{3x+4}{8} = \frac{5x-(3x+4)}{8} = \frac{5x-3x-4}{8} = \frac{2x-4}{8} = \frac{\overset{1}{\not 2}(x-2)}{\underset{4}{\not 8}} = \frac{x-2}{4}$$

$$\frac{10}{3y} - \frac{1+3y}{3y} =$$

$\frac{y+8}{y+3}$

---

47. Following the example, reduce the other answer to lowest terms.

$$\frac{x}{x^2-1} - \frac{1}{x^2-1} = \frac{x-1}{x^2-1} = \frac{\overset{1}{\cancel{x-1}}}{(x+1)\cancel{(x-1)}} = \frac{1}{x+1}$$

$$\frac{a^2}{a+b} - \frac{b^2}{a+b} =$$

$\frac{3-y}{y}$

---

48. Following the example, reduce the other answer to lowest terms.

$$\frac{x^2-2}{x^2-x-12} - \frac{2x+13}{x^2-x-12} = \frac{(x^2-2)-(2x+13)}{x^2-x-12}$$

$$= \frac{x^2-2-2x-13}{x^2-x-12}$$

$$= \frac{x^2-2x-15}{x^2-x-12}$$

$$= \frac{\cancel{(x+3)}(x-5)}{\cancel{(x+3)}(x-4)} = \frac{x-5}{x-4}$$

$$\frac{5y-4}{3y^2+5y-2} - \frac{2y-3}{3y^2+5y-2} =$$

$a - b$

$\frac{1}{y+2}$

49. When one denominator is the additive inverse of the other, we can get like denominators by multiplying one expression by $\frac{-1}{-1}$. An example is shown.

$$\frac{x}{x - 3} + \frac{5}{3 - x} = \frac{x}{x - 3} + \left(\frac{-1}{-1}\right)\left(\frac{5}{3 - x}\right)$$

$$= \frac{x}{x - 3} + \frac{-5}{x - 3}$$

$$= \frac{x - 5}{x - 3}$$

Do this addition. Reduce to lowest terms.

$$\frac{y^2}{y - 4} + \frac{16}{4 - y} =$$

$y + 4$

50. We multiplied one expression by $\frac{-1}{-1}$ to get like denominators below. Do the other subtraction.

$$\frac{7a}{a - 3b} - \frac{4a - 5}{3b - a} = \frac{7a}{a - 3b} - \left(\frac{-1}{-1}\right)\left(\frac{4a - 5}{3b - a}\right)$$

$$= \frac{7a}{a - 3b} - \frac{5 - 4a}{a - 3b}$$

$$= \frac{7a - (5 - 4a)}{a - 3b}$$

$$= \frac{7a - 5 + 4a}{a - 3b}$$

$$= \frac{11a - 5}{a - 3b}$$

$$\frac{2x + 5}{x - 1} - \frac{3x - 1}{1 - x} =$$

$\frac{5x + 4}{x - 1}$

4-6 FINDING LOWEST COMMON DENOMINATORS

To add rational expressions with unlike denominators, we use the lowest common denominator (LCD). We will discuss a method for finding lowest common denominators in this section.

---

51. To find the lowest common denominator (LCD) for unlike denominators, we begin by factoring the denominators into primes. For example:

$$\frac{7}{12} + \frac{11}{30} \qquad \begin{aligned} 12 &= (2)(2)(3) \\ 30 &= (2)(3)(5) \end{aligned}$$

Then to get the LCD, we use each factor the greatest number of times it appears in any denominator. Therefore, we get the following LCD for the addition above:

$$\text{LCD} = (2)(2)(3)(5) = 60$$

Note: Since 12 has 2 as a factor twice and 30 has 2 as a factor only once, the LCD has 2 as a factor twice (the greatest number of times it occurs).

We factored each denominator below. Use the method above to find the LCD.

$$\frac{4}{15} + \frac{7}{18} \qquad \begin{aligned} 15 &= (3)(5) \\ 18 &= (2)(3)(3) \end{aligned} \qquad \text{LCD} = \underline{\hspace{3cm}}$$

---

52. The following principle is used after factoring each denominator.

Answer to 51: (2)(3)(3)(5) = 90

> To get the LCD, use each factor the greatest number of times it appears in any denominator.

We used the principle above to find the LCD for the three denominators below.

$$\frac{5}{6} + \frac{4}{9} + \frac{17}{21} \qquad \begin{aligned} 6 &= (2)(3) \\ 9 &= (3)(3) \\ 21 &= (3)(7) \end{aligned} \qquad \text{LCD} = (2)(3)(3)(7) = 126$$

Following the example, find the LCD for this addition.

$$\frac{3}{4} + \frac{9}{10} + \frac{1}{12} \qquad \begin{aligned} 4 &= \underline{\hspace{2cm}} \\ 10 &= \underline{\hspace{2cm}} \\ 12 &= \underline{\hspace{2cm}} \end{aligned} \qquad \text{LCD} = \underline{\hspace{3cm}}$$

Answer to 52:

4 = (2)(2)
10 = (2)(5)
12 = (2)(2)(3)

LCD = (2)(2)(3)(5) = 60

53. The same method is used when one or more denominators contains a variable. An example is shown.

$$\frac{2}{3y^2} + \frac{5}{6y} \qquad \begin{aligned} 3y^2 &= (3)(y)(y) \\ 6y &= (2)(3)(y) \end{aligned} \qquad \text{LCD} = (2)(3)(y)(y) = 6y^2$$

Find the LCD for each addition.

a) $\frac{7}{12y} + \frac{3}{16y}$ b) $\frac{3}{5t} + \frac{1}{4t^3}$

LCD = ______________ LCD = ______________

Answers to 53: a) $(2)(2)(2)(2)(3)(y) = 48y$ b) $(5)(2)(2)(t)(t)(t) = 20t^3$

54. We found the LCD for the addition below.

$$\frac{3}{8xy^2} + \frac{5}{6x^3y} \qquad \begin{aligned} 8xy^2 &= (2)(2)(2)(x)(y)(y) \\ 6x^3y &= (2)(3)(x)(x)(x)(y) \end{aligned}$$

$$\text{LCD} = (2)(2)(2)(3)(x)(x)(x)(y)(y) = 24x^3y^2$$

Find the LCD for each addition.

a) $\frac{1}{3a^2b} + \frac{2}{2ab}$ b) $\frac{3}{4x^2y^2} + \frac{1}{10x^3y} + \frac{5}{6y^3}$

LCD = ______________ LCD = ______________

Answers to 54: a) $6a^2b$ b) $60x^3y^3$

55. To find the LCD for the addition below, we began by factoring the denominators.

$$\frac{3}{x^2 - 2x} + \frac{5}{3x - 6} = \frac{3}{x(x - 2)} + \frac{5}{3(x - 2)}$$

$$\text{LCD} = 3x(x - 2)$$

Find the LCD for each addition.

a) $\frac{4}{t^2 + 2t} + \frac{7}{5t^2}$ b) $\frac{1}{10y - 5} + \frac{7}{2y^2 - y}$

LCD = ______________ LCD = ______________

Answers to 55: a) $5t^2(t + 2)$ b) $5y(2y - 1)$

56. In the addition below, the denominators do not have a common factor. Therefore, the LCD is the product of the denominators.

$$\frac{4}{x + 3} + \frac{3}{x - 4} \qquad \text{LCD} = (x + 3)(x - 4)$$

Find the LCD for each addition.

a) $\frac{3}{2a} + \frac{7}{5b}$ b) $\frac{6}{7t} + \frac{5}{t - 1}$

LCD = ______________ LCD = ______________

Answers to 56: a) $10ab$ b) $7t(t - 1)$

57. We found the LCD for the addition below.

$$\frac{x + 1}{x^2 - 4} + \frac{x - 1}{x^2 + 5x + 6} = \frac{x + 1}{(x + 2)(x - 2)} + \frac{x - 1}{(x + 2)(x + 3)}$$

$$\text{LCD} = (x + 2)(x - 2)(x + 3)$$

Find the LCD for each addition.

a) $\frac{y - 3}{y + 3} + \frac{2y + 1}{y^2 + 8y + 15}$

LCD = ______________

b) $\frac{x - 5}{x^2 + 3x - 10} + \frac{3x + 7}{x^2 - 10x + 16}$

LCD = ______________

---

a) $(y + 3)(y + 5)$

b) $(x-2)(x+5)(x-8)$

58. We can use the same method to find the LCD when an addition contains three denominators. An example is shown.

$$\frac{x}{x + 5} + \frac{3x}{x^2 - 25} + \frac{x - 12}{x^2 - 3x - 10}$$

$$x + 5 = (x + 5)$$

$$x^2 - 25 = (x + 5)(x - 5)$$

$$x^2 - 3x - 10 = (x + 2)(x - 5)$$

$$\text{LCD} = (x + 5)(x - 5)(x + 2)$$

Find the LCD for each addition.

a) $\frac{3y}{2y + 4} + \frac{y + 1}{3y - 6} + \frac{y - 1}{y^2 - 4}$

LCD = ______________

b) $\frac{3}{x^2y} + \frac{9}{5x^2 - 5xy} + \frac{7}{xy^3}$

LCD = ______________

---

a) $6(y + 2)(y - 2)$

b) $5x^2y^3(x - y)$

4-7 ADDITION AND SUBTRACTION WITH UNLIKE DENOMINATORS

In this section, we will discuss the procedure for adding and subtracting rational expressions with unlike denominators.

---

59. To add or subtract rational expressions with unlike denominators, we use the following three steps.

1. Find the lowest common denominator.
2. Convert each rational expression to an equivalent expression with the LCD as the denominator.
3. Then add or subtract in the usual way.

Continued on following page.

59. Continued

We used the three steps for the addition below. The LCD is 6x. To convert the rational expressions to expressions with 6x as the denominator, we multiplied $\frac{1}{2x}$ by $\frac{3}{3}$ and $\frac{4}{3x}$ by $\frac{2}{2}$.

$$\frac{1}{2x} + \frac{4}{3x} = \left(\frac{1}{2x}\right)\left(\frac{3}{3}\right) + \left(\frac{4}{3x}\right)\left(\frac{2}{2}\right)$$

$$= \frac{3}{6x} + \frac{8}{6x}$$

$$= \frac{11}{6x}$$

Following the example, complete these.

a) $\frac{5}{6y} + \frac{1}{4y} =$ b) $\frac{3}{t} - \frac{4}{t^2} =$

---

a) $\frac{13}{12y}$, from: $\frac{10}{12y} + \frac{3}{12y}$

b) $\frac{3t - 4}{t^2}$, from: $\frac{3t}{t^2} - \frac{4}{t^2}$

60. The LCD for the addition below is (x + 1)(x - 2). To get the LCD for each expression, we multiplied $\frac{2}{x + 1}$ by $\frac{x - 2}{x - 2}$ and $\frac{3}{x - 2}$ by $\frac{x + 1}{x + 1}$. Notice how we simplified the numerator of the sum. Do the other addition.

$$\frac{2}{x + 1} + \frac{3}{x - 2} = \left(\frac{2}{x + 1}\right)\left(\frac{x - 2}{x - 2}\right) + \left(\frac{3}{x - 2}\right)\left(\frac{x + 1}{x + 1}\right)$$

$$= \frac{2(x - 2) + 3(x + 1)}{(x + 1)(x - 2)}$$

$$= \frac{2x - 4 + 3x + 3}{(x + 1)(x - 2)}$$

$$= \frac{5x - 1}{(x + 1)(x - 2)}$$

$$\frac{y}{y + 3} + \frac{5}{2y} =$$

$\frac{2y^2 + 5y + 15}{2y(y + 3)}$

61. To find the LCD below, we began by factoring the denominators. The LCD is $2x(x - 3)$. Do the other addition.

$$\frac{5}{x^2 - 3x} + \frac{x}{2x - 6} = \frac{5}{x(x - 3)} + \frac{x}{2(x - 3)}$$

$$= \left[\frac{5}{x(x - 3)}\right]\left(\frac{2}{2}\right) + \left[\frac{x}{2(x - 3)}\right]\left(\frac{x}{x}\right)$$

$$= \frac{5(2) + x(x)}{2x(x - 3)}$$

$$= \frac{x^2 + 10}{2x(x - 3)}$$

$$\frac{1}{6y - 3} + \frac{4}{10y - 5} =$$

$$\frac{17}{15(2y - 1)}$$

62. To find the LCD below, we began by factoring the denominators. The LCD is $(x + 3)(x + 4)(x - 2)$. Do the other addition.

$$\frac{2x}{x^2 + 7x + 12} + \frac{3}{x^2 + x - 6} = \frac{2x}{(x + 3)(x + 4)} + \frac{3}{(x + 3)(x - 2)}$$

$$= \left[\frac{2x}{(x + 3)(x + 4)}\right]\left(\frac{x - 2}{x - 2}\right) + \left[\frac{3}{(x + 3)(x - 2)}\right]\left(\frac{x + 4}{x + 4}\right)$$

$$= \frac{2x(x - 2) + 3(x + 4)}{(x + 3)(x + 4)(x - 2)}$$

$$= \frac{2x^2 - 4x + 3x + 12}{(x + 3)(x + 4)(x - 2)}$$

$$= \frac{2x^2 - x + 12}{(x + 3)(x + 4)(x - 2)}$$

$$\frac{y}{y^2 - 1} + \frac{5}{y^2 + 3y - 4} =$$

63. The LCD for the subtraction below is (x - 2)(x + 1). Notice how we wrote parentheses around $x^2 - 3x + 2$ to make sure we subtracted that whole numerator. Do the other subtraction.

$$\frac{x+3}{x-2} - \frac{x-1}{x+1} = \left(\frac{x+3}{x-2}\right)\left(\frac{x+1}{x+1}\right) - \left(\frac{x-1}{x+1}\right)\left(\frac{x-2}{x-2}\right)$$

$$= \frac{(x+3)(x+1) - (x-1)(x-2)}{(x-2)(x+1)}$$

$$= \frac{x^2 + 4x + 3 - (x^2 - 3x + 2)}{(x-2)(x+1)}$$

$$= \frac{x^2 + 4x + 3 - x^2 + 3x - 2}{(x-2)(x+1)}$$

$$= \frac{7x + 1}{(x-2)(x+1)}$$

$$\frac{y-1}{y+3} - \frac{y+4}{2y} =$$

$$\frac{y^2 + 9y + 5}{(y+1)(y-1)(y+4)}$$

---

64. To find the LCD below, we began by factoring the denominators. The LCD is (x + 2)(x - 2). Notice how we wrote parentheses around 5x - 10 to make sure we subtracted that whole numerator. Do the other subtraction.

$$\frac{7}{x^2-4} - \frac{5}{x+2} = \frac{7}{(x+2)(x-2)} - \frac{5}{x+2}$$

$$= \frac{7}{(x+2)(x-2)} - \left(\frac{5}{x+2}\right)\left(\frac{x-2}{x-2}\right)$$

$$= \frac{7 - 5(x-2)}{(x+2)(x-2)}$$

$$= \frac{7 - (5x - 10)}{(x+2)(x-2)}$$

$$= \frac{7 - 5x + 10}{(x+2)(x-2)}$$

$$= \frac{17 - 5x}{(x+2)(x-2)}$$

Continued on following page.

$$\frac{y^2 - 9y - 12}{2y(y+3)}$$

64. Continued

$$\frac{5a}{a^2 - b^2} - \frac{4}{a - b} =$$

$$\frac{a - 4b}{(a + b)(a - b)}$$

65. To find the LCD below, we began by factoring both denominators. The LCD is (x - 3)(x - 3)(x + 3). Do the other subtraction.

$$\frac{5x}{(x - 3)^2} - \frac{2}{x^2 - 9} = \frac{5x}{(x - 3)(x - 3)} - \frac{2}{(x + 3)(x - 3)}$$

$$= \left[\frac{5x}{(x - 3)(x - 3)}\right]\left(\frac{x + 3}{x + 3}\right) - \left[\frac{2}{(x + 3)(x - 3)}\right]\left(\frac{x - 3}{x - 3}\right)$$

$$= \frac{5x(x + 3) - 2(x - 3)}{(x - 3)(x - 3)(x + 3)}$$

$$= \frac{5x^2 + 15x - (2x - 6)}{(x - 3)(x - 3)(x + 3)}$$

$$= \frac{5x^2 + 15x - 2x + 6}{(x - 3)(x - 3)(x + 3)}$$

$$= \frac{5x^2 + 13x + 6}{(x - 3)(x - 3)(x + 3)}$$

$$\frac{a}{a^2 - b^2} - \frac{b}{(a + b)^2} =$$

66. The expression below contains both an addition and a subtraction. The LCD is $4x^2$. Simplify the other expression.

$$\frac{1}{2} + \frac{3}{4x} - \frac{2}{x^2} = \frac{1}{2}\left(\frac{2x^2}{2x^2}\right) + \frac{3}{4x}\left(\frac{x}{x}\right) - \frac{2}{x^2}\left(\frac{4}{4}\right)$$

$$= \frac{2x^2 + 3x - 8}{4x^2}$$

$$\frac{2}{3} - \frac{1}{5y} + \frac{4}{y^2} =$$

$$\frac{a^2 + b^2}{(a + b)(a + b)(a - b)}$$

67. The expression below also contains both an addition and a subtraction. The LCD is $(x + 1)(x - 1)$. Simplify the other expression.

$$\frac{10y^2 - 3y + 60}{15y^2}$$

$$\frac{3}{x + 1} + \frac{x + 4}{x^2 - 1} - \frac{2}{x - 1} = \frac{3}{x + 1}\left(\frac{x - 1}{x - 1}\right) + \frac{x + 4}{(x + 1)(x - 1)} - \frac{2}{x - 1}\left(\frac{x + 1}{x + 1}\right)$$

$$= \frac{3(x - 1) + x + 4 - 2(x + 1)}{(x + 1)(x - 1)}$$

$$= \frac{3x - 3 + x + 4 - (2x + 2)}{(x + 1)(x - 1)}$$

$$= \frac{3x - 3 + x + 4 - 2x - 2}{(x + 1)(x - 1)}$$

$$= \frac{2x - 1}{(x + 1)(x - 1)}$$

$$\frac{4}{2y + 3} - \frac{y}{y - 1} + \frac{5y^2 + 3}{2y^2 + y - 3} =$$

$$\frac{3y^2 + y - 1}{(2y + 3)(y - 1)}$$

4-8 COMPLEX FRACTIONS

In this section, we will discuss the procedure for simplifying complex fractions.

---

68. Any rational expression with a fraction in its numerator or its denominator or both is called a <u>complex fraction</u>. Some examples of complex fractions are:

$$\frac{\frac{x}{5}}{\frac{3}{4}} \qquad \frac{1 + \frac{4}{x}}{\frac{3}{2}} \qquad \frac{\frac{x}{x - y}}{\frac{1}{x} + \frac{1}{y}}$$

To simplify a complex fraction, we begin by adding or subtracting (if necessary) to get a single rational expression in both numerator and denominator. Then we divide by multiplying the numerator by the reciprocal of the denominator. An example is shown. Simplify the other complex fraction.

$$\frac{1 + \frac{4}{x}}{\frac{3}{2}} = \frac{1\left(\frac{x}{x}\right) + \frac{4}{x}}{\frac{3}{2}} \qquad\qquad \frac{\frac{y}{3} + 2}{\frac{7}{6}} =$$

$$= \frac{\frac{x + 4}{x}}{\frac{3}{2}}$$

$$= \left(\frac{x + 4}{x}\right)\left(\frac{2}{3}\right)$$

$$= \frac{2(x + 4)}{3x}$$

---

69. Following the example, simplify the other complex fraction.

Answer to 68: $\frac{2(y + 6)}{7}$

$$\frac{\frac{x + y}{x}}{\frac{1}{x} + \frac{1}{y}} = \frac{\frac{x + y}{x}}{\left(\frac{1}{x}\right)\left(\frac{y}{y}\right) + \left(\frac{1}{y}\right)\left(\frac{x}{x}\right)} \qquad\qquad \frac{t - \frac{1}{t}}{t + \frac{1}{t}} =$$

$$= \frac{\frac{x + y}{x}}{\frac{y + x}{xy}}$$

$$= \left(\frac{\cancel{x + y}}{\cancel{x}}\right)\left(\frac{\cancel{x}y}{\cancel{y + x}}\right)$$

$$= y$$

Answer to 69: $\frac{t^2 - 1}{t^2 + 1}$

70. Notice how we factored $(x^2 - y^2)$ to simplify the expression below. Simplify the other expression.

$$\frac{\frac{x}{y} - \frac{y}{x}}{\frac{1}{y} + \frac{1}{x}} = \frac{\left(\frac{x}{y}\right)\left(\frac{x}{x}\right) - \left(\frac{y}{x}\right)\left(\frac{y}{y}\right)}{\left(\frac{1}{y}\right)\left(\frac{x}{x}\right) + \left(\frac{1}{x}\right)\left(\frac{y}{y}\right)}$$

$$= \frac{\frac{x^2}{xy} - \frac{y^2}{xy}}{\frac{x}{xy} + \frac{y}{xy}}$$

$$= \frac{\frac{x^2 - y^2}{xy}}{\frac{x + y}{xy}}$$

$$= \left(\frac{x^2 - y^2}{\cancel{xy}}\right)\left(\frac{\cancel{xy}}{x + y}\right)$$

$$= \frac{\cancel{(x + y)}(x - y)}{\cancel{x + y}}$$

$$= x - y$$

$$\frac{1 + \frac{1}{m}}{1 - \frac{1}{m^2}} =$$

$$\frac{m}{m - 1}$$

71. Following the example, simplify the other expression.

$$\frac{x - 3 + \frac{x - 3}{x + 2}}{x + 4 - \frac{4x + 23}{x + 2}} = \frac{(x - 3)\left(\frac{x + 2}{x + 2}\right) + \frac{x - 3}{x + 2}}{(x + 4)\left(\frac{x + 2}{x + 2}\right) - \frac{4x + 23}{x + 2}}$$

$$= \frac{\frac{x^2 - x - 6 + x - 3}{x + 2}}{\frac{x^2 + 6x + 8 - (4x + 23)}{x + 2}}$$

$$= \frac{\frac{x^2 - 9}{x + 2}}{\frac{x^2 + 6x + 8 - 4x - 23}{x + 2}}$$

$$= \frac{\frac{x^2 - 9}{x + 2}}{\frac{x^2 + 2x - 15}{x + 2}}$$

$$= \frac{x^2 - 9}{x + 2} \cdot \frac{x + 2}{x^2 + 2x - 15}$$

$$= \frac{(x + 3)\cancel{(x - 3)}}{\cancel{x + 2}} \cdot \frac{\cancel{x + 2}}{\cancel{(x - 3)}(x + 5)}$$

$$= \frac{x + 3}{x + 5}$$

$$\frac{\frac{4y}{4y + 1} + \frac{1}{y}}{\frac{2}{4y + 1} + 2} =$$

$$\frac{2y + 1}{4y}$$

72. We simplified the same expression below with two different methods.

Method 1

We simplified the numerator and denominator and then divided.

$$\frac{\frac{2}{y}-\frac{1}{3y}}{\frac{1}{2y}+\frac{5}{4y}} = \frac{\left(\frac{2}{y}\right)\left(\frac{3}{3}\right)-\frac{1}{3y}}{\left(\frac{1}{2y}\right)\left(\frac{2}{2}\right)+\frac{5}{4y}}$$

$$= \frac{\frac{6}{3y}-\frac{1}{3y}}{\frac{2}{4y}+\frac{5}{4y}}$$

$$= \frac{\frac{5}{3y}}{\frac{7}{4y}}$$

$$= \left(\frac{5}{3\not{y}}\right)\left(\frac{4\not{y}}{7}\right)$$

$$= \frac{20}{21}$$

Method 2

We multiplied both numerator and denominator by 12y, the LCD of all four denominators.

$$\frac{\frac{2}{y}-\frac{1}{3y}}{\frac{1}{2y}+\frac{5}{4y}} = \frac{12y\left(\frac{2}{y}-\frac{1}{3y}\right)}{12y\left(\frac{1}{2y}+\frac{5}{4y}\right)}$$

$$= \frac{12\not{y}\left(\frac{2}{\not{y}}\right)-\overset{4}{\cancel{12y}}\left(\frac{1}{\cancel{3y}}\right)}{\overset{6}{\cancel{12y}}\left(\frac{1}{\cancel{2y}}\right)+\overset{3}{\cancel{12y}}\left(\frac{5}{\cancel{4y}}\right)}$$

$$= \frac{24-4}{6+15}$$

$$= \frac{20}{21}$$

Use Method 2 to simplify the expression below. The LCD of all four denominators is 6b.

$$\frac{\frac{a}{b}-\frac{3}{2b}}{\frac{a}{3b}+\frac{1}{2b}} =$$

$\frac{6a-9}{2a+3}$

SELF-TEST 14 (pages 192-206)

Add or subtract. Report each answer in lowest terms.

1. $\frac{2y}{5} - \frac{y - 7}{5}$

2. $\frac{p}{p^2 - 1} - \frac{1}{p^2 - 1}$

3. $\frac{3x - 4}{x - 2} - \frac{5x - 1}{2 - x}$

4. $\frac{1}{6x} + \frac{3}{10x}$

5. $\frac{t}{t - 2} + \frac{4}{3t}$

6. $\frac{y + 1}{y - 2} - \frac{y - 1}{y + 3}$

7. $\frac{x}{3x^2 + 5x - 2} + \frac{5}{x + 2}$

8. $\frac{1}{x + 2} + \frac{4x + 5}{x^2 - 4} - \frac{3}{x - 2}$

Simplify each expression.

9. $\dfrac{\frac{1}{a} + \frac{1}{b}}{\frac{a}{b} - \frac{b}{a}}$

10. $\dfrac{\frac{4d}{4d + 1} + \frac{1}{d}}{\frac{2}{4d + 1} + 2}$

ANSWERS:

1. $\frac{y + 7}{5}$
2. $\frac{1}{p + 1}$
3. $\frac{8x - 5}{x - 2}$
4. $\frac{7}{15x}$
5. $\frac{3t^2 + 4t - 8}{3t(t - 2)}$
6. $\frac{7y + 1}{(y - 2)(y + 3)}$
7. $\frac{16x - 5}{(3x - 1)(x + 2)}$
8. $\frac{2x - 3}{(x + 2)(x - 2)}$
9. $\frac{1}{a - b}$
10. $\frac{2d + 1}{4d}$

4-9 SOLVING FRACTIONAL EQUATIONS

To solve equations containing one or more rational expressions, we begin by clearing the fraction or fractions. We will discuss the method in this section.

---

73. In an earlier chapter, we solved equations containing fractions with numerical denominators. To do so, we began by clearing the fraction or fractions by multiplying both sides by the LCD of all denominators. For example, we multiplied both sides by 12 below. Solve the other equation.

$$\frac{m}{4} + \frac{2m}{3} = 1 \qquad\qquad \frac{3x}{2} - \frac{x}{3} = 1$$

$$12\left(\frac{m}{4} + \frac{2m}{3}\right) = 12(1)$$

$$\overset{3}{\cancel{12}}\left(\frac{m}{\cancel{4}}\right) + \overset{4}{\cancel{12}}\left(\frac{2m}{\cancel{3}}\right) = 12$$

$$3m + 8m = 12$$

$$11m = 12$$

$$m = \frac{12}{11}$$

The solution set is $\left\{\frac{12}{11}\right\}$.

---

74. The same method is used when one or more denominators contains a variable. For example, we multiplied both sides by 2y below. Solve the other equation.

$x = \frac{6}{7}$

$$\frac{5}{2y} = 4 \qquad\qquad 1 = \frac{3}{7x}$$

$$\cancel{2y}\left(\frac{5}{\cancel{2y}}\right) = 2y(4)$$

$$5 = 8y$$

$$y = \frac{5}{8}$$

---

75. To solve the equation below, we began by multiplying both sides by <u>x</u>. Solve the other equation.

$x = \frac{3}{7}$

$$\frac{x - 1}{x} = 4 \qquad\qquad 1 = \frac{4 - y}{3y}$$

$$\cancel{x}\left(\frac{x - 1}{\cancel{x}}\right) = x(4)$$

$$x - 1 = 4x$$

$$-1 = 3x$$

$$x = -\frac{1}{3}$$

76. To solve the equation below, we began by multiplying both sides by $(x + 4)$. Solve the other equation.

$$\frac{9}{x+4} = 2 \qquad\qquad 3 = \frac{y}{y-2}$$

$$\cancel{x+4}\left(\frac{9}{\cancel{x+4}}\right) = 2(x + 4)$$

$$9 = 2x + 8$$

$$1 = 2x$$

$$x = \frac{1}{2}$$

y = 1

---

77. A value that makes a denominator equal 0 cannot be the solution of a fractional equation. For example:

$x = 0$ cannot be the solution of $\frac{x - 3}{2x} = 1$

$y = 4$ cannot be the solution of $\frac{y}{y - 4} = 2$

Therefore, when you finish solving a fractional equation, always check to see whether the proposed solution makes a denominator equal 0. If so, it is not a solution.

y = 3

---

78. To clear the fraction below, we multiplied both sides by $3x$. Notice that we multiplied by the distributive principle on the left side. Solve the other equation.

$$2 - \frac{45}{3x} = 5 \qquad\qquad 4 = \frac{3}{y} + 1$$

$$3x\left(2 - \frac{45}{3x}\right) = 3x(5)$$

$$3x(2) - \overset{1}{\cancel{3x}}\left(\frac{45}{\cancel{3x}}\right) = 15x$$

$$6x - 45 = 15x$$

$$-45 = 9x$$

$$x = -5$$

---

y = 1, from:

$4y = 3 + y$

79. To clear the fractions below, we multiplied both sides by 5x, the LCD of both denominators. Solve the other equation.

$$\frac{x + 2}{x} = \frac{3}{5} \qquad\qquad \frac{t - 9}{t} = \frac{1}{2}$$

$$5\cancel{x}\left(\frac{x + 2}{\cancel{x}}\right) = \cancel{5}x\left(\frac{3}{\cancel{5}}\right)$$

$$5(x + 2) = x(3)$$

$$5x + 10 = 3x$$

$$2x = -10$$

$$x = \frac{-10}{2} = -5$$

---

t = 18

80. To clear the fractions below, we multiplied both sides by 5(x + 5). Solve the other equation.

$$\frac{x - 3}{x + 5} = \frac{3}{5} \qquad\qquad \frac{5}{y} = \frac{1}{3y + 4}$$

$$5\cancel{(x + 5)}\left(\frac{x - 3}{\cancel{x + 5}}\right) = \cancel{5}(x + 5)\left(\frac{3}{\cancel{5}}\right)$$

$$5(x - 3) = (x + 5)(3)$$

$$5x - 15 = 3x + 15$$

$$2x = 30$$

$$x = 15$$

$y = -\frac{10}{7}$

81. To clear the fractions below, we multiplied both sides by (x + 1)(x - 4). Solve the other equation.

$$\frac{3}{x + 1} = \frac{2}{x - 4} \qquad\qquad \frac{4}{y - 1} = \frac{5}{y + 3}$$

$$\cancel{(x + 1)}(x - 4)\left(\frac{3}{\cancel{x + 1}}\right) = (x + 1)\cancel{(x - 4)}\left(\frac{2}{\cancel{x - 4}}\right)$$

$$(x - 4)(3) = (x + 1)(2)$$

$$3x - 12 = 2x + 2$$

$$x = 14$$

---

y = 17

82. Notice how the squared terms are eliminated in the solution below. Solve the other equation.

$$\frac{x-1}{x-2} = \frac{x-3}{x+2}$$

$$\frac{y-3}{y-1} = \frac{y+4}{y+1}$$

$$(\cancel{x-2})(x+2)\left(\frac{x-1}{\cancel{x-2}}\right) = (x-2)(\cancel{x+2})\left(\frac{x-3}{\cancel{x+2}}\right)$$

$$(x+2)(x-1) = (x-2)(x-3)$$

$$x^2 + x - 2 = x^2 - 5x + 6$$

$$6x = 8$$

$$x = \frac{8}{6} = \frac{4}{3}$$

---

83. To clear the fractions below, we multiplied both sides by 6x, the LCD of all three denominators. Solve the other equation.

$y = \frac{1}{5}$

$$\frac{1}{2} - \frac{1}{3} = \frac{1}{x}$$

$$\frac{1}{4} + \frac{1}{6} = \frac{1}{y}$$

$$6x\left(\frac{1}{2} - \frac{1}{3}\right) = 6\cancel{x}\left(\frac{1}{\cancel{x}}\right)$$

$$\overset{3}{\cancel{6}}x\left(\frac{1}{\cancel{2}}\right) - \overset{2}{\cancel{6}}x\left(\frac{1}{\cancel{3}}\right) = 6(1)$$

$$3x - 2x = 6$$

$$x = 6$$

---

84. To clear the fractions below, we multiplied both sides by 2x. Solve the other equation.

$y = \frac{12}{5}$

$$\frac{3}{x} = \frac{4}{x} - \frac{1}{2}$$

$$\frac{1}{3y} + \frac{5}{y} = 2$$

$$2\cancel{x}\left(\frac{3}{\cancel{x}}\right) = 2x\left(\frac{4}{x} - \frac{1}{2}\right)$$

$$2(3) = 2\cancel{x}\left(\frac{4}{\cancel{x}}\right) - \cancel{2}x\left(\frac{1}{\cancel{2}}\right)$$

$$6 = 8 - x$$

$$-2 = -x$$

$$x = 2$$

$y = \frac{8}{3}$

85. Since $x^2 - 4 = (x + 2)(x - 2)$, the LCD below is $(x + 2)(x - 2)$. Therefore, to clear the fractions, we multiplied both sides by $(x + 2)(x - 2)$.

$$\frac{3}{x+2} + \frac{4x}{x^2 - 4} = \frac{1}{x-2}$$

$$(x+2)(x-2)\left[\frac{3}{x+2} + \frac{4x}{x^2-4}\right] = \left(\frac{1}{\cancel{x-2}}\right)(x+2)\cancel{(x-2)}$$

$$\cancel{(x+2)}(x-2)\left(\frac{3}{\cancel{x+2}}\right) + \cancel{(x+2)}\cancel{(x-2)}\left(\frac{4x}{\cancel{x^2-4}}\right) = 1(x+2)$$

$$3x - 6 + 4x = x + 2$$

$$7x - 6 = x + 2$$

$$6x = 8$$

$$x = \frac{8}{6} = \frac{4}{3}$$

Using the same steps, solve this equation.

$$\frac{4m - 1}{m^2 + 3m - 10} = \frac{1}{m+5} + \frac{2}{m-2}$$

---

m = 9

86. When clearing fractions, we can get an equation that has two solutions. An example is shown. Solve the other equation.

$$2x + \frac{3}{x} = 5$$

$$x\left(2x + \frac{3}{x}\right) = 5(x)$$

$$x(2x) + \cancel{x}\left(\frac{3}{\cancel{x}}\right) = 5x$$

$$2x^2 + 3 = 5x$$

$$2x^2 - 5x + 3 = 0$$

$$(2x - 3)(x - 1) = 0$$

$$x = \frac{3}{2} \text{ and } 1$$

$$y + \frac{5}{y} = 6$$

87. When both sides of an equation are multiplied by an expression containing a variable, we can get an equation with a solution that is not a solution of the original equation. As an example, we solved the equation below.

$$\frac{x-1}{x-6} = \frac{5}{x-6}$$

$$\cancel{x-6}\left(\frac{x-1}{\cancel{x-6}}\right) = \cancel{x-6}\left(\frac{5}{\cancel{x-6}}\right)$$

$$x - 1 = 5$$

$$x = 6$$

The proposed solution is $x = 6$. However, 6 is a value that makes both denominators equal 0. Therefore:

a) Is $x = 6$ the solution of the equation? ________

b) Does the equation have a solution? ________

y = 1 and 5, from:

$y^2 - 6y + 5 = 0$

---

88. To clear the fractions below, we also multiplied by an expression containing a variable.

$$\frac{x^2}{x-4} = \frac{16}{x-4}$$

$$\cancel{x-4}\left(\frac{x^2}{\cancel{x-4}}\right) = \cancel{x-4}\left(\frac{16}{\cancel{x-4}}\right)$$

$$x^2 = 16$$

$$x^2 - 16 = 0$$

$$(x + 4)(x - 4) = 0$$

$$x = -4 \text{ and } 4$$

The two proposed solutions are -4 and 4. We checked each proposed solution in the original equation below.

Checking x = -4	Checking x = 4
$\frac{x^2}{x-4} = \frac{16}{x-4}$	$\frac{x^2}{x-4} = \frac{16}{x-4}$
$\frac{(-4)^2}{(-4)-4} = \frac{16}{(-4)-4}$	$\frac{(4)^2}{(4)-4} = \frac{16}{(4)-4}$
$\frac{16}{-8} = \frac{16}{-8}$	$\frac{16}{0} = \frac{16}{0}$

a) Is -4 a solution of the original equation? ________

b) Is 4 a solution of the original equation? ________

a) No

b) No

---

a) Yes

b) No, because it leads to a division by 0.

4-10 FORMULAS

In this section, we will do some evaluations and rearrangements with fractional formulas. Some rearrangements that require multiplying or factoring by the distributive principle are also included.

---

89. When it makes sense to use the same letter for more than one variable in a formula, subscripts are used. Either letters or numbers can be used as the subscripts. For example, the formula below shows the relationship between total resistance of two resistors connected in parallel. The letter "t" and the numbers 1 and 2 are used as subscripts.

$$\frac{1}{R_t} = \frac{1}{R_1} + \frac{1}{R_2}$$

where: R_t = total resistance
R_1 = first resistance
R_2 = second resistance

The abbreviation "sub" is used when naming a letter with a subscript. For example:

R_t is called "R sub t".

R_1 is called "R sub 1".

Write a letter with a subscript for each of these:

a) P sub 2 = ______ b) v sub f = ______ c) d sub 1 = ______

---

90. To do the evaluation below, we had to solve a proportion. Do the other evaluation.

Answers to 89: a) P_2 b) v_f c) d_1

In the formula below, find d_1 when $F_1 = 12$, $F_2 = 20$, and $d_2 = 40$.

$$\frac{F_1}{F_2} = \frac{d_1}{d_2}$$

$$\frac{12}{20} = \frac{d_1}{40}$$

$$\overset{2}{\cancel{40}}\left(\frac{12}{\cancel{20}}\right) = \cancel{40}\left(\frac{d_1}{\cancel{40}}\right)$$

$$24 = d_1$$

In the formula below, find P_1 when $P_2 = 12$, $V_2 = 12$, and $V_1 = 18$.

$$\frac{P_1}{P_2} = \frac{V_2}{V_1}$$

Answer to 90: $P_1 = 8$

91. To do the evaluation below, we had to solve a fractional equation. Do the other evaluation.

In the formula below, find R_1 when $R_t = 10$ and $R_2 = 30$.

$$\frac{1}{R_t} = \frac{1}{R_1} + \frac{1}{R_2}$$

$$\frac{1}{10} = \frac{1}{R_1} + \frac{1}{30}$$

$$\overset{3}{\cancel{30}}R_1\left(\frac{1}{\cancel{10}}\right) = 30R_1\left(\frac{1}{R_1} + \frac{1}{30}\right)$$

$$3R_1 = 30\cancel{R_1}\left(\frac{1}{\cancel{R_1}}\right) + \cancel{30}R_1\left(\frac{1}{\cancel{30}}\right)$$

$$3R_1 = 30 + R_1$$

$$2R_1 = 30$$

$$R_1 = 15$$

In the formula below, find d when $D = 90$ and $f = 30$.

$$\frac{1}{D} + \frac{1}{d} = \frac{1}{f}$$

---

92. To do the evaluation below, we also had to solve a fractional equation. Do the other evaluation.

Answer to 91: $d = 45$

Find n when $I = 5$, $E = 35$, $R = 30$ and $r = 4$.

$$I = \frac{nE}{R + nr}$$

$$5 = \frac{n(35)}{30 + n(4)}$$

$$(30 + 4n)(5) = \cancel{(30 + 4n)}\left(\frac{35n}{\cancel{30 + 4n}}\right)$$

$$150 + 20n = 35n$$

$$150 = 15n$$

$$n = 10$$

Find p when $I = 4$, $T = 25$, $M = 40$, and $r = 5$.

$$I = \frac{pT}{M + pr}$$

---

93. To solve for R below, we began by multiplying both sides by R to clear the fraction. Solve for a in the other formula.

Answer to 92: $p = 32$

$$I = \frac{E}{R}$$

$$R(I) = \cancel{R}\left(\frac{E}{\cancel{R}}\right)$$

$$IR = E$$

$$R = \frac{E}{I}$$

$$V = \frac{4st}{a}$$

94. We solved for m below. Solve for a_1 in the other formula.

$$S = \frac{H}{m(t_1 - t_2)} \qquad S_n = \frac{a_1(r^n - 1)}{r - 1}$$

$$m(t_1 - t_2)S = m(t_1 - t_2)\left[\frac{H}{m(t_1 - t_2)}\right]$$

$$mS(t_1 - t_2) = H$$

$$m = \frac{H}{S(t_1 - t_2)}$$

$a = \frac{4st}{V}$

---

95. After clearing the fraction below, we had to use the addition axiom to solve for v_2. Solve for F_i in the other formula.

$$a = \frac{v_2 - v_1}{t} \qquad w = \frac{F_o + F_i}{t}$$

$$t(a) = t\left(\frac{v_2 - v_1}{t}\right)$$

$$at = v_2 - v_1$$

$$at + v_1 = v_2 - v_1 + v_1$$

$$v_2 = at + v_1$$

$a_1 = \frac{S_n(r - 1)}{r^n - 1}$

---

96. To solve for P_1 below, we began by multiplying both sides by P_2V_1 to clear the fractions. Solve for T_1 in the other formula.

$$\frac{P_1}{P_2} = \frac{V_2}{V_1} \qquad \frac{V_1}{V_2} = \frac{T_1}{T_2}$$

$$P_2V_1\left(\frac{P_1}{P_2}\right) = P_2V_1\left(\frac{V_2}{V_1}\right)$$

$$P_1V_1 = P_2V_2$$

$$P_1 = \frac{P_2V_2}{V_1}$$

$F_i = wt - F_o$

---

97. To solve for T_2 below, we began by multiplying both sides by T_1T_2. Solve for d in the other formula.

$$\frac{P_1V_1}{T_1} = \frac{P_2V_2}{T_2} \qquad \frac{D}{d} = \frac{F}{f}$$

$$T_1T_2\left(\frac{P_1V_1}{T_1}\right) = T_1T_2\left(\frac{P_2V_2}{T_2}\right)$$

$$P_1T_2V_1 = P_2T_1V_2$$

$$T_2 = \frac{P_2T_1V_2}{P_1V_1}$$

$T_1 = \frac{V_1T_2}{V_2}$

$d = \frac{Df}{F}$

98. Sometimes we have to factor by the distributive principle when rearranging a formula. Two factorings of that type are shown.

$$ab + ac = a(b + c)$$

$$CV - CT = C(V - T)$$

In $E = IR_1 + IR_2$, the variable I appears in both terms on the right side. To solve for I, we begin by factoring by the distributive principle as we have done below. Use the same method to solve for S in the other formula.

$E = IR_1 + IR_2$ | $MS - QS = R$

$E = I(R_1 + R_2)$

$I = \dfrac{E}{R_1 + R_2}$

---

99. To solve for S below, we isolated both S terms on one side before factoring. Solve for P_2 in the other formula.

Answer: $S = \dfrac{R}{M - Q}$

$MS = R - QS$ | $P_1P_2 = I - RP_2$

$MS + QS = R$

$S(M + Q) = R$

$S = \dfrac{R}{M + Q}$

---

100. To solve for $\underline{t}$ below, we isolated both $\underline{t}$ terms on one side before factoring. Solve for T in the other formula.

Answer: $P_2 = \dfrac{I}{P_1 + R}$

$bt + cm = dt$ | $RV + ST = QT$

$cm = dt - bt$

$cm = t(d - b)$

$t = \dfrac{cm}{d - b}$

---

101. To solve for $\underline{c}$ below, we multiplied both sides by $\underline{abc}$ to clear the fractions and then factored by the distributive principle. Solve for $\underline{f}$ in the other formula.

Answer: $T = \dfrac{RV}{Q - S}$

$\dfrac{1}{a} + \dfrac{1}{b} = \dfrac{1}{c}$ | $\dfrac{1}{D} + \dfrac{1}{d} = \dfrac{1}{f}$

$abc\left(\dfrac{1}{a} + \dfrac{1}{b}\right) = abc\left(\dfrac{1}{c}\right)$

$\not{a}bc\left(\dfrac{1}{\not{a}}\right) + a\not{b}c\left(\dfrac{1}{\not{b}}\right) = ab\not{c}\left(\dfrac{1}{\not{c}}\right)$

$bc + ac = ab$

$c(b + a) = ab$

$c = \dfrac{ab}{b + a}$

102. To solve for d below, we had to isolate the d terms before factoring. Solve for C_t in the other formula.

$$\frac{1}{D} + \frac{1}{d} = \frac{1}{f} \qquad\qquad \frac{1}{C_1} = \frac{1}{C_t} - \frac{1}{C_2}$$

$$Ddf\left(\frac{1}{D} + \frac{1}{d}\right) = Ddf\left(\frac{1}{f}\right)$$

$$\not{D}df\left(\frac{1}{\not{D}}\right) + D\not{d}f\left(\frac{1}{\not{d}}\right) = Dd\not{f}\left(\frac{1}{\not{f}}\right)$$

$$df + Df = Dd$$

$$Df = Dd - df$$

$$Df = d(D - f)$$

$$d = \frac{Df}{D - f}$$

Answer: $f = \frac{Dd}{d + D}$

---

103. To solve for B below, we had to isolate the B terms before factoring B - AB. Solve for D in the other formula.

$$A = \frac{B}{B + 1} \qquad\qquad C = \frac{DF}{D + F}$$

$$A(B + 1) = \cancel{(B + 1)}\left(\frac{B}{\cancel{B + 1}}\right)$$

$$AB + A = B$$

$$A = B - AB$$

$$A = B(1 - A)$$

$$B = \frac{A}{1 - A}$$

Answer: $C_t = \frac{C_1 C_2}{C_1 + C_2}$

---

104. We solved for r below. Solve for v in the other formula.

$$\frac{E}{e} = \frac{R + r}{r} \qquad\qquad \frac{F}{f} = \frac{v + v_o}{v - v_f}$$

$$\not{e}r\left(\frac{E}{\not{e}}\right) = e\not{r}\left(\frac{R + r}{\not{r}}\right)$$

$$rE = e(R + r)$$

$$rE = eR + er$$

$$rE - er = eR$$

$$r(E - e) = eR$$

$$r = \frac{eR}{E - e}$$

Answer: $D = \frac{CF}{F - C}$

Answer: $v = \frac{Fv_f + fv_o}{F - f}$

SELF-TEST 15 (pages 207-218)

Solve each equation.

1. $\frac{x - 5}{x} = \frac{3}{2}$

2. $\frac{1}{y - 1} = \frac{3}{y + 3}$

3. $\frac{1}{2y} - 3 = \frac{5}{y}$

4. $3x - \frac{2}{x} = 5$

5. In the formula below, find P_2 when $P_1 = 36$, $V_2 = 40$, and $V_1 = 30$.

$$\frac{P_1}{P_2} = \frac{V_2}{V_1}$$

6. Solve for h.

$$\frac{B}{b} = \frac{H}{h}$$

7. Solve for a.

$$\frac{1}{a} + \frac{1}{b} = \frac{1}{c}$$

8. Solve for M.

$$S = \frac{MT}{M + T}$$

ANSWERS:

1. $x = -10$
2. $y = 3$
3. $y = -\frac{3}{2}$
4. $x = -\frac{1}{3}$ and 2
5. $P_2 = 27$
6. $h = \frac{bH}{B}$
7. $a = \frac{bc}{b - c}$
8. $M = \frac{ST}{T - S}$

4-11 WORD PROBLEMS

In this section, we will solve some word problems involving fractional equations.

105. To solve the problem below, we set up and solved a fractional equation. We let x equal the smaller number and x + 9 equal the larger number. Solve the other problem.

One number is 9 more than another. The quotient of the larger divided by the smaller is 4. Find the two numbers.

$$\frac{x+9}{x} = 4$$

$$\not{x}\left(\frac{x+9}{\not{x}}\right) = x(4)$$

$$x + 9 = 4x$$

$$9 = 3x$$

$$x = 3$$

Therefore, the two numbers are 3 and 12 (which is x + 9).

One number is 4 more than another. The quotient of the larger divided by the smaller is $\frac{5}{4}$. Find the two numbers.

smaller is 16, larger is 20, from:

$$\frac{x+4}{x} = \frac{5}{4}$$

106. To solve the problem below, we let x equal the original numerator and x + 2 equal the original denominator. Solve the other problem.

The denominator of a fraction is 2 more than the numerator. If 5 is subtracted from both the numerator and denominator, the resulting fraction is $\frac{3}{4}$. Find the original fraction.

$$\frac{x-5}{(x+2)-5} = \frac{3}{4}$$

$$\frac{x-5}{x-3} = \frac{3}{4}$$

$$4\cancel{(x-3)}\left(\frac{x-5}{\cancel{x-3}}\right) = \cancel{4}(x-3)\left(\frac{3}{\cancel{4}}\right)$$

$$4(x-5) = (x-3)(3)$$

$$4x - 20 = 3x - 9$$

$$x = 11$$

The original fraction is:

$$\frac{x}{x+2} = \frac{11}{13}$$

In a certain fraction, the denominator is 6 larger than the numerator. If 4 is added to both the numerator and denominator, the result is $\frac{3}{5}$. Find the original fraction.

107. To solve the problem below, we let x equal the original numerator and 3x equal the original denominator. Solve the other problem.

The denominator of a fraction is 3 times the numerator. If 4 is added to the numerator and subtracted from the denominator, the result equals $\frac{5}{7}$. Find the original fraction.

The denominator of a fraction is twice the numerator. If 6 is added to the numerator and subtracted from the denominator, the result equals 5. Find the original fraction.

$$\frac{x+4}{3x-4} = \frac{5}{7}$$

$$7(3x-4)\left(\frac{x+4}{3x-4}\right) = 7(3x-4)\left(\frac{5}{7}\right)$$

$$7(x+4) = (3x-4)(5)$$

$$7x + 28 = 15x - 20$$

$$x = 6$$

The original fraction is:

$$\frac{x}{3x} = \frac{6}{18}$$

$\frac{5}{11}$, from:

$$\frac{x+4}{(x+6)+4} = \frac{3}{5}$$

---

108. Following the example, solve the other problem.

One-third of a number is 2 more than one-fourth of the same number. Find the number.

One-half of a number is 10 more than one-sixth of the same number. Find the number.

$$\frac{1}{3}(x) = \frac{1}{4}(x) + 2$$

$$\frac{x}{3} = \frac{x}{4} + 2$$

$$12\left(\frac{x}{3}\right) = 12\left(\frac{x}{4} + 2\right)$$

$$4x = 12\left(\frac{x}{4}\right) + 12(2)$$

$$4x = 3x + 24$$

$$x = 24$$

The number is 24.

$\frac{4}{8}$, from:

$$\frac{x+6}{2x-6} = 5$$

30, from:

$$\frac{x}{2} = \frac{x}{6} + 10$$

109. For the problem below, we let the consecutive integers be x and x + 1. Solve the other problem.

Find two consecutive integers such that one-third of the smaller plus one-half of the larger is equal to 13.

Find two consecutive integers such that two-thirds of the smaller plus one-fifth of the larger is equal to 21.

$$\frac{1}{3}(x) + \frac{1}{2}(x + 1) = 13$$

$$\frac{x}{3} + \frac{x + 1}{2} = 13$$

$$6\left(\frac{x}{3} + \frac{x + 1}{2}\right) = 6(13)$$

$$\overset{2}{\cancel{6}}\left(\frac{x}{\cancel{3}}\right) + \overset{3}{\cancel{6}}\left(\frac{x + 1}{\cancel{2}}\right) = 78$$

$$2x + 3x + 3 = 78$$

$$5x = 75$$

$$x = 15$$

The two consecutive integers are 15 and 16.

The two consecutive integers are 24 and 25, from:

$$\frac{2x}{3} + \frac{x + 1}{5} = 21$$

---

110. A proportion is an equation of the form $\frac{a}{b} = \frac{c}{d}$. To solve the problem below, we set up a proportion with units, solved the same proportion without units, and then stated the solution in terms of the original wording of the problem.

A train travels 110 miles in 2 hours. At that rate, how many miles would it travel in 5 hours?

$$\frac{110 \text{ miles}}{2 \text{ hours}} = \frac{x \text{ miles}}{5 \text{ hours}}$$

$$\frac{110}{2} = \frac{x}{5}$$

$$(\cancel{2})(5)\left(\frac{110}{\cancel{2}}\right) = (2)(\cancel{5})\left(\frac{x}{\cancel{5}}\right)$$

$$550 = 2x$$

$$x = 275$$

Therefore, the train would travel ________ miles in 5 hours.

275

111. When solving a proportion, we usually multiply both sides by the given denominators rather than trying to find the LCD. For example, we multiplied both sides by (30)(42) below. Solve the other proportion.

$$\frac{10}{30} = \frac{x}{42}$$

$$\frac{18}{15} = \frac{x}{50}$$

$$(\not{30})(42)\left(\frac{10}{\not{30}}\right) = \left(\frac{x}{\not{42}}\right)(30)(\not{42})$$

$$420 = 30x$$

$$x = \frac{420}{30}$$

$$x = 14$$

---

112. To clear the fractions below, we multiplied both sides by 44x. Solve the other proportion.

x = 60

$$\frac{8}{44} = \frac{10}{x}$$

$$\frac{11}{12} = \frac{77}{x}$$

$$\not{44}x\left(\frac{8}{\not{44}}\right) = \left(\frac{10}{\not{x}}\right)(44\not{x})$$

$$8x = 440$$

$$x = 55$$

---

113. When using a proportion to solve a problem, the following steps should be used to set up the proportion correctly.

x = 84

1. Set up the known ratio on the left side.
2. Then make sure that the units in the ratio on the right side correspond to the units on the left side.

For example, we set up a proportion correctly below. Notice that the ratio is $\boxed{\frac{\text{gallons}}{\text{miles}}}$ on both sides.

If a car uses 2 gallons of gasoline to travel 30 miles, how many gallons will it need to travel 75 miles?

$$\frac{2 \text{ gallons}}{30 \text{ miles}} = \frac{x \text{ gallons}}{75 \text{ miles}}$$

Use a proportion to solve each problem.

a) On a map, 6 inches represents 200 miles. How many miles would be represented by 15 inches?

b) A woman saved $60 in 4 weeks. At that rate, how long would it take her to save $210?

114. Proportions can be used for conversions of units. An example is shown.

There are approximately 28 grams in 1 ounce. Approximately how many grams are there in 4.5 ounces?

$$\frac{28 \text{ grams}}{1 \text{ ounce}} = \frac{x \text{ grams}}{4.5 \text{ ounces}} \qquad \frac{28}{1} = \frac{x}{4.5}$$

$$4.5\left(\frac{28}{1}\right) = \overset{1}{\cancel{4.5}}\left(\frac{x}{\cancel{4.5}}\right)$$

$$126 = x$$

Therefore, there are approximately 126 grams in 4.5 ounces.

Using the same steps, solve these.

a) There are 36 inches in 1 yard. How many inches are there in 5.5 yards?

b) There are approximately 1.6 kilometers in 1 mile. Therefore, 24 kilometers is approximately how many miles?

a) 500 miles, from: $\frac{200}{6} = \frac{x}{15}$

b) 14 weeks, from: $\frac{60}{4} = \frac{210}{x}$

---

a) 198 inches, from: $\frac{36}{1} = \frac{x}{5.5}$

b) Approximately 15 miles, from: $\frac{1.6}{1} = \frac{24}{x}$

4-12 WORK PROBLEMS

In this section, we will solve some problems involving the time needed for two people, working together, to complete a job. Each problem involves a fractional equation.

---

115. If a job can be done in $\underline{x}$ hours (or some other unit of time), then $\frac{1}{x}$ of the job can be done in 1 hour. For example:

If a job can be done in 3 hours, $\frac{1}{3}$ of the job can be done in 1 hour.

If a job can be done in 4 hours, $\frac{1}{4}$ of the job can be done in 1 hour.

The relationship above is used to solve the problem below.

Problem: John can paint a room in 3 hours. Mike can paint the same room in 4 hours. How long would it take them, working together, to paint the room?

Continued on following page.

115. Continued

In terms of the part of the job that can be done in 1 hour, we know these facts:

1. Working alone, John can paint $\frac{1}{3}$ of the room in 1 hour and Mike can paint $\frac{1}{4}$ of the room in 1 hour.
2. If together they can paint the room in <u>x</u> hours, together they can paint $\frac{1}{x}$ of the room in 1 hour.

Therefore, we can set up the following equation based on the parts that can be done in 1 hour.

$$\begin{pmatrix}\text{Part John}\\ \text{can paint}\\ \text{in 1 hour}\end{pmatrix} + \begin{pmatrix}\text{Part Mike}\\ \text{can paint}\\ \text{in 1 hour}\end{pmatrix} = \begin{pmatrix}\text{Part both}\\ \text{can paint}\\ \text{in 1 hour}\end{pmatrix}$$

$$\downarrow \quad\quad \downarrow \quad\quad \downarrow \quad\quad \downarrow \quad\quad \downarrow$$

$$\frac{1}{3} + \frac{1}{4} = \frac{1}{x}$$

To clear the fraction, we multiply both sides by 12x. We get:

$$12x\left(\frac{1}{3} + \frac{1}{4}\right) = 12\cancel{x}\left(\frac{1}{\cancel{x}}\right)$$

$$\overset{4}{\cancel{12}}x\left(\frac{1}{\cancel{3}}\right) + \overset{3}{\cancel{12}}x\left(\frac{1}{\cancel{4}}\right) = 12(1)$$

$$4x + 3x = 12$$

$$7x = 12$$

$$x = \frac{12}{7} = 1\frac{5}{7}$$

Therefore, working together, John and Mike can paint the room in ________ hours.

Answer: $1\frac{5}{7}$ hours

---

116. Following the example, solve the other problem.

A man can plant his garden in 8 hours. His wife can plant the same garden in 6 hours. How long would it take them if they worked together?

$$\frac{1}{8} + \frac{1}{6} = \frac{1}{x}$$

$$24x\left(\frac{1}{8} + \frac{1}{6}\right) = 24\cancel{x}\left(\frac{1}{\cancel{x}}\right)$$

$$\overset{3}{\cancel{24}}x\left(\frac{1}{\cancel{8}}\right) + \overset{4}{\cancel{24}}x\left(\frac{1}{\cancel{6}}\right) = 24(1)$$

$$3x + 4x = 24$$

$$7x = 24$$

$$x = \frac{24}{7} = 3\frac{3}{7}$$

Together they can plant the garden in $3\frac{3}{7}$ hours.

A swimming pool can be filled in 4 hours by pipe A alone and in 6 hours by pipe B alone. How long would it take to fill the tank if both pipes were used together?

117. Use the same method for these.

a) Sue can do a job in 6 days, but Mary needs only 5 days. How long would it take them working together?

b) Pete can dig a trench in 8 hours, but his son needs 10 hours. How long would it take them working together?

$2\frac{2}{5}$ hours, from:

$\frac{1}{4} + \frac{1}{6} = \frac{1}{x}$

---

118. To translate the problem below to an equation, we used x for the number of hours Bill needs and 2x for the number of hours Jim needs.

Problem: Working together, Bill and Jim can complete a job in 7 hours. Working alone, it would take Jim twice as long as it would Bill. How long would it take each of them working alone?

$$\frac{1}{x} + \frac{1}{2x} = \frac{1}{7}$$

$$14x\left(\frac{1}{x} + \frac{1}{2x}\right) = \overset{2}{\cancel{14}}x\left(\frac{1}{\cancel{7}}\right)$$

$$14\cancel{x}\left(\frac{1}{\cancel{x}}\right) + \overset{7}{\cancel{14x}}\left(\frac{1}{\cancel{2x}}\right) = 2x$$

$$14 + 7 = 2x$$

$$21 = 2x$$

$$x = \frac{21}{2} = 10\frac{1}{2}$$

Working alone, Bill would take $10\frac{1}{2}$ hours and Jim would take $2\left(10\frac{1}{2}\right) = 21$ hours.

Using the same method, solve this one.

Problem: Two pipes carry water to the same tank. Working alone, Pipe A can fill the tank three times as fast as Pipe B. Together, the pipes can fill the tank in 6 hours. How long would it take each of them working alone?

a) $\frac{30}{11} = 2\frac{8}{11}$ days

b) $\frac{40}{9} = 4\frac{4}{9}$ hours

Pipe A alone takes 8 hours and Pipe B alone takes 24 hours.

4-13 MOTION PROBLEMS

In this section, we will solve some motion problems based on the formula d = rt. All of the problems involve a fractional equation.

---

119. We use the formula d = rt for the problem below. In that formula, d = distance traveled, r = rate (or speed), and t = time traveled.

<u>Problem</u>: One car travels 10 miles an hour faster than another. While one car travels 270 miles, the other travels 330 miles. Find the speed of each car.

We diagrammed the problem below.

Car 1

270 miles ———— x mph →

Car 2

330 miles ———— (x + 10) mph →

The data is summarized in the chart below. The time <u>t</u> for each is equal. We used $t = \frac{d}{r}$ to get $\frac{270}{x}$ and $\frac{330}{x + 10}$.

	Distance	Rate	Time
Car 1	270	x	$t = \frac{270}{x}$
Car 2	330	x + 10	$t = \frac{330}{x + 10}$

Since the two times are equal, we can set up and solve the following fractional equation.

$$\frac{270}{x} = \frac{330}{x + 10}$$

$$\cancel{x}(x + 10)\left(\frac{270}{\cancel{x}}\right) = x\cancel{(x + 10)}\left(\frac{330}{\cancel{x + 10}}\right)$$

$$270x + 2700 = 330x$$

$$2700 = 60x$$

$$x = 45$$

Therefore, the speed of Car 1 is 45 mph and the speed of Car 2 is 45 + 10 = 55 mph. These values check since $\frac{270}{45} = 6$ and $\frac{330}{55} = 6$. They both travel 6 hours at those speeds.

Using a diagram and a chart, do this one.

<u>Problem</u>: A freight train travels 20 mph slower than a passenger train. While the freight train travels 180 miles, the passenger train travels 270 miles. Find the speed of each train.

120. Since the two times are equal, we use a fractional equation to solve the problem below.

Problem: A river has a current of 3 mph. A boat takes as long to go 21 miles downstream as to go 12 miles upstream. What is the speed of the boat in still water?

We diagrammed the problem below.

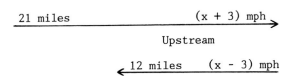

Downstream

21 miles (x + 3) mph →

Upstream

← 12 miles (x - 3) mph

We organized the data in the chart below. The time t for each trip is equal. We used $t = \frac{d}{r}$ to get $\frac{21}{x + 3}$ and $\frac{12}{x - 3}$.

	Distance	Rate	Time
Downstream	21	x + 3	$t = \frac{21}{x + 3}$
Upstream	12	x - 3	$t = \frac{12}{x - 3}$

Since the two times are equal, we can set up and solve the equation below.

$$\frac{21}{x + 3} = \frac{12}{x - 3}$$

$$\cancel{(x + 3)}(x - 3)\left(\frac{21}{\cancel{x + 3}}\right) = (x + 3)\cancel{(x - 3)}\left(\frac{12}{\cancel{x - 3}}\right)$$

$$21x - 63 = 12x + 36$$

$$9x = 99$$

$$x = 11$$

Therefore, the speed of the boat is 11 mph in still water. This checks since $\frac{21}{14} = \frac{3}{2}$ and $\frac{12}{8} = \frac{3}{2}$. Both trips take $1\frac{1}{2}$ hours.

Using a diagram and a chart, solve this one.

Problem: An airplane takes as long to fly 450 miles with the wind as it does to fly 360 miles against the wind. If the wind is blowing at 20 mph, what is the speed of the plane without a wind?

The speed of the freight train is 40 mph; the speed of the passenger train is 60 mph.

From either:

$$\frac{180}{x} = \frac{270}{x + 20}$$

or $$\frac{180}{x - 20} = \frac{270}{x}$$

121. We also use a fractional equation to solve the problem below.

Problem: A jet flies at an average speed of 500 mph without a wind. If it takes as long to fly 896 miles with the wind as it does to fly 854 miles against the wind, how strong is the wind?

We diagrammed the problem below.

With the wind

896 miles (500 + x) mph →

Against the wind

← 854 miles (500 - x) mph

We organized the data in the chart below.

	Distance	Rate	Time
With the wind	896	500 + x	$t = \frac{896}{500 + x}$
Against the wind	854	500 - x	$t = \frac{854}{500 - x}$

Since the two times are equal, we can set up and solve this equation where x is the speed of the wind.

$$\frac{896}{500 + x} = \frac{854}{500 - x}$$

$$\cancel{(500 + x)}(500 - x)\left(\frac{896}{\cancel{500 + x}}\right) = (500 + x)\cancel{(500 - x)}\left(\frac{854}{\cancel{500 - x}}\right)$$

$$448{,}000 - 896x = 427{,}000 + 854x$$

$$21{,}000 = 1750x$$

$$x = 12$$

Therefore, the wind is blowing at 12 mph. This checks since $\frac{896}{512} = \frac{7}{4}$ and $\frac{854}{488} = \frac{7}{4}$. Both flights take $1\frac{3}{4}$ hours.

Using a diagram and a chart, solve this one.

Problem: Joe can row 5 miles per hour in still water. It takes as long to row 14 miles upstream as it does to row 26 miles downstream. How fast is the current?

180 mph, from:

$$\frac{450}{x + 20} = \frac{360}{x - 20}$$

1.5 mph, from:

$$\frac{14}{5 - x} = \frac{26}{5 + x}$$

SELF TEST 16 (pages 219-229)

(pages 219-229)

Solve each problem.

1. One number is 16 more than another. If the larger is divided by the smaller, we get 5. Find the two numbers.

2. Find two consecutive integers such that one-half of the smaller plus three-fifths of the larger is equal to 16.

3. The denominator of a fraction is 4 more than the numerator. If 5 is subtracted from both the numerator and denominator, the resulting fraction is $\frac{1}{2}$. Find the original fraction.

4. A train travels 235 miles in 5 hours. At that rate, how many miles would it travel in 8 hours?

5. Joan can type a report in 4 hours, but Lori needs 5 hours. How long would it take them if they worked together?

6. One car travels 6 miles an hour faster than another. While one car travels 230 miles, the other travels 260 miles. Find the speed of each car.

ANSWERS:

1. 4 and 20
2. 14 and 15
3. $\frac{9}{13}$
4. 376 miles
5. $2\frac{2}{9}$ hours
6. 46 mph and 52 mph

SUPPLEMENTARY PROBLEMS - CHAPTER 4

Assignment 13

Reduce to lowest terms.

1. $\frac{3ab}{ab}$ 2. $\frac{m}{5m^2}$ 3. $\frac{4x^3y^2}{7xy^5}$ 4. $\frac{12cd^5}{8c^2d^3}$

5. $\frac{4x - 8}{3x - 6}$ 6. $\frac{2y^2 - 2y}{4y^2 + 4y}$ 7. $\frac{10x + 5}{15}$ 8. $\frac{4t}{2t - 8}$

9. $\frac{t^2 - 6t + 8}{t^2 - t - 12}$ 10. $\frac{y^2 + 7y + 10}{y^2 - 4}$ 11. $\frac{a - 7}{7 - a}$ 12. $\frac{c^2 - d^2}{d - c}$

Multiply. Report each product in lowest terms.

13. $\left(\frac{x}{3}\right)\left(\frac{2}{y}\right)$ 14. $\left(\frac{8}{5y}\right)\left(\frac{3y}{4}\right)$ 15. $\left(\frac{5x^2}{6}\right)\left(\frac{3}{2x}\right)$

16. $\left(\frac{1}{ay^3}\right)\left(\frac{5ay^2}{6}\right)$ 17. $\left(\frac{x - 3}{2x}\right)\left(\frac{2x^2 - 2x}{x - 3}\right)$ 18. $\left(\frac{y - 3}{y + 1}\right)\left(\frac{y^2 - 1}{y^2 - 5y + 6}\right)$

19. $\left(\frac{ab - b^2}{7a}\right)\left(\frac{7a + 7b}{a^2b - b^3}\right)$ 20. $\left(\frac{x^2 + 4x - 5}{10}\right)\left(\frac{5x}{2x^2 + x - 3}\right)$ 21. $\left(\frac{t + 1}{t - 2}\right)\left(\frac{t^2 - 5t + 6}{(t + 1)^2}\right)$

22. $\left(\frac{p^2 - 3p}{p^2 + 2p - 15}\right)\left(\frac{p^2 - 25}{3p}\right)$ 23. $\left(\frac{x^2 - 16}{x^3 + 64}\right)\left(\frac{x^2 - 4x + 16}{x^2 - 8x + 16}\right)$ 24. $\left(\frac{p^3 - q^3}{p^2 - q^2}\right)\left(\frac{3p^2 + 2pq - q^2}{p^2 + pq + q^2}\right)$

Divide. Report each quotient in lowest terms.

25. $\frac{\frac{x^2}{9}}{\frac{x}{3}}$ 26. $\frac{\frac{c^2d}{t^3}}{\frac{c^4d}{t}}$ 27. $\frac{\frac{a}{a + b}}{\frac{b}{a + b}}$ 28. $\frac{\frac{x^2 - 36}{3x + 9}}{\frac{x + 6}{6}}$

29. $\frac{\frac{12y^2}{(x + y)^2}}{\frac{16y}{x + y}}$ 30. $\frac{\frac{x^2 - 5x}{x^2 - 2x - 15}}{\frac{x^2 + 4x}{x^2 + 5x + 4}}$ 31. $\frac{\frac{a^2 - 5a + 4}{a^2 - a - 12}}{\frac{a^2 - 2a + 1}{a^2 + 6a + 9}}$ 32. $\frac{\frac{c^2 - d^2}{2c^2 - cd - d^2}}{\frac{2c + 2d}{2c - d}}$

33. $\frac{\frac{d^3 - 64}{d^2 - 16}}{\frac{d^2 - 2d + 1}{d + 4}}$ 34. $\frac{\frac{x^2 - y^2}{x^3 + y^3}}{\frac{x^2 - 2xy + y^2}{x^2 - xy + y^2}}$ 35. $\frac{\frac{p^2 - q^2}{p + q}}{p}$ 36. $\frac{\frac{(k - 3)^2}{5}}{k - 3}$

Assignment 14

Add or subtract. Report each answer in lowest terms.

1. $\frac{2x}{3} + \frac{1}{3}$ 2. $\frac{3t + 2}{t - 1} + \frac{t - 7}{t - 1}$ 3. $\frac{3x}{10} + \frac{3x}{10}$

4. $\frac{y^2 + y}{y^2 - y} + \frac{y^2 - 6y}{y^2 - y}$ 5. $\frac{2p}{p + 3} - \frac{7}{p + 3}$ 6. $\frac{5}{6x} - \frac{1}{6x}$

7. $\frac{3d - 5}{d - 3} - \frac{2d - 7}{d - 3}$ 8. $\frac{7}{2x} - \frac{1 - 4x}{2x}$ 9. $\frac{y}{y^2 - 4} - \frac{2}{y^2 - 4}$

10. $\frac{3t + 5}{t^2 - t - 6} - \frac{2t + 3}{t^2 - t - 6}$ 11. $\frac{m}{m - 4} + \frac{3}{4 - m}$ 12. $\frac{6p}{2p - q} - \frac{p - 3}{q - 2p}$

Find the LCD only. (Do not add or subtract.)

13. $\frac{7}{24} + \frac{1}{18}$

14. $\frac{11}{12} + \frac{1}{6} + \frac{8}{15}$

15. $\frac{5}{8x} - \frac{1}{6x}$

16. $\frac{3}{4x} - \frac{1}{2y}$

17. $\frac{5}{6t^2} + \frac{3}{10t}$

18. $\frac{2}{3y} - \frac{1}{3y - 9}$

19. $\frac{x}{4x + 2} + \frac{1}{2x^2 + x}$

20. $\frac{y - 1}{y^2 - 1} - \frac{y + 1}{y^2 + 3y - 4}$

21. $\frac{m - 5}{(m + 2)^2} + \frac{m + 4}{m^2 - 3m - 10}$

22. $\frac{1}{3} + \frac{3}{4x} - \frac{5}{2x^2}$

23. $\frac{y}{y + 3} - \frac{3}{y - 1} + \frac{y + 5}{y^2 + 2y - 3}$

24. $\frac{3m}{4m + 2} - \frac{m - 1}{6m - 3} + \frac{2m + 1}{4m^2 - 1}$

Add or subtract.

25. $\frac{3}{4x} + \frac{1}{10x}$

26. $\frac{m}{m - 1} + \frac{3}{2m}$

27. $\frac{y + 1}{y - 2} + \frac{y - 7}{y + 3}$

28. $\frac{x}{3x - 6} + \frac{5}{x^2 - 2x}$

29. $\frac{2y}{y^2 - y - 6} + \frac{1}{y^2 - 9}$

30. $\frac{3}{2x} - \frac{2}{3x}$

31. $\frac{p + 3}{p - 2} - \frac{p + 1}{p - 3}$

32. $\frac{5x}{x^2 - 16} - \frac{2}{x - 4}$

33. $\frac{x}{(x - y)^2} - \frac{y}{x^2 - y^2}$

34. $\frac{1}{4} + \frac{3}{2x} - \frac{2}{x^2}$

35. $\frac{5}{y - 2} - \frac{y + 3}{y^2 - 4} + \frac{1}{y + 2}$

36. $\frac{4}{3a - 2} + \frac{5a^2 - 6}{3a^2 + a - 2} - \frac{2}{a + 1}$

Simplify each expression.

37. $\dfrac{1 + \frac{3}{x}}{\frac{2}{5}}$

38. $\dfrac{\frac{y}{4} + 3}{\frac{7}{8}}$

39. $\dfrac{\frac{3}{y} - \frac{1}{2y}}{\frac{1}{y} + \frac{4}{3y}}$

40. $\dfrac{1 - \frac{1}{t}}{1 + \frac{1}{t}}$

41. $\dfrac{\frac{a}{b} - \frac{b}{a}}{\frac{1}{a} + \frac{1}{b}}$

42. $\dfrac{\frac{x}{y} - \frac{1}{3y}}{\frac{x}{3y} + \frac{1}{6y}}$

43. $\dfrac{\frac{b}{b + 1} + 1}{\frac{2b + 1}{b - 1}}$

44. $\dfrac{\frac{x + y}{y} + \frac{y}{x - y}}{\frac{y}{x - y}}$

45. $\dfrac{a + 5 + \frac{a + 5}{a - 3}}{a - 4 - \frac{2}{a - 3}}$

Assignment 15

Solve each equation.

1. $\frac{2x}{3} + \frac{x}{6} = 1$

2. $\frac{3x}{2} - \frac{4x}{5} = 2$

3. $\frac{7}{3x} = 4$

4. $2 = \frac{9}{4y}$

5. $\frac{y - 1}{y} = 2$

6. $1 = \frac{5 - y}{2y}$

7. $\frac{10}{x + 4} = 2$

8. $5 = \frac{1}{t - 1}$

9. $6 = \frac{10}{x} - 1$

10. $3 - \frac{1}{2x} = 5$

11. $\frac{t - 3}{t} = \frac{2}{3}$

12. $\frac{m + 5}{3m - 1} = \frac{1}{2}$

13. $\frac{2}{y} = \frac{5}{y - 4}$

14. $\frac{2}{x + 1} = \frac{4}{x - 4}$

15. $\frac{p + 1}{p + 2} = \frac{p - 2}{p - 3}$

16. $\frac{x - 1}{x - 4} = \frac{3}{x - 4}$

17. $\frac{1}{8} + \frac{1}{6} = \frac{1}{x}$

18. $\frac{5}{6} - \frac{2}{3} = \frac{1}{y}$

19. $\frac{1}{y} = \frac{5}{2y} - \frac{1}{2}$

20. $\frac{2}{m} + \frac{3}{2m} = \frac{7}{6}$

21. $\frac{3}{4p} - \frac{2}{p} = \frac{5}{12}$

22. $m - \frac{20}{m} = 1$

23. $3x + \frac{4}{x} = 7$

24. $\frac{x^2}{x - 3} = \frac{9}{x - 3}$

25. $\frac{1}{x + 3} + \frac{1}{x^2 - 9} = \frac{2}{x - 3}$

26. $\frac{1}{x + 2} + \frac{1}{x - 4} = \frac{1}{x^2 - 2x - 8}$

Do each evaluation.

27. In $\frac{F_1}{F_2} = \frac{d_1}{d_2}$, find F_1 when $F_2 = 60$, $d_1 = 54$, and $d_2 = 90$.

28. In $\frac{P_1}{P_2} = \frac{V_2}{V_1}$, find V_1 when $P_1 = 20$, $P_2 = 15$, and $V_2 = 48$.

29. In $\frac{1}{R_t} = \frac{1}{R_1} + \frac{1}{R_2}$, find R_2 when $R_t = 20$ and $R_1 = 60$.

30. In $\frac{1}{D} + \frac{1}{d} = \frac{1}{f}$, find D when $d = 90$ and $f = 60$.

31. In $I = \frac{nE}{R + nr}$, find R when $I = 4$, $n = 2$, $E = 50$ and $r = 10$.

32. In $I = \frac{pT}{M + pr}$, find p when $I = 5$, $T = 80$, $M = 24$ and $r = 8$.

Do each rearrangement.

33. Solve for d_2.

$$F_1 = \frac{d_1 F_2}{d_2}$$

34. Solve for H.

$$P = \frac{M}{H(t_2 - t_1)}$$

35. Solve for C_o.

$$v = \frac{C_o + C_i}{t}$$

36. Solve for P_1.

$$\frac{P_1 V_1}{T_1} = \frac{P_2 V_2}{T_2}$$

37. Solve for F.

$$\frac{D}{d} = \frac{F}{f}$$

38. Solve for P_2.

$$\frac{P_1}{P_2} = \frac{V_2}{V_1}$$

39. Solve for T_2.

$$\frac{V_1}{V_2} = \frac{T_1}{T_2}$$

40. Solve for A.

$$B = AC + AD$$

41. Solve for K.

$$KT = AB + KR$$

42. Solve for t.

$$\frac{1}{m} + \frac{1}{p} = \frac{1}{t}$$

43. Solve for R.

$$\frac{1}{F} + \frac{1}{R} = \frac{1}{S}$$

44. Solve for D.

$$C = \frac{D}{D + 1}$$

45. Solve for T.

$$H = \frac{PT}{P + T}$$

46. Solve for t.

$$\frac{B}{b} = \frac{D + t}{t}$$

47. Solve for a.

$$\frac{D}{d} = \frac{a + a_o}{a + a_f}$$

Assignment 16

Solve each problem.

1. One number is 10 more than another. The quotient of the larger divided by the smaller is 6. Find the two numbers.

2. One number is 4 less than another. The quotient of the smaller divided by the larger is $\frac{4}{5}$. Find the two numbers.

3. The denominator of a fraction is 3 more than the numerator. If 5 is subtracted from both the numerator and denominator, the resulting fraction is $\frac{1}{2}$. Find the original fraction.

4. The denominator of a fraction is 3 times the numerator. If 5 is added to the numerator and subtracted from the denominator, the result equals $\frac{3}{5}$. Find the original fraction.

5. One-fourth of a number is 2 more than one-fifth of the same number. Find the number.

6. One-half of a number is 6 more than one-third of the same number. Find the number.

7. Find two consecutive integers such that four-fifths of the smaller plus one-half of the larger is equal to 20.

8. Find two consecutive integers such that one-fourth of the smaller plus two-thirds of the larger is equal to 19.

9. A car travels 135 miles in 3 hours. At that rate, how long would it take to travel 225 miles?

10. A hospital charges a patient $6.60 for 12 pills. At that rate, what would the charge be for 30 pills?

11. There are 36 inches in 1 yard. How many inches are there in 2.75 yards?

12. There are 28 grams in 1 ounce. Therefore, 126 grams equal how many ounces?

13. A man can cut 2 acres of grass in 4 hours on a riding mower. His son can cut the 2 acres of grass in 6 hours with a walking mower. How long would it take them if they worked together?

14. Peggy can type a report in 4 hours, but Joan needs 5 hours. How long would it take them working together?

15. Dan can lay some sod in 6 hours, but his son needs 8 hours. How long would it take them working together?

16. Working together, Mike and Joe can complete a job in 6 hours. It would take Mike 9 more hours working alone than it would take Joe. How long would it take each of them working alone?

17. One car travels 8 miles an hour faster than another. While one car travels 230 miles, the other travels 270 miles. Find the speed of each car.

18. One airplane travels 50 miles per hour faster than another. While one travels 1125 miles, the other travels 1000 miles. Find the speed of each plane.

19. An airplane takes as long to fly 380 miles with the wind as it does to fly 320 miles against the wind. If the wind is blowing at 12 mph, what is the speed of the plane without a wind?

20. Sue can row 3 miles per hour in still water. It takes as long to row 6 miles upstream as it does to row 18 miles downstream. How fast is the current?

5 Exponents and Radicals

In this chapter, we will discuss the basic operations with radicals, a method for solving radical equations, and some evaluations and rearrangements with radical formulas. We will review scientific notation, define powers with rational exponents, and extend the laws of exponents to powers of that type. We will define complex numbers and discuss the basic operations with complex numbers.

5-1 RADICAL EXPRESSIONS

In this section, we will discuss radical expressions involving various roots.

1. A <u>square root</u> of a number <u>a</u> is a number <u>b</u> whose square is <u>a</u>. That is, <u>b</u> is a square root of <u>a</u> if $b^2 = a$.

   Any positive number has two square roots, one positive and one negative. For example:

   4 is a square root of 16, since $(4)^2 = 16$.
   -4 is a square root of 16, since $(-4)^2 = 16$.

   The number 0 has only one square root. It is 0, since $(0)^2 = 0$.

   Complete: a) The <u>positive</u> square root of 36 is _______.

   b) The <u>negative</u> square root of 100 is _______.

   c) The two square roots of 49 are _______ and _______.

   d) The square root of 0 is _______.

a) 6
b) -10
c) 7 and -7
d) 0

2. Though negative square roots are called <u>negative</u> square roots, positive square roots are called <u>principal</u> square roots.

The symbol $\sqrt{\quad}$, called a <u>radical sign</u>, is used for <u>principal</u> square roots. That is:

$\sqrt{25} = 5$ $\sqrt{81} = 9$ $\sqrt{400} = 20$

The symbol $-\sqrt{\quad}$ is used for <u>negative</u> square roots. That is:

$-\sqrt{9} = -3$ $-\sqrt{1} = -1$ $-\sqrt{225} = -15$

Complete these:

a) $\sqrt{1}$ = _____ b) $-\sqrt{64}$ = _____ c) $-\sqrt{4}$ = _____

---

3. Negative numbers do not have real number square roots because the square of any real number is positive. For example, -16 does not have a real number square root.

$\sqrt{-16} \neq 4$, because $(4)^2 = 16$.

$\sqrt{-16} \neq -4$, because $(-4)^2 = 16$.

Complete these:

a) $\sqrt{-81}$ = _____ b) $-\sqrt{81}$ = _____ c) $\sqrt{0}$ = _____

Answers (frame 2): a) 1 b) -8 c) -2

---

4. Except for perfect squares like 25 or 64, the square roots of numbers are non-terminating decimal numbers. We can use a calculator to find square roots. Using a calculator, we get:

$\sqrt{19} = 4.3588989$

$\sqrt{84} = 9.1651514$

A table of square roots for numbers from 1 to 100 is given on the inside of the back cover. In the table, the square roots of numbers that are not perfect squares are rounded <u>to three decimal places</u>. For example, the table has these entries which are rounded from the values above.

N	$\sqrt{N}$
19	4.359
84	9.165

Though the entries above are not exact, we will treat them as if they were exact. That is:

We will say: $\sqrt{19} = 4.359$ $\sqrt{84} = 9.165$

Using the table, complete these:

a) $\sqrt{28}$ = _______ b) $-\sqrt{41}$ = _______ c) $\sqrt{98}$ = _______

Answers (frame 3): a) -81 has no real number square root. b) -9 c) 0

Answers (frame 4): a) 5.292 b) -6.403 c) 9.899

5. A cube root of a number a is a number b whose cube is a. That is, b is a cube root of a if $b^3 = a$. Every real number has only one cube root. The cube root of a positive number is positive. The cube root of a negative number is negative. For example:

3 is the cube root of 27, since $3^3 = 27$.

-4 is the cube root of -64, since $(-4)^3 = -64$.

The symbol $\sqrt[3]{}$ is used for cube roots. Complete these.

a) $\sqrt[3]{8} =$ ______ b) $\sqrt[3]{-27} =$ ______ c) $\sqrt[3]{125} =$ ______

---

6. The general symbol for principal roots is given below.

$\sqrt[k]{a} = b$ means $b^k = a$

The number k is called the index. It indicates the root. When the index is 2, it is not written. For example:

In $\sqrt{64}$, the index is 2. It indicates a square root.

In $\sqrt[3]{27}$, the index is 3. It indicates a cube root.

In $\sqrt[6]{64}$, the index is 6. It indicates a sixth root.

As a help in problems with roots, some powers are given in the table below.

Number	Cube	Fourth Power	Fifth Power	Sixth Power
2	8	16	32	64
3	27	81	243	729
4	64	256	1024	4096
5	125	625	3125	15,625
6	216	1296	7776	46,656

We used the table to complete these:

$\sqrt[4]{625} = 5$, since $5^4 = 625$

$\sqrt[5]{243} = 3$, since $3^5 = 243$

Complete these:

a) $\sqrt[3]{216} =$ ____ b) $\sqrt[5]{1024} =$ ____ c) $\sqrt[6]{729} =$ ____

Answers to frame 5: a) 2 b) -3 c) 5

---

7. When the index is an odd number, it indicates an odd root. Every real number has just one odd root. If the number is positive, the root is positive. If the number is negative, the root is negative. For example:

$\sqrt[3]{125} = 5$, since $5^3 = 125$

$\sqrt[5]{-32} = -2$, since $(-2)^5 = -32$

Complete these:

a) $\sqrt[3]{-64} =$ ______ b) $\sqrt[5]{3125} =$ ______ c) $\sqrt[5]{-243} =$ ______

Answers to frame 6: a) 6 b) 4 c) 3

8.	When the index is an <u>even</u> number, it indicates an even root. Every positive real number has two even roots, one positive and one negative. Negative real numbers do not have real number even roots. For example: $\sqrt[4]{625} = 5$, since $5^4 = 625$ $-\sqrt[6]{729} = -3$, since $(-3)^6 = 729$ $\sqrt[6]{-64}$ There is no real number root. Complete these: a) $-\sqrt[4]{256}$ = _____ b) $\sqrt[6]{15,625}$ = _____ c) $\sqrt[4]{-16}$ = _____	a) -4 b) 5 c) -3
9.	Each root below is either "1" or -1. $-\sqrt{1} = -1$ $\sqrt[3]{-1} = -1$ $\sqrt[4]{1} = 1$ Complete these: a) $\sqrt[5]{1}$ = _____ b) $-\sqrt[6]{1}$ = _____ c) $\sqrt[4]{-1}$ = _____	a) -4 b) 5 c) No real number root
10.	When a "-" appears in front of a radical with an even index, it indicates the negative root. When a "-" appears in front of a radical with an odd index, it indicates the <u>additive inverse</u> of the one root. For example: $-\sqrt[4]{81} = -3$ $-\sqrt[3]{125} = -(5) = -5$ $-\sqrt[5]{-32} = -(-2) = 2$ Complete these: a) $-\sqrt[5]{1024}$ = _____ b) $-\sqrt[6]{64}$ = _____ c) $-\sqrt[3]{-1}$ = _____	a) 1 b) -1 c) No real number root
11.	Each expression below is called a radical expression. $\sqrt{17}$ $\sqrt{x}$ $\sqrt[3]{5y^2}$ $\sqrt[4]{x^2 - 4}$ The expression under the radical is called the <u>radicand</u>. For example: In $\sqrt{17}$, the radicand is 17. In $\sqrt[3]{5y^2}$, the radicand is $5y^2$. In $\sqrt[4]{x^2 - 4}$, the radicand is __________.	a) -4 b) -2 c) 1
12.	In $\sqrt{x^2}$, the radicand is a perfect square. We can substitute either a positive or a negative value for <u>x</u>. If $x = 4$, $\sqrt{x^2} = \sqrt{4^2} = \sqrt{16} = 4$. If $x = -4$, $\sqrt{x^2} = \sqrt{(-4)^2} = \sqrt{16} = 4$.	$x^2 - 4$

Continued on following page.

12. Continued

Since $\sqrt{x^2}$ is a principal square root, it represents a positive number in both cases. Therefore:

If $\underline{x}$ is positive, $\sqrt{x^2}$ equals $\underline{x}$.

If $\underline{x}$ is negative, $\sqrt{x^2}$ equals the additive inverse of $\underline{x}$.

In both cases, $\sqrt{x^2}$ equals the absolute value of $\underline{x}$. Therefore:

For any real number $\underline{x}$, $\sqrt{x^2} = \lvert x \rvert$

That is: $\sqrt{5^2} = \lvert 5 \rvert = 5$ $\qquad$ $\sqrt{(-3)^2} = \lvert -3 \rvert = 3$

Complete these:

a) $\sqrt{10^2}$ = _____ $\qquad$ b) $\sqrt{(-1)^2}$ = _____ $\qquad$ c) $\sqrt{(-8)^2}$ = _____

---

13. When $\underline{k}$ is any even number, $\sqrt[k]{x^k} = \lvert x \rvert$. The absolute value symbol is needed. For example:

$\sqrt[4]{7^4} = \lvert 7 \rvert = 7$ $\qquad$ $\sqrt[6]{(-4)^6} = \lvert -4 \rvert = 4$

Complete these:

a) $\sqrt[4]{(-2)^4}$ = _____ $\qquad$ b) $\sqrt[6]{9^6}$ = _____ $\qquad$ c) $\sqrt[8]{(-1)^8}$ = _____

Answers to 12: a) $\lvert 10 \rvert = 10$ b) $\lvert -1 \rvert = 1$ c) $\lvert -8 \rvert = 8$

---

14. In $\sqrt[3]{x^3}$, the radicand is a perfect cube. We can substitute either a positive or negative value for $\underline{x}$.

If $x = 2$, $\sqrt[3]{x^3} = \sqrt[3]{2^3} = \sqrt[3]{8} = 2$

If $x = -2$, $\sqrt[3]{x^3} = \sqrt[3]{(-2)^3} = \sqrt[3]{-8} = -2$

The absolute value symbol is not needed when finding $\sqrt[3]{x^3}$. If $\underline{x}$ is positive, the cube root is positive. If $\underline{x}$ is negative, the cube root is negative. That is:

$\sqrt[3]{5^3} = 5$ $\qquad$ $\sqrt[3]{(-7)^3} = -7$

Complete these:

a) $\sqrt[3]{6^3}$ = _____ $\qquad$ b) $\sqrt[3]{(-4)^3}$ = _____ $\qquad$ c) $\sqrt[3]{(-1)^3}$ = _____

Answers to 13: a) $\lvert -2 \rvert = 2$ b) $\lvert 9 \rvert = 9$ c) $\lvert -1 \rvert = 1$

---

15. When $\underline{k}$ is any odd number, $\sqrt[k]{x^k} = x$. The absolute value symbol is not needed. For example:

$\sqrt[5]{3^5} = 3$ $\qquad$ $\sqrt[7]{(-4)^7} = -4$

Complete these:

a) $\sqrt[5]{(-9)^5}$ = _____ $\qquad$ b) $\sqrt[7]{10^7}$ = _____ $\qquad$ c) $\sqrt[9]{(-1)^9}$ = _____

Answers to 14: a) 6 b) -4 c) -1

Answers to 15: a) -9 $\qquad$ b) 10 $\qquad$ c) -1

16. If both positive and negative values are substituted for x:

when k is even, $\sqrt[k]{x^k} = |x|$

when k is odd, $\sqrt[k]{x^k} = x$

If only positive values are substituted for x, the absolute value sign is not needed when k is even. In that case,

when k is even or odd, $\sqrt[k]{x^k} = x$

In this chapter, we will assume that only positive values are substituted for x. With that assumption:

$\sqrt[3]{x^3} = x$ $\sqrt[4]{x^4} = x$ $\sqrt[6]{x^6} =$ ______

x

5-2 MULTIPLYING AND SIMPLIFYING RADICALS

In this section, we will discuss the procedure for multiplying radicals. We will also show how some radicals can be simplified by factoring out perfect powers.

Note: In this chapter, we will assume that all radicands represent positive numbers.

17. To multiply radicals with the same index, we multiply their radicands. That is:

$$\sqrt[k]{x} \cdot \sqrt[k]{y} = \sqrt[k]{xy}$$

The following two multiplications show that the definition above makes sense.

$\sqrt{25} \cdot \sqrt{4} = \sqrt{(25)(4)}$ $\sqrt[3]{8} \cdot \sqrt[3]{27} = \sqrt[3]{(8)(27)}$

$5 \cdot 2 = \sqrt{100}$ $2 \cdot 3 = \sqrt[3]{216}$

$10 = 10$ $6 = 6$

When multiplying radicals, we simplify the radicand of the product. For example:

$$\sqrt{2} \cdot \sqrt{3} = \sqrt{(2)(3)} = \sqrt{6}$$

$$\sqrt[3]{7x} \cdot \sqrt[3]{2x} = \sqrt[3]{(7x)(2x)} = \sqrt[3]{14x^2}$$

a) $\sqrt[4]{y+5} \cdot \sqrt[4]{y-5} = \sqrt[4]{(y+5)(y-5)} =$ ______________

b) $\sqrt[3]{\frac{a}{7}} \cdot \sqrt[3]{\frac{2}{b}} = \sqrt[3]{\left(\frac{a}{7}\right)\left(\frac{2}{b}\right)} =$ ______________

18. We can factor a radical by reversing the procedure for multiplying. That is:

$$\sqrt{6} = \sqrt{2} \cdot \sqrt{3} \qquad \sqrt[3]{35} = \sqrt[3]{5} \cdot \sqrt[3]{7}$$

When we factor and get a radicand that is a perfect power, we can simplify. For example:

$$\sqrt{18} = \sqrt{9} \cdot \sqrt{2} = 3\sqrt{2}$$

$$\sqrt[3]{40} = \sqrt[3]{8} \cdot \sqrt[3]{5} = 2\sqrt[3]{5}$$

The process above is called <u>simplifying by factoring out a perfect power</u>. Simplify these by that method.

a) $\sqrt{8}$ = ____________ c) $\sqrt[3]{54}$ = ____________

b) $\sqrt{75}$ = ____________ d) $\sqrt[4]{160}$ = ____________

a) $\sqrt[4]{y^2 - 25}$

b) $\sqrt[3]{\frac{2a}{7b}}$

---

19. We simplified each radical below by factoring out a perfect power.

$$\sqrt{50x} = \sqrt{25} \cdot \sqrt{2x} = 5\sqrt{2x}$$

$$\sqrt[3]{27ab} = \sqrt[3]{27} \cdot \sqrt[3]{ab} = 3\sqrt[3]{ab}$$

Simplify these:

a) $\sqrt{81y}$ = ____________ c) $\sqrt[3]{54m}$ = ____________

b) $\sqrt{20xy}$ = ____________ d) $\sqrt[4]{256cd}$ = ____________

a) $2\sqrt{2}$

b) $5\sqrt{3}$

c) $3\sqrt[3]{2}$

d) $2\sqrt[4]{10}$

---

20. The definition of roots can be extended to include the roots of powers. For example:

$$\sqrt{2^6} = 2^3\text{, since } (2^3)^2 = 2^6$$

$$\sqrt[3]{x^{12}} = x^4\text{, since } (x^4)^3 = x^{12}$$

$$\sqrt[4]{y^{20}} = y^5\text{, since } (y^5)^4 = y^{20}$$

To find the root of a power, we divide the exponent of the power by the index. That is:

$$\sqrt{2^6} = 2^{\frac{6}{2}} = 2^3$$

$$\sqrt[3]{x^{12}} = x^{\frac{12}{3}} = x^4$$

$$\sqrt[4]{y^{20}} = y^{\frac{20}{4}} = y^5$$

A power whose exponent is exactly divisible by the index of the radical is called a <u>perfect power</u>. Complete the division to find the roots of these perfect powers.

a) $\sqrt[5]{t^{10}} = t^{\frac{10}{5}}$ = ________ b) $\sqrt[8]{m^{56}} = m^{\frac{56}{8}}$ = ________

a) $9\sqrt{y}$

b) $2\sqrt{5xy}$

c) $3\sqrt[3]{2m}$

d) $4\sqrt[4]{cd}$

---

a) t^2 b) m^7

21. The root of a perfect power can be a power whose exponent is "1". That is:

$$\sqrt[4]{x^4} = x^{\frac{4}{4}} = x^1 \text{ or } x$$

Divide to find each root.

a) $\sqrt[3]{d^{18}}$ = ________ c) $\sqrt[5]{y^{50}}$ = ________

b) $\sqrt{p^2}$ = ________ d) $\sqrt[7]{m^7}$ = ________

---

22. We used the same method to find the root below.

$$\sqrt[3]{(x-5)^6} = (x-5)^{\frac{6}{3}} = (x-5)^2$$

Find each root.

a) $\sqrt[5]{(y+1)^5}$ = ________ b) $\sqrt{(a-b)^{10}}$ = ________

Answers to 21: a) d^6 b) p c) y^{10} d) m

---

23. We simplified each radical below by factoring out perfect powers.

$$\sqrt{4x^2} = \sqrt{4}\cdot\sqrt{x^2} = 2x$$

$$\sqrt[3]{5y^{15}} = \sqrt[3]{5}\cdot\sqrt[3]{y^{15}} = y^5\sqrt[3]{5}$$

$$\sqrt[4]{32m^{12}} = \sqrt[4]{16}\cdot\sqrt[4]{2}\cdot\sqrt[4]{m^{12}} = 2m^3\sqrt[4]{2}$$

Simplify these.

a) $\sqrt{25x^6}$ = ________ c) $\sqrt[3]{27(t-1)^3}$ = ________

b) $\sqrt{18t^8}$ = ________ d) $\sqrt[5]{7t^5}$ = ________

Answers to 22: a) $(y+1)^1$ or $y+1$ b) $(a-b)^5$

---

24. Two more simplifications are shown below.

$$\sqrt{3x^{10}y^4} = \sqrt{3}\cdot\sqrt{x^{10}}\cdot\sqrt{y^4} = x^5y^2\sqrt{3}$$

$$\sqrt[3]{40ax^9} = \sqrt[3]{8}\cdot\sqrt[3]{5a}\cdot\sqrt[3]{x^9} = 2x^3\sqrt[3]{5a}$$

Simplify these.

a) $\sqrt{x^4y^2}$ = ________ c) $\sqrt[4]{16a^4b^8c^4}$ = ________

b) $\sqrt[3]{cd^6}$ = ________ d) $\sqrt{32at^8}$ = ________

Answers to 23: a) $5x^3$ b) $3t^4\sqrt{2}$ c) $3(t-1)$ d) $t\sqrt[5]{7}$

---

25. When the exponent of a power is not exactly divisible by the index, it is not a perfect power. But if the exponent is greater than the index, we can factor to get a perfect power and another power. For example:

$$\sqrt{x^5} = \sqrt{x^4}\cdot\sqrt{x^1} = x^2\sqrt{x}$$

$$\sqrt[3]{y^{11}} = \sqrt[3]{y^9}\cdot\sqrt[3]{y^2} = y^3\sqrt[3]{y^2}$$

Continued on following page.

Answers to 24: a) x^2y b) $d^2\sqrt[3]{c}$ c) $2ab^2c$ d) $4t^4\sqrt{2a}$

25. Continued

Notice that we factored out the largest possible perfect powers. That is, in the top example, the exponent 4 is the largest multiple of the index 2. In the bottom example, the exponent 9 is the largest multiple of the index 3. Simplify these by factoring out the largest perfect power.

a) $\sqrt{x^3}$ = ______ b) $\sqrt[3]{y^7}$ = ______ c) $\sqrt[4]{t^{10}}$ = ______

---

26. We simplified each radical below by factoring out the largest perfect powers.

$$\sqrt{12x^7} = \sqrt{4} \cdot \sqrt{x^6} \cdot \sqrt{3x} = 2x^3\sqrt{3x}$$

$$\sqrt[3]{64y^8} = \sqrt[3]{64} \cdot \sqrt[3]{y^6} \cdot \sqrt[3]{y^2} = 4y^2\sqrt[3]{y^2}$$

Simplify these.

a) $\sqrt{9x^9}$ = ______ b) $\sqrt[4]{32t^7}$ = ______

Answers to 25: a) $x\sqrt{x}$ b) $y^2\sqrt[3]{y}$ c) $t^2\sqrt[4]{t^2}$

---

27. We simplified below by factoring out the largest perfect powers.

$$\sqrt{5x^2y^3} = \sqrt{x^2} \cdot \sqrt{y^2} \cdot \sqrt{5y} = xy\sqrt{5y}$$

$$\sqrt[3]{24a^2b^7} = \sqrt[3]{8} \cdot \sqrt[3]{b^6} \cdot \sqrt[3]{3a^2b} = 2b^2\sqrt[3]{3a^2b}$$

Simplify these.

a) $\sqrt{49a^4b^5}$ = ______ b) $\sqrt[3]{54x^2y^3z^4}$ = ______

Answers to 26: a) $3x^4\sqrt{x}$ b) $2t\sqrt[4]{2t^3}$

---

28. Notice how we factored to simplify each radical below.

$$\sqrt{x^2 + 8x + 16} = \sqrt{(x+4)^2} = x + 4$$

$$\sqrt[3]{8a^3b - 8a^3c} = \sqrt[3]{8a^3(b - c)} = 2a\sqrt[3]{b - c}$$

Simplify these.

a) $\sqrt{y^2 + 10y + 25}$ = ______ b) $\sqrt[4]{c^4x + c^4y}$ = ______

Answers to 27: a) $7a^2b^2\sqrt{b}$ b) $3yz\sqrt[3]{2x^2z}$

---

29. After multiplying below, we simplified each product.

$$\sqrt{2} \cdot \sqrt{10} = \sqrt{20} = 2\sqrt{5}$$

$$\sqrt{3a^3} \cdot \sqrt{12b^5} = \sqrt{36a^3b^5} = 6ab^2\sqrt{ab}$$

$$\sqrt[3]{12x^3} \cdot \sqrt[3]{4y^2} = \sqrt[3]{48x^3y^2} = 2x\sqrt[3]{6y^2}$$

Multiply and simplify each product.

a) $\sqrt{5} \cdot \sqrt{10}$ = ______

b) $\sqrt{20a^4} \cdot \sqrt{5b^7}$ = ______

c) $\sqrt[3]{9x^2y^3} \cdot \sqrt[3]{6z^5}$ = ______

Answers to 28: a) $y + 5$ b) $c\sqrt[4]{x + y}$

30. We simplified the product below.

$$\sqrt{(x + y)^3} \cdot \sqrt{x + y} = \sqrt{(x + y)^4} = (x + y)^2$$

Multiply and simplify each product.

a) $\sqrt[3]{(m - 1)^3} \cdot \sqrt[3]{(m - 1)^4} =$ __________

b) $\sqrt{5(b + 7)} \cdot \sqrt{4(b + 7)^4} =$ __________

Answers: a) $5\sqrt{2}$ b) $10a^2b^3\sqrt{b}$ c) $3yz\sqrt[3]{2x^2z^2}$

---

31. To indicate a multiplication of a non-radical and a radical, we write the non-radical in front of the radical. That is:

$2\sqrt{5}$ means: multiply 2 and $\sqrt{5}$

To simplify the expressions below, we multiplied the non-radical factors.

$$2 \cdot 4\sqrt{7} = 8\sqrt{7}$$

$$y \cdot 6\sqrt[3]{y} = 6y\sqrt[3]{y}$$

Simplify these.

a) $3x \cdot 5\sqrt{x} =$ __________ b) $t^2 \cdot t^3\sqrt[4]{t^3} =$ __________

Answers: a) $(m - 1)^2\sqrt[3]{(m - 1)}$ b) $2(b+7)^2\sqrt{5(b + 7)}$

---

32. To simplify the expressions below, we multiplied the radical factors and then simplified the product if possible.

$$7\sqrt{3} \cdot \sqrt{2} = 7\sqrt{6}$$

$$a\sqrt[3]{x^2} \cdot \sqrt[3]{5x^2} = a\sqrt[3]{5x^4} = ax\sqrt[3]{5x}$$

Simplify these.

a) $9\sqrt{y^3} \cdot \sqrt{y^3} =$ __________ b) $at\sqrt[3]{t^2} \cdot \sqrt[3]{a^3t^2} =$ __________

Answers: a) $15x\sqrt{x}$ b) $t^5\sqrt[4]{t^3}$

---

33. To simplify the expressions below, we multiplied both the non-radical and the radical factors.

$$2\sqrt{3} \cdot 5\sqrt{6} = 2 \cdot 5 \cdot \sqrt{3} \cdot \sqrt{6} = 10\sqrt{18} = 30\sqrt{2}$$

$$a\sqrt[3]{b} \cdot a\sqrt[3]{b^4} = a \cdot a \cdot \sqrt[3]{b} \cdot \sqrt[3]{b^4} = a^2\sqrt[3]{b^5} = a^2b\sqrt[3]{b^2}$$

Simplify these.

a) $3\sqrt{5x} \cdot x\sqrt{2} =$ __________

b) $2p\sqrt[4]{q^3} \cdot p^2\sqrt[4]{q^2} =$ __________

Answers: a) $9y^3$ b) $a^2t^2\sqrt[3]{t}$

Answers: a) $3x\sqrt{10x}$ b) $2p^3q\sqrt[4]{q}$

5-3 DIVIDING RADICALS

In this section, we will discuss the procedure for dividing radicals. We will assume that all radicands represent positive numbers.

34. To divide radicals with the same index, we divide their radicands. That is:

$$\frac{\sqrt[k]{x}}{\sqrt[k]{y}} = \sqrt[k]{\frac{x}{y}}$$

The following two divisions show that the definition above makes sense.

$$\frac{\sqrt{36}}{\sqrt{4}} = \sqrt{\frac{36}{4}} \qquad \frac{\sqrt[3]{64}}{\sqrt[3]{8}} = \sqrt[3]{\frac{64}{8}}$$

$$\frac{6}{2} = \sqrt{9} \qquad \frac{4}{2} = \sqrt[3]{8}$$

$$3 = 3 \qquad 2 = 2$$

When dividing radicals, we simplify the quotient as much as possible. For example:

$$\frac{\sqrt{50}}{\sqrt{2}} = \sqrt{\frac{50}{2}} = \sqrt{25} = 5$$

$$\frac{\sqrt[3]{32}}{\sqrt[3]{2}} = \sqrt[3]{\frac{32}{2}} = \sqrt[3]{16} = 2\sqrt[3]{2}$$

Complete these.

a) $\frac{\sqrt{150}}{\sqrt{3}} =$ ________ b) $\frac{\sqrt[3]{128}}{\sqrt[3]{2}} =$ ________

35. In each division below, we simplified the quotient as much as possible.

a) $5\sqrt{2}$ b) 4

$$\frac{\sqrt{18x^5}}{\sqrt{2x^3}} = \sqrt{\frac{18x^5}{2x^3}} = \sqrt{9x^2} = 3x$$

$$\frac{\sqrt[4]{14a^5b^5}}{\sqrt[4]{2a}} = \sqrt[4]{\frac{14a^5b^5}{2a}} = \sqrt[4]{7a^4b^5} = ab\sqrt[4]{7b}$$

Simplify each quotient as much as possible.

a) $\frac{\sqrt{6p^6q^8}}{\sqrt{2pq}} =$ ________________

b) $\frac{\sqrt[3]{24x^4y^8}}{\sqrt[3]{3xy^2}} =$ ________________

a) $p^2q^3\sqrt{3pq}$

b) $2xy^2$

36. In each example below, we divided the non-radicals and the radicals.

$$\frac{15\sqrt[3]{x^4}}{3\sqrt[3]{x}} = \frac{15}{3}\sqrt[3]{\frac{x^4}{x}} = 5\sqrt[3]{x^3} = 5x$$

$$\frac{4\sqrt{xy^4}}{8\sqrt{xy}} = \frac{4}{8}\sqrt{\frac{xy^4}{xy}} = \frac{1}{2}\sqrt{y^3} = \frac{y}{2}\sqrt{y}$$

Complete these.

a) $\frac{10\sqrt[4]{12x}}{2\sqrt[4]{4x}}$ = __________

b) $\frac{8\sqrt{10a^2b^3}}{12\sqrt{5ab}}$ = __________

---

37. Notice how we got "1" as the numerator in the second example below.

a) $5\sqrt[4]{3}$

b) $\frac{2b}{3}\sqrt{2a}$

$$\frac{7\sqrt[3]{16x^2}}{\sqrt[3]{2x}} = 7\sqrt[3]{\frac{16x^2}{2x}} = 7\sqrt[3]{8x} = 7 \cdot 2\sqrt[3]{x} = 14\sqrt[3]{x}$$

$$\frac{\sqrt{32a^2b^2}}{2\sqrt{2ab}} = \frac{1}{2}\sqrt{\frac{32a^2b^2}{2ab}} = \frac{1}{2}\sqrt{16ab} = \frac{1}{2} \cdot 4\sqrt{ab} = 2\sqrt{ab}$$

Complete these.

a) $\frac{3\sqrt{10p^5q^8}}{\sqrt{2p^2q^2}}$ = __________

b) $\frac{\sqrt[3]{54cd}}{3\sqrt[3]{2}}$ = __________

---

38. To simplify a radical containing a fraction, we can reverse the procedure for dividing radicals. That is, we can convert the radical to a division of two radicals and then simplify both terms. For example:

a) $3pq^3\sqrt{5p}$

b) $\sqrt[3]{cd}$

$$\sqrt{\frac{1}{25}} = \frac{\sqrt{1}}{\sqrt{25}} = \frac{1}{5}$$

$$\sqrt[3]{\frac{8a^6}{b^3}} = \frac{\sqrt[3]{8a^6}}{\sqrt[3]{b^3}} = \frac{2a^2}{b}$$

Simplify these.

a) $\sqrt[3]{\frac{64}{27}}$ = ______ b) $\sqrt{\frac{49m^4}{81}}$ = ______

---

39. Two more simplifications are shown below.

a) $\frac{4}{3}$ b) $\frac{7m^2}{9}$

$$\sqrt{\frac{18y}{49}} = \frac{\sqrt{18y}}{\sqrt{49}} = \frac{3\sqrt{2y}}{7}$$

$$\sqrt[3]{\frac{125a^3}{64b^5}} = \frac{\sqrt[3]{125a^3}}{\sqrt[3]{64b^5}} = \frac{5a}{4b\sqrt[3]{b^2}}$$

Continued on following page.

39. Continued

Simplify these.

a) $\sqrt[3]{\frac{16c^4}{27d^6}}$ = ____________ b) $\sqrt{\frac{49}{20m}}$ = ____________

---

40. Simplify each radical.

a) $\sqrt{\frac{7}{4t^6}}$ = ____________ c) $\sqrt[3]{\frac{1}{ab^6}}$ = ____________

b) $\sqrt[3]{\frac{64}{3ax}}$ = ____________ d) $\sqrt{\frac{64x^3}{9y^2}}$ = ____________

Answers to 39: a) $\frac{2c\sqrt[3]{2c}}{3d^2}$ b) $\frac{7}{2\sqrt{5m}}$

---

41. After multiplying radicals with fractional radicands, we simplify the product. For example:

$$\sqrt{\frac{5}{2}} \cdot \sqrt{\frac{4}{5}} = \sqrt{\frac{20}{10}} = \sqrt{2}$$

$$\sqrt[3]{\frac{8}{3}} \cdot \sqrt[3]{\frac{x}{9}} = \sqrt[3]{\frac{8x}{27}} = \frac{2\sqrt[3]{x}}{3}$$

Simplify each product.

a) $\sqrt[3]{250} \cdot \sqrt[3]{\frac{1}{2}}$ = ____________ b) $\sqrt{\frac{y^2}{3}} \cdot \sqrt{\frac{y^3}{12}}$ = ____________

Answers to 40: a) $\frac{\sqrt{7}}{2t^3}$ b) $\frac{4}{\sqrt[3]{3ax}}$ c) $\frac{1}{b^2\sqrt[3]{a}}$ d) $\frac{8x\sqrt{x}}{3y}$

Answers to 41: a) 5 b) $\frac{y^2\sqrt{y}}{6}$

SELF-TEST 17 (pages 234-246)

Find each root.

1. $-\sqrt{49}$ =	2. $\sqrt[3]{-64}$ =	3. $\sqrt[4]{16}$ =	4. $-\sqrt[5]{6^5}$ =

Simplify.

5. $\sqrt{80x^4}$ =	6. $\sqrt[3]{27a^4b^8}$ =	7. $\sqrt[4]{16x - 16y}$ =

Multiply and simplify each product.

8. $\sqrt{2r^3} \cdot \sqrt{32s^4}$ =

9. $\sqrt[3]{9x^2y^2} \cdot \sqrt[3]{6xy^2}$ =

10. $\sqrt[4]{(a + b)^3} \cdot \sqrt[4]{(a + b)^2}$ =

11. $2\sqrt{5x} \cdot 4\sqrt{3x}$ =

Divide and simplify each quotient.

12. $\frac{\sqrt{60}}{\sqrt{3}}$ =

13. $\frac{4\sqrt[3]{12x^5}}{6\sqrt[3]{2x}}$ =

14. $\sqrt{\frac{27y}{64}}$ =

15. $\sqrt[3]{\frac{5a^5}{8b^6}}$ =

ANSWERS:

1. -7	5. $4x^2\sqrt{5}$	8. $8rs^2\sqrt{r}$	12. $2\sqrt{5}$
2. -4	6. $3ab^2\sqrt[3]{ab^2}$	9. $3xy\sqrt[3]{2y}$	13. $\frac{2x}{3}\sqrt[3]{6x}$
3. 2	7. $2\sqrt[4]{x-y}$	10. $(a+b)\sqrt[4]{a+b}$	14. $\frac{3\sqrt{3y}}{8}$
4. -6		11. $8x\sqrt{15}$	15. $\frac{a\sqrt[3]{5a^2}}{2b^2}$

5-4 ADDING AND SUBTRACTING RADICALS

In this section, we will discuss the procedure for adding and subtracting radicals. We will assume that all radicands represent positive numbers.

---

42. Two radical terms are called like terms if they contain like radicals. Like radicals are radicals with the same index and radicand. For example:

$2\sqrt{5}$ and $7\sqrt{5}$ are like terms because both contain $\sqrt{5}$.

$8\sqrt[3]{2x}$ and $5\sqrt[3]{2x}$ are like terms because both contain $\sqrt[3]{2x}$.

We use the distributive principle to add or subtract like radical terms. For example:

$$2\sqrt{5} + 7\sqrt{5} = (2+7)\sqrt{5} = 9\sqrt{5}$$

$$8\sqrt[3]{2x} - 5\sqrt[3]{2x} = (8-5)\sqrt[3]{2x} = 3\sqrt[3]{2x}$$

Notice that using the distributive principle is the same as adding or subtracting the coefficients of the radicals. Use that method for these.

a) $5\sqrt[3]{y} + 6\sqrt[3]{y} =$ ____________ b) $4\sqrt{3} - 9\sqrt{3} =$ ____________

Answers: a) $11\sqrt[3]{y}$ b) $-5\sqrt{3}$

43. If the coefficient of a radical is not explicitly shown, its coefficient is a "1". Therefore:

$$\sqrt{x} + 5\sqrt{x} = 1\sqrt{x} + 5\sqrt{x} = 6\sqrt{x}$$

$$4\sqrt[3]{7} - \sqrt[3]{7} = 4\sqrt[3]{7} - 1\sqrt[3]{7} = 3\sqrt[3]{7}$$

Complete these.

a) $9\sqrt{t} + \sqrt{t} =$ ____________ c) $\sqrt[3]{y} - 3\sqrt[3]{y} =$ ____________

b) $\sqrt[3]{3x} + \sqrt[3]{3x} =$ ____________ d) $\sqrt{10} - \sqrt{10} =$ ____________

Answers: a) $10\sqrt{t}$ b) $2\sqrt[3]{3x}$ c) $-2\sqrt[3]{y}$ d) 0

44. The same procedure is used when the radicand is a binomial. For example:

$$5\sqrt{x - 1} + 4\sqrt{x - 1} = 9\sqrt{x - 1}$$

$$8\sqrt[3]{a + b} - \sqrt[3]{a + b} = 7\sqrt[3]{a + b}$$

Complete these.

a) $\sqrt{y + 7} + 3\sqrt{y + 7} =$ ________________

b) $12\sqrt[4]{p + q} - 5\sqrt[4]{p + q} =$ ________________

Answers: a) $4\sqrt{y + 7}$ b) $7\sqrt[4]{p + q}$

45. Two radical terms cannot be combined into one term <u>if they do not contain like radicals</u>. For example:

$4\sqrt{3} + 5\sqrt{7}$ cannot be combined.

$8\sqrt[3]{x} - 3\sqrt[4]{x}$ cannot be combined.

Add or subtract if possible.

a) $\sqrt[3]{y} + 2\sqrt[3]{y} =$ __________

b) $5\sqrt{2} + \sqrt{3} =$ __________

c) $2\sqrt{7} - \sqrt{7} =$ __________

d) $8\sqrt[3]{x} - 5\sqrt[5]{x} =$ __________

Answers: a) $3\sqrt[3]{y}$ b) Not possible c) $\sqrt{7}$ d) Not possible

46. Following the example, use the distributive principle to combine like terms.

$$5\sqrt{3} - 2x\sqrt{3} + 6\sqrt{3} = (5 - 2x + 6)\sqrt{3} = (11 - 2x)\sqrt{3}$$

$6y\sqrt[3]{x^2} + 4\sqrt[3]{x^2} - y\sqrt[3]{x^2} =$ ________________________

Answer: $(5y + 4)\sqrt[3]{x^2}$

47. We simplified the expression below by combining the two like terms. Simplify the other expression.

$$5\sqrt[3]{2x} + 3\sqrt[3]{2x} - 7\sqrt[4]{2x} = 8\sqrt[3]{2x} - 7\sqrt[4]{2x}$$

$9\sqrt{5} - \sqrt{7} - 4\sqrt{5} =$ ________________________

Answer: $5\sqrt{5} - \sqrt{7}$

48. Sometimes we can add or subtract after simplifying by factoring out perfect powers. For example:

$$4\sqrt{20} + \sqrt{45} = 4 \cdot 2\sqrt{5} + 3\sqrt{5} = 8\sqrt{5} + 3\sqrt{5} = 11\sqrt{5}$$

Complete these.

a) $4\sqrt{75} - 2\sqrt{27} =$ ________________________

b) $2\sqrt[3]{16x} + \sqrt[3]{54x} =$ ________________________

Answers: a) $14\sqrt{3}$ b) $7\sqrt[3]{2x}$

49. We could add below after factoring out perfect powers.

$$\sqrt{5a^3} + \sqrt{45a} = a\sqrt{5a} + 3\sqrt{5a} = (a + 3)\sqrt{5a}$$

Complete these.

a) $2\sqrt[3]{24x^4} - 5\sqrt[3]{3x} =$ ____________________

b) $4\sqrt{y^3} - 3\sqrt{y} + 5\sqrt{y^4} =$ ____________________

---

50. Notice how we factored out perfect powers below and then added. Use the same steps for the subtraction.

$$\sqrt{x^3 + x^2} + \sqrt{9x + 9} = \sqrt{x^2(x + 1)} + \sqrt{9(x + 1)}$$
$$= x\sqrt{x + 1} + 3\sqrt{x + 1}$$
$$= (x + 3)\sqrt{x + 1}$$

$$\sqrt{12y - 12} - \sqrt{3y - 3} =$$

Answers to 49: a) $(4x - 5)\sqrt[3]{3x}$ b) $(4y - 3)\sqrt{y} + 5y^2$

---

51. A radical containing a binomial with two perfect-square terms <u>cannot be broken up into an addition or subtraction of two radicals</u>. For example:

$$\sqrt{36 + 64} \neq \sqrt{36} + \sqrt{64}$$

Since: $\sqrt{100} \neq 6 + 8$

$$10 \neq 14$$

Two more examples of the above statement are:

$$\sqrt{x^2 + 9} \neq \sqrt{x^2} + \sqrt{9}$$
$$\sqrt{a^2 - b^2} \neq \sqrt{a^2} - \sqrt{b^2}$$

Which of the following are true? __________

a) $\sqrt{9 + 16} = \sqrt{9} + \sqrt{16}$ c) $\sqrt{49 - 25} = \sqrt{24}$

b) $\sqrt{t^2 - 100} = \sqrt{t^2} - \sqrt{100}$ d) $\sqrt{p^2 + q^2} = \sqrt{p^2} + \sqrt{q^2}$

Answer to 50: $\sqrt{3y - 3}$

---

52. If the radicand is a <u>multiplication</u> with two perfect-square factors, it <u>can</u> be simplified. For example:

$$\sqrt{9x^2} = \sqrt{9} \cdot \sqrt{x^2} = 3x$$
$$\sqrt{a^2b^2} = \sqrt{a^2} \cdot \sqrt{b^2} = ab$$

Continued on following page.

Answer to 51: Only (c)

52. Continued

If the radicand is a binomial with two perfect-square terms, it cannot be simplified. For example:

$$\sqrt{x^2 + 9} \neq \sqrt{x^2} + \sqrt{9} \quad \text{or} \quad x + 3$$

$$\sqrt{a^2 - b^2} \neq \sqrt{a^2} - \sqrt{b^2} \quad \text{or} \quad a - b$$

Simplify if possible:

a) $\sqrt{y^2 - 25}$ = ________ c) $\sqrt{p^2 + q^2}$ = ________

b) $\sqrt{25y^2}$ = ________ d) $\sqrt{p^2q^2}$ = ________

---

a) Not possible b) 5y c) Not possible d) pq

5-5 MULTIPLICATIONS INVOLVING BINOMIALS

In this section, we will discuss multiplications involving binomials that contain one or two radicals. We will assume that all radicands represent positive numbers.

---

53. We used the distributive principle for the multiplication below.

$$\sqrt{2}(x - \sqrt{7}) = \sqrt{2} \cdot x - \sqrt{2} \cdot \sqrt{7}$$
$$= x\sqrt{2} - \sqrt{14}$$

Complete these.

a) $\sqrt{3}(y + \sqrt{5})$ = ________ b) $\sqrt{7}(\sqrt{2} - \sqrt{3})$ = ________

---

54. If we multiply a square root radical by itself, the product is the radicand. For example:

a) $y\sqrt{3} + \sqrt{15}$

b) $\sqrt{14} - \sqrt{21}$

$$\sqrt{7} \cdot \sqrt{7} = \sqrt{49} = 7$$

$$\sqrt{2x} \cdot \sqrt{2x} = \sqrt{4x^2} = 2x$$

Notice in the multiplication below that $\sqrt{3} \cdot \sqrt{3} = 3$

$$\sqrt{3}(5 + \sqrt{3}) = \sqrt{3} \cdot 5 + \sqrt{3} \cdot \sqrt{3} = 5\sqrt{3} + 3$$

Complete these.

a) $\sqrt{7}(\sqrt{7} - 2)$ = ________ b) $\sqrt{2}(\sqrt{3} + \sqrt{2})$ = ________

---

a) $7 - 2\sqrt{7}$

b) $\sqrt{6} + 2$

55. Notice how we simplified the product below. Do the other multiplication.

$\sqrt{2}(3\sqrt{5} - 2\sqrt{6}) = \sqrt{2} \cdot 3\sqrt{5} - \sqrt{2} \cdot 2\sqrt{6}$ $\qquad$ $\sqrt{3}(2\sqrt{7} + 4\sqrt{6}) =$

$= 3\sqrt{10} - 2\sqrt{12}$

$= 3\sqrt{10} - 2 \cdot 2\sqrt{3}$

$= 3\sqrt{10} - 4\sqrt{3}$

---

56. Notice how we simplified the product below. Do the other multiplication.

Answer: $2\sqrt{21} + 12\sqrt{2}$

$\sqrt[3]{5}(\sqrt[3]{25} + 7\sqrt[3]{6}) = \sqrt[3]{5} \cdot \sqrt[3]{25} + \sqrt[3]{5} \cdot 7\sqrt[3]{6}$ $\qquad$ $\sqrt[3]{4}(3\sqrt[3]{16} - \sqrt[3]{2}) =$

$= \sqrt[3]{125} + 7\sqrt[3]{30}$

$= 5 + 7\sqrt[3]{30}$

---

57. Squaring a square root radical is the same as multiplying it by itself. Therefore, its square is the radicand. For example:

Answer: 10

$(\sqrt{5})^2 = 5$, since $(\sqrt{5})^2 = \sqrt{5} \cdot \sqrt{5} = \sqrt{25} = 5$

$(\sqrt{3y})^2 = 3y$, since $(\sqrt{3y})^2 = \sqrt{3y} \cdot \sqrt{3y} = \sqrt{9y^2} = 3y$

The multiplication below fits the pattern $(a + b)(a - b) = a^2 - b^2$. Notice that $(\sqrt{6})^2 = 6$ and $(\sqrt{2})^2 = 2$. Complete the other multiplications.

$(\sqrt{6} + \sqrt{2})(\sqrt{6} - \sqrt{2}) = (\sqrt{6})^2 - (\sqrt{2})^2 = 6 - 2 = 4$

a) $(\sqrt{8} + \sqrt{5})(\sqrt{8} - \sqrt{5}) =$ ____________

b) $(\sqrt{3} + \sqrt{7})(\sqrt{3} - \sqrt{7}) =$ ____________

---

58. Following the example, complete the other multiplications.

Answers: a) 3 b) -4

$(\sqrt{x} + \sqrt{y})(\sqrt{x} - \sqrt{y}) = (\sqrt{x})^2 - (\sqrt{y})^2 = x - y$

a) $(\sqrt{p} - \sqrt{q})(\sqrt{p} + \sqrt{q}) =$ ____________

b) $(3 + \sqrt{5})(3 - \sqrt{5}) =$ ____________

---

59. Earlier we saw that $(ab)^2 = a^2b^2$. Therefore, to square $2\sqrt{5}$ below, we squared both 2 and $\sqrt{5}$.

Answers: a) $p - q$ b) 4

$$(2\sqrt{5})^2 = 2^2 \cdot (\sqrt{5})^2 = 4 \cdot 5 = 20$$

Notice how we squared $3\sqrt{10}$ in the multiplication below. Complete the other multiplication.

$(3\sqrt{10} + \sqrt{70})(3\sqrt{10} - \sqrt{70}) = (3\sqrt{10})^2 - (\sqrt{70})^2$ $\qquad$ $(\sqrt{40} - 5\sqrt{2})(\sqrt{40} + 5\sqrt{2}) =$

$= 3^2 \cdot (\sqrt{10})^2 - 70$

$= 9 \cdot 10 - 70$

$= 90 - 70$

$= 20$

60. We used the FOIL method for the multiplication below. Complete the other multiplication. $(\sqrt{5}+\sqrt{3})(\sqrt{2}-\sqrt{7}) = \sqrt{5}\cdot\sqrt{2}-\sqrt{5}\cdot\sqrt{7}+\sqrt{3}\cdot\sqrt{2}-\sqrt{3}\cdot\sqrt{7}$ $= \sqrt{10}-\sqrt{35}+\sqrt{6}-\sqrt{21}$ $(\sqrt{x}+\sqrt{7})(\sqrt{y}+\sqrt{2}) =$	-10
61. Notice how we simplified the product below. Do the other multiplication. $(3-\sqrt{a})(2-\sqrt{a}) = 6-3\sqrt{a}-2\sqrt{a}+(\sqrt{a})^2$ $= 6-5\sqrt{a}+a$ $(4-\sqrt{2})(3+\sqrt{2}) =$	$\sqrt{xy}+\sqrt{2x}+\sqrt{7y}+\sqrt{14}$
62. Following the example, do the other multiplication. $(\sqrt{5}+2)(4\sqrt{5}+3) = 4(\sqrt{5})^2+3\sqrt{5}+8\sqrt{5}+6$ $= 4\cdot5+11\sqrt{5}+6$ $= 20+11\sqrt{5}+6$ $= 26+11\sqrt{5}$ $(10\sqrt{3}+7)(\sqrt{3}-3) =$	$10+\sqrt{2}$
63. Notice how we simplified below. Do the other multiplication. $(2\sqrt{5}-6\sqrt{2})(3\sqrt{5}+9\sqrt{2}) = 6(\sqrt{5})^2+18\sqrt{5}\cdot\sqrt{2}-18\sqrt{2}\cdot\sqrt{5}-54(\sqrt{2})^2$ $= 6\cdot5+18\sqrt{10}-18\sqrt{10}-54\cdot2$ $= 30-108$ $= -78$ $(5\sqrt{2}+2\sqrt{3})(3\sqrt{2}-4\sqrt{3}) =$	$9-23\sqrt{3}$

64. To square the binomial below, we used the principle $(a + b)^2 = a^2 + 2ab + b^2$.

$$(3 + \sqrt{5})^2 = 3^2 + 2(3)(\sqrt{5}) + (\sqrt{5})^2$$
$$= 9 + 6\sqrt{5} + 5$$
$$= 14 + 6\sqrt{5}$$

$(\sqrt{7} + 2)^2 =$

$6 - 14\sqrt{6}$

$11 + 4\sqrt{7}$

5-6 RATIONALIZING DENOMINATORS

When a fraction contains a radical in its denominator, we can convert it to an equivalent form without a radical in its denominator. The process of doing so is called rationalizing the denominator. We will discuss that process in this section. We will assume that all radicands represent positive numbers.

---

65. To rationalize the denominator below, we multiplied the fraction by $\frac{\sqrt{7}}{\sqrt{7}}$. Notice that we got rid of the radical in the denominator by multiplying it by itself.

$$\frac{5}{\sqrt{7}} = \frac{5}{\sqrt{7}}\left(\frac{\sqrt{7}}{\sqrt{7}}\right) = \frac{5\sqrt{7}}{7}$$

$\sqrt{7}$ is not a rational number (the denominator $\sqrt{7}$)

7 is a rational number (the denominator 7)

Rationalize each denominator.

a) $\frac{1}{\sqrt{15}} =$ ____________________

b) $\frac{3}{\sqrt{2y}} =$ ____________________

---

66. Following the example, rationalize the other denominators.

$$\frac{x}{2\sqrt{y}} = \frac{x}{2\sqrt{y}}\left(\frac{\sqrt{y}}{\sqrt{y}}\right) = \frac{x\sqrt{y}}{2y}$$

a) $\frac{3}{8\sqrt{5}} =$ ____________________

b) $\frac{1}{2\sqrt{7x}} =$ ____________________

a) $\frac{\sqrt{15}}{15}$

b) $\frac{3\sqrt{2y}}{2y}$

---

a) $\frac{3\sqrt{5}}{40}$ b) $\frac{\sqrt{7x}}{14x}$

67. When a denominator is rationalized, sometimes the new fraction can be reduced to lowest terms. For example:

$$\frac{5}{2\sqrt{5}} = \frac{5}{2\sqrt{5}}\left(\frac{\sqrt{5}}{\sqrt{5}}\right) = \frac{\cancel{5}\sqrt{5}}{2(\cancel{5})} = \frac{\sqrt{5}}{2}$$

Rationalize each denominator and then reduce to lowest terms.

a) $\frac{3}{\sqrt{3}}$ = ______________

b) $\frac{c}{7\sqrt{c}}$ = ______________

---

68. After simplifying a radical containing a fraction, we can frequently rationalize the denominator of the new fraction. For example:

$$\sqrt{\frac{c^2x}{d^2y}} = \frac{c\sqrt{x}}{d\sqrt{y}} = \frac{c\sqrt{x}}{d\sqrt{y}}\left(\frac{\sqrt{y}}{\sqrt{y}}\right) = \frac{c\sqrt{xy}}{dy}$$

Simplify and then rationalize the denominator of the new fraction.

a) $\sqrt{\frac{8}{27}}$ = ______________

b) $\sqrt{\frac{m^2}{bt^4}}$ = ______________

Answers to 67: a) $\sqrt{3}$ b) $\frac{\sqrt{c}}{7}$

---

69. Even when we cannot simplify a radical containing a fraction, we can still rationalize its denominator. For example:

$$\sqrt{\frac{5}{7}} = \frac{\sqrt{5}}{\sqrt{7}} = \frac{\sqrt{5}}{\sqrt{7}}\left(\frac{\sqrt{7}}{\sqrt{7}}\right) = \frac{\sqrt{35}}{7}$$

Rationalize each denominator.

a) $\sqrt{\frac{3}{10}}$ = ______________

b) $\sqrt{\frac{5}{2x}}$ = ______________

Answers to 68: a) $\frac{2\sqrt{6}}{9}$ b) $\frac{m\sqrt{b}}{bt^2}$

---

70. The <u>conjugate</u> of a binomial containing a square root radical is a binomial with the same terms but with the opposite sign. For example:

The conjugate of $2 + \sqrt{3}$ is $2 - \sqrt{3}$.

The conjugate of $5 - \sqrt{7}$ is $5 + \sqrt{7}$.

The conjugate of $\sqrt{5} + \sqrt{2}$ is $\sqrt{5} - \sqrt{2}$.

Continued on following page.

Answers to 69: a) $\frac{\sqrt{30}}{10}$ b) $\frac{\sqrt{10x}}{2x}$

70. Continued

To rationalize the denominator below, we multiplied the fraction by $\frac{5+\sqrt{2}}{5 \div \sqrt{2}}$. Notice that we got rid of the radical in the denominator by multiplying it by its conjugate.

$$\frac{3}{5-\sqrt{2}} = \frac{3}{5-\sqrt{2}}\left(\frac{5+\sqrt{2}}{5+\sqrt{2}}\right) = \frac{3(5+\sqrt{2})}{25-2} = \frac{15+3\sqrt{2}}{23}$$

Rationalize this denominator.

$$\frac{1}{3+\sqrt{5}} =$$

---

71. After rationalizing the denominator below, we reduced to lowest terms. Rationalize the other denominator.

$\frac{3-\sqrt{5}}{4}$

$$\frac{4}{3+\sqrt{7}} = \frac{4}{3+\sqrt{7}}\left(\frac{3-\sqrt{7}}{3-\sqrt{7}}\right) = \frac{4(3-\sqrt{7})}{9-7} = \frac{\overset{2}{\cancel{4}}(3-\sqrt{7})}{\cancel{2}} = 6 - 2\sqrt{7}$$

$$\frac{6}{\sqrt{3}-1} =$$

---

72. After rationalizing the denominator below, we reduced to lowest terms. Rationalize the other denominator.

$3\sqrt{3} + 3$

$$\frac{2}{\sqrt{3}-\sqrt{5}} = \frac{2}{\sqrt{3}-\sqrt{5}}\left(\frac{\sqrt{3}+\sqrt{5}}{\sqrt{3}+\sqrt{5}}\right)$$

$$= \frac{2(\sqrt{3}+\sqrt{5})}{3-5}$$

$$= \frac{2(\sqrt{3}+\sqrt{5})}{-2}$$

$$= \left(\frac{2}{-2}\right)(\sqrt{3}+\sqrt{5})$$

$$= (-1)(\sqrt{3}+\sqrt{5})$$

$$= -\sqrt{3}-\sqrt{5}$$

$$\frac{6}{\sqrt{3}+\sqrt{6}} =$$

---

73. Following the example, rationalize the other denominator.

$-2\sqrt{3} + 2\sqrt{6}$

$$\frac{\sqrt{5}+\sqrt{2}}{\sqrt{5}-\sqrt{2}} = \frac{\sqrt{5}+\sqrt{2}}{\sqrt{5}-\sqrt{2}}\left(\frac{\sqrt{5}+\sqrt{2}}{\sqrt{5}+\sqrt{2}}\right) = \frac{(\sqrt{5})^2 + 2\sqrt{5}\cdot\sqrt{2} + (\sqrt{2})^2}{5-2}$$

$$= \frac{5 + 2\sqrt{10} + 2}{3}$$

$$= \frac{7 + 2\sqrt{10}}{3}$$

Continued on following page.

73. Continued

$$\frac{\sqrt{a} - \sqrt{b}}{\sqrt{a} + \sqrt{b}} =$$

---

74. To rationalize the denominator below, we multiplied by $\frac{\sqrt[3]{2}}{\sqrt[3]{2}}$ to get a perfect cube as the radicand in the denominator. Rationalize the other denominators.

$$\frac{3}{\sqrt[3]{4}} = \frac{3}{\sqrt[3]{4}}\left(\frac{\sqrt[3]{2}}{\sqrt[3]{2}}\right) = \frac{3\sqrt[3]{2}}{\sqrt[3]{8}} = \frac{3\sqrt[3]{2}}{2}$$

a) $\frac{5}{\sqrt[3]{9}} =$

b) $\frac{7}{\sqrt[3]{2}} =$

Answer: $\frac{a - 2\sqrt{ab} + b}{a - b}$

---

75. Following the example, rationalize the other denominator.

$$\sqrt[3]{\frac{11}{25}} = \frac{\sqrt[3]{11}}{\sqrt[3]{25}}\left(\frac{\sqrt[3]{5}}{\sqrt[3]{5}}\right) = \frac{\sqrt[3]{55}}{\sqrt[3]{125}} = \frac{\sqrt[3]{55}}{5}$$

$$\sqrt[3]{\frac{x^6}{y}} =$$

Answers: a) $\frac{5\sqrt[3]{3}}{3}$ b) $\frac{7\sqrt[3]{4}}{2}$

---

76. To rationalize the denominator below, we multiplied by $\frac{\sqrt[3]{3x^2}}{\sqrt[3]{3x^2}}$ to get a perfect cube as the radicand in the denominator. Rationalize the other denominator.

$$\sqrt[3]{\frac{5y^5}{9x^4}} = \frac{y\sqrt[3]{5y^2}}{x\sqrt[3]{9x}}\left(\frac{\sqrt[3]{3x^2}}{\sqrt[3]{3x^2}}\right) = \frac{y\sqrt[3]{15x^2y^2}}{x\sqrt[3]{27x^3}} = \frac{y\sqrt[3]{15x^2y^2}}{x(3x)} = \frac{y\sqrt[3]{15x^2y^2}}{3x^2}$$

$$\sqrt[3]{\frac{a^9}{2b^2}} =$$

Answer: $\frac{x^2\sqrt[3]{y^2}}{y}$

Answer: $\frac{a\sqrt[3]{4b}}{2b}$

SELF-TEST 18 (pages 247-257)

Add or subtract.

1. $\sqrt{x + 1} + 2\sqrt{x + 1}$

2. $5\sqrt[3]{y} - \sqrt[3]{y}$

3. $8\sqrt{3} - \sqrt{5} - 3\sqrt{3}$

4. $\sqrt[3]{7a^4} + \sqrt[3]{56a}$

Multiply. Simplify each product.

5. $\sqrt{5}(x - \sqrt{2})$

6. $\sqrt[3]{3}(\sqrt[3]{9} + 2\sqrt[3]{16}$

7. $(\sqrt{7} + \sqrt{3})(\sqrt{7} - \sqrt{3})$

8. $(5\sqrt{3} + 4\sqrt{2})(3\sqrt{3} - 2\sqrt{2})$

Rationalize each denominator. Report each answer in lowest terms.

9. $\dfrac{3}{4\sqrt{3}}$

10. $\sqrt{\dfrac{x^4}{ay^2}}$

11. $\dfrac{3}{\sqrt{5} - \sqrt{2}}$

12. $\sqrt[3]{\dfrac{a^3}{8b^2}}$

ANSWERS:

1. $3\sqrt{x + 1}$
2. $4\sqrt[3]{y}$
3. $5\sqrt{3} - \sqrt{5}$
4. $(a + 2)\sqrt[3]{7a}$
5. $x\sqrt{5} - \sqrt{10}$
6. $3 + 4\sqrt[3]{6}$
7. 4
8. $29 + 2\sqrt{6}$
9. $\dfrac{\sqrt{3}}{4}$
10. $\dfrac{x^2\sqrt{a}}{ay}$
11. $\sqrt{5} + \sqrt{2}$
12. $\dfrac{a\sqrt[3]{b}}{2b}$

5-7 RADICAL EQUATIONS

In this section, we will discuss a method for solving radical equations.

---

77. A radical equation is an equation in which the variable appears in a radicand. For example:

$\sqrt{x + 1} = x - 4$ $\sqrt[3]{5x} + 2 = 7$

To solve a radical equation, we must get rid of the radical. To do so, we use the POWER PRINCIPLE FOR EQUATIONS. That principle says:

> **The Power Principle For Equations**
>
> If we raise both sides of a true equation to the same power, the new equation is true. That is:
>
> If $a = b$
>
> Then $a^n = b^n$

The equation below contains a square root radical on one side. To get rid of the radical, we squared both sides. Remember that the square of a square root radical is the radicand. That is, $(\sqrt{2x})^2 = 2x$. Use the same steps to solve the other equation.

$\sqrt{2x} = 4$

$(\sqrt{2x})^2 = (4)^2$

$2x = 16$

$x = 8$

$\sqrt{x - 1} = 9$

Answer: $(\sqrt{x - 1})^2 = (9)^2$

$x - 1 = 81$

$x = 82$

---

78. When the radical is isolated on one side, only two steps are needed to solve a radical equation. They are:

1. Use the power principle to eliminate the radical.

2. Then solve the resulting non-radical equation.

We used the two steps to solve the equation on the following page and then checked the solution. Solve the other equation and check your solution.

Continued on following page.

78. Continued

$$\sqrt{\frac{3x}{4}} = 6 \qquad 1 = \sqrt{\frac{7}{2y}}$$

$$\left(\sqrt{\frac{3x}{4}}\right)^2 = (6)^2$$

$$\frac{3x}{4} = 36$$

$$3x = 144$$

$$x = 48$$

Check

$$\sqrt{\frac{3x}{4}} = 6$$

$$\sqrt{\frac{3(\overset{12}{\cancel{48}})}{\cancel{4}}} = 6$$

$$\sqrt{36} = 6$$

$$6 = 6$$

---

79. Some radical equations have no solution. For example, we get 25 as a solution below, but 25 does not check.

$$\sqrt{x} = -5$$

$$(\sqrt{x})^2 = (-5)^2$$

$$x = 25$$

Check

$$\sqrt{x} = -5$$

$$\sqrt{25} = -5$$

$$5 \neq -5$$

The equation above has no solution because a principal square root cannot be negative. The same applies to the equation below. Solve it and show that the obtained solution does not check.

$$\sqrt{x+1} = -2$$

Check

Answer to 78: $y = \frac{7}{2}$

Check:

$$1 = \sqrt{\frac{7}{\cancel{2}\left(\frac{7}{\cancel{2}}\right)}}$$

$$1 = \sqrt{\frac{7}{7}}$$

$$1 = \sqrt{1}$$

$$1 = 1$$

Answer to 79: $x = 3$ does not check, since:

$$\sqrt{3+1} \neq -2$$

$$\sqrt{4} \neq -2$$

$$2 \neq -2$$

80. When solving a radical equation, it is important to check the solution because some radical equations do not have solutions. Following the example, solve and check the other equation below.

$$\frac{1}{3} = \sqrt{\frac{1}{d}}$$

$$\left(\frac{1}{3}\right)^2 = \left(\sqrt{\frac{1}{d}}\right)^2$$

$$\frac{1}{9} = \frac{1}{d}$$

$$9d\left(\frac{1}{9}\right) = 9d\left(\frac{1}{d}\right)$$

$$d = 9$$

Check

$$\frac{1}{3} = \sqrt{\frac{1}{d}}$$

$$\frac{1}{3} = \sqrt{\frac{1}{9}}$$

$$\frac{1}{3} = \frac{1}{3}$$

$$\sqrt{\frac{2p}{9}} = \frac{4}{3}$$

81. When the radical is not isolated, we isolate the radical before using the power principle. An example is shown. Use the same steps to solve the other equation.

$$\sqrt{T + 10} + 5 = 10$$

$$\sqrt{T + 10} = 5$$

$$(\sqrt{T + 10})^2 = 5^2$$

$$T + 10 = 25$$

$$T = 15$$

$$\sqrt{y + 1} - 3 = 4$$

$p = 8$

Check:

$$\sqrt{\frac{2(8)}{9}} = \frac{4}{3}$$

$$\sqrt{\frac{16}{9}} = \frac{4}{3}$$

$$\frac{4}{3} = \frac{4}{3}$$

82. The equation below contains a radical on each side. We got rid of both radicals in one step by squaring both sides. Solve and check the other equation.

$$\sqrt{3x - 7} = \sqrt{x + 1}$$

$$(\sqrt{3x - 7})^2 = (\sqrt{x + 1})^2$$

$$3x - 7 = x + 1$$

$$2x = 8$$

$$x = 4$$

Check

$$\sqrt{3x - 7} = \sqrt{x + 1}$$

$$\sqrt{3(4) - 7} = \sqrt{4 + 1}$$

$$\sqrt{12 - 7} = \sqrt{5}$$

$$\sqrt{5} = \sqrt{5}$$

$$\sqrt{5y - 9} = \sqrt{4y - 2}$$

$y = 48$

83. Notice how we squared $y - 1$ below. Notice also how we factored and used the principle of zero products. We got two solutions.

$$y - 1 = \sqrt{4y - 7}$$

$$(y - 1)^2 = (\sqrt{4y - 7})^2$$

$$y^2 - 2y + 1 = 4y - 7$$

$$y^2 - 6y + 8 = 0$$

$$(y - 2)(y - 4) = 0$$

$y - 2 = 0$	$y - 4 = 0$
$y = 2$	$y = 4$

The two solutions are 2 and 4. We checked 2 below. Check 4 in the space provided.

Checking 2	Checking 4
$y - 1 = \sqrt{4y - 7}$	$y - 1 = \sqrt{4y - 7}$
$2 - 1 = \sqrt{4(2) - 7}$	
$1 = \sqrt{8 - 7}$	
$1 = \sqrt{1}$	
$1 = 1$	

$y = 7$

Check:

$\sqrt{5(7)-9} = \sqrt{4(7)-2}$

$\sqrt{35 - 9} = \sqrt{28 - 2}$

$\sqrt{26} = \sqrt{26}$

---

84. Sometimes when we get two solutions, one of them does not satisfy the original radical equation. A solution of that type is called an <u>extraneous</u> solution. For example, we got two solutions below.

$$\sqrt{x + 3} = x + 1$$

$$(\sqrt{x + 3})^2 = (x + 1)^2$$

$$x + 3 = x^2 + 2x + 1$$

$$x^2 + x - 2 = 0$$

$$(x - 1)(x + 2) = 0$$

$x - 1 = 0$	$x + 2 = 0$
$x = 1$	$x = -2$

The two solutions for $x^2 + x - 2 = 0$ are "1" and -2. We checked both solutions in the original radical equation below.

Checking "1"	Checking -2
$\sqrt{1 + 3} = 1 + 1$	$\sqrt{-2 + 3} \neq -2 + 1$
$\sqrt{4} = 2$	$\sqrt{1} \neq -1$
$2 = 2$	$1 \neq -1$

Of the two solutions, only "1" satisfies the original equation. Therefore, -2 is an <u>extraneous</u> solution. -2 is not a solution of the original equation.

$4 - 1 = \sqrt{4(4) - 7}$

$3 = \sqrt{9}$

$3 = 3$

85. Whenever you get two solutions for a radical equation, <u>the only way to detect an extraneous solution is to check both solutions in the original equation</u>. Solve the equation below and check both solutions.

$$m - 5 = \sqrt{3m - 5}$$

a) The two obtained solutions are ________ and ________.

b) Is either solution an extraneous solution? __________

---

a) 3 and 10

b) Yes, 3 is extraneous

---

86. When an equation contains two radical terms, we use the following steps to solve it.

1. Isolate one of the radicals.

2. Use the power principle to eliminate that radical.

3. If a radical remains, isolate it and use the power principle a second time.

We used the steps above to solve the equation below. Notice that the power principle was used twice. Notice also how the pattern $(a - b)^2 = a^2 - 2ab + b^2$ is used to square $9 - \sqrt{x}$. The steps are:

$$(9 - \sqrt{x})^2 = (9)^2 - 2(9)(\sqrt{x}) + (\sqrt{x})^2$$

$$= 81 - 18\sqrt{x} + x$$

Use the same steps to solve the other equation.

$$\sqrt{x - 9} + \sqrt{x} = 9$$

$$\sqrt{x - 9} = 9 - \sqrt{x}$$

$$(\sqrt{x - 9})^2 = (9 - \sqrt{x})^2$$

$$x - 9 = 81 - 18\sqrt{x} + x$$

$$-90 = -18\sqrt{x}$$

$$\sqrt{x} = \frac{-90}{-18}$$

$$\sqrt{x} = 5$$

$$x = 25$$

$$\sqrt{x + 16} - \sqrt{x} = 2$$

87. We used the power principle twice to solve the equation below. The pattern $(a - b)^2 = a^2 - 2ab + b^2$ was also used to square $4 - \sqrt{x - 4}$. The steps are:

x = 9

$$(4 - \sqrt{x - 4})^2 = (4)^2 - 2(4)(\sqrt{x - 4}) + (\sqrt{x - 4})^2$$
$$= 16 - 8\sqrt{x - 4} + (x - 4)$$

Use the same steps to solve the other equation.

$$\sqrt{x + 4} + \sqrt{x - 4} = 4 \qquad\qquad \sqrt{x + 6} - \sqrt{x - 6} = 2$$

$$\sqrt{x + 4} = 4 - \sqrt{x - 4}$$

$$(\sqrt{x + 4})^2 = (4 - \sqrt{x - 4})^2$$

$$x + 4 = 16 - 8\sqrt{x - 4} + (x - 4)$$

$$-8 = -8\sqrt{x - 4}$$

$$\sqrt{x - 4} = \frac{-8}{-8}$$

$$\sqrt{x - 4} = 1$$

$$x - 4 = 1$$

$$x = 5$$

---

88. We used the power principle twice to solve the equation below. Notice how we used the pattern $(ab)^2 = a^2b^2$ to square $2\sqrt{2x - 1}$. The steps are:

x = 10

$$(2\sqrt{2x - 1})^2 = 2^2(\sqrt{2x - 1})^2 = 4(2x - 1) = 8x - 4$$

Use the same steps to solve the other equation.

$$\sqrt{3x + 1} - \sqrt{2x - 1} = 1 \qquad\qquad \sqrt{3y + 4} - \sqrt{2y - 4} = 2$$

$$\sqrt{3x + 1} = 1 + \sqrt{2x - 1}$$

$$(\sqrt{3x + 1})^2 = (1 + \sqrt{2x - 1})^2$$

$$3x + 1 = 1 + 2\sqrt{2x - 1} + (2x - 1)$$

$$x + 1 = 2\sqrt{2x - 1)}$$

$$(x + 1)^2 = (2\sqrt{2x - 1})^2$$

$$x^2 + 2x + 1 = 4(2x - 1)$$

$$x^2 + 2x + 1 = 8x - 4$$

$$x^2 - 6x + 5 = 0$$

$$(x - 1)(x - 5) = 0$$

$$x - 1 = 0 \quad \Big| \quad x - 5 = 0$$
$$x = 1 \quad \Big| \quad x = 5$$

Both roots check.

89. If we cube a cube root radical, we get the radicand. For example:

$$(\sqrt[3]{2x})^3 = 2x, \quad \text{since } (\sqrt[3]{2x})^3 = \sqrt[3]{2x} \cdot \sqrt[3]{2x} \cdot \sqrt[3]{2x} = \sqrt[3]{8x^3} = 2x$$

Therefore, to eliminate the cube root radical below, we cubed each side. Solve the other equation.

$$\sqrt[3]{x - 3} = 2 \qquad\qquad \sqrt[3]{2y} = 4$$

$$(\sqrt[3]{x - 3})^3 = (2)^3$$

$$x - 3 = 8$$

$$x = 11$$

y = 4 and 20

---

90. If we raise a fourth root radical to the fourth power, we get the radicand. For example.

$$(\sqrt[4]{3x})^4 = 3x, \quad \text{since } (\sqrt[4]{3x})^4 = \sqrt[4]{3x} \cdot \sqrt[4]{3x} \cdot \sqrt[4]{3x} \cdot \sqrt[4]{3x} = \sqrt[4]{81x^4} = 3x$$

Therefore to eliminate the fourth root radical below, we raised each side to the fourth power. Solve the other equation.

$$\sqrt[4]{3x + 4} = 2 \qquad\qquad \sqrt[4]{9y} = 3$$

$$(\sqrt[4]{3x + 4})^4 = (2)^4$$

$$3x + 4 = 16$$

$$3x = 12$$

$$x = 4$$

y = 32

---

91. Before using the power principle, we had to isolate the radical below. Solve the other equation.

$$\sqrt[4]{x + 7} - 2 = 0 \qquad\qquad \sqrt[3]{2x - 3} - 1 = 2$$

$$\sqrt[4]{x + 7} = 2$$

$$(\sqrt[4]{x + 7})^4 = (2)^4$$

$$x + 7 = 16$$

$$x = 9$$

y = 9

---

x = 15

5-8 RADICAL FORMULAS

In this section, we will do some evaluations and rearrangements with formulas containing radicals.

---

92. To do the evaluations below, we substituted, simplified, and then found the square root.

In $v = \sqrt{2as}$, find $\underline{v}$ when $a = 3$ and $s = 6$.

$$v = \sqrt{2as} = \sqrt{2(3)(6)} = \sqrt{36} = 6$$

In $t = \sqrt{\frac{2s}{g}}$, find $\underline{t}$ when $s = 64$ and $g = 32$.

$$t = \sqrt{\frac{2s}{g}} = \sqrt{\frac{2(64)}{32}} = \sqrt{\frac{128}{32}} = \sqrt{4} = 2$$

Following the example, do these evaluations.

a) In $R = \sqrt{Z^2 - X^2}$, find R when $Z = 10$ and $X = 8$.

R = ______________________

b) In $N = K\sqrt{\frac{V}{H}}$, find N when $K = 7$, $V = 18$, and $H = 2$.

N = ______________________

---

a) $R = 6$

b) $N = 21$

93. To perform the evaluation below, we had to solve a radical equation. Do the other evaluation.

In the formula below, find A when $s = 10$.

$$s = \sqrt{A}$$

$$10 = \sqrt{A}$$

$$10^2 = (\sqrt{A})^2$$

$$A = 100$$

In the formula below, find V when $s = 5$.

$$s = \sqrt[3]{V}$$

$V = 125$

94. We also had to solve a radical equation to do the evaluation below. Complete the other evaluation.

In the formula below, find $\underline{s}$ when $v = 8$ and $a = 2$.

$$v = \sqrt{2as}$$

$$8 = \sqrt{2(2)s}$$

$$8 = \sqrt{4s}$$

$$64 = 4s$$

$$s = 16$$

In the formula below, find R when $E = 9$ and $P = 3$.

$$E = \sqrt{PR}$$

$R = 27$

95. Following the example, complete the other evaluation.

In the formula below, find R when I = 6 and P = 72.

$$I = \sqrt{\frac{P}{R}}$$

$$6 = \sqrt{\frac{72}{R}}$$

$$36 = \frac{72}{R}$$

$$36R = 72$$

$$R = 2$$

In the formula below, find s when t = 10 and g = 3.

$$t = \sqrt{\frac{2s}{g}}$$

---

96. Following the example, complete the other evaluation.

s = 150

In the formula below, find g_p when $g_o = 4$, $r_o = 60$, and $r_p = 12$.

$$\sqrt{\frac{g_p}{g_o}} = \frac{r_o}{r_p}$$

$$\sqrt{\frac{g_p}{4}} = \frac{60}{12}$$

$$\sqrt{\frac{g_p}{4}} = 5$$

$$\frac{g_p}{4} = 25$$

$$g_p = 100$$

In the formula below, find S when T = 45, V = 30, and D = 10.

$$\sqrt{\frac{T}{S}} = \frac{V}{D}$$

---

97. To solve for P in the formula below, we squared both sides to eliminate the radical and then proceeded in the usual way. Solve for g in the other formula.

S = 5

$$I = \sqrt{\frac{P}{R}}$$

$$(I)^2 = \left(\sqrt{\frac{P}{R}}\right)^2$$

$$I^2 = \frac{P}{R}$$

$$R(I^2) = \cancel{R}\left(\frac{P}{\cancel{R}}\right)$$

$$P = I^2R$$

$$t = \sqrt{\frac{2s}{g}}$$

$$g = \frac{2s}{t^2}$$

98. We solved for s below. Solve for b in the other formula.

$$v = \sqrt{2as} \qquad a = \sqrt{b + c}$$

$$v^2 = 2as$$

$$s = \frac{v^2}{2a}$$

$b = a^2 - c$

99. We solved for b_o below. Solve for A_1 in the other formula.

$$\sqrt{\frac{b_p}{b_o}} = \frac{d_o}{d_p} \qquad \sqrt{\frac{A_1}{A_2}} = \frac{B_2}{B_1}$$

$$\frac{b_p}{b_o} = \frac{d_o^2}{d_p^2}$$

$$b_o d_p^2\left(\frac{b_p}{b_o}\right) = b_o d_p^2\left(\frac{d_o^2}{d_p^2}\right)$$

$$b_p d_p^2 = b_o d_o^2$$

$$b_o = \frac{b_p d_p^2}{d_o^2}$$

$$A_1 = \frac{A_2 B_2^2}{B_1^2}$$

100. To solve for V below, we isolated the radical before squaring. Solve for g in the other formula.

$$N = K\sqrt{\frac{V}{H}} \qquad T = 2\pi\sqrt{\frac{\ell}{g}}$$

$$\frac{N}{K} = \sqrt{\frac{V}{H}}$$

$$\frac{N^2}{K^2} = \frac{V}{H}$$

$$K^2 H\left(\frac{N^2}{K^2}\right) = K^2 H\left(\frac{V}{H}\right)$$

$$HN^2 = K^2 V$$

$$V = \frac{HN^2}{K^2}$$

$$g = \frac{4\pi^2 \ell}{T^2}$$

SELF-TEST 19 (pages 258-268)

(pages 258-268)

Solve each equation.

1. $\sqrt{3x} = 6$

2. $\sqrt{\frac{2x}{3}} = 4$

3. $\sqrt{5x - 9} = \sqrt{x + 3}$

4. $\sqrt{y - 1} = y - 3$

5. $\sqrt{2x - 5} - \sqrt{x - 3} = 1$

6. $\sqrt[3]{5x + 7} = 3$

7. In $E = \sqrt{PR}$, find P when E = 10 and R = 5.

8. In $\sqrt{\frac{E}{F}} = \frac{H}{P}$, find F when E = 20, H = 18, and P = 9.

9. Solve for s. $t = \sqrt{\frac{2s}{g}}$

10. Solve for D. $A = B\sqrt{\frac{C}{D}}$

ANSWERS:

1. x = 12
2. x = 24
3. x = 3
4. y = 5 (not y = 2)
5. x = 3 and 7
6. x = 4
7. P = 20
8. F = 5
9. $s = \frac{gt^2}{2}$
10. $D = \frac{B^2C}{A^2}$

5-9 RATIONAL EXPONENTS

In this section we will define rational (or fractional) exponents, show that the laws of exponents can be used with them, and show how they can be used to simplify various radical expressions. We will assume that all radicands represent positive numbers.

---

101. We know that $\sqrt{a} \cdot \sqrt{a} = a$. And if the law of exponents still applies then $a^{\frac{1}{2}} \cdot a^{\frac{1}{2}} = a^{\frac{1}{2}+\frac{1}{2}} = a^1 = a$. Therefore:

$$a^{\frac{1}{2}} = \sqrt{a}$$

We know that $\sqrt[3]{a} \cdot \sqrt[3]{a} \cdot \sqrt[3]{a} = a$. And if the law of exponents still applies, then $a^{\frac{1}{3}} \cdot a^{\frac{1}{3}} \cdot a^{\frac{1}{3}} = a^{\frac{1}{3}+\frac{1}{3}+\frac{1}{3}} = a^1 = a$. Therefore:

$$a^{\frac{1}{3}} = \sqrt[3]{a}$$

Generalizing from the examples above, we get the definition below. Notice that the denominator of the exponent is the index of the radical.

Definition: $a^{\frac{1}{n}} = \sqrt[n]{a}$

Examples: $5^{\frac{1}{2}} = \sqrt{5}$ $y^{\frac{1}{3}} = \sqrt[3]{y}$ $7^{\frac{1}{5}} = \sqrt[5]{7}$

Convert each power to a radical.

a) $x^{\frac{1}{2}}$ = ________ b) $3^{\frac{1}{4}}$ = ________ c) $t^{\frac{1}{8}}$ = ________

---

102. We used a law of exponents and the definition in the last frame to convert $a^{\frac{2}{3}}$ and $a^{\frac{5}{4}}$ to radicals below.

Answers to 101: a) $\sqrt{x}$ b) $\sqrt[4]{3}$ c) $\sqrt[8]{t}$

$$a^{\frac{2}{3}} = (a^2)^{\frac{1}{3}} = \sqrt[3]{a^2}$$

$$a^{\frac{5}{4}} = (a^5)^{\frac{1}{4}} = \sqrt[4]{a^5}$$

Generalizing from the examples above, we get the definition below. Notice again that n, the denominator of the exponent, is the index of the radical.

Definition: $a^{\frac{m}{n}} = \sqrt[n]{a^m}$

Examples: $3^{\frac{7}{2}} = \sqrt{3^7}$ $x^{\frac{5}{3}} = \sqrt[3]{x^5}$ $t^{\frac{4}{7}} = \sqrt[7]{t^4}$

Continued on following page.

102. Continued

Convert each power to a radical.

a) $5^{\frac{3}{2}}$ = ________ b) $x^{\frac{3}{4}}$ = ________ c) $y^{\frac{10}{9}}$ = ________

---

103. We converted $a^{\frac{2}{3}}$ and $a^{\frac{5}{4}}$ to a different radical form below.

$$a^{\frac{2}{3}} = \left(a^{\frac{1}{3}}\right)^2 = (\sqrt[3]{a})^2$$

$$a^{\frac{5}{4}} = \left(a^{\frac{1}{4}}\right)^5 = (\sqrt[4]{a})^5$$

Generalizing from the examples above, we extended our definition of $a^{\frac{m}{n}}$ below.

Definition: $a^{\frac{m}{n}} = \sqrt[n]{a^m} = (\sqrt[n]{a})^m$

Examples: $7^{\frac{3}{2}} = \sqrt{7^3}$ or $(\sqrt{7})^3$

$x^{\frac{2}{5}} = \sqrt[5]{x^2}$ or $(\sqrt[5]{x})^2$

Convert each power to a radical.

a) $t^{\frac{5}{2}}$ = ________ b) $5^{\frac{3}{4}}$ = ________ c) $y^{\frac{7}{9}}$ = ________

Answers to 102: a) $\sqrt{5^3}$ b) $\sqrt[4]{x^3}$ c) $\sqrt[9]{y^{10}}$

---

104. To convert a radical to a power with a fractional exponent, we can reverse our definition.

Definition: $\sqrt[n]{a^m} = (\sqrt[n]{a})^m = a^{\frac{m}{n}}$

We used the definition for some conversions below. Notice that the index of the radical is the denominator of the fractional exponent.

$$\sqrt{2^5} = (\sqrt{2})^5 = 2^{\frac{5}{2}}$$

$$\sqrt[4]{t^3} = (\sqrt[4]{t})^3 = t^{\frac{3}{4}}$$

Convert each radical to a power.

a) $\sqrt{7^3}$ = _______ b) $\sqrt[5]{y^2}$ = _______ c) $(\sqrt[4]{m})^9$ = _______

Answers to 103: a) $\sqrt{t^5}$ or $(\sqrt{t})^5$ b) $\sqrt[4]{5^3}$ or $(\sqrt[4]{5})^3$ c) $\sqrt[9]{y^7}$ or $(\sqrt[9]{y})^7$

Answers to 104: a) $7^{\frac{3}{2}}$ b) $y^{\frac{2}{5}}$ c) $m^{\frac{9}{4}}$

105. In each conversion below, the numerator of the fractional exponent is "1".

$$\sqrt{7} = \sqrt{7^1} = 7^{\frac{1}{2}} \qquad \sqrt[6]{x} = \sqrt[6]{x^1} = x^{\frac{1}{6}}$$

Convert each radical to a power.

a) $\sqrt{y}$ = ________ b) $\sqrt[3]{6}$ = ________ c) $\sqrt[5]{m}$ = ________

---

a) $y^{\frac{1}{2}}$

b) $6^{\frac{1}{3}}$

c) $m^{\frac{1}{5}}$

106. By converting to a radical, we simplified each expression below.

$$36^{\frac{1}{2}} = \sqrt{36} = 6 \qquad 64^{\frac{1}{3}} = \sqrt[3]{64} = 4$$

Simplify these.

a) $100^{\frac{1}{2}}$ = ________ b) $81^{\frac{1}{4}}$ = ________

---

a) 10 b) 3

107. By converting to a radical, we simplified each expression below. Notice that we converted to $(\sqrt[k]{a})^m$ rather than $\sqrt[k]{a^m}$.

$$16^{\frac{3}{2}} = (\sqrt{16})^3 = 4^3 = 64$$

$$1000^{\frac{2}{3}} = (\sqrt[3]{1000})^2 = 10^2 = 100$$

Simplify these.

a) $4^{\frac{5}{2}}$ = ________ c) $32^{\frac{3}{5}}$ = ________

b) $64^{\frac{2}{3}}$ = ________ d) $27^{\frac{4}{3}}$ = ________

---

a) 32 c) 8

b) 16 d) 81

108. The definition of a power with a negative exponent is restated below.

$$a^{-n} = \frac{1}{a^n} \qquad (a \neq 0)$$

The definition also applies to powers with negative fractional exponents. For example:

$$5^{-\frac{1}{2}} = \frac{1}{5^{\frac{1}{2}}} \qquad x^{-\frac{4}{3}} = \frac{1}{x^{\frac{4}{3}}}$$

We used the definition above to simplify the expression below. Simplify the other expressions.

$$8^{-\frac{4}{3}} = \frac{1}{8^{\frac{4}{3}}} = \frac{1}{(\sqrt[3]{8})^4} = \frac{1}{2^4} = \frac{1}{16}$$

a) $49^{-\frac{1}{2}}$ = ________

b) $81^{-\frac{3}{4}}$ = ________

109. Three laws of exponents are given in the table below.

Laws of Exponents		
Multiplication	$a^m \cdot a^n = a^{m+n}$	
Division	$\dfrac{a^m}{a^n} = a^{m-n}$	$(a \neq 0)$
Power	$(a^m)^n = a^{mn}$	

The laws above apply to powers with fractional exponents. For example:

$$x^{\frac{1}{2}} \cdot x^{\frac{1}{4}} = x^{\frac{1}{2}+\frac{1}{4}} = x^{\frac{3}{4}}$$

$$\frac{x^{\frac{3}{4}}}{x^{\frac{1}{2}}} = x^{\frac{3}{4}-\frac{1}{2}} = x^{\frac{1}{4}}$$

$$(x^{\frac{1}{2}})^{\frac{3}{4}} = x^{(\frac{1}{2})(\frac{3}{4})} = x^{\frac{3}{8}}$$

Use the laws of exponents for these:

a) $5^{\frac{2}{3}} \cdot 5^{\frac{1}{2}} =$ ________

c) $\dfrac{3^{\frac{9}{8}}}{3^{\frac{5}{8}}} =$ ________

b) $(7^{\frac{3}{5}})^2 =$ ________

d) $(d^{\frac{1}{2}})^{\frac{4}{3}} =$ ________

a) $\frac{1}{7}$ b) $\frac{1}{27}$

---

110. The laws also apply to powers with negative fractional exponents. Use them to complete these.

a) $x^{-\frac{1}{5}} \cdot x^{\frac{3}{5}} =$ ________

c) $\dfrac{7^{\frac{1}{4}}}{7^{\frac{1}{2}}} =$ ________

b) $(5^{-2})^{\frac{1}{6}} =$ ________

d) $(y^{-\frac{1}{3}})^{-\frac{6}{5}} =$ ________

a) $5^{\frac{7}{6}}$ c) $3^{\frac{1}{2}}$

b) $7^{\frac{6}{5}}$ d) $d^{\frac{2}{3}}$

---

111. Notice how we used a fractional exponent to simplify each radical below.

$$\sqrt[4]{x^2} = x^{\frac{2}{4}} = x^{\frac{1}{2}} = \sqrt{x}$$

$$\sqrt[6]{y^4} = y^{\frac{4}{6}} = y^{\frac{2}{3}} = \sqrt[3]{y^2}$$

Simplify these.

a) $\sqrt[6]{x^2} =$ ____________

b) $\sqrt[10]{y^6} =$ ____________

a) $x^{\frac{2}{5}}$ c) $7^{-\frac{1}{4}}$

b) $5^{-\frac{1}{3}}$ d) $y^{\frac{2}{5}}$

---

a) $\sqrt[3]{x}$ b) $\sqrt[5]{y^3}$

112. Notice how we used $(a^m b^n)^p = a^{mp}b^{np}$ to simplify each radical below.

$$\sqrt[4]{9x^2} = \sqrt[4]{3^2x^2} = (3^2x^2)^{\frac{1}{4}} = 3^{\frac{2}{4}} \cdot x^{\frac{2}{4}} = 3^{\frac{1}{2}} \cdot x^{\frac{1}{2}} = (3x)^{\frac{1}{2}} = \sqrt{3x}$$

$$\sqrt[6]{a^2b^4} = (a^2b^4)^{\frac{1}{6}} = a^{\frac{2}{6}} \cdot b^{\frac{4}{6}} = a^{\frac{1}{3}} \cdot b^{\frac{2}{3}} = (ab^2)^{\frac{1}{3}} = \sqrt[3]{ab^2}$$

Simplify these.

a) $\sqrt[6]{25m^2}$ = ____________________

b) $\sqrt[8]{x^4y^6}$ = ____________________

Answers: a) $\sqrt[3]{5m}$ b) $\sqrt[4]{x^2y^3}$

113. Notice how we used fractional exponents to write each multiplication below as a single radical.

$$\sqrt{y} \cdot \sqrt[3]{y} = y^{\frac{1}{2}} \cdot y^{\frac{1}{3}} = y^{\frac{3}{6} + \frac{2}{6}} = y^{\frac{5}{6}} = \sqrt[6]{y^5}$$

$$\sqrt[3]{2} \cdot \sqrt{5} = 2^{\frac{1}{3}} \cdot 5^{\frac{1}{2}} = 2^{\frac{2}{6}} \cdot 5^{\frac{3}{6}} = (2^2 \cdot 5^3)^{\frac{1}{6}} = \sqrt[6]{2^2 \cdot 5^3} = \sqrt[6]{500}$$

Write each of these as a single radical.

a) $\sqrt[4]{x} \cdot \sqrt[8]{x^3}$ = ____________________

b) $\sqrt{5} \cdot \sqrt[4]{7}$ = ____________________

Answers: a) $\sqrt[8]{x^5}$ b) $\sqrt[4]{175}$

114. Notice how we were able to factor a perfect power out of $\sqrt[12]{x^{17}}$ below. Simplify the other multiplication.

$$\sqrt[3]{x^2} \cdot \sqrt[4]{x^3} = x^{\frac{2}{3}} \cdot x^{\frac{3}{4}} = x^{\frac{8}{12} + \frac{9}{12}} = x^{\frac{17}{12}} = \sqrt[12]{x^{17}} = x\sqrt[12]{x^5}$$

$\sqrt{y} \cdot \sqrt[3]{y^2}$ = ____________________

Answer: $y\sqrt[6]{y}$, from $\sqrt[6]{y^7}$

115. Following the example, simplify the other multiplication.

$$\sqrt[8]{x^3} \cdot \sqrt{x} \cdot \sqrt[4]{x^3} = x^{\frac{3}{8}} \cdot x^{\frac{1}{2}} \cdot x^{\frac{3}{4}} = x^{\frac{3}{8} + \frac{4}{8} + \frac{6}{8}} = x^{\frac{13}{8}} = \sqrt[8]{x^{13}} = x\sqrt[8]{x^5}$$

$\sqrt{y} \cdot \sqrt[3]{y} \cdot \sqrt[4]{y^3}$ = ____________________

Answer: $y\sqrt[12]{y^7}$, from $\sqrt[12]{y^{19}}$

116. Following the example, simplify the other division.

$$\frac{\sqrt[5]{x^4}}{\sqrt{x}} = \frac{x^{\frac{4}{5}}}{x^{\frac{1}{2}}} = x^{\frac{8}{10} - \frac{5}{10}} = x^{\frac{3}{10}} = \sqrt[10]{x^3}$$

$$\frac{\sqrt[6]{y^5}}{\sqrt[3]{y^2}} =$$

Answer: $\sqrt[6]{y}$

117. Following the example, simplify the other division.

$$\frac{\sqrt[4]{x}}{\sqrt[3]{x}} = \frac{x^{\frac{1}{4}}}{x^{\frac{1}{3}}} = x^{\frac{3}{12} - \frac{4}{12}} = x^{-\frac{1}{12}} = \frac{1}{x^{\frac{1}{12}}} = \frac{1}{\sqrt[12]{x}}$$

$$\frac{\sqrt[3]{y}}{\sqrt{y}} =$$

$\dfrac{1}{\sqrt[6]{y}}$

5-10 ADDING AND SUBTRACTING COMPLEX NUMBERS

In this section, we will define imaginary and complex numbers. We will also show the procedures for adding and subtracting complex numbers.

118. We have seen that the square roots of negative numbers do not equal real numbers. However, the square root of any negative numbers can be written as a real number times $\sqrt{-1}$. For example:

$$\sqrt{-9} = \sqrt{9(-1)} = \sqrt{9} \cdot \sqrt{-1} = 3\sqrt{-1}$$

$$\sqrt{-36} = \sqrt{36(-1)} = \sqrt{36} \cdot \sqrt{-1} = 6\sqrt{-1}$$

$$\sqrt{-47} = \sqrt{47(-1)} = \sqrt{47} \cdot \sqrt{-1}$$

We use $\underline{i}$ for $\sqrt{-1}$. And since $i = \sqrt{-1}$, $i^2 = -1$. That is:

<u>Definitions</u>

$i = \sqrt{-1}$

$i^2 = -1$

Therefore: Instead of $3\sqrt{-1}$, we write 3i.

a) Instead of $6\sqrt{-1}$, we write _________.

b) Instead of $\sqrt{47} \cdot \sqrt{-1}$, we write _________.

a) 6i

b) $\sqrt{47}i$

119. To contrast them with real numbers, numbers containing i as a factor are called imaginary numbers. An imaginary number is a number that can be written in the form bi, where b is a real number and $i = \sqrt{-1}$.

$3i$, $6i$, and $\sqrt{47}i$ are called imaginary numbers.

Note: The term "imaginary" was probably a bad choice. It suggests that the numbers do not exist or that they are useless. However, they do exist and they are very useful in physical science and engineering.

Write each of these as an imaginary number.

a) $\sqrt{-4}$ = _______ b) $\sqrt{-100}$ = _______ c) $\sqrt{-83}$ = _________

---

120. We can add, subtract, multiply, divide, and square imaginary numbers. When doing so, the letter "i" is handled like any other letter. For example:

$2i + 5i = 7i$

$6i - 8i = -2i$

$7(4i) = 28i$

$(4i)^2 = 16i^2 = 16(-1) = -16$

$\frac{3i}{6i} = \left(\frac{3}{6}\right)\left(\frac{i}{i}\right) = \frac{1}{2}(1) = \frac{1}{2}$

Complete these.

a) $4i - 9i$ = _________ c) $(8i)^2$ = _________

b) $-3(10i)$ = _________ d) $\frac{24i}{8i}$ = _________

Answers to 119: a) $2i$ b) $10i$ c) $\sqrt{83}i$

---

121. We used the reverse of the rule below to convert the square roots of negative numbers to imaginary numbers.

$$\sqrt{x} \cdot \sqrt{y} = \sqrt{xy}$$

The rule applies when one of either x or y is negative. For example:

$$\sqrt{4} \cdot \sqrt{-9} = \sqrt{-36} = 6i$$

$$\sqrt{4} \cdot \sqrt{-9} = 2(3i) = 6i$$

However, the rule does not apply when both x and y are negative. For example:

$$\sqrt{-4} \cdot \sqrt{-9} = \sqrt{36} = 6$$

$$\sqrt{-4} \cdot \sqrt{-9} = 2i(3i) = 6i^2 = 6(-1) = -6$$

Note: The products 6 and -6 are not equal.

Use imaginary numbers for these.

a) $\sqrt{-16} \cdot \sqrt{9}$ = _____ b) $\sqrt{-4} \cdot \sqrt{-16}$ = _____ c) $\sqrt{-9} \cdot \sqrt{-1}$ = _____

Answers to 120: a) $-5i$ b) $-30i$ c) -64 d) 3

122.	A <u>complex number</u> is a <u>number</u> of the form <u>a + bi</u>, where <u>a</u> and <u>b</u> are real numbers and $i = \sqrt{-1}$. Some examples are shown. 4 + 5i 8 - 3i -2 + 7i As you can see, a complex number is an addition of a real number and an imaginary number. For example: In 4 + 5i: the real part is 4. the imaginary part is 5i. In -2 + 7i: a) the real part is ________. b) the imaginary part is ________.	a) 12i b) -8 c) -3
123.	<u>When the real part is zero</u>, the complex number is a <u>pure imaginary number</u>. Two examples are shown. 0 + 5i 0 - 9i When the imaginary part is zero, the complex number is a <u>real number</u>. Two examples are shown. (Note: 0i = 0) 4 + 0i -8 + 0i When either <u>a</u> or <u>b</u> in a complex number is 0, we can write the complex number without the "0" term. For example: Instead of 0 + 5i, we can write 5i. Instead of 4 + 0i, we can write _____.	a) -2 b) 7i
124.	To add complex numbers, we add their real parts and their imaginary parts. For example: (2 + 3i) + (6 + 4i) = (2 + 6) + (3i + 4i) = 8 + 7i (-5 - 7i) + (3 + 4i) = (-5 + 3) + (-7i + 4i) = -2 - 3i Do these additions. a) (7 + 2i) + (3 - 8i) = ______________ b) (-9 - 3i) + (4 - i) = ______________	4
125.	Since the sum of 5 + 4i and -5 - 4i is 0 + 0i which equals 0, those two complex numbers are additive inverses of each other. That is: The additive inverse of 5 + 4i is -5 - 4i. The additive inverse of -5 - 4i is 5 + 4i. Write the additive inverse of each complex number. a) 7 - 3i __________ b) -9 + 2i __________	a) 10 - 6i b) -5 - 4i
		a) -7 + 3i b) 9 - 2i

126. To subtract complex numbers, we add the additive inverse of the second complex number to the first complex number. For example:

additive inverse ↓

$$(6 - 2i) - (3 - 9i) = (6 - 2i) + (-3 + 9i) = 3 + 7i$$

$$(-4 + 7i) - (-8 + 9i) = (-4 + 7i) + (8 - 9i) = 4 - 2i$$

Do these subtractions.

a) $(7 - 9i) - (5 - 6i) =$ ______

b) $(6 - i) - (-2 - 2i) =$ ______

---

127. When subtracting below, we got a real number and a pure imaginary number.

$$(7 + 5i) - (4 + 5i) = 3 + 0i = 3$$

$$(8 - 9i) - (8 - 3i) = 0 - 6i = -6i$$

Do these subtractions.

a) $(-4 + 3i) - (-4 + 7i) =$ ______

b) $(-8 - i) - (-1 - i) =$ ______

Answers to 126: a) $2 - 3i$ b) $8 + i$

Answers to 127: a) $-4i$ b) -7

5-11 MULTIPLYING AND DIVIDING COMPLEX NUMBERS

In this section, we will discuss the procedures for multiplying and dividing complex numbers.

---

128. To multiply a complex number by a real number, we use the distributive principle. For example:

$$7(4 - 3i) = 7(4) - 7(3i) = 28 - 21i$$

Complete these.

a) $5(2 + i) =$ ______ b) $10(-6 - 4i) =$ ______

---

129. To multiply a complex number by an imaginary number, we use the distributive principle, substitute -1 for i^2, collect like terms, and write the product in $a + bi$ form. An example is shown. Do the other multiplication.

$$\begin{aligned} 3i(5 + 4i) &= 3i(5) + 3i(4i) \\ &= 15i + 12i^2 \\ &= 15i + 12(-1) \\ &= 15i - 12 \\ &= -12 + 15i \end{aligned}$$

$4i(-2 + 7i) =$

Answers to 128: a) $10 + 5i$ b) $-60 - 40i$

130. Following the example, do the other multiplication. $2i(10 - i) = 2i(10) - 2i(i)$ $= 20i - 2i^2$ $= 20i - 2(-1)$ $= 20i + 2$ $= 2 + 20i$ $-6i(-1 + 5i) =$	-28 - 8i
131. To multiply two complex numbers, we use the FOIL method, substitute -1 for i^2, collect like terms, and write the product in <u>a + bi</u> form. For example: $(3 + 4i)(5 + 2i) = \underset{F}{3(5)} + \underset{O}{3(2i)} + \underset{I}{4i(5)} + \underset{L}{4i(2i)}$ $= 15 + 6i + 20i + 8i^2$ $= 15 + 6i + 20i + 8(-1)$ $= 15 + 26i - 8$ $= 7 + 26i$ Using the same method, do this multiplication. $(1 + 6i)(2 + 3i) =$	30 + 6i
132. Another multiplication is shown below. $(3 - 2i)(1 + 4i) = 3(1) + 3(4i) - 2i(1) - 2i(4i)$ $= 3 + 12i - 2i - 8i^2$ $= 3 + 12i - 2i - 8(-1)$ $= 3 + 10i + 8$ $= 11 + 10i$ Use the same method for this multiplication. $(1 - 6i)(2 + 5i) =$	-16 + 15i

133. Do each multiplication. a) $(5 - 3i)(4 - 2i) =$ b) $(7 + i)(5 - i) =$	$32 - 7i$
134. Another multiplication is shown below. Do the other multiplication. $(-3 - 2i)(-6 + i) = -3(-6) - 3(i) - 2i(-6) - 2i(i)$ $= 18 - 3i + 12i - 2i^2$ $= 18 - 3i + 12i - 2(-1)$ $= 18 + 9i + 2$ $= 20 + 9i$ $(-7 - i)(5 - 4i) =$	a) $14 - 22i$ b) $36 - 2i$
135. We squared one complex number. Square the other complex number. $(4 + 3i)^2 = 4^2 + 2(4)(3i) + (3i)^2$ $= 16 + 24i + 9i^2$ $= 16 + 24i - 9$ $= 7 + 24i$ $(7 - 2i)^2 =$	$-39 + 23i$
	$45 - 28i$

136. The two complex numbers a + bi and a - bi are called conjugates of each other. For example:

The conjugate of $5 + 3i$ is $5 - 3i$.

The conjugate of $-4 - 7i$ is $-4 + 7i$.

Write the conjugate of each complex number.

a) $8 + 2i$ _______ b) $6 - i$ _______ c) $-9 + 10i$ _______

---

137. When we multiply a complex number and its conjugate, we get a real number. For example:

$$(8 + 5i)(8 - 5i) = 8^2 - (5i)^2 = 64 - 25i^2 = 64 - (-25) = 89$$

$$(-4 - i)(-4 + i) = (-4)^2 - i^2 = 16 - (-1) = 17$$

Complete these:

a) $(2 - 7i)(2 + 7i) =$

b) $(10 + i)(10 - i) =$

c) $(-1 + 6i)(-1 - 6i) =$

Answers to 136: a) $8 - 2i$ b) $6 + i$ c) $-9 - 10i$

---

138. Any imaginary number also has a conjugate. For example:

The conjugate of $5i$ is $-5i$, since $5i = 0 + 5i$ and its conjugate $0 - 5i = -5i$.

If we multiply an imaginary number by its conjugate, we also get a real number. For example:

$$5i(-5i) = -25i^2 = -25(-1) = 25$$

Do each multiplication.

a) $-9i(9i) =$ _________ b) $i(-i) =$ _________

Answers to 137: a) 53 b) 101 c) 37

---

139. To divide a complex number by a real number, we divide each part by the real number. For example:

$$\frac{10 - 15i}{5} = \frac{10}{5} - \frac{15i}{5} = 2 - 3i$$

Do each division.

a) $\frac{12 + 8i}{2} =$ _________ b) $\frac{-9 - 3i}{3} =$ _________

Answers to 138: a) 81 b) 1

Answers to 139: a) $6 + 4i$ b) $-3 - i$

140. When dividing by a real number, a and b in the quotient can be fractions. They should be reduced to lowest terms. For example:

$$\frac{6 - 8i}{10} = \frac{6}{10} - \frac{8}{10}i = \frac{3}{5} - \frac{4}{5}i$$

Do each division.

a) $\frac{5 + 3i}{7} =$ ______________ b) $\frac{-4 - i}{8} =$ ______________

---

141. To divide a complex number by an imaginary number, we multiply both terms by the conjugate of the imaginary number. By doing so, we get a real number in the denominator. For example:

$$\frac{2 - 8i}{4i} = \frac{2 - 8i}{4i}\left(\frac{-4i}{-4i}\right) = \frac{-8i + 32i^2}{-16i^2} = \frac{-8i - 32}{16} = \frac{-32 - 8i}{16} = -2 - \frac{1}{2}i$$

Do each division.

a) $\frac{5 + 2i}{3i} =$

b) $\frac{1 - i}{i} =$

Answers to 140: a) $\frac{5}{7} + \frac{3}{7}i$ b) $-\frac{1}{2} - \frac{1}{8}i$

---

142. To divide a complex number by a complex number, we multiply both the numerator and denominator by the conjugate of the denominator. By doing so, we get a real number in the denominator. An example is shown. Do the other division.

$$\frac{3 + 4i}{5 - i} = \frac{3 + 4i}{5 - i}\left(\frac{5 + i}{5 + i}\right) = \frac{15 + 23i + 4i^2}{25 - i^2} = \frac{11 + 23i}{26} = \frac{11}{26} + \frac{23}{26}i$$

$\frac{6 - i}{3 + 2i} =$

Answers to 141: a) $\frac{2}{3} - \frac{5}{3}i$ b) $-1 - i$

---

143. Another example is shown. Do the other division.

$$\frac{2 - 4i}{1 + 3i} = \frac{2 - 4i}{1 + 3i}\left(\frac{1 - 3i}{1 - 3i}\right) = \frac{2 - 10i + 12i^2}{1 - 9i^2} = \frac{-10 - 10i}{10} = -1 - i$$

$\frac{6 + 2i}{1 - i} =$

Answer to 142: $\frac{16}{13} - \frac{15}{13}i$

---

144. Do each division.

a) $\frac{4 - 6i}{3 - i} =$

b) $\frac{10 - 5i}{1 + 2i} =$

Answer to 143: $2 + 4i$

Answers to 144: a) $\frac{9}{5} - \frac{7}{5}i$ b) $-5i$, from $\frac{0 - 25i}{5}$

145. We know that $i^2 = -1$. We can use that fact to find higher powers of $\underline{i}$. For example:

$$i^3 = i^2 \cdot i = (-1)i = -i$$

$$i^4 = i^2 \cdot i^2 = (-1)(-1) = 1$$

The first four powers of $\underline{i}$ are given in the table below.

$$i^1 = i$$
$$i^2 = -1$$
$$i^3 = -i$$
$$i^4 = 1$$

We can use those values to find higher powers of i. For example:

$$i^5 = i^4 \cdot i = 1 \cdot i = i$$

$$i^{10} = (i^4)^2 \cdot i^2 = (1)^2 \cdot (-1) = 1(-1) = -1$$

We can also use those values to simplify expressions like those below.

$$(4i)^3 = 64i^3 = 64(-i) = -64i$$

$$(2i)^4 = 16i^4 = 16(1) = 16$$

Simplify each expression.

a) $(9i)^2$ = _______ b) $(2i)^3$ = _______ c) $(3i)^4$ = _______

a) -81

b) -8i

c) 81

SELF-TEST 20 (pages 269-283)

Convert to a radical.

1. $y^{\frac{3}{2}}$

Convert to a power.

2. $\sqrt[6]{x^5}$

Simplify.

3. $8^{\frac{2}{3}}$

4. $9^{-\frac{1}{2}}$

Use the laws of exponents to simplify.

5. $x^{\frac{1}{3}} \cdot x^{\frac{1}{2}}$

6. $\dfrac{y^{\frac{3}{4}}}{y^{-\frac{1}{2}}}$

7. $(7^{-3})^{\frac{1}{6}}$

Use fractional exponents to simplify.

8. $\sqrt[8]{4x^4}$

9. $\sqrt[3]{m^2} \cdot \sqrt{m}$

10. $\dfrac{\sqrt[3]{t}}{\sqrt[4]{t}}$

Do each operation.

11. $(5 - 2i) + (-3 - i)$

12. $(1 + 2i) - (2 - i)$

13. $-3i(-4 + 2i)$

14. $(6 - 5i)(3 - 4i)$

15. $(2 - 9i)(2 + 9i)$

16. $\dfrac{6 + 5i}{3i}$

17. $\dfrac{3 - 2i}{1 - 3i}$

ANSWERS:

1. $\sqrt{y^3}$
2. $x^{\frac{5}{6}}$
3. 4
4. $\frac{1}{3}$
5. $x^{\frac{5}{6}}$
6. $y^{\frac{5}{4}}$
7. $7^{-\frac{1}{2}}$
8. $\sqrt[4]{2x^2}$
9. $m\sqrt[6]{m}$
10. $\sqrt[12]{t}$
11. $2 - 3i$
12. $-1 + 3i$
13. $6 + 12i$
14. $-2 - 39i$
15. 85
16. $\frac{5}{3} - 2i$
17. $\frac{9}{10} + \frac{7}{10}i$

SUPPLEMENTARY PROBLEMS - CHAPTER 5

Assignment 17

Find each real root. Use the table when needed.

1. $\sqrt{49}$ 2. $-\sqrt{16}$ 3. $\sqrt{-64}$ 4. $\sqrt{37}$ 5. $-\sqrt{89}$

6. $\sqrt[3]{64}$ 7. $\sqrt[3]{-125}$ 8. $-\sqrt[4]{16}$ 9. $\sqrt[4]{1296}$ 10. $\sqrt[5]{-1}$

Simplify.

11. $\sqrt{50}$ 12. $\sqrt[3]{24x}$ 13. $\sqrt[6]{y^{18}}$ 14. $\sqrt{(x + y)^8}$

15. $\sqrt[5]{32t^5}$ 16. $\sqrt[3]{4y^{12}}$ 17. $\sqrt[4]{16x^4y^8}$ 18. $\sqrt{18ab^6}$

19. $\sqrt{m^7}$ 20. $\sqrt[5]{y^{13}}$ 21. $\sqrt{72x^3}$ 22. $\sqrt[3]{27y^5}$

23. $\sqrt{81x^2y^5}$ 24. $\sqrt[3]{16a^2b^4}$ 25. $\sqrt{x^2 - 6x + 9}$ 26. $\sqrt[3]{b^3x + b^3y}$

Multiply and simplify each product.

27. $\sqrt{3} \cdot \sqrt{6}$ 28. $\sqrt[3]{4x^3} \cdot \sqrt[3]{4y^5}$ 29. $\sqrt[4]{(x + 1)^2} \cdot \sqrt[4]{(x + 1)^3}$

30. $2x \cdot 3\sqrt{x^3}$ 31. $5\sqrt[3]{y^2} \cdot \sqrt[3]{y^5}$ 32. $3\sqrt{6x} \cdot x\sqrt{2}$

Divide and simplify each quotient.

33. $\frac{\sqrt{40}}{\sqrt{5}}$ 34. $\frac{\sqrt[3]{81}}{\sqrt[3]{3}}$ 35. $\frac{\sqrt{32x^3y^4}}{\sqrt{2xy}}$ 36. $\frac{\sqrt[4]{20m^6}}{\sqrt[4]{10m}}$

37. $\frac{8\sqrt[3]{10x^2}}{4\sqrt[3]{5x}}$ 38. $\frac{4\sqrt{6x^3y^4}}{10\sqrt{2x^2y^2}}$ 39. $\frac{5\sqrt{18d^2}}{\sqrt{2d}}$ 40. $\frac{\sqrt[3]{24ab}}{2\sqrt[3]{3}}$

Simplify.

41. $\sqrt{\frac{4}{9}}$ 42. $\sqrt{\frac{36x^2}{25}}$ 43. $\sqrt[3]{\frac{1}{8}}$ 44. $\sqrt[3]{\frac{16y}{27}}$

45. $\sqrt{\frac{64}{18t}}$ 46. $\sqrt[3]{\frac{27x^5}{64y^4}}$ 47. $\sqrt{\frac{7x^3}{16y^2}}$ 48. $\sqrt[3]{\frac{1}{bd^9}}$

Assignment 18

Add or subtract.

1. $7\sqrt{3} + 2\sqrt{3}$ 2. $4\sqrt[3]{7} - 9\sqrt[3]{7}$ 3. $\sqrt{x} + 3\sqrt{x}$

4. $6\sqrt[4]{9} - \sqrt[4]{9}$ 5. $\sqrt{y} - \sqrt{y}$ 6. $5\sqrt{a - 3} + 8\sqrt{a - 3}$

7. $a\sqrt{t} - b\sqrt{t}$ 8. $2x\sqrt{y} - 3\sqrt{y} - x\sqrt{y}$ 9. $4\sqrt[3]{x^2} - \sqrt[3]{x^2} + \sqrt[4]{x^2}$

10. $3\sqrt{18} + \sqrt{8}$ 11. $2\sqrt[3]{81x} - \sqrt[3]{24x}$ 12. $\sqrt[3]{2a^4} + \sqrt[3]{54a}$

13. $5\sqrt{x^3} - \sqrt{x} + 3\sqrt{x^2}$ 14. $\sqrt{8y - 4} - \sqrt{2y - 1}$ 15. $\sqrt{x^5 + x^4} + \sqrt{25x + 25}$

Multiply and simplify each product.

16. $\sqrt{5}(x - \sqrt{2})$ 17. $\sqrt{3}(\sqrt{2} + \sqrt{5})$ 18. $\sqrt{2}(\sqrt{2} - 5)$

19. $\sqrt{7}(\sqrt{3} - \sqrt{7})$ 20. $\sqrt{5}(3\sqrt{2} + 2\sqrt{7})$ 21. $\sqrt[3]{7}(5\sqrt[3]{2} - \sqrt[3]{3})$

22. $(\sqrt{10} + \sqrt{5})(\sqrt{10} - \sqrt{5})$
23. $(\sqrt{2} - \sqrt{6})(\sqrt{2} + \sqrt{6})$
24. $(\sqrt{a} + \sqrt{b})(\sqrt{a} - \sqrt{b})$
25. $(4 - \sqrt{11})(4 + \sqrt{11})$
26. $(2\sqrt{30} + \sqrt{90})(2\sqrt{30} - \sqrt{90})$
27. $(\sqrt{50} - 4\sqrt{5})(\sqrt{50} + 4\sqrt{5})$
28. $(\sqrt{2} + \sqrt{7})(\sqrt{5} - \sqrt{3})$
29. $(5 - \sqrt{3})(2 + \sqrt{3})$
30. $(2\sqrt{5} + 3\sqrt{2})(3\sqrt{5} - 4\sqrt{2})$
31. $(10\sqrt{5} + 9)(\sqrt{5} - 4)$
32. $(4 + \sqrt{3})^2$
33. $(\sqrt{5} - 6)^2$

Rationalize each denominator. Report each answer in lowest terms.

34. $\frac{2}{\sqrt{5x}}$
35. $\frac{7}{3\sqrt{2}}$
36. $\frac{1}{5\sqrt{3y}}$
37. $\frac{7}{\sqrt{7}}$
38. $\frac{x}{9\sqrt{x}}$
39. $\sqrt{\frac{16}{27}}$
40. $\sqrt{\frac{x^2}{5y^4}}$
41. $\sqrt{\frac{3}{5}}$
42. $\sqrt{\frac{5}{7x}}$
43. $\frac{1}{4 + \sqrt{7}}$
44. $\frac{4}{\sqrt{5} - 1}$
45. $\frac{8}{\sqrt{5} - \sqrt{3}}$
46. $\frac{\sqrt{x} + \sqrt{y}}{\sqrt{x} - \sqrt{y}}$
47. $\frac{4}{\sqrt[3]{9}}$
48. $\sqrt[3]{\frac{5}{2}}$
49. $\sqrt[3]{\frac{3x^3}{25y^4}}$

Assignment 19

Solve each equation.

1. $\sqrt{x + 5} = 3$
2. $\sqrt{\frac{m}{2}} = 5$
3. $1 = \sqrt{\frac{5}{7x}}$
4. $\sqrt{y - 9} = -4$
5. $\sqrt{2V - 1} + 3 = 8$
6. $\sqrt{4x - 9} = \sqrt{3x + 1}$
7. $\sqrt{t + 5} = \sqrt{5t - 7}$
8. $w + 3 = \sqrt{9w + 7}$
9. $\sqrt{2x + 5} = x + 1$
10. $\sqrt{y - 3} + \sqrt{y} = 3$
11. $\sqrt{x + 9} - \sqrt{x} = 1$
12. $\sqrt{x + 10} + \sqrt{x - 10} = 10$
13. $\sqrt{x + 8} - \sqrt{x - 8} = 2$
14. $\sqrt{3x + 10} - \sqrt{2x - 1} = 2$
15. $\sqrt{4y + 1} - \sqrt{y - 2} = 3$
16. $\sqrt[3]{m + 50} = 4$
17. $\sqrt[4]{8t} = 2$
18. $\sqrt[3]{5x - 8} - 2 = 1$

Do each evaluation.

19. In $t = \sqrt{\frac{2s}{g}}$, find $\underline{t}$ when $s = 144$ and $g = 32$.

20. In $a = \sqrt{c^2 - b^2}$, find $\underline{a}$ when $c = 5$ and $b = 3$.

21. In $N = K\sqrt{\frac{V}{H}}$, find N when $K = 5$, $V = 48$, and $H = 12$.

22. In $s = \sqrt[3]{V}$, find V when $s = 4$.

23. In $v = \sqrt{2as}$, find $\underline{a}$ when $v = 20$ and $s = 25$.

24. In $I = \sqrt{\frac{P}{R}}$, find R when $P = 100$ and $I = 5$.

25. In $t = \sqrt{\frac{2s}{g}}$, find $\underline{s}$ when $t = 2$ and $g = 32$.

26. In $\sqrt{\frac{a}{b}} = \frac{c}{d}$, find $\underline{a}$ when $b = 3$, $c = 80$, and $d = 20$.

Do each rearrangement.

27. Solve for A.
$s = \sqrt{A}$

28. Solve for V.
$s = \sqrt[3]{V}$

29. Solve for T.
$P = \sqrt{ST}$

30. Solve for $\underline{p}$.
$m = \sqrt{p + q}$

31. Solve for V.

$$M = \sqrt{\frac{T}{V}}$$

32. Solve for a.

$$h = \sqrt{\frac{3a}{b}}$$

33. Solve for E.

$$\sqrt{\frac{E}{F}} = \frac{H}{P}$$

34. Solve for d.

$$a = b\sqrt{\frac{c}{d}}$$

Assignment 20

Convert each power to a radical.

1. $5^{\frac{1}{2}}$ 2. $x^{\frac{1}{5}}$ 3. $7^{\frac{2}{3}}$ 4. $10^{\frac{3}{2}}$ 5. $x^{\frac{7}{8}}$

Convert each radical to a power.

6. $\sqrt{3}$ 7. $\sqrt[4]{8}$ 8. $\sqrt{y^5}$ 9. $\sqrt[4]{2^3}$ 10. $\sqrt[6]{m^5}$

Evaluate by converting to a radical.

11. $49^{\frac{1}{2}}$ 12. $125^{\frac{1}{3}}$ 13. $8^{-\frac{1}{3}}$ 14. $4^{\frac{3}{2}}$ 15. $8^{-\frac{2}{3}}$

Use the laws of exponents to simplify.

16. $5^{\frac{1}{2}} \cdot 5^{\frac{1}{3}}$ 17. $x^{-\frac{2}{3}} \cdot x^{\frac{7}{4}}$ 18. $\left(3^{\frac{2}{5}}\right)^2$ 19. $\left(y^{\frac{7}{8}}\right)^{\frac{4}{3}}$

20. $(m^3)^{-\frac{1}{2}}$ 21. $\dfrac{2^{\frac{7}{8}}}{2^{\frac{3}{8}}}$ 22. $\dfrac{7^{\frac{1}{3}}}{7^{\frac{2}{3}}}$ 23. $\dfrac{m^{-\frac{1}{4}}}{m^{-\frac{1}{2}}}$

Simplify by converting to powers.

24. $\sqrt[6]{x^4}$ 25. $\sqrt[8]{m^2}$ 26. $\sqrt[4]{25y^2}$ 27. $\sqrt[6]{9a^4b^2}$

Use powers to write as a single radical.

28. $\sqrt[3]{6} \cdot \sqrt{2}$ 29. $\sqrt{5} \cdot \sqrt[4]{4}$ 30. $\sqrt[6]{x} \cdot \sqrt[3]{x^2}$ 31. $\sqrt{y} \cdot \sqrt[5]{y^4}$

32. $\sqrt[4]{m^3} \cdot \sqrt{m} \cdot \sqrt[8]{m^7}$ 33. $\dfrac{\sqrt[3]{t^2}}{\sqrt[6]{t}}$ 34. $\dfrac{\sqrt[8]{5^7}}{\sqrt{5}}$ 35. $\dfrac{\sqrt[4]{x}}{\sqrt{x}}$

Express in terms of i.

36. $\sqrt{-4}$ 37. $\sqrt{-81}$ 38. $\sqrt{-100}$ 39. $\sqrt{-7}$ 40. $\sqrt{-37}$

Do each addition and subtraction.

41. $(5 + 3i) + (2 + 6i)$ 42. $(-1 - 7i) + (8 + 4i)$ 43. $(9 + i) + (2 - 3i)$

44. $(4 - i) + (-4 + 5i)$ 45. $(9 + 7i) - (5 + 6i)$ 46. $(3 - 4i) - (1 - 8i)$

47. $(2 - 9i) - (3 + i)$ 48. $(-5 - 8i) - (-1 - 8i)$ 49. $(-4 + i) - (-2 - i)$

Simplify.

50. $2i(3 - 5i)$ 51. $4i(-1 + i)$ 52. $-6i(-2 - 7i)$

53. $(3 + 2i)(4 + 5i)$ 54. $(1 - 5i)(8 + 2i)$ 55. $(-8 + i)(6 - i)$

56. $(-2 - i)(3 - 5i)$ 57. $(6 + i)^2$ 58. $(5 - 3i)^2$

59. $(3 + 7i)(3 - 7i)$ 60. $(9 - i)(9 + i)$ 61. $(-4 + 2i)(-4 - 2i)$

62. $-4i(4i)$ 63. $(8i)^2$ 64. $(5i)^3$ 65. $(4i)^4$

Do each division.

66. $\frac{4 + 12i}{6i}$ 67. $\frac{5 - i}{2i}$ 68. $\frac{2 + 3i}{4 - i}$

69. $\frac{2 + 4i}{1 + i}$ 70. $\frac{1 + 6i}{3 + 2i}$ 71. $\frac{8 - 4i}{1 + 2i}$

6 Quadratic Equations

In this chapter, we will define quadratic equations and solve them by factoring, the square root method, completing the square, and the quadratic formula. Some word problems and some evaluations and rearrangements with formulas are included, with a special section for the Pythagorean Theorem. Equations quadratic in form and quadratic inequalities are solved.

6-1 SOLVING BY FACTORING

In this section, we will define quadratic equations and review the factoring method for solving quadratic equations.

---

1. A quadratic equation is a polynomial equation whose highest-degree term is second-degree. Quadratic equations are also called second-degree equations.

   The Standard Form Of A Quadratic Equation is:

   $$ax^2 + bx + c = 0$$

   where a, b, and c are real numbers and $a \neq 0$.

Continued on following page.

1. Continued

In Chapter 3, we solved quadratic equations by the factoring method. An example is shown below. Notice that we factored the left side and then used the principle of zero products. Solve the other equation.

$$5d^2 + 3d - 2 = 0 \qquad\qquad 2x^2 - x - 10 = 0$$

$$(5d - 2)(d + 1) = 0$$

$5d - 2 = 0$	$d + 1 = 0$
$5d = 2$	$d = -1$
$d = \frac{2}{5}$	

Both $\frac{2}{5}$ and -1 check.

---

2. Though a cannot be 0 in a quadratic equation, b and c can be 0. In the equation below, b = 0. Therefore, there is no bx term. Following the example, solve the other equation.

$x = \frac{5}{2}$ and -2

$$9x^2 - 4 = 0 \qquad\qquad 36y^2 - 1 = 0$$

$$(3x + 2)(3x - 2) = 0$$

$3x + 2 = 0$	$3x - 2 = 0$
$3x = -2$	$3x = 2$
$x = -\frac{2}{3}$	$x = \frac{2}{3}$

---

3. In the equations below, there is no c term because c = 0. For equations of that type, one root is always 0. Solve the other equation.

$y = \frac{1}{6}$ and $-\frac{1}{6}$

$$x^2 + 3x = 0 \qquad\qquad 4y^2 - 5y = 0$$

$$x(x + 3) = 0$$

$x = 0$	$x + 3 = 0$
	$x = -3$

---

4. Before using the factoring method, we must get all terms on one side with 0 on the other side. An example is shown. Solve the other equation.

$y = 0$ and $\frac{5}{4}$

$$x^2 = 5x + 14 \qquad\qquad 3y^2 + 2 = 5y$$

$$x^2 - 5x - 14 = 0$$

$$(x + 2)(x - 7) = 0$$

$x + 2 = 0$	$x - 7 = 0$
$x = -2$	$x = 7$

5. Following the example, solve the other equation.

$$5x^2 = 4x$$
$$5x^2 - 4x = 0$$
$$x(5x - 4) = 0$$
$$x = 0 \quad\Big|\quad 5x - 4 = 0$$
$$5x = 4$$
$$x = \frac{4}{5}$$

$$16y^2 = 25$$

$y = \frac{2}{3}$ and 1

$y = \frac{5}{4}$ and $-\frac{5}{4}$

6-2 SOLVING BY THE SQUARE ROOT METHOD

In this section, we will show how quadratic equations of the form $x^2 = k$ and $(x + h)^2 = k$ can be solved by the square root method.

---

6. We can use the square root principle to solve equations of the form $x^2 = k$. That principle is stated below.

> The Square Root Principle For Equations
>
> $$x^2 = k$$
> $$\sqrt{x^2} = \pm\sqrt{k}$$
> $$x = \pm\sqrt{k}$$
>
> Note: $x = \pm\sqrt{k}$ means: $x = \sqrt{k}$ and $x = -\sqrt{k}$.

We used the square root principle to solve $x^2 = 64$ below. Use the same method to solve the other equation.

$$x^2 = 64$$
$$\sqrt{x^2} = \pm\sqrt{64}$$
$$x = \pm\sqrt{64}$$
$$x = \sqrt{64} \text{ and } -\sqrt{64}$$
$$x = 8 \text{ and } -8$$

$$y^2 = 81$$

---

y = 9 and -9

7. To solve the equation below, we used both axioms to isolate x^2 and then used the square root principle. Solve the other equation.

$$25x^2 - 49 = 0$$

$$25x^2 = 49$$

$$x^2 = \frac{49}{25}$$

$$x = \pm\sqrt{\frac{49}{25}}$$

$$x = \frac{7}{5} \text{ and } -\frac{7}{5}$$

$$1 = \frac{16t^2}{9}$$

$t = \frac{3}{4}$ and $-\frac{3}{4}$

8. The solution of the equation below is $y = \pm\sqrt{7}$. We can either leave the solution in radical form or use the square root table (or a calculator) to report the solution in decimal form. The expression ± 2.646 means 2.646 and -2.646. Solve the other equation.

$$y^2 - 7 = 0$$

$$y^2 = 7$$

$$y = \pm\sqrt{7}$$

$$y = \pm 2.646$$

$$2p^2 = 30$$

$p = \pm\sqrt{15}$

$= \pm 3.873$

9. Notice how we rationalized the denominator of the solution below and then used the square root table to write the solution in decimal form. Solve the other equation. Round the decimal solution to two decimal places.

$$3x^2 - 11 = 0$$

$$3x^2 = 11$$

$$x^2 = \frac{11}{3}$$

$$x = \pm\sqrt{\frac{11}{3}}$$

$$x = \pm\frac{\sqrt{11}}{\sqrt{3}}\left(\frac{\sqrt{3}}{\sqrt{3}}\right)$$

$$x = \pm\frac{\sqrt{33}}{3}$$

$$x = \pm\frac{5.745}{3}$$

$$x = \pm 1.915$$

$$\frac{5t^2}{2} = 4$$

$t = \pm\frac{\sqrt{40}}{5}$

$= \pm 1.265$

10. The solutions for the equation below are imaginary numbers. Solve the other equation.

$$x^2 + 9 = 0$$
$$x^2 = -9$$
$$x = \pm\sqrt{-9}$$
$$x = \pm 3i, \text{ or}$$
$$x = 3i \text{ and } -3i$$

$$2y^2 + 72 = 0$$

$y = \pm 6i$, or

$y = 6i$ and $-6i$

11. Following the example, solve the other equation.

$$9x^2 + 4 = 0$$
$$9x^2 = -4$$
$$x^2 = -\frac{4}{9}$$
$$x = \pm\sqrt{-\frac{4}{9}}$$
$$x = \pm\frac{2}{3}i, \text{ or}$$
$$x = \frac{2}{3}i \text{ and } -\frac{2}{3}i$$

$$25m^2 + 49 = 0$$

$m = \pm\frac{7}{5}i$, or

$m = \frac{7}{5}i$ and $-\frac{7}{5}i$

12. The square root method can also be used with equations of the form $(x + h)^2 = k$. For example, we used it to solve the equation below.

$$(x + 2)^2 = 25$$
$$\sqrt{(x + 2)^2} = \pm\sqrt{25}$$
$$x + 2 = \pm\sqrt{25}$$

$x + 2 = 5$	$x + 2 = -5$
$x = 3$	$x = -7$

We checked 3 as a root below. Check -7 as a root.

$$(x + 2)^2 = 25$$
$$(3 + 2)^2 = 25$$
$$5^2 = 25$$
$$25 = 25$$

$$(x + 2)^2 = 25$$

$$(-7 + 2)^2 = 25$$
$$(-5)^2 = 25$$
$$25 = 25$$

13. Following the example, solve the other equation.

$$(2x - 1)^2 = 9 \qquad\qquad (3y - 7)^2 = 4$$

$$2x - 1 = \pm\sqrt{9}$$

$2x - 1 = 3$	$2x - 1 = -3$
$2x = 4$	$2x = -2$
$x = 2$	$x = -1$

$y = 3$ and $\frac{5}{3}$

14. Notice how we left the solution below in radical form. Solve the other equation.

$$(2x + 3)^2 = 5 \qquad\qquad (3y - 5)^2 = 11$$

$$2x + 3 = \pm\sqrt{5}$$

$2x + 3 = \sqrt{5}$	$2x + 3 = -\sqrt{5}$
$2x = -3 + \sqrt{5}$	$2x = -3 - \sqrt{5}$
$x = \frac{-3 + \sqrt{5}}{2}$	$x = \frac{-3 - \sqrt{5}}{2}$

$y = \frac{5 + \sqrt{11}}{3}$ and $\frac{5 - \sqrt{11}}{3}$

15. The solutions for the equation below are complex numbers. Solve the other equation.

$$(y - 2)^2 = -36 \qquad\qquad (x + 1)^2 = -9$$

$$y - 2 = \pm\sqrt{-36}$$

$$y - 2 = \pm 6i$$

$y - 2 = 6i$	$y - 2 = -6i$
$y = 2 + 6i$	$y = 2 - 6i$

$x = -1 + 3i$ and $-1 - 3i$

6-3 SOLVING BY COMPLETING THE SQUARE

The factoring and square root methods are not general methods. That is, they cannot be used to solve all quadratic equations. In this section, we will solve quadratic equations by a general method called completing the square.

---

16. The perfect square trinomial below is the square of a binomial. The coefficient of the x^2 term is "1".

$$x^2 + 10x + 25 = (x + 5)^2$$

Continued on following page.

16. Continued

Notice the relationship between 25 and 10. 25 is the square of half of 10. That is:

$$\frac{1}{2}(10) = 5 \quad \text{and} \quad 5^2 = 25$$

We can use the relationship above to add a number to $x^2 + 12x$ to make it a perfect square trinomial. That is, we take half of 12 and square it.

Since $\frac{1}{2}(12) = 6$ and $6^2 = 36$, we add 36 to $x^2 + 12x$ and get:

$$x^2 + 12x + 36 = (x + 6)^2$$

The procedure above is called completing the square. By taking half the coefficient of x and squaring it, complete the square for each of these.

a) $x^2 + 8x +$ ______ $= (x +$ ______$)^2$

b) $x^2 - 18x +$ ______ $= (x -$ ______$)^2$

---

17. To complete a square, we sometimes add a fraction. For example, to complete the square for $x^2 + 5x$ below, we took half of 5 and squared it.

Since $\frac{1}{2}(5) = \frac{5}{2}$ and $\left(\frac{5}{2}\right)^2 = \frac{25}{4}$, we add $\frac{25}{4}$ to $x^2 + 5x$ and get:

$$x^2 + 5x + \frac{25}{4} = \left(x + \frac{5}{2}\right)^2$$

By taking half the coefficient of x and squaring it, complete the square for each of these.

a) $x^2 + 3x +$ ______ $= (x +$ ______$)^2$

b) $x^2 - 7x +$ ______ $= (x -$ ______$)^2$

Answers to Frame 16:

a) $x^2 + 8x + 16 = (x + 4)^2$

b) $x^2 - 18x + 81 = (x - 9)^2$

---

18. We could solve the equation below by the factoring method. However, we will solve it by a method called completing the square. In that method, we write the equation in the form $(x + h)^2 = k$ and then use the square root principle. To begin, we remove the 12 from the left side.

$$x^2 + 8x + 12 = 0$$

$$x^2 + 8x = -12$$

Continued on following page.

Answers to Frame 17:

a) $x^2 + 3x + \frac{9}{4} = \left(x + \frac{3}{2}\right)^2$

b) $x^2 - 7x + \frac{49}{4} = \left(x - \frac{7}{2}\right)^2$

18. Continued

If we add 4^2 or 16 to both sides, we complete the square on the left side. We can then write the equation in the form $(x + h)^2 = k$ and solve it.

$$x^2 + 8x + 16 = -12 + 16$$
$$(x + 4)^2 = 4$$
$$x + 4 = \pm\sqrt{4}$$
$$x + 4 = \pm 2$$

$x + 4 = 2$, $x = -2$ | $x + 4 = -2$, $x = -6$

We checked -2 as a root below. Check -6 as a root.

$$x^2 + 8x + 12 = 0$$
$$(-2)^2 + 8(-2) + 12 = 0$$
$$4 + (-16) + 12 = 0$$
$$0 = 0$$

$$x^2 + 8x + 12 = 0$$

---

19. To complete the square below, we added $\left(\frac{5}{2}\right)^2$ or $\frac{25}{4}$ to both sides. We left the solutions in radical form. Use the same method to solve the other equation.

$$x^2 + 5x - 7 = 0$$
$$x^2 + 5x = 7$$
$$x^2 + 5x + \frac{25}{4} = 7 + \frac{25}{4}$$
$$\left(x + \frac{5}{2}\right)^2 = \frac{28}{4} + \frac{25}{4}$$
$$\left(x + \frac{5}{2}\right)^2 = \frac{53}{4}$$
$$x + \frac{5}{2} = \pm\sqrt{\frac{53}{4}}$$
$$x + \frac{5}{2} = \pm\frac{\sqrt{53}}{2}$$

$x + \frac{5}{2} = \frac{\sqrt{53}}{2}$ | $x + \frac{5}{2} = -\frac{\sqrt{53}}{2}$

$x = -\frac{5}{2} + \frac{\sqrt{53}}{2}$ | $x = -\frac{5}{2} - \frac{\sqrt{53}}{2}$

$x = \frac{-5 + \sqrt{53}}{2}$ | $x = \frac{-5 - \sqrt{53}}{2}$

$$y^2 - 3y - 8 = 0$$

Answer to 18:

$$(-6)^2 + 8(-6) + 12 = 0$$
$$36 + (-48) + 12 = 0$$
$$0 = 0$$

20. To use completing the square, the coefficient of the x^2 term must be "1". If it is not "1", we can make it "1" by multiplying both sides by its reciprocal. For example, we multiplied both sides below by $\frac{1}{2}$. Then to complete the square, we added $\frac{9}{16}$ to both sides, since $\frac{1}{2}\left(-\frac{3}{2}\right) = \left(-\frac{3}{4}\right)$ and $\left(-\frac{3}{4}\right)^2 = \frac{9}{16}$. Use the same method to solve the other equation.

$y = \frac{3 + \sqrt{41}}{2}$, and

$y = \frac{3 - \sqrt{41}}{2}$

$$2x^2 - 3x - 3 = 0 \qquad\qquad 3x^2 - 2x - 4 = 0$$

$$\frac{1}{2}(2x^2 - 3x - 3) = \frac{1}{2}(0)$$

$$x^2 - \frac{3}{2}x - \frac{3}{2} = 0$$

$$x^2 - \frac{3}{2}x = \frac{3}{2}$$

$$x^2 - \frac{3}{2}x + \frac{9}{16} = \frac{3}{2} + \frac{9}{16}$$

$$\left(x - \frac{3}{4}\right)^2 = \frac{24}{16} + \frac{9}{16}$$

$$\left(x - \frac{3}{4}\right)^2 = \frac{33}{16}$$

$$x - \frac{3}{4} = \pm\sqrt{\frac{33}{16}}$$

$$x - \frac{3}{4} = \pm\frac{\sqrt{33}}{4}$$

$x - \frac{3}{4} = \frac{\sqrt{33}}{4}$	$x - \frac{3}{4} = -\frac{\sqrt{33}}{4}$
$x = \frac{3}{4} + \frac{\sqrt{33}}{4}$	$x = \frac{3}{4} - \frac{\sqrt{33}}{4}$
$x = \frac{3 + \sqrt{33}}{4}$	$x = \frac{3 - \sqrt{33}}{4}$

$x = \frac{1 + \sqrt{13}}{3}$ and $\frac{1 - \sqrt{13}}{3}$

6-4 THE QUADRATIC FORMULA

Though completing the square is a general method that will solve any quadratic equation, it is not a very efficient method. However, we can use completing the square to solve the standard quadratic equation $ax^2 + bx + c = 0$. By doing so, we can derive a general formula, called the quadratic formula, which can be used to solve any quadratic equation.

21. The standard form of a quadratic equation is:

$$ax^2 + bx + c = 0 \text{ , where } a \neq 0$$

We can solve the standard equation by completing the square. To do so, we begin by multiplying both sides by $\frac{1}{a}$ and then getting rid of $\frac{c}{a}$ on the left side.

$$\frac{1}{a}(ax^2 + bx + c) = \frac{1}{a}(0)$$

$$x^2 + \frac{b}{a}x + \frac{c}{a} = 0$$

$$x^2 + \frac{b}{a}x = -\frac{c}{a}$$

Half of $\frac{b}{a} = \frac{1}{2}\left(\frac{b}{a}\right) = \frac{b}{2a}$, and $\left(\frac{b}{2a}\right)^2 = \frac{b^2}{4a^2}$. We add $\frac{b^2}{4a^2}$ to both sides.

$$x^2 + \frac{b}{a}x + \frac{b^2}{4a^2} = -\frac{c}{a} + \frac{b^2}{4a^2}$$

$$\left(x + \frac{b}{2a}\right)^2 = -\frac{4ac}{4a^2} + \frac{b^2}{4a^2}$$

$$\left(x + \frac{b}{2a}\right)^2 = \frac{b^2 - 4ac}{4a^2}$$

$$x + \frac{b}{2a} = \pm\sqrt{\frac{b^2 - 4ac}{4a^2}}$$

$$x + \frac{b}{2a} = \pm\frac{\sqrt{b^2 - 4ac}}{2a}$$

$$x = -\frac{b}{2a} \pm \frac{\sqrt{b^2 - 4ac}}{2a}$$

The two solutions of the standard equation are:

$$x = -\frac{b}{2a} + \frac{\sqrt{b^2 - 4ac}}{2a} \qquad x = -\frac{b}{2a} - \frac{\sqrt{b^2 - 4ac}}{2a}$$

$$x = \frac{-b + \sqrt{b^2 - 4ac}}{2a} \qquad x = \frac{-b - \sqrt{b^2 - 4ac}}{2a}$$

The two solutions are usually stated together as we have done below.

<u>The Quadratic Formula</u> is: $x = \frac{-b \pm \sqrt{b^2 - 4ac}}{2a}$ $(a \neq 0)$

22. To use the quadratic formula, you must be able to identify a, b, and c in a quadratic equation. An example is shown. Identify a, b, and c in the other equation.

$5y^2 - 4y - 1 = 0$

$a = 5$

$b = -4$

$c = -1$

$t^2 - 2t + 30 = 0$

a = ______

b = ______

c = ______

---

23. Before identifying a, b, and c, we must write the equation in standard form. An example is shown. Identify a, b, and c in the other equation.

$a = 1$

$b = -2$

$c = 30$

$7d^2 - 8 = 2d$

$7d^2 - 2d - 8 = 0$

$a = 7$

$b = -2$

$c = -8$

$3x^2 = x - 5$

a = ______

b = ______

c = ______

---

24. We used the quadratic formula to solve one equation below.

$a = 3$

$b = -1$

$c = 5$

$2x^2 + 3x - 20 = 0$

$a = 2,\ b = 3,\ c = -20$

$$x = \frac{-b \pm \sqrt{b^2 - 4ac}}{2a}$$

$$x = \frac{-3 \pm \sqrt{3^2 - 4(2)(-20)}}{2(2)}$$

$$= \frac{-3 \pm \sqrt{9 + 160}}{4}$$

$$= \frac{-3 \pm \sqrt{169}}{4}$$

$$= \frac{-3 \pm 13}{4}$$

The two solutions are:

$$x = \frac{-3 + 13}{4} = \frac{10}{4} = \frac{5}{2}$$

and

$$x = \frac{-3 - 13}{4} = \frac{-16}{4} = -4$$

$3y^2 + 2y - 5 = 0$

$y = -\frac{5}{3}$ and 1

25. We used the quadratic formula to solve one equation below. Though solutions are frequently left in radical form (like $x = 2 \pm 2\sqrt{2}$), we can also use the square root table to report answers in decimal form. Solve the other equation.

$x^2 - 4x - 4 = 0$ $\qquad\qquad$ $m^2 - 4m + 1 = 0$

$a = 1,\ b = -4,\ c = -4$

$$x = \frac{-b \pm \sqrt{b^2 - 4ac}}{2a}$$

$$x = \frac{-(-4) \pm \sqrt{(-4)^2 - 4(1)(-4)}}{2(1)}$$

$$= \frac{4 \pm \sqrt{16 + 16}}{2}$$

$$= \frac{4 \pm \sqrt{32}}{2}$$

$$= \frac{4 \pm 4\sqrt{2}}{2}$$

$$= \frac{4}{2} \pm \frac{4\sqrt{2}}{2}$$

$$= 2 \pm 2\sqrt{2}$$

Substituting 1.414 for $\sqrt{2}$, we get:

$$x = 2 \pm 2(1.414)$$

$$= 2 \pm 2.828$$

The two solutions are:

$$x = 2 + 2.828 = 4.828$$

and

$$x = 2 - 2.828 = -0.828$$

$m = 2 \pm \sqrt{3}$

or

$m = 3.732$ and 0.268

26. Before using the quadratic formula below, we wrote the equation in standard form. Solve the other equation. Round to two decimal places.

$$y^2 + 7 = 9y$$
$$y^2 - 9y + 7 = 0$$
$$a = 1,\ b = -9,\ c = 7$$

$$2x^2 = 6x + 1$$

$$y = \frac{-(-9) \pm \sqrt{(-9)^2 - 4(1)(7)}}{2(1)}$$

$$= \frac{9 \pm \sqrt{81 - 28}}{2}$$

$$= \frac{9 \pm \sqrt{53}}{2}$$

Substituting 7.28 for $\sqrt{53}$, we get:

$$y = \frac{9 \pm 7.28}{2}$$

The two solutions are:

$$y = \frac{9 + 7.28}{2} = \frac{16.28}{2} = 8.14$$

$$y = \frac{9 - 7.28}{2} = \frac{1.72}{2} = 0.86$$

---

27. We get complex numbers as the solutions below. We left the solution in radical form. Solve the other equation.

$x = 3.16$ and -0.16

$$x^2 - 3x + 4 = 0$$
$$a = 1,\ b = -3,\ c = 4$$

$$3y^2 - 5y + 3 = 0$$

$$x = \frac{-(-3) \pm \sqrt{(-3)^2 - 4(1)(4)}}{2(1)}$$

$$= \frac{3 \pm \sqrt{9 - 16}}{2}$$

$$= \frac{3 \pm \sqrt{-7}}{2}$$

$$= \frac{3 \pm \sqrt{7}i}{2}$$

The two solutions are:

$$x = \frac{3 + \sqrt{7}i}{2}, \text{ and}$$

$$x = \frac{3 - \sqrt{7}i}{2}$$

$$y = \frac{5 \pm \sqrt{11}i}{6}$$

28. Using the factoring method, we got -4 as a root twice for the equation below. Though there is only one root, sometimes it is called a double root, or a root of multiplicity two, so that all quadratic equations can be considered to have two roots.

$$x^2 + 8x + 16 = 0$$

$$(x + 4)(x + 4) = 0$$

$x + 4 = 0$	$x + 4 = 0$
$x = -4$	$x = -4$

We used the quadratic formula to solve the same equation below. Notice that $b^2 - 4ac = 0$ when an equation has a double root.

$$x^2 + 8x + 16 = 0$$

$$a = 1,\ b = 8,\ c = 16$$

$$x = \frac{-8 \pm \sqrt{8^2 - 4(1)(16)}}{2(1)}$$

$$= \frac{-8 \pm \sqrt{64 - 64}}{2}$$

$$= \frac{-8 \pm \sqrt{0}}{2}$$

The two solutions are:

$$x = \frac{-8 + 0}{2} = \frac{-8}{2} = -4 \quad \text{and} \quad x = \frac{-8 - 0}{2} = \frac{-8}{2} = -4$$

---

29. The following suggestions can be used to pick a method for solving quadratic equations.

To Solve A Quadratic Equation

1. Use the square root method for equations of the form $x^2 = k$.
2. Try the factoring method for all other equations.
3. If factoring is impossible or difficult, use the quadratic formula. It solves all quadratics.

SELF-TEST 21 (pages 288-302)

Solve by the factoring method.

1. $3x^2 - 2x - 5 = 0$

2. $y^2 = 7y$

Solve by the square root method.

3. $16x^2 - 9 = 0$

4. $y^2 + 25 = 0$

Complete the square.

5. $x^2 + 12x +$ ______ $= (x +$ ______$)^2$

6. $y^2 - 3y +$ ______ $= (y -$ ______$)^2$

Solve by completing the square.

7. $x^2 + 10x - 4 = 0$

Solve with the quadratic formula.

8. $8x^2 - 2x - 3 = 0$

9. $2y^2 + y - 5 = 0$

ANSWERS:

1. $x = \frac{5}{3}$ and -1
2. $y = 0$ and 7
3. $x = \frac{3}{4}$ and $-\frac{3}{4}$
4. $y = 5i$ and $-5i$
5. $x^2 + 12x + 36 = (x + 6)^2$
6. $y^2 - 3y + \frac{9}{4} = \left(y - \frac{3}{2}\right)^2$
7. $x = -5 \pm \sqrt{29}$
8. $x = \frac{3}{4}$ and $-\frac{1}{2}$
9. $y = \dfrac{-1 \pm \sqrt{41}}{4}$

   or

   $y = 1.351$ and -1.851

6-5 WRITING EQUATIONS IN STANDARD FORM

Before solving quadratic equations, we write them in standard form. In this section, we will write equations in standard form. Only a few equations are solved.

---

30. The standard form of quadratic equations is $ax^2 + bx + c = 0$. In the standard form, a is always positive. Therefore, the equation below is not in standard form. To write it in standard form, we multiplied both sides by -1 to make a positive. Write the other equation in standard form.

$$-5x^2 + 7x - 3 = 0 \qquad\qquad -y^2 - 4y + 9 = 0$$

$$-1(-5x^2 + 7x - 3) = -1(0)$$

$$5x^2 - 7x + 3 = 0$$

---

31. Following the example, write the other equation in standard form.

Answer to 30: $y^2 + 4y - 9 = 0$

$$(x + 2)(x - 3) = 5x - 1 \qquad\qquad (2y - 3)(y - 4) = y + 7$$

$$x^2 - x - 6 = 5x - 1$$

$$x^2 - 6x - 5 = 0$$

---

32. To write the equation below in standard form, we began by clearing the fractions. To do so, we multiplied both sides by x^2. Write the other equation in standard form.

Answer to 31: $2y^2 - 12y + 5 = 0$

$$3 + \frac{5}{x} = \frac{4}{x^2} \qquad\qquad \frac{7}{y^2} = \frac{2}{y} + 1$$

$$x^2\left(3 + \frac{5}{x}\right) = \cancel{x^2}\left(\frac{4}{\cancel{x^2}}\right)$$

$$x^2(3) + \cancel{x^2}\left(\frac{5}{\cancel{x}}\right) = 4$$

$$3x^2 + 5x = 4$$

$$3x^2 + 5x - 4 = 0$$

---

33. To write the equation below in standard form, we also began by clearing the fractions. To do so, we multiplied both sides by $x(x + 4)$. Write the other equation in standard form.

Answer to 32: $y^2 + 2y - 7 = 0$

$$\frac{x + 4}{x} = \frac{1}{x + 4} \qquad\qquad \frac{1}{y - 5} = \frac{y - 5}{y}$$

$$\cancel{x}(x + 4)\left(\frac{x + 4}{\cancel{x}}\right) = \left(\frac{1}{\cancel{x + 4}}\right)(x)\cancel{(x + 4)}$$

$$(x + 4)(x + 4) = (1)(x)$$

$$x^2 + 8x + 16 = x$$

$$x^2 + 7x + 16 = 0$$

34. To write the equation below in standard form, we began by multiplying both sides by $(x + 2)(x - 2)$.

$y^2 - 11y + 25 = 0$

$$\frac{5}{x+2} + \frac{5}{x-2} = 3$$

$$(x+2)(x-2)\left(\frac{5}{x+2} + \frac{5}{x-2}\right) = 3(x+2)(x-2)$$

$$\cancel{(x+2)}(x-2)\left(\frac{5}{\cancel{x+2}}\right) + (x+2)\cancel{(x-2)}\left(\frac{5}{\cancel{x-2}}\right) = 3(x+2)(x-2)$$

$$5(x-2) + 5(x+2) = 3(x+2)(x-2)$$

$$5x - 10 + 5x + 10 = 3(x^2 - 4)$$

$$10x = 3x^2 - 12$$

$$3x^2 - 10x - 12 = 0$$

Following the example, write this equation in standard form.

$$\frac{10}{x-3} + \frac{10}{x+3} = 1$$

---

35. Let's solve: $x + 3 = \frac{2}{x+3}$

$x^2 - 20x - 9 = 0$

a) Write the equation in standard form.

$$x + 3 = \frac{2}{x+3}$$

b) Use the quadratic formula. Report the solutions as decimal numbers.

36. Let's solve: $\frac{2}{x + 1} + \frac{2}{x - 1} = 3$

a) Write the equation in standard form.

$$\frac{2}{x + 1} + \frac{2}{x - 1} = 3$$

b) Use the quadratic formula. Report the solutions as decimal numbers rounded to hundredths.

a) $x^2 + 6x + 7 = 0$

b) $x = -1.586$ and -4.414

---

a) $3x^2 - 4x - 3 = 0$ b) $x = 1.87$ and -0.54

6-6 WORD PROBLEMS

In this section, we will solve some word problems involving quadratic equations.

---

37. We need a quadratic equation to solve the problem below.

<u>Problem</u>: Working together, Mike and Dan can complete a job in 6 hours. It would take Dan 9 hours longer than Mike to do the job alone. How long would it take each to do the job working alone?

We let <u>x</u> equal the time it takes Mike working alone and <u>x + 9</u> equal the time it takes Dan working alone.

$\frac{1}{x}$ = the part of the job Mike can do in 1 hour

$\frac{1}{x + 9}$ = the part of the job Dan can do in 1 hour

$\frac{1}{6}$ = the part of the job both can do in 1 hour

Continued on following page.

37. Continued

Therefore, we can set up and solve the following equation.

$$\frac{1}{x} + \frac{1}{x + 9} = \frac{1}{6}$$

$$6x(x + 9)\left(\frac{1}{x} + \frac{1}{x + 9}\right) = 6x(x + 9)\left(\frac{1}{6}\right)$$

$$6\cancel{x}(x + 9)\left(\frac{1}{\cancel{x}}\right) + 6x\cancel{(x + 9)}\left(\frac{1}{\cancel{x + 9}}\right) = \cancel{6}x(x + 9)\left(\frac{1}{\cancel{6}}\right)$$

$$6(x + 9) + 6x = x(x + 9)$$

$$6x + 54 + 6x = x^2 + 9x$$

$$12x + 54 = x^2 + 9x$$

$$0 = x^2 - 3x - 54$$

$$0 = (x - 9)(x + 6)$$

$$x = 9 \quad \text{or} \quad x = -6$$

Since -6 does not make sense in this problem, Mike would take 9 hours working alone and Dan would take 9 + 9 or 18 hours working alone. The answer checks since $\frac{1}{9} + \frac{1}{18} = \frac{2}{18} + \frac{1}{18} = \frac{3}{18} = \frac{1}{6}$.

Use the same method to solve this problem.

Two hoses are used to fill the same swimming pool. Together they can fill the pool in 2 hours. Working alone, the larger hose can fill the pool in 3 hours less than the smaller hose. Find the time it takes each hose working alone.

6 hours for the smaller hose, 3 hours for the larger hose

38. To solve the problem below, we let <u>x</u> equal the width and <u>x + 1</u> equal the length. We used the quadratic formula. Solve the other problem.

The length of a rectangle is 1 ft more than the width. The area of the rectangle is 15.75 ft^2. Find the length and width.

The width of a rectangle is 2m less than the length. If the area is $1.25m^2$, find the length and width.

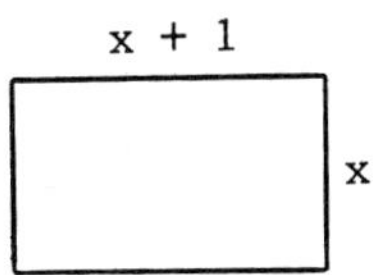

$$A = LW$$

$$15.75 = (x + 1)(x)$$

$$15.75 = x^2 + x$$

$$x^2 + x - 15.75 = 0$$

$$x = \frac{-1 \pm \sqrt{1^2 - 4(1)(-15.75)}}{2(1)}$$

$$= \frac{-1 \pm \sqrt{1 + 63}}{2}$$

$$= \frac{-1 \pm \sqrt{64}}{2}$$

$$= \frac{-1 \pm 8}{2}$$

$$x = \frac{-1 + 8}{2} = \frac{7}{2} = 3.5$$

and

$$x = \frac{-1 - 8}{2} = \frac{-9}{2} = -4.5$$

Since a negative length does not make sense, W = 3.5 ft and L = 4.5 ft.

---

39. We also need a quadratic equation to solve this problem.

<u>Problem</u>: The dimensions of a picture frame are 16 in by 20 in. The area of the picture is 192 in^2. Find the width of the frame.

We made a drawing on the next page. In the drawing, <u>x</u> equals the width of the frame. The length of the picture is 20 - 2x and the width of the picture is 16 - 2x.

L = 2.5m, W = 0.5m

Continued on following page.

39. Continued

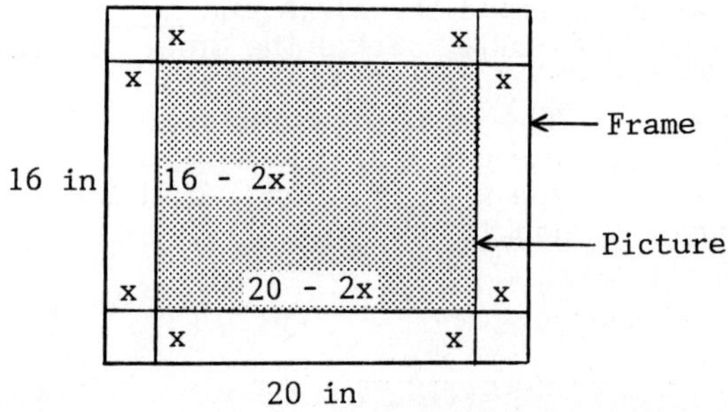

Using the formula A = LW, we can set up the following equation for the area of the picture.

$$(20 - 2x)(16 - 2x) = 192$$

$$320 - 40x - 32x + 4x^2 = 192$$

$$4x^2 - 72x + 128 = 0$$

$$x^2 - 18x + 32 = 0 \qquad \text{(Multiplying by } \tfrac{1}{4}\text{)}$$

$$(x - 2)(x - 16) = 0$$

$$x = 2 \quad \text{and} \quad x = 16$$

16 in. cannot be the width of the frame since we would get negative values if we subtracted (2)(16) from either 16 or 20. Therefore, the width of the frame is 2 in. The solution checks since $[20 - 2(2)][16 - 2(2)] = 16(12) = 192$.

Use the same method to solve this problem.

The dimensions of a picture frame are 18 in. by 30 in. The area of the picture is 220 in^2. Find the width of the frame.

The width of the frame is 4 in.

---

40. The solution below is based on the formula d = rt. We used the factoring method.

Problem: A river has a 2 mph current. A boat travels 30 miles upstream and 30 miles downstream in a total of 8 hours. What is the speed of the boat in still water?

Continued on following page.

40. Continued

The information in the problem is given in the diagram below.

Upstream
30 miles (x-2) mph →

Downstream
← 30 miles (x+2) mph

We summarized the information in the problem in the chart below.

	Distance	Rate	Time
Upstream	30 miles	x - 2	t_1
Downstream	30 miles	x + 2	t_2

Since the total time is 8 hours, we know this fact:

$$t_1 + t_2 = 8$$

Rearranging d = rt to solve for t, we get: $t = \frac{d}{r}$. Therefore:

$$t_1 = \frac{30}{x - 2} \quad \text{and} \quad t_2 = \frac{30}{x + 2}$$

Substituting those values in $t_1 + t_2 = 8$, we get:

$$\frac{30}{x - 2} + \frac{30}{x + 2} = 8$$

Clearing the fractions and solving, we get:

$$(x - 2)(x + 2)\left(\frac{30}{x - 2} + \frac{30}{x + 2}\right) = 8(x - 2)(x + 2)$$

$$\cancel{(x - 2)}(x + 2)\left(\frac{30}{\cancel{x - 2}}\right) + (x - 2)\cancel{(x + 2)}\left(\frac{30}{\cancel{x + 2}}\right) = 8(x^2 - 4)$$

$$30x + 60 + 30x - 60 = 8x^2 - 32$$

$$60x = 8x^2 - 32$$

$$8x^2 - 60x - 32 = 0$$

$$2x^2 - 15x - 8 = 0 \qquad \text{(Multiplying by } \tfrac{1}{4}\text{)}$$

$$(2x + 1)(x - 8) = 0$$

$$x = -\frac{1}{2} \quad \text{and} \quad x = 8$$

Since $-\frac{1}{2}$ mph does not make sense, the speed of the boat in still water is 8 mph. Therefore, the speed upstream would be 6 mph and the speed downstream would be 10 mph. These values check since it would take 5 hours upstream (from $\frac{30}{6}$) and 3 hours downstream (from $\frac{30}{10}$) for a total of 8 hours.

Continued on following page.

40. Continued

Use the same method for this problem.

A river has a current of 1 mph. A boat travels 24 miles upstream and 24 miles downstream in a total of 7 hours. What is the speed of the boat in still water?

the speed in still water is 7 mph

6-7 FORMULAS

In this section, we will do some evaluations and rearrangements with formulas containing a squared variable.

41. In the evaluation below, we squared before multiplying. Do the other evaluation.

In $s = \frac{1}{2}at^2$, find s when a = 12 and t = 4.

$$s = \frac{1}{2}(12)(4^2) = \frac{1}{2}(12)(16) = 6(16) = 96$$

a) In $A = 3.14r^2$, find A when r = 10.

A = ______________________

b) In $P = \frac{E^2}{R}$, find P when E = 9 and R = 10.

P = ______________________

a) A = 314

b) P = 8.1

42. If an object is dropped, the distance d it falls in t seconds (disregarding air resistance) is given by the formula $d = 16t^2$. We used the formula to solve one problem below. Notice that we had to solve a quadratic equation. Solve the other problem.

How long would it take an object to fall 80 ft?

$$d = 16t^2$$

$$80 = 16t^2$$

$$t^2 = \frac{80}{16}$$

$$t^2 = 5$$

$$t = \pm\sqrt{5}$$

$$t = \pm 2.236$$

Since a negative time interval does not make sense, it takes 2.236 seconds to drop 80 ft.

How long would it take an object to fall 320 ft?

---

43. When solving for a variable that is squared in a formula, we report only the positive root because the negative root does not ordinarily make sense. An example is shown. Do the other evaluation.

4.472 seconds

In the formula below, find s when A = 150.

$$A = 6s^2$$

$$150 = 6s^2$$

$$s^2 = \frac{150}{6} = 25$$

$$s = \sqrt{25} = 5$$

In the formula below, find t when s = 100 and a = 10.

$$s = \frac{1}{2}at^2$$

---

44. Following the example, do the other evaluation.

$t = \sqrt{20} = 4.472$

In the formula below, find s when N = 1000, k = 10, $Q_1 = 60$, and $Q_2 = 80$.

$$N = \frac{kQ_1Q_2}{s^2}$$

$$1000 = \frac{10(60)(80)}{s^2}$$

$$1000s^2 = 600(80)$$

$$s^2 = \frac{48000}{1000}$$

$$s^2 = 48$$

$$s = 6.928$$

In the formula below, find r when F = 10, G = 20, $m_1 = 5$, and $m_2 = 9$.

$$F = \frac{Gm_1m_2}{r^2}$$

45. The formula $s = v_0t + 16t^2$ shows the relationship between distance traveled (s), initial velocity (v_0) and time traveled (t) for an object that is thrown vertically downward. We used the formula for one evaluation below. Do the other evaluation.

$r = \sqrt{90} = 9.487$

If the initial velocity is 64 ft per second, find the amount of time needed to travel 80 ft.

If the initial velocity is 48 ft per second, find the amount of time needed to travel 160 ft.

$$s = v_0t + 16t^2$$

$$80 = 64t + 16t^2$$

$$\frac{1}{16}(80) = \frac{1}{16}(64t + 16t^2)$$

$$5 = 4t + t^2$$

$$t^2 + 4t - 5 = 0$$

$$(t - 1)(t + 5) = 0$$

$$t = 1 \text{ and } -5$$

Since a negative time interval does not make sense, it takes 1 second to travel 80 ft.

---

46. When rearranging formulas to solve for a variable that is squared, negative values do not ordinarily make sense. Therefore, when using the square root principle to solve for v below, we only used the principle square root of 2as. Solve for s in the other two formulas.

$t = 2$ seconds

$$v^2 = 2as$$

$$\sqrt{v^2} = \sqrt{2as}$$

$$v = \sqrt{2as}$$

a) $s^2 = A$

b) $s^2 = \frac{A}{6}$

---

47. When solving for a variable that is squared, be sure to take the square root of all of the other side. An example is shown. Solve for c in the other formula.

a) $s = \sqrt{A}$

b) $s = \sqrt{\frac{A}{6}}$

$$a^2 = \frac{b + c}{d}$$

$$a = \sqrt{\frac{b + c}{d}}$$

$$c^2 = a^2 + b^2$$

---

48. To solve for c below, we isolated c^2 first and then used the square root principle. Solve for r in the other formula.

$c = \sqrt{a^2 + b^2}$

$$E = mc^2$$

$$c^2 = \frac{E}{m}$$

$$c = \sqrt{\frac{E}{m}}$$

$$A = 4\pi r^2$$

49. To solve for t below, we began by multiplying both sides by 2 to get rid of the $\frac{1}{2}$. Solve for r in the other formula.

$$s = \frac{1}{2}at^2 \qquad\qquad V = \frac{1}{3}\pi r^2 h$$

$$2s = 2\left(\frac{1}{2}at^2\right)$$

$$2s = at^2$$

$$t^2 = \frac{2s}{a}$$

$$t = \sqrt{\frac{2s}{a}}$$

$r = \sqrt{\frac{A}{4\pi}}$ or $r = \frac{1}{2}\sqrt{\frac{A}{\pi}}$

---

50. To solve for r below, we isolated r^2 first. Solve for s in the other formula.

$$F = \frac{Gm_1m_2}{r^2} \qquad\qquad N = \frac{kQ_1Q_2}{s^2}$$

$$Fr^2 = \cancel{r^2}\left(\frac{Gm_1m_2}{\cancel{r^2}}\right)$$

$$Fr^2 = Gm_1m_2$$

$$r^2 = \frac{Gm_1m_2}{F}$$

$$r = \sqrt{\frac{Gm_1m_2}{F}}$$

$r = \sqrt{\frac{3V}{\pi h}}$

---

51. To solve for X below, we isolated X^2 first. Solve for b in the other formula.

$$Z^2 = X^2 + R^2 \qquad\qquad c^2 = a^2 + b^2$$

$$Z^2 + (-R^2) = X^2 + R^2 + (-R^2)$$

$$Z^2 - R^2 = X^2$$

$$X = \sqrt{Z^2 - R^2}$$

$s = \sqrt{\frac{kQ_1Q_2}{N}}$

$b = \sqrt{c^2 - a^2}$

52. To solve for t below, we wrote the formula in standard form and then used the quadratic formula. Solve for V in the other formula.

$$s = gt + 16t^2 \qquad\qquad h = V^2 + kV$$

$$16t^2 + gt - s = 0$$

$$a = 16,\ b = g,\ c = -s$$

$$t = \frac{-g \pm \sqrt{g^2 - 4(16)(-s)}}{2(16)}$$

$$= \frac{-g \pm \sqrt{g^2 + 64s}}{32}$$

$$V = \frac{-k \pm \sqrt{k^2 + 4h}}{2}$$

53. We used the quadratic formula to solve for r below. Notice how we factored a 2 out of the radical and then reduced to lowest terms. Solve for r in the other formula.

$$A = 2\pi r^2 + 2\pi rh \qquad\qquad S = \pi r^2 + 2\pi rh$$

$$2\pi r^2 + 2\pi rh - A = 0$$

$$a = 2\pi,\ b = 2\pi h,\ c = -A$$

$$r = \frac{-2\pi h \pm \sqrt{(2\pi h)^2 - 4(2\pi)(-A)}}{2(2\pi)}$$

$$= \frac{-2\pi h \pm \sqrt{4\pi^2h^2 + 8\pi A}}{4\pi}$$

$$= \frac{-2\pi h \pm \sqrt{4(\pi^2h^2 + 2\pi A)}}{4\pi}$$

$$= \frac{-2\pi h \pm 2\sqrt{\pi^2h^2 + 2\pi A}}{4\pi}$$

$$= \frac{\not{2}(-\pi h \pm \sqrt{\pi^2h^2 + 2\pi A})}{\underset{2}{\not{4}}\pi}$$

$$= \frac{-\pi h \pm \sqrt{\pi^2h^2 + 2\pi A}}{2\pi}$$

$$r = \frac{-\pi h \pm \sqrt{\pi^2h^2 + \pi S}}{\pi}$$

SELF-TEST 22 (pages 303-315)

Write in standard form and then solve.

1. $(10 - x)(8 - x) = 35$

2. $\frac{1}{x} + \frac{1}{x+3} = \frac{1}{4}$

3. The length of a rectangle is 1 ft more than the width. The area of the rectangle is 3.75 ft^2. Find the length and width.

4. A river has a current of 3 mph. A boat travels 20 miles upstream and 20 miles downstream in a total of 7 hours. What is the speed of the boat in still water?

5. In $F = \frac{m_1 m_2}{rd^2}$, find d when F = 8, $m_1 = 8$, $m_2 = 10$, and r = 2.

6. In $s = v_0 t + 16t^2$, find t when s = 96 and $v_0 = 80$.

7. Solve for m.

$$V = \frac{1}{2}bm^2$$

8. Solve for Q.

$$P^2 = Q^2 + R^2$$

9. Solve for h.

$$c = h^2 + dh$$

ANSWERS:

1. $x^2 - 18x + 45 = 0$
   x = 3 and 15

2. $x^2 - 5x - 12 = 0$

   $x = \frac{5 \pm \sqrt{73}}{2}$

   or

   x = 6.772 and -1.772

3. W = 1.5 ft, L = 2.5 ft

4. 7 mph

5. d = 2.236

6. t = 1

7. $m = \sqrt{\frac{2V}{b}}$

8. $Q = \sqrt{P^2 - R^2}$

9. $h = \frac{-d \pm \sqrt{d^2 + 4c}}{2}$

6-8 THE PYTHAGOREAN THEOREM

The Pythagorean Theorem is a formula relating the hypotenuse and legs of a right triangle. We will discuss that formula in this section and use it to solve some problems.

---

54. A right triangle is a triangle that contains a right angle (90°). To label the sides of a right triangle, we frequently use the letters a, b, and c.

c is the hypotenuse, the side opposite the right angle.

a and b are the legs.

The Pythagorean Theorem says this: In any right triangle, the square of the hypotenuse is equal to the sum of the squares of the two legs. That is:

$$c^2 = a^2 + b^2$$

We used the formula to find the length of the hypotenuse of one right triangle below. Use it to find the length of the hypotenuse of the other triangle.

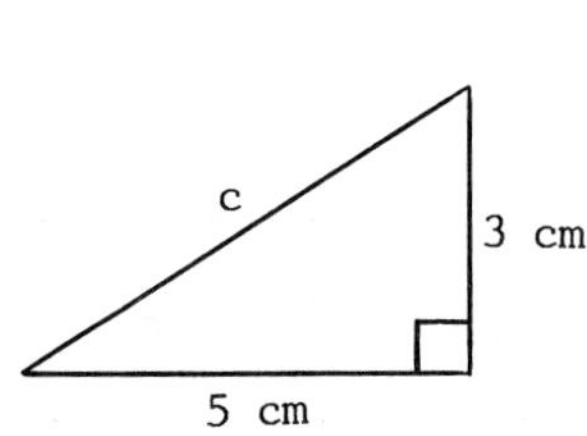

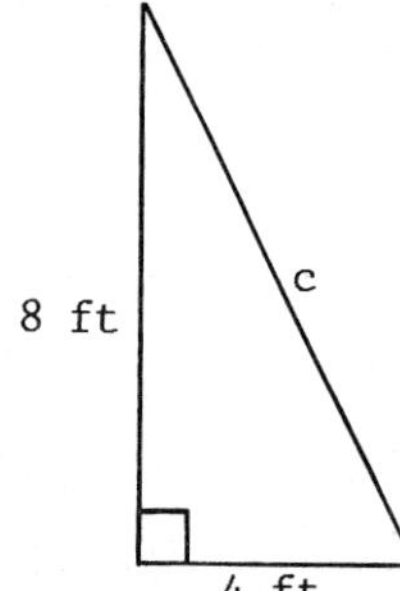

$$c^2 = a^2 + b^2$$
$$c^2 = 5^2 + 3^2$$
$$c^2 = 25 + 9$$
$$c^2 = 34$$
$$c = \sqrt{34} = 5.831 \text{ cm}$$

---

55. The Pythagorean Theorem was used to find the length of leg a in the right triangle below. Use it to find the length of leg b in the other right triangle.

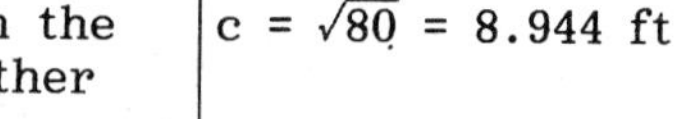
$c = \sqrt{80} = 8.944$ ft

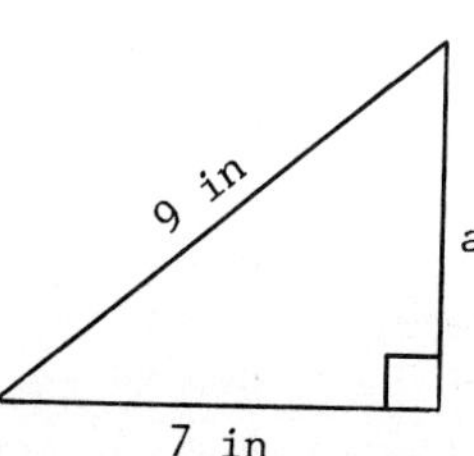

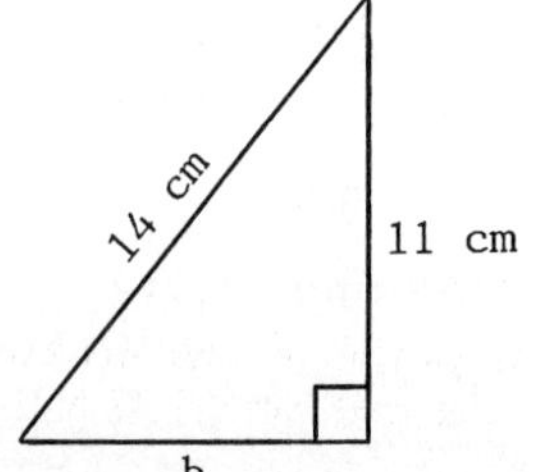

Continued on following page.

55. Continued

$c^2 = a^2 + b^2$

$9^2 = a^2 + 7^2$

$81 = a^2 + 49$

$a^2 = 32$

$a = \sqrt{32} = 5.657$ in

---

56. The distance between the opposite corners of a rectangle is called the diagonal. The diagonal is the hypotenuse of a right triangle. We used the Pythagorean Theorem to find the diagonal of one rectangle below. Use it to find the width of the other rectangle.

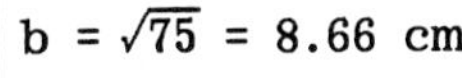
$b = \sqrt{75} = 8.66$ cm

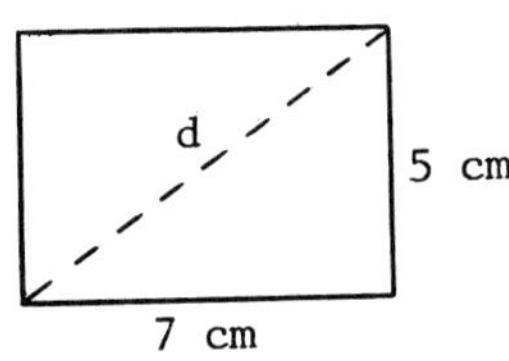

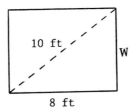

10 ft

W

8 ft

$d^2 = 7^2 + 5^2$

$d^2 = 49 + 25$

$d^2 = 74$

$d = \sqrt{74} = 8.602$ cm

---

57. We used the Pythagorean Theorem to solve the problem below.

W = 6 ft

Problem: A 13-foot ladder is leaning against a house. The distance from the top of the ladder to the ground is 7 feet more than the distance from the bottom of the ladder to the house. How far is the bottom of the ladder from the house?

Letting x equal the distance from the bottom of the ladder to the house and $x + 7$ equal the distance from the top of the ladder to the ground, we sketched the problem at the right.

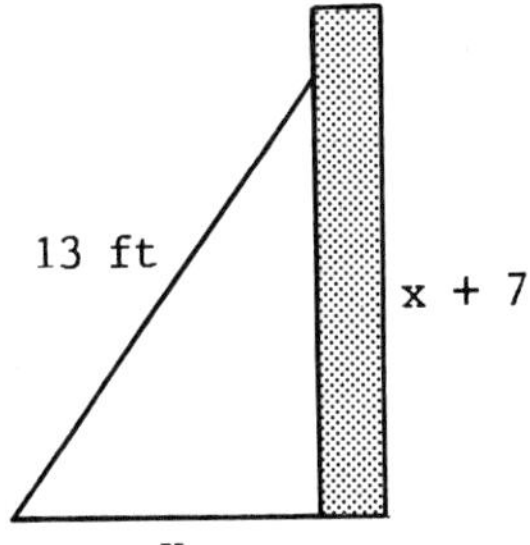

Using the Pythagorean Theorem, we get:

$13^2 = x^2 + (x + 7)^2$

$169 = x^2 + x^2 + 14x + 49$

$0 = 2x^2 + 14x - 120$

$0 = x^2 + 7x - 60$ (Multiplying by $\frac{1}{2}$)

$(x + 12)(x - 5) = 0$

$x = -12$ and $x = 5$

Since distances are positive, we can ignore -12 ft. Therefore, the distance from the bottom of the ladder to the house is 5 ft.

Continued on following page.

57. Continued

Using the same method, solve this problem.

A 20-foot wire is stretched from the ground to the top of a building. The height of the building is 4 feet more than the distance (d) from the building. Find the distance d and the height of the building.

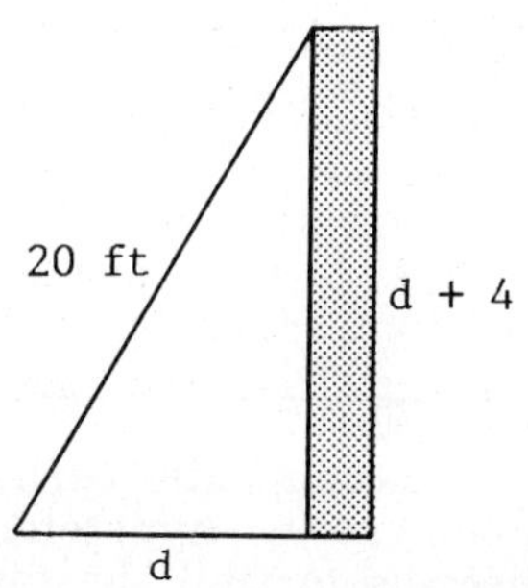

$d = 12$ ft,
the height $= 16$ ft

58. We used the Pythagorean Theorem to solve this problem.

Problem: Two cars left the same place at the same time. One headed due north, the other due east. When they were exactly 25 miles apart, the car headed due north had gone 5 miles farther. How far had each car traveled?

Letting x equal the distance traveled by the car heading due east and $x + 5$ equal the distance traveled by the car heading due north, we sketched the problem at the right. The triangle is a right triangle.

North
25 miles
x + 5
East
x

Using the Pythagorean Theorem, we get:

$$25^2 = x^2 + (x + 5)^2$$

$$625 = x^2 + x^2 + 10x + 25$$

$$0 = 2x^2 + 10x - 600$$

$$0 = x^2 + 5x - 300 \qquad \text{(Multipling by } \tfrac{1}{2}\text{)}$$

$$(x + 20)(x - 15) = 0$$

$$x = -20 \text{ and } x = 15$$

Since distances are positive, we can ignore $x = -20$ miles. Therefore, the car going east traveled 15 miles and the car going north traveled 20 miles.

Continued on following page.

58. Continued

Using the same method, solve this problem.

In the diagram at the right, A, B, and C represent towns. The roads meeting at B form a right angle. The distance from B to C is 2 miles more than the distance from A to B. Find the distance from A to B and the distance from B to C.

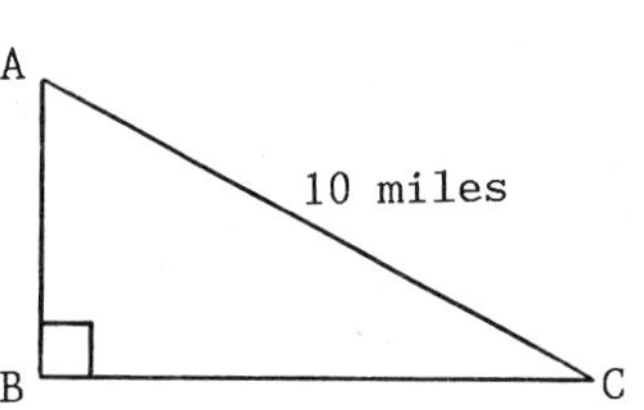

6 miles from A to B
8 miles from B to C

6-9 THE DISCRIMINANT

In the quadratic formula, the radicand $b^2 - 4ac$ is called the discriminant because it determines the nature of the solutions. We will discuss the discriminant in this section.

---

59. The solutions of the quadratic equation $ax^2 + bx + c = 0$ are given by the quadratic formula. They are:

$$x = \frac{-b \pm \sqrt{b^2 - 4ac}}{2a}$$

As we will show by examples, the value of the radicand $b^2 - 4ac$ determines the nature of the solutions. For that reason, $b^2 - 4ac$ is called the discriminant.

For each equation below, we will compute the discriminant and state the nature of the solution.

Example 1: $2x^2 - 3x - 5 = 0$

$a = 2,\ b = -3,\ c = -5$

$b^2 - 4ac = (-3)^2 - 4(2)(-5) = 9 + 40 = 49$

The discriminant 49 is a perfect square. Since $\sqrt{49}$ is a rational number, the two solutions are real, rational numbers.

Example 2: $3x^2 + 6x - 2 = 0$

$a = 3,\ b = 6,\ c = -2$

$b^2 - 4ac = 6^2 - 4(3)(-2) = 36 + 24 = 60$

The discriminant 60 is not a perfect square. Since $\sqrt{60}$ is an irrational number, the two solutions are real, irrational numbers.

Continued on following page.

59. Continued

Example 3: $9x^2 + 12x + 4 = 0$

$a = 9, \; b = 12, \; c = 4$

$b^2 - 4ac = 12^2 - 4(9)(4) = 144 - 144 = 0$

Since the discriminant is 0, there is only one solution. It is a real, rational number.

Note: When $b^2 - 4ac = 0$, the one real root is sometimes called a double root, or a root of multiplicity two, so that all quadratic equations can be considered to have two roots.

Example 4: $5x^2 - 2x + 1 = 0$

$a = 5, \; b = -2, \; c = 1$

$b^2 - 4ac = (-2)^2 - 4(5)(1) = 4 - 20 = -16$

Since the discriminant is negative, its square root is an imaginary number. Therefore, the two solutions are complex numbers. They are complex conjugates.

---

60. The information in the last frame is summarized in the table below.

Discriminant: $b^2 - 4ac$	Nature of Solutions
Positive, perfect square	Two real, rational solutions
Positive, not a perfect square	Two real, irrational solutions
Zero	One real, rational solution
Negative	Two complex solutions (complex conjugates)

By computing their discriminants, determine the nature of the solutions of each equation below.

a) $2x^2 - 7x - 3 = 0$

b) $x^2 - 10x + 25 = 0$

---

61. Compute their discriminants to determine the nature of the solutions of each equation below.

a) $3x^2 + x + 5 = 0$

b) $x^2 - 2x - 15 = 0$

Answers to 60:

a) two real, irrational solutions ($b^2 - 4ac = 73$)

b) one real, rational solution ($b^2 - 4ac = 0$)

Answers to 61:

a) two complex solutions ($b^2 - 4ac = -59$)

b) two real, rational solutions ($b^2 - 4ac = 64$)

6-10 EQUATIONS QUADRATIC IN FORM

Some equations that are not quadratic can be solved like quadratics by making a substitution. We will discuss the method in this section.

---

62. The equation below is not a quadratic equation.

$$(x^2 - 2)^2 - 7(x^2 - 2) + 10 = 0$$

However, by letting $a = x^2 - 2$ and substituting $\underline{a}$ for $x^2 - 2$, we get a quadratic equation that we can solve.

$$a^2 - 7a + 10 = 0$$
$$(a - 2)(a - 5) = 0$$
$$a = 2 \quad \text{and} \quad a = 5$$

Now we can substitute $x^2 - 2$ for $\underline{a}$ and solve for $\underline{x}$. We get:

$a = 2$	$a = 5$
$x^2 - 2 = 2$	$x^2 - 2 = 5$
$x^2 = 4$	$x^2 = 7$
$x = \pm 2$	$x = \pm\sqrt{7}$

The solution set is $\{2, -2, \sqrt{7}, -\sqrt{7}\}$. All four solutions check.

Use the same method to solve this equation.

$$(y^2 - 5)^2 - 2(y^2 - 5) - 8 = 0$$

Answer: $y = \pm 3$ and $\pm\sqrt{3}$

---

63. The equation below is not a quadratic equation.

$$(x^2 + 2x)^2 - 18(x^2 + 2x) + 45 = 0$$

However, by letting $a = x^2 + 2x$ and substituting $\underline{a}$ for $x^2 + 2x$, we get a quadratic equation that we can solve.

$$a^2 - 18a + 45 = 0$$
$$(a - 3)(a - 15) = 0$$
$$a = 3 \quad \text{and} \quad a = 15$$

Continued on following page.

63. Continued

Now we can substitute $x^2 + 2x$ for $\underline{a}$, and solve for $\underline{x}$. We get:

$$a = 3 \qquad\qquad a = 15$$

$$x^2 + 2x = 3 \qquad\qquad x^2 + 2x = 15$$

$$x^2 + 2x - 3 = 0 \qquad\qquad x^2 + 2x - 15 = 0$$

$$(x + 3)(x - 1) = 0 \qquad\qquad (x + 5)(x - 3) = 0$$

$$x = -3 \text{ and } x = 1 \qquad\qquad x = -5 \text{ and } x = 3$$

The solution set is $\{-5, -3, 1, 3\}$. All four solutions check.

Use the same method to solve this equation.

$$(2y^2 - y)^2 - 4(2y^2 - y) + 3 = 0$$

$y = -\frac{1}{2}, 1, \frac{3}{2}$, and -1

64. The equation below is also not a quadratic equation.

$$x^4 - 9x^2 + 20 = 0$$

However, by letting $a = x^2$, we can substitute $\underline{a}$ for x^2 and a^2 for x^4. We then get a quadratic equation that we can solve.

$$a^2 - 9a + 20 = 0$$

$$(a - 4)(a - 5) = 0$$

$$a = 4 \text{ and } a = 5$$

Now we can substitute x^2 for $\underline{a}$ and solve for $\underline{x}$. We get:

$$a = 4 \quad \text{and} \quad a = 5$$

$$x^2 = 4 \quad \text{and} \quad x^2 = 5$$

$$x = \pm 2 \quad \text{and} \quad x = \pm\sqrt{5}$$

The solution set is $\{-\sqrt{5}, -2, 2, \sqrt{5}\}$. All four solutions check.

Use the same method to solve this equation.

$$y^4 - 11y^2 + 18 = 0$$

65. To solve the equation below, we substituted a for x^2 and a^2 for x^4. Solve the other equation.

$$9x^4 - 13x^2 + 4 = 0$$

$$9a^2 - 13a + 4 = 0$$

$$(9a - 4)(a - 1) = 0$$

$$a = \frac{4}{9} \quad \text{and} \quad a = 1$$

$$x^2 = \frac{4}{9} \quad \text{and} \quad x^2 = 1$$

$$x = \pm\frac{2}{3} \quad \text{and} \quad x = \pm 1$$

$$4y^4 - 17y^2 + 4 = 0$$

Answer: $y = \pm\sqrt{2}$ and ± 3

---

66. Notice that we only got two solutions for the equation below. Solve the other equation.

$$x^4 - 8x^2 + 16 = 0$$

$$a^2 - 8a + 16 = 0$$

$$(a - 4)(a - 4) = 0$$

$$a = 4 \quad \text{and} \quad a = 4$$

$$x^2 = 4 \quad \text{and} \quad x^2 = 4$$

$$x^2 = \pm 2 \quad \text{and} \quad x^2 = \pm 2$$

$$9y^4 - 12y^2 + 4 = 0$$

Answer: $y = \pm\frac{1}{2}$ and ± 2

Answer: $y = \pm\sqrt{\frac{2}{3}}$

6-11 QUADRATIC INEQUALITIES

In this section, we will show a method for solving quadratic inequalities and graphing the solutions.

---

67. The inequality below is a quadratic inequality.

$$x^2 - x - 6 > 0$$

To solve it, we begin by factoring the left side.

$$(x + 2)(x - 3) > 0$$

The product $(x + 2)(x - 3)$ is greater than 0 when it is positive. The product is positive when both factors are positive or both factors are negative.

Continued on following page.

67. Continued

<u>The factors are positive when they are greater than 0.</u>

$x + 2 > 0$ $\qquad$ $x - 3 > 0$

$x > -2$ $\qquad$ $x > 3$

<u>The factors are negative when they are less than 0.</u>

$x + 2 < 0$ $\qquad$ $x - 3 < 0$

$x < -2$ $\qquad$ $x < 3$

That is, $x + 2$ is positive when $x > -2$ and negative when $x < -2$. $x - 3$ is positive when $x > 3$ and negative when $x < 3$. These facts are shown in the chart below. The critical values are -2 and 3.

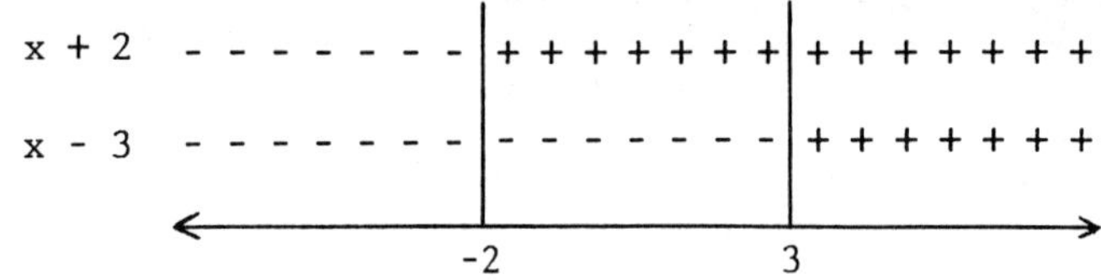

From the chart, we can see that both factors are negative when $x < -2$ and both factors are positive when $x > 3$. Therefore, the solution is:

$x < -2$ or $x > 3$

The solution is graphed below.

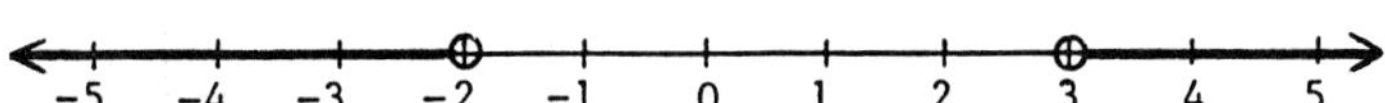

Using the same method, solve this inequality and graph the solution.

$$x^2 + 2x - 3 > 0$$

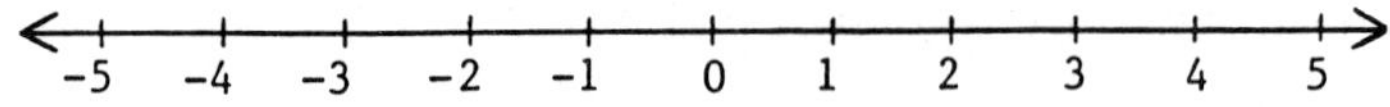

$x < -3$ or $x > 1$

-3 $\qquad$ 1

68. The inequality below is also a quadratic inequality. We factored the left side.

$$x^2 - x - 12 < 0$$

$$(x + 3)(x - 4) < 0$$

The product $(x + 3)(x - 4)$ is less than 0 when it is negative. The product is negative when the factors have opposite signs.

$x + 3$ is negative when $x < -3$ and positive when $x > -3$.

$x - 4$ is negative when $x < 4$ and positive when $x > 4$.

Those facts are shown in the chart below. The critical values are -3 and 4.

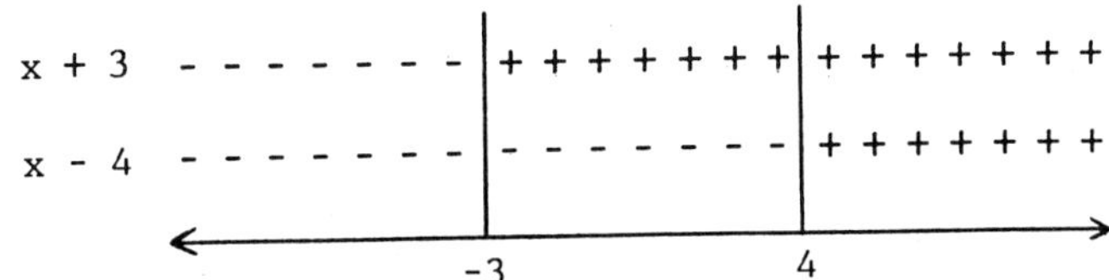

From the chart, we can see that the factors have opposite signs when x is between -3 and 4. Therefore, the solution is:

$$-3 < x < 4$$

The solution is graphed below.

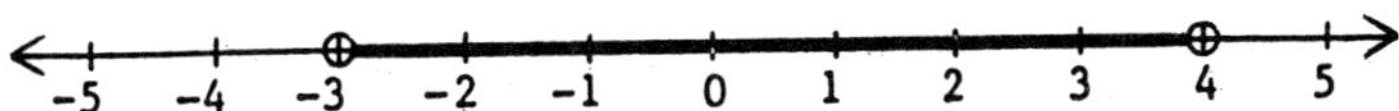

Using the same method, solve this inequality and graph the solution.

$$x^2 - x - 2 < 0$$

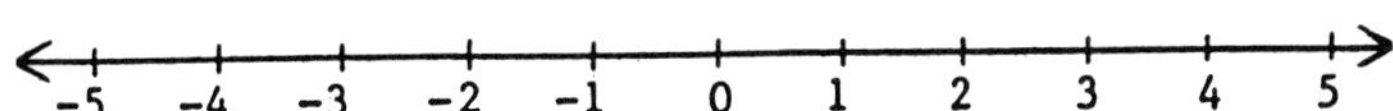

$-1 < x < 2$

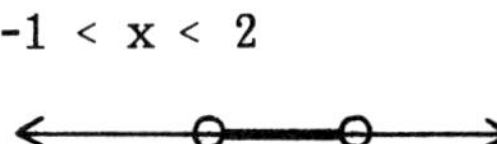

---

69. To solve the inequality below, we began by factoring the left side.

$$2x^2 - x - 6 \geq 0$$

$$(2x + 3)(x - 2) \geq 0$$

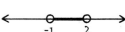

Continued on following page.

69. Continued

The product $(2x + 3)(x - 2)$ equals 0 when $x = -\frac{3}{2}$ or $x = 2$. The product is greater than 0 when both factors are positive or both factors are negative.

$2x + 3$ is positive when $x > -\frac{3}{2}$ and negative when $x < -\frac{3}{2}$.

$x - 2$ is positive when $x > 2$ and negative when $x < 2$.

Those facts are shown in the chart below.

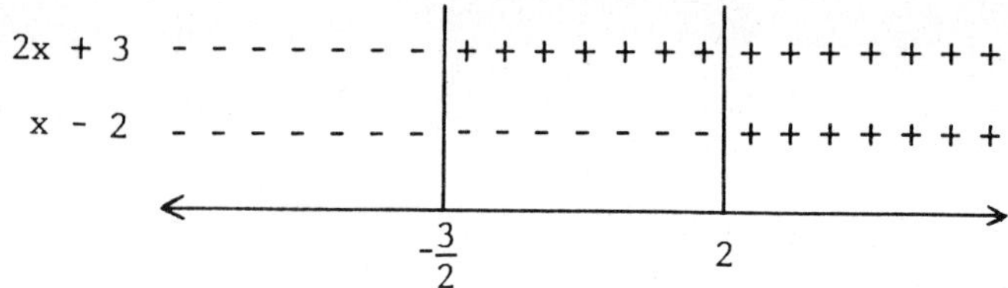

From the chart, we can see that both factors are negative when $x < -\frac{3}{2}$ and both factors are positive when $x > 2$. Therefore, the solution is:

$$x \leq -\frac{3}{2} \qquad \text{or} \qquad x \geq 2$$

The solution is graphed below. Notice the solid points at $-\frac{3}{2}$ and 2.

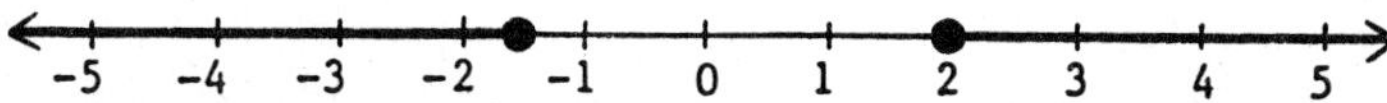

Using the chart above, write the solution for this inequality and then graph it.

$$2x^2 - x - 6 \leq 0$$

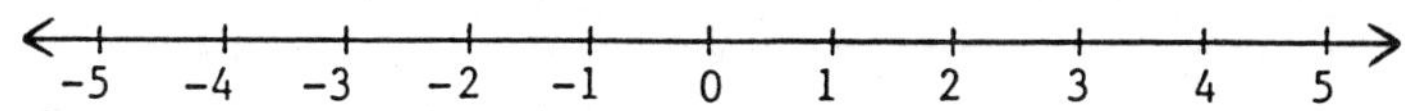

$-\frac{3}{2} \leq x \leq 2$

$-\frac{3}{2}$ 2

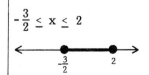

70. To solve the inequality below, we began by factoring the left side.

$$2x^2 + 7x + 3 \le 0$$
$$(2x + 1)(x + 3) \le 0$$

The product $(2x + 1)(x + 3)$ equals 0 when $x = -\frac{1}{2}$ and $x = -3$. The product is less than 0 when both factors have opposite signs.

$2x + 1$ is negative when $x < -\frac{1}{2}$ and positive when $x > -\frac{1}{2}$.

$x + 3$ is negative when $x < -3$ and positive when $x > -3$.

Those facts are shown in the chart below. The critical values are $-\frac{1}{2}$ and -3.

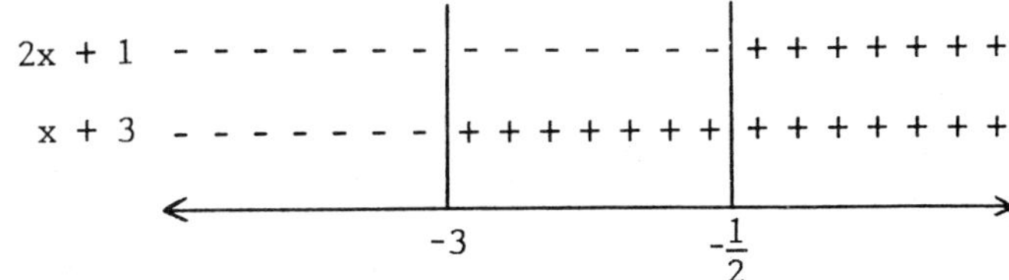

From the chart, we can see that the factors have opposite signs when x is between -3 and $-\frac{1}{2}$. Therefore, the solution is:

$$-3 \le x \le -\frac{1}{2}$$

The solution is graphed below.

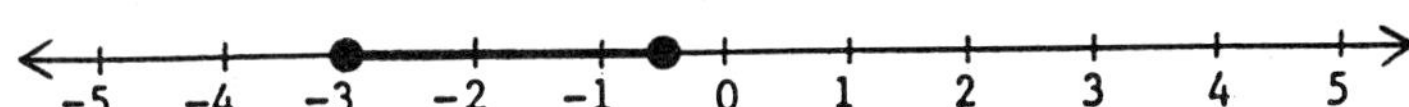

Using the chart above, write the solution for this inequality and then graph it.

$$2x^2 + 7x + 3 \ge 0$$

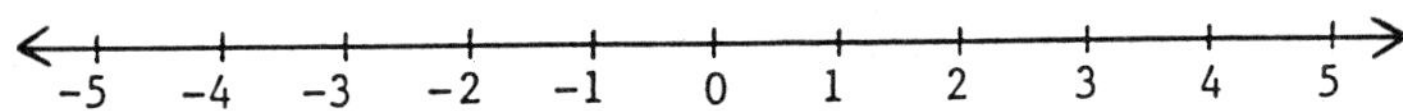

$x \le -3$ or $x \ge -\frac{1}{2}$

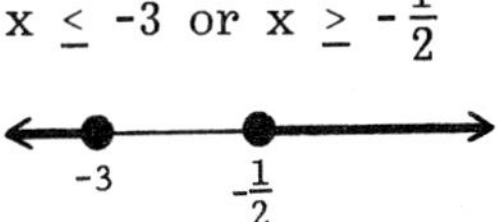

71. Solve each inequality and graph each solution.

a) $x^2 - 5x + 4 > 0$

b) $2x^2 - 9x + 7 \le 0$

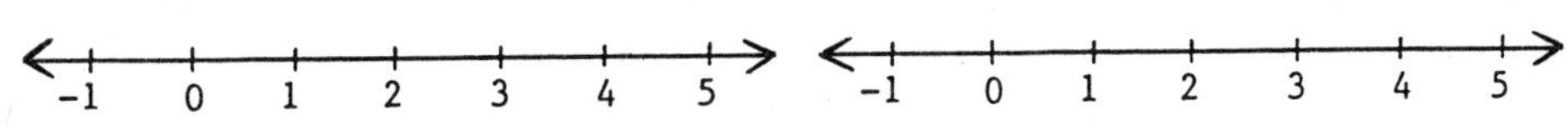

72. Though the inequality below is not a quadratic inequality, the same general method can be used to solve it.

$$\frac{x + 3}{x - 1} > 0$$

For the fraction to be greater than 0 or positive, both x + 3 and x - 1 must be either positive or negative.

x + 3 is positive when x > -3 and negative when x < -3.

x - 1 is positive when x > 1 and negative when x < 1.

Those facts are shown in the chart below. The critical values are -3 and 1.

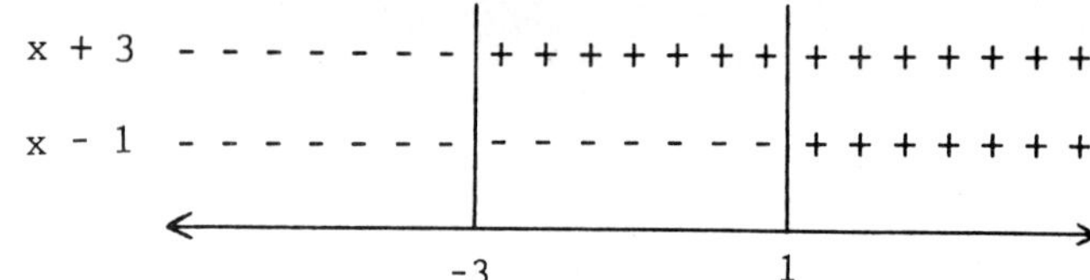

From the chart, we can see that x + 3 and x - 1 have the same signs when x is less than -3 or greater than "1". Therefore, the solution is:

$$x < -3 \quad \text{or} \quad x > 1$$

We graphed the solution below.

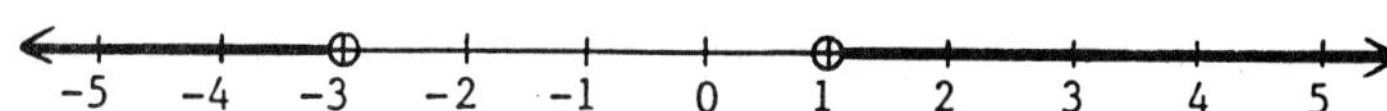

Using the same chart, write the solution for this inequality and graph the solution.

$$\frac{x + 3}{x - 1} < 0$$

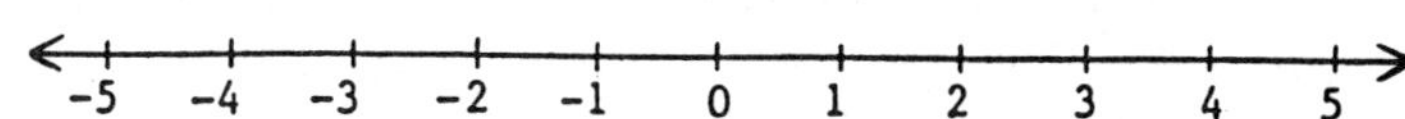

a) $x < 1$ or $x > 4$

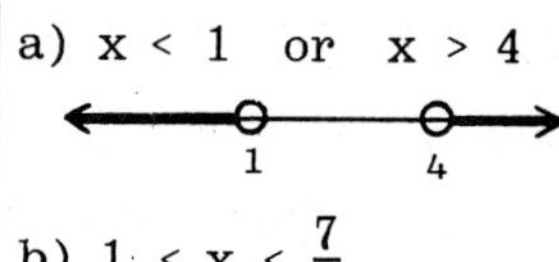

b) $1 \le x \le \frac{7}{2}$

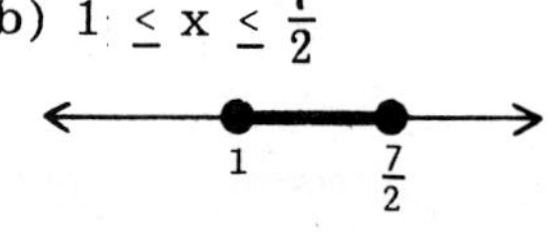

$-3 < x < 1$

-3 1

SELF-TEST 23 (pages 316-329)

1. Find the diagonal of a rectangle if its length is 5 ft and its width is 3 ft.

2. Find the second leg of a right triangle if the first leg is 6 cm and the hypotenuse is 11 cm.

Solve each equation.

3. $(2x^2 + x)^2 - 7(2x^2 + x) + 6 = 0$

4. $y^4 - 7y^2 + 12 = 0$

Use the discriminant to determine the nature of the solutions of each equation.

5. $x^2 - 6x + 3 = 0$

6. $6x^2 - x - 2 = 0$

7. $3x^2 + 2x + 1 = 0$

Solve each inequality and graph the solution.

8. $x^2 + x - 2 \geq 0$

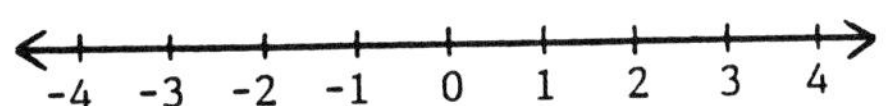

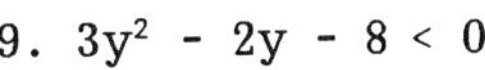

9. $3y^2 - 2y - 8 < 0$

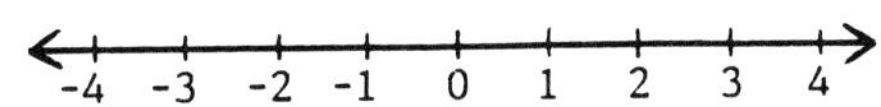

ANSWERS:

1. $d = 5.831$
2. $\text{leg} = 9.22$
3. $x = \frac{1}{2}, -1, \frac{3}{2}, -2$
4. $y = \pm 2$ and $\pm\sqrt{3}$
5. two real, irrational solutions ($b^2 - 4ac = 24$)
6. two real, rational solutions ($b^2 - 4ac = 49$)
7. two complex solutions ($b^2 - 4ac = -8$)
8. $x \leq -2$ or $x \geq 1$

-2 1

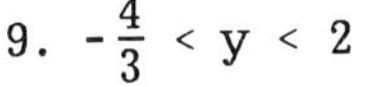

9. $-\frac{4}{3} < y < 2$

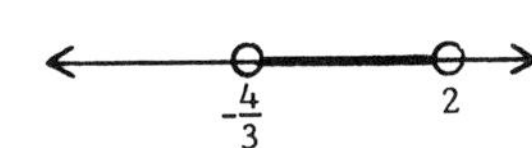

SUPPLEMENTARY PROBLEMS - CHAPTER 6

Assignment 21

Solve by factoring.

1. $x^2 - 3x - 40 = 0$
2. $3x^2 - 8x + 5 = 0$
3. $5t^2 + 7t - 6 = 0$
4. $m^2 - 81 = 0$
5. $x^2 - 12x = 0$
6. $5d^2 + 6d = 0$
7. $6t^2 = 11t - 3$
8. $25m^2 = 1$
9. $b^2 = 5b$

Solve by the square root method.

10. $x^2 = 36$
11. $m^2 - 1 = 0$
12. $49y^2 - 36 = 0$
13. $1 = \frac{4b^2}{9}$
14. $h^2 - 5 = 0$
15. $\frac{5x^2}{3} = 2$
16. $3y^2 + 27 = 0$
17. $16p^2 + 1 = 0$
18. $(x + 4)^2 = 36$
19. $(3y - 1)^2 = 16$
20. $(y - 7)^2 = -25$
21. $(2b + 5)^2 = -49$

Complete the square.

22. $x^2 + 6x +$ ______ $= (x +$ ______$)^2$
23. $y^2 - 14y +$ ______ $= (y -$ ______$)^2$
24. $d^2 + d +$ ______ $= (d +$ ______$)^2$
25. $m^2 - 5m +$ ______ $= (m -$ ______$)^2$

Solve by completing the square.

26. $x^2 - 8x + 7 = 0$
27. $x^2 - 4x - 11 = 0$
28. $y^2 + 10y + 20 = 0$
29. $t^2 - 3t + 1 = 0$
30. $3y^2 + 4y - 4 = 0$
31. $2b^2 - 5b - 1 = 0$

Solve with the quadratic formula.

32. $x^2 - 2x - 8 = 0$
33. $6t^2 + t - 2 = 0$
34. $4w^2 - w - 3 = 0$
35. $3d^2 - 5d - 1 = 0$
36. $5p^2 - 8p + 2 = 0$
37. $2F^2 + F - 7 = 0$
38. $h^2 + 2h - 1 = 0$
39. $x^2 - 3x + 5 = 0$
40. $5y^2 + 2y + 4 = 0$

Assignment 22

Write in standard form.

1. $-x^2 + 5x - 7 = 0$
2. $y^2 = 9 - 3y$
3. $5t - 4 = t^2 + 2t$
4. $(x + 1)(x - 4) = 7x - 2$
5. $(3y - 1)(y - 2) = y + 9$
6. $2 + \frac{1}{x} = \frac{9}{x^2}$
7. $\frac{m + 5}{m} = \frac{1}{m - 5}$
8. $x + 2 = \frac{3}{x + 2}$
9. $\frac{4}{x + 3} + \frac{4}{x - 3} = 2$

Write in standard form and then solve.

10. $5x - 2 = 3x^2$
11. $t(2t - 5) = 3$
12. $(8 - x)(6 - x) = 15$
13. $1 + \frac{2}{x} = \frac{7}{x^2}$
14. $x - 2 = \frac{7}{x - 2}$
15. $\frac{3}{y + 2} + \frac{3}{y - 2} = 1$

Solve each problem.

16. Working together, Joan and Judy can complete a job in 4 hours. It would take Joan 6 hours longer than Judy to do the job alone. How long would it take each to do the job working alone?

17. When both are working, two pipes can fill a tank in 8 hours. The larger pipe, working alone, can fill the tanks in 12 hours less time than the smaller pipe. How long would it take each pipe working alone to fill the tank?

18. The width of a rectangle is 3 ft less than the length. If the area is 6.75 ft^2, find the length and width.

19. The length of a rectangle is 2m more than the width. If the area is 11.25m^2, find the length and width.

20. The dimensions of a picture frame are 20 in by 26 in. The area of the picture is 280 in^2. Find the width of the frame.

21. The dimensions of a picture frame are 10 in by 12 in. The area of the picture is 80 in^2. Find the width of the frame.

22. A river has a 2 mph current. A boat travels 16 miles upstream and 16 miles downstream in a total of 6 hours. What is the speed of the boat in still water?

23. A river has a current of 3 mph. A boat travels 20 miles upstream and 20 miles downstream in a total of 7 hours. Find the speed of the boat in still water.

Do these evaluations.

24. In $A = 6s^2$, find A when s = 10.

25. In $s = \frac{1}{2}at^2$, find s when a = 60 and t = 3.

26. In $d = 16t^2$, find t when d = 64.

27. In $A = 3.14r^2$, find r when A = 6.28.

28. In $E = \frac{1}{2}mv^2$, find v when E = 27 and m = 6.

29. In $F = \frac{m_1m_2}{rd^2}$, find d when F = 4, $m_1 = 10$, $m_2 = 20$, and r = 2.

30. In $Z^2 = R^2 + X^2$, find R when Z = 13 and X = 5.

31. In $P = ht^2 - dt$, find t when P = 10, h = 2, and d = 1.

Do these rearrangements.

32. Solve for P.

$P^2 = Q^2 + R^2$

33. Solve for a.

$a^2 = \frac{b}{c}$

34. Solve for r.

$A = \pi r^2$

35. Solve for v.

$E = \frac{1}{2}mv^2$

36. Solve for d.

$F = \frac{m_1m_2}{rd^2}$

37. Solve for G.

$B^2 = F^2 + G^2$

38. Solve for s.

$s^2 + as = b$

39. Solve for r.

$A = \pi r^2 + \pi rs$

Assignment 23

Solve each problem.

1. Find the hypotenuse of a right triangle if its legs are 8 cm and 5 cm.

2. Find the diagonal of a square whose side is 3 ft.

3. Find the second leg of a right triangle if the first leg is 6 in and the hypotenuse is 10 in.

4. Find the length of a rectangle if the diagonal is 9m and the width is 7m.

5. A 15 ft wire is stretched from the ground to the top of a garage. The height of the garage is 3 ft more than the distance of the wire from the bottom of the garage. Find the height of the garage.

6. Two cars left the same place at the same time. One headed due south, the other due west. When they were exactly 50 miles apart, the car headed due south has gone 10 miles farther. How far had each car traveled?

Use the discriminant to determine the nature of the solutions for each equation.

7. $3x^2 + 2x - 8 = 0$ 8. $x^2 + 8x + 16 = 0$ 9. $2y^2 - 4y - 7 = 0$

10. $4m^2 - 3m + 2 = 0$ 11. $t^2 - 4t - 9 = 0$ 12. $p^2 + 5p + 10 = 0$

13. $5x^2 - 9x + 4 = 0$ 14. $4x^2 - 20x + 25 = 0$ 15. $3d^2 - d - 7 = 0$

Solve each equation.

16. $(x^2 - 4)^2 - 3(x^2 - 4) - 10 = 0$ 17. $(x^2 - 1)^2 - (x^2 - 1) - 2 = 0$

18. $(x^2 + 2x)^2 - 11(x^2 + 2x) + 24 = 0$ 19. $(3y^2 - y)^2 - 6(3y^2 - y) + 8 = 0$

20. $t^4 - 5t^2 + 6 = 0$ 21. $m^4 - 8m^2 + 7 = 0$

22. $2x^4 - 11x^2 + 12 = 0$ 23. $3y^4 - 28y^2 + 9 = 0$ 24. $x^4 - 10x^2 + 25 = 0$

Solve each inequality and graph the solution.

25. $x^2 - 2x - 3 > 0$

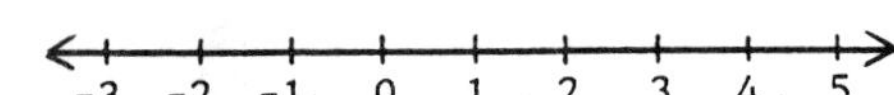

26. $y^2 + y - 2 < 0$

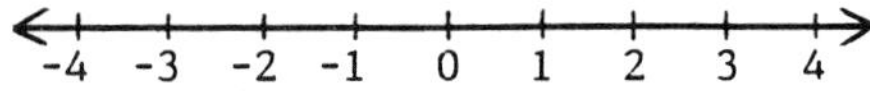

27. $m^2 + m - 6 \geq 0$

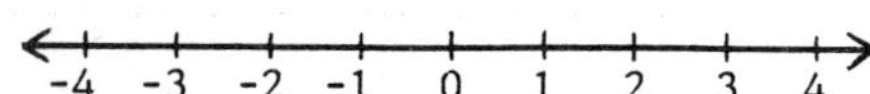

28. $b^2 - 3b - 4 \leq 0$

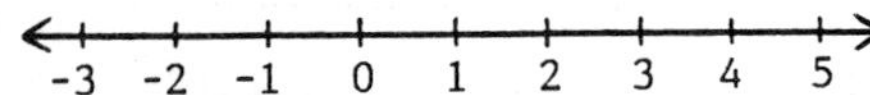

29. $2x^2 - 5x - 3 > 0$

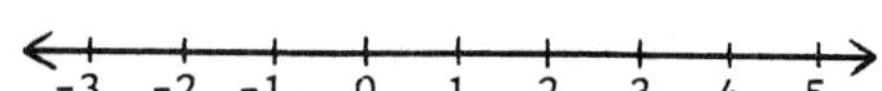

30. $2t^2 + t - 6 \leq 0$

-4 -3 -2 -1 0 1 2 3 4

31. $2x^2 - 7x + 3 < 0$

-3 -2 -1 0 1 2 3 4 5

32. $d^2 + 5d + 4 \geq 0$

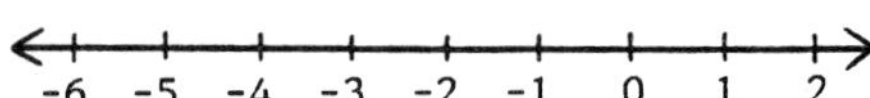

33. $\dfrac{x + 2}{x - 2} > 0$

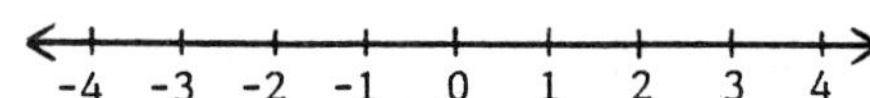

34. $\dfrac{x + 1}{x - 4} \leq 0$

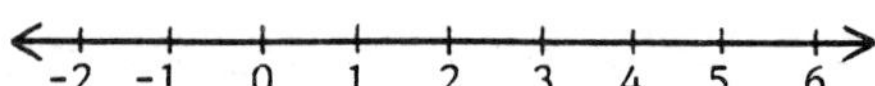

Graphing and Linear Functions

7

In this chapter, we will introduce the coordinate system and discuss the methods for graphing linear equations and linear inequalities. Slope, slope-intercept form, point-slope form, and parallel and perpendicular lines are discussed. Functions are also discussed, as well as the following types of variation: direct, inverse, direct square, inverse square, joint, and combined.

7-1 LINEAR EQUATIONS

In this section, we will discuss linear equations and their solutions.

---

1. A linear equation is an equation with no more than two variables that is either in the form or can be put in the form:

$$Ax + By = C$$

where A, B, and C can be any real number, but A and B cannot both be 0.

The following equations are linear equations because they are in the form $Ax + By = C$.

$2x + 3y = 6$ $\qquad\qquad$ $5x - y = 0$

Continued on following page.

1. Continued

The following equations <u>are</u> linear equations because they can be put in the form $Ax + By = C$.

$y = 3x - 1$ $\qquad$ $y = 6x$

In a linear equation, each term containing a variable is a first-degree term. The following equations <u>are not</u> linear equations because they contain a second-degree term.

$y = x^2$ $\qquad$ $y = 2x^2 + 1$

Which of the following are linear equations? ______________

a) $y = 2x^2$ b) $y = 2x$ c) $5x - y = 1$ d) $y = 5x^2 - 1$

---

2. A solution of a linear equation is a pair of values that satisfies the equation. For example, $x = 3$ and $y = 7$ is a solution of $y = 2x + 1$ since:

Answer to frame 1: (b) and (c)

$$y = 2x + 1$$
$$7 = 2(3) + 1$$
$$7 = 6 + 1$$
$$7 = 7$$

Usually $x = 3$ and $y = 7$ is written (3,7), with the x-value written first. Therefore, the <u>order</u> in which the numbers are written makes a difference. That is:

(3,7) means: $x = 3, y = 7$

(7,3) means: $x = 7, y = 3$

Since the order in which the numbers are written makes a difference, (3,7) and (7,3) are called <u>ordered pairs</u>.

a) Write $x = -1$, $y = 5$ as an ordered pair. ______________

b) Write $y = -4$, $x = 9$ as an ordered pair. ______________

---

3. To decide whether an ordered pair is a solution of a linear equation, we substitute the values and see whether they satisfy the equation. For example:

Answers to frame 2:

a) (-1,5)

b) (9,-4)

(4,3) is a solution of $2x - y = 5$, since $2(4) - 3 = 5$.

(-1,5) is not a solution of $y = 3x$, since $5 \neq 3(-1)$.

a) Is (10,4) a solution of $x - y = 3$? __________

b) Is (2,-1) a solution of $3x + 2y = 4$? __________

---

4. Any linear equation has many solutions. For example, each ordered pair below is a solution of $y = 2x$.

Answers to frame 3:

a) No, since: $10 - 4 \neq 3$

b) Yes, since: $3(2) + 2(-1) = 4$

(5,10)	(-1,-2)
(1.5,3)	(-2.5,-5)
$\left(\frac{1}{2},1\right)$	$\left(-\frac{7}{2},-7\right)$
(0,0)	(-40,-80)

How many more solutions are there for $y = 2x$?

5. We completed the solutions (4,) and (,0) for $y = 2x + 3$ below.

To complete (4,), we substituted 4 for x and found the corresponding value of y by evaluating.

$$y = 2x + 3$$
$$y = 2(4) + 3$$
$$y = 8 + 3$$
$$y = 11$$

The ordered pair is (4,11).

To complete (,0), we substituted 0 for y and found the corresponding value of x by solving an equation.

$$y = 2x + 3$$
$$0 = 2x + 3$$
$$-3 = 2x$$
$$x = -\frac{3}{2}$$

The ordered pair is $\left(-\frac{3}{2},0\right)$.

Complete the solutions (-2,) and (,-5) for $y = 2x + 3$ below.

a) (-2,)

b) (,-5)

An infinite number

---

6. We completed the solutions (2,) and (,-8) for $4x - y = 12$ below.

To complete (2,), we substituted 2 for x and got:

$$4x - y = 12$$
$$4(2) - y = 12$$
$$8 - y = 12$$
$$-y = 4$$
$$y = -4$$

The ordered pair is (2,-4).

To complete (,-8), we substituted -8 for y and got:

$$4x - y = 12$$
$$4x - (-8) = 12$$
$$4x + 8 = 12$$
$$4x = 4$$
$$x = 1$$

The ordered pair is (1,-8).

Complete (0,) and (,8) for $4x - y = 12$ below.

a) (0,)

b) (,8)

a) (-2,-1)

b) (-4,-5)

---

a) (0,-12)

b) (5,8)

7-2 THE COORDINATE SYSTEM

In this section, we will introduce the rectangular (or Cartesian) coordinate system that is used to graph equations.

---

7. The rectangular coordinate system is shown at the right. It consists of a horizontal number line used for x-values and a vertical number line used for y-values.

The horizontal number line is called the horizontal axis or x-axis.

The vertical number line is called the vertical axis or y-axis.

The two number lines together are called coordinate axes.

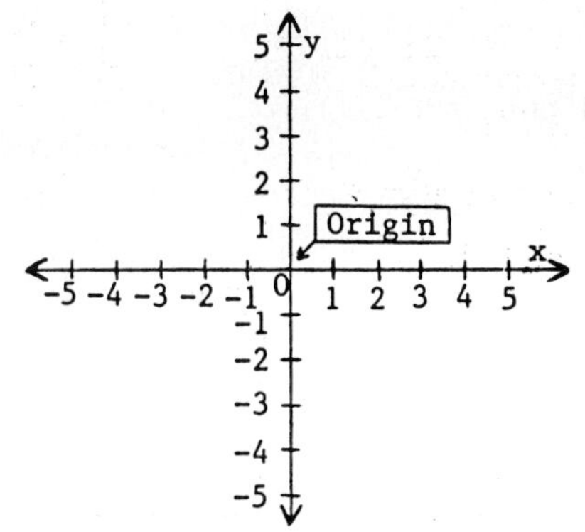

The two coordinate axes intercept at 0 on each. That point is called the ______________.

---

origin

8. Each point on the coordinate system represents an ordered pair. To find the ordered pair for point A:

1. We drew an arrow down to the horizontal axis. The arrow points to x = 3.
2. We drew an arrow over to the vertical axis. The arrow points to y = 4.
3. Therefore, point A represents (3,4).

Similarly: Point B represents (,).

Point C represents (,).

Point D represents (,).

B (-2,3)

C (-5,-4)

D (1,-2)

9. When scaling the axes at the right, we counted by 5's on the x-axis and by 10's on the y-axis. Therefore, point A represents (10,30).

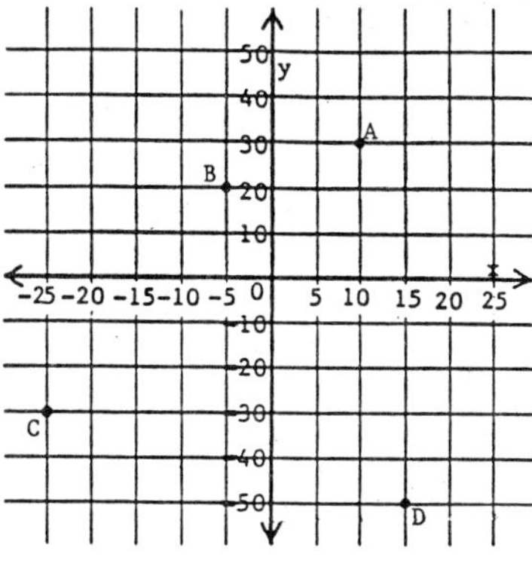

10 and 30 are called the coordinates of point A. 10 is called the x-coordinate. 30 is called the y-coordinate.

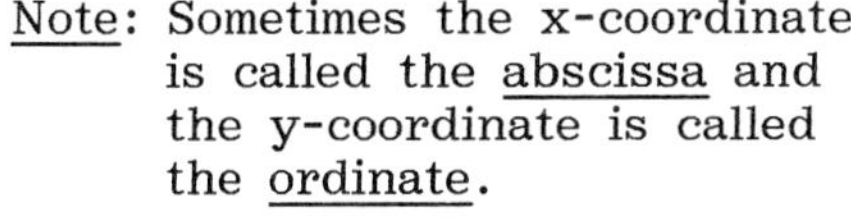

Note: Sometimes the x-coordinate is called the abscissa and the y-coordinate is called the ordinate.

Write the coordinates of the other three points.

B ______

C ______

D ______

B (-5,20)

C (-25,-30)

D (15,-50)

10. Points A and B at the right lie on the horizontal axis. The y-coordinate for all points on that axis is 0. Using that fact, write the coordinates of A and B below.

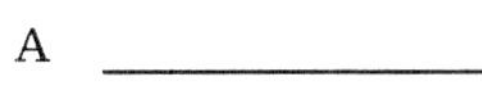

A ______

B ______

Points C and D at the right lie on the vertical axis. The x-coordinate for all points on that axis is 0. Using that fact, write the coordinates of C and D below.

C ______

D ______

The origin lies on both axes. Write the coordinates of the origin below.

Origin ______

A (-4,0)
B (2,0)

C (0,3)
D (0,-5)

Origin (0,0)

11. Point A at the right does not lie at an intersection of coordinate lines. Therefore, we have to estimate its coordinates. We get:

$\left(2\frac{1}{2}, 7\right)$

Estimate the coordinates of these points.

B ______________

C ______________

D ______________

E ______________

F ______________

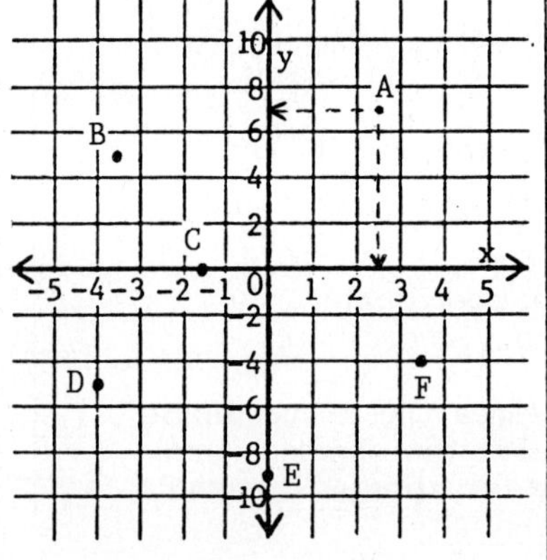

---

12. Points A and B at the right do not lie at an intersection of coordinate lines. Estimating their coordinates, we get:

A (25,45)

B (-32,18)

Estimate the coordinates of these points.

C ______________

D ______________

E ______________

F ______________

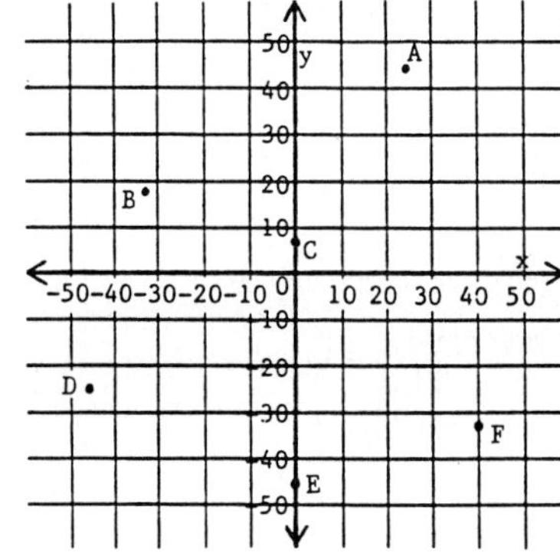

Your answers should be close to these:

B $\left(-3\frac{1}{2}, 5\right)$

C $\left(-1\frac{1}{2}, 0\right)$

D (-4,-5)

E (0,-9)

F $\left(3\frac{1}{2}, -4\right)$

---

13. The coordinate axes divide the coordinate system into four parts called quadrants. We labeled the four quadrants in Figure 1 below. Notice that they are numbered in a counter-clockwise direction, beginning with the upper right quadrant.

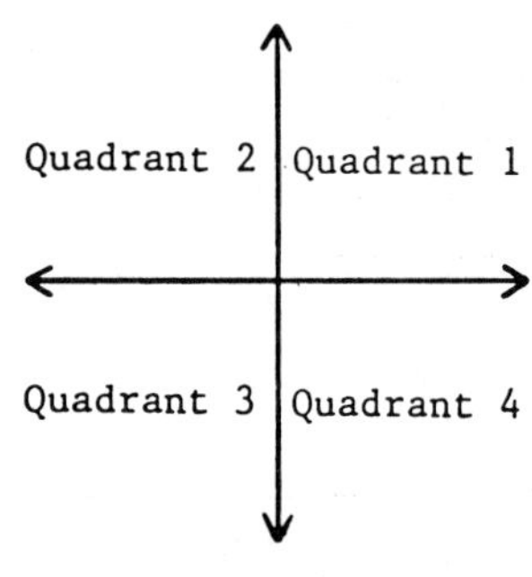

Figure 1

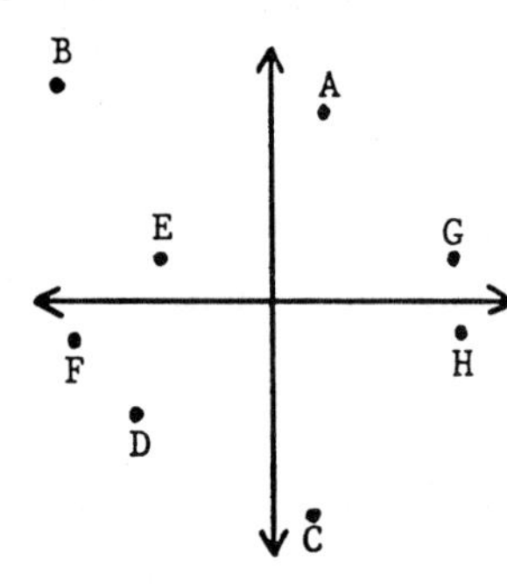

Figure 2

On the coordinate system in Figure 2 above, there are two points labeled in each quadrant.

a) Points B and E lie in Quadrant _____.

b) Points C and H lie in Quadrant _____.

Your answers should be close to these:

C (0,7)

D (-46,-25)

E (0,-45)

F (40,-32)

14. To locate or <u>plot</u> a point on the coordinate system, we simply reverse the procedure for reading the coordinates of a point. For example, we have plotted points A and B at the right by drawing arrows from the axes.

A (2,4)

B (-4,-3)

Plot and label the four points below on the same coordinate system.

C (3,1) E (-2,5)

D (3,-2) F (-2,-4)

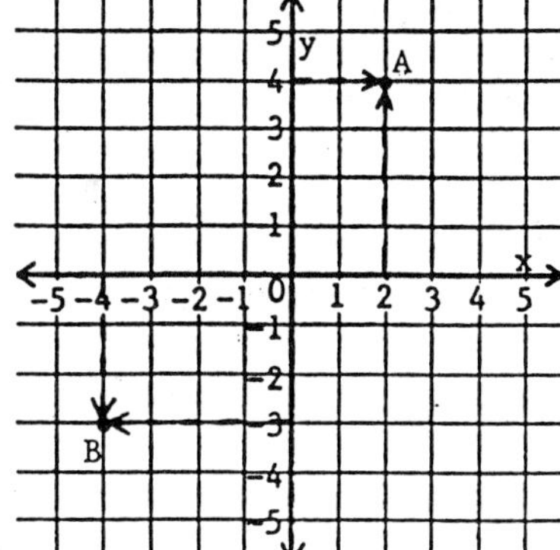

a) Quadrant 2

b) Quadrant 4

---

15. The correct plotting of the points from the last frame is shown at the right.

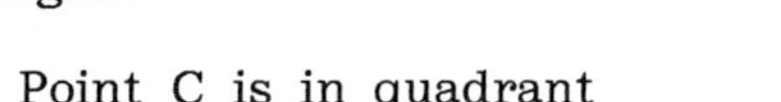
Point C is in quadrant ______.

Point D is in quadrant ______.

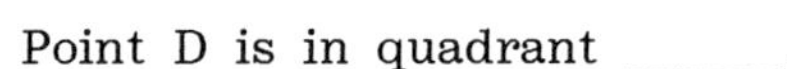
Point E is in quadrant ______.

Point F is in quadrant ______.

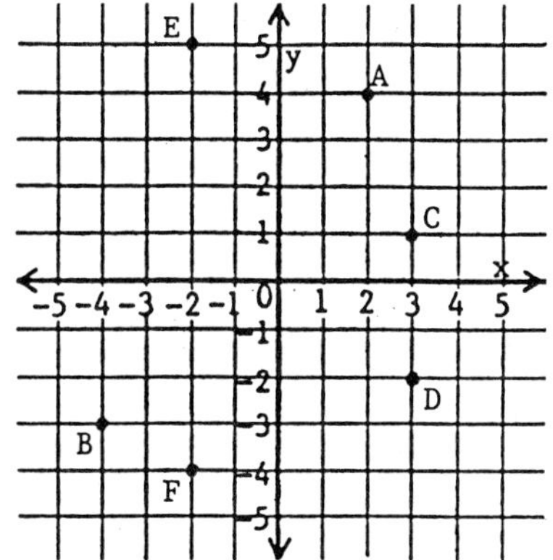

See graph at left for answer.

---

16. All four points below lie on an axis. Plot them at the right.

A (4,0)

B (0,4)

C (-3,0)

D (0,-3)

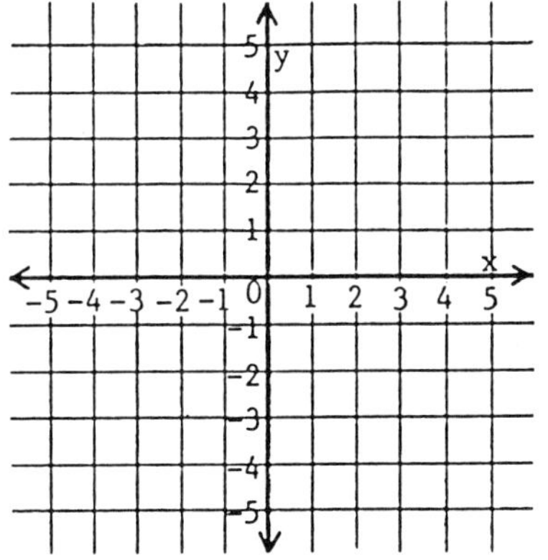

C (quadrant 1)

D (quadrant 4)

E (quadrant 2)

F (quadrant 3)

<u>Answer to Frame 16</u>:

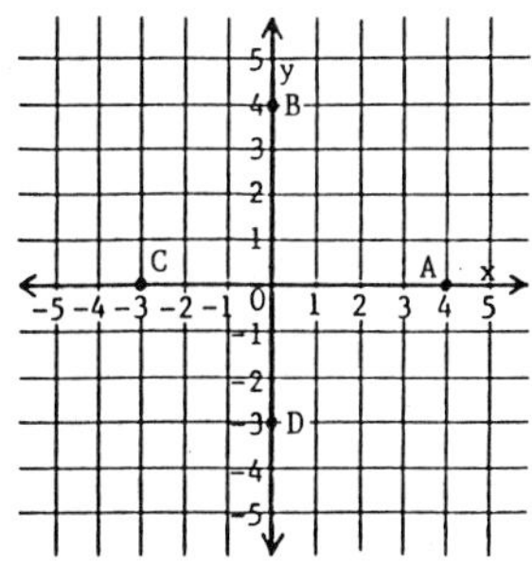

17. To plot the following points at the right, we had to estimate their positions.

A (50,90)

B (-75,45)

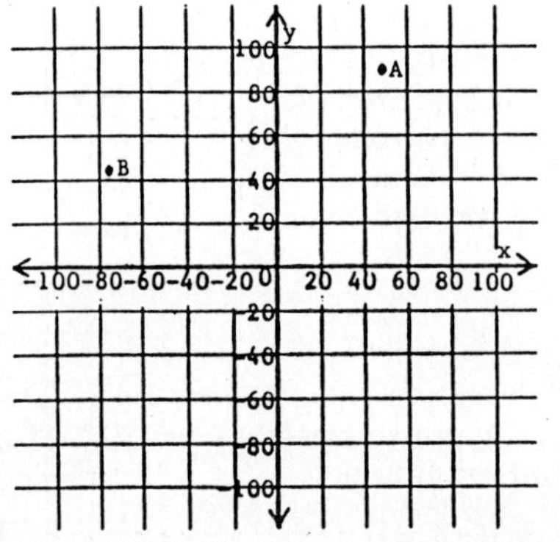

Plot and label these points.

C (0,32)

D (-47,-85)

E (30,-68)

F (73,0)

Answer to Frame 17:

Your plotted points should be approximately like those below.

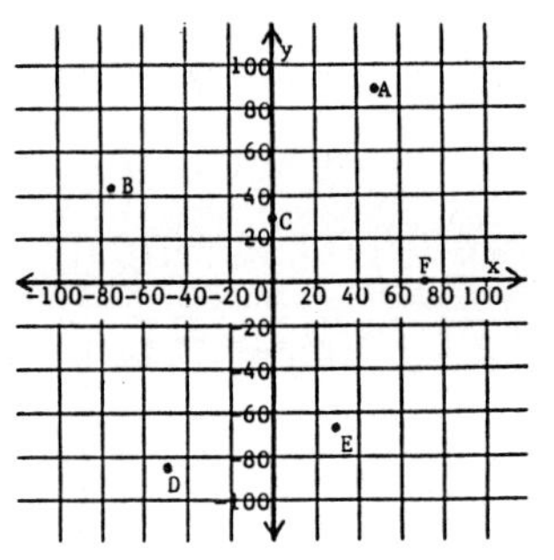

7-3 GRAPHING LINEAR EQUATIONS

The graph of any linear equation is a straight line. We will discuss the method for graphing linear equations in this section.

---

18. The graph of a linear equation is a drawing of its solutions on the coordinate system. Since the graph of any linear equation is a straight line, we can graph it by plotting a few points. Suppose we want to graph the equation below.

$$y = x + 3$$

To find some solutions to use for graphing, we substitute values for $\underline{x}$ and find the corresponding values for $\underline{y}$. We substitute a few positive values, 0, and a few negative values for $\underline{x}$. For example, substituting 5, 3, 0, -3, and -5, we get the solutions on the following page. They are listed both as ordered pairs and in a solution-table.

Continued on following page.

18. Continued

	x	y
(5,8)	5	8
(3,6)	3	6
(0,3)	0	3
(-3,0)	-3	0
(-5,-2)	-5	-2

Are all possible solutions for y = x + 3 listed in the table? _________

---

19. The three steps needed to graph the following equation are discussed below.

$$y = x + 1$$

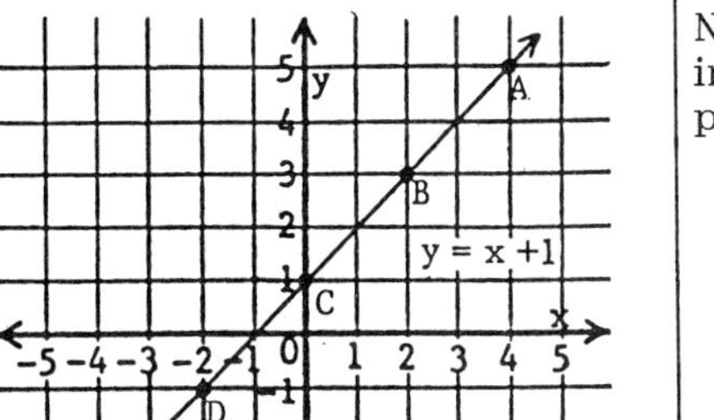

Step 1: Make up a solution-table. Notice that we avoid points like (5,6) that lie off the graph.

	x	y
A	4	5
B	2	3
C	0	1
D	-2	-1
E	-4	-3

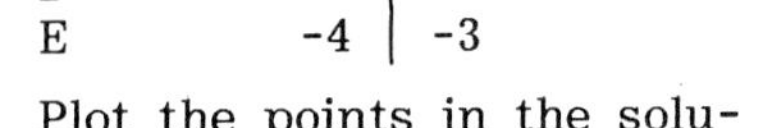

Step 2: Plot the points in the solution-table.

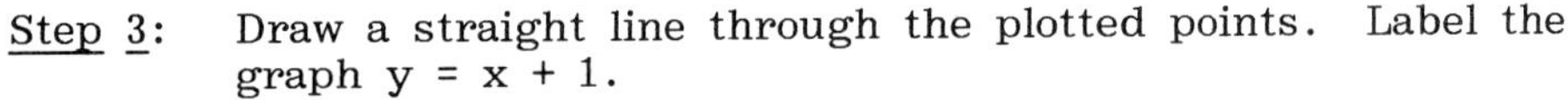

Step 3: Draw a straight line through the plotted points. Label the graph y = x + 1.

The straight line shown is the graph of y = x + 1.

No. There are an infinite number of possible solutions.

---

20. We used the solution-table shown to graph x + y = 1 in two different ways below.

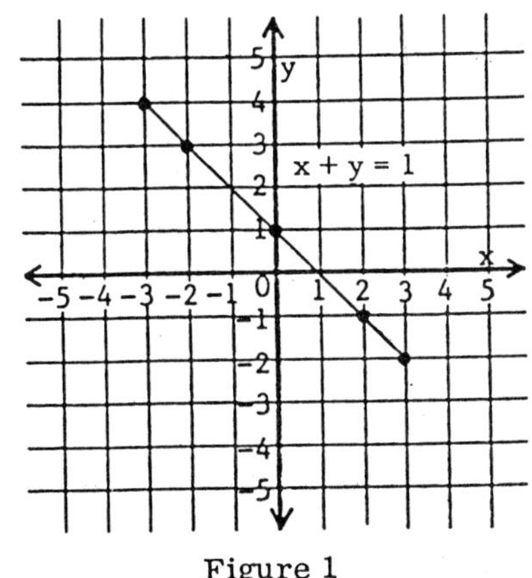

Figure 1

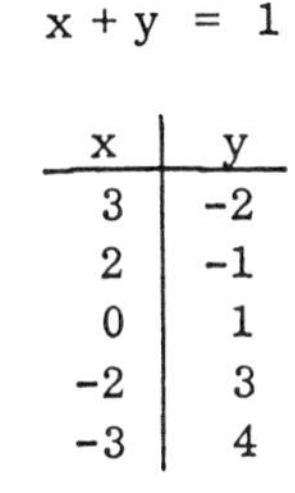

x + y = 1

x	y
3	-2
2	-1
0	1
-2	3
-3	4

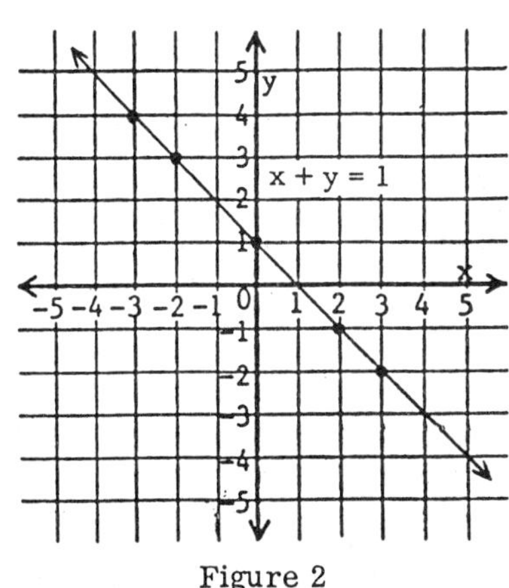

Figure 2

Note: In Figure 1, we stopped the graphed line at the last plotted point on each end.

In Figure 2, we extended the graphed line to the edge of the coordinate system shown and put arrowheads at each end of it.

Continued on following page.

20. Continued

Figure 2 is the correct graphing of $x + y = 1$ for these reasons:

1) The line should be extended to the edge of the coordinate system to show that there are other solutions, like (5,-4) and (-4,5), beyond (-3,4) and (3,-2).
2) Arrowheads should be put at each end of the graphed line to show that there are solutions, like (10,-9) and (-20,21), beyond the edge of the coordinate system shown.

---

21. Since two points determine a straight line, we only need to plot two points to graph a line. However, we always plot a third point as a check. Complete the table below and graph the equation.

$$y = -3x$$

x	y
2	
0	
-2	

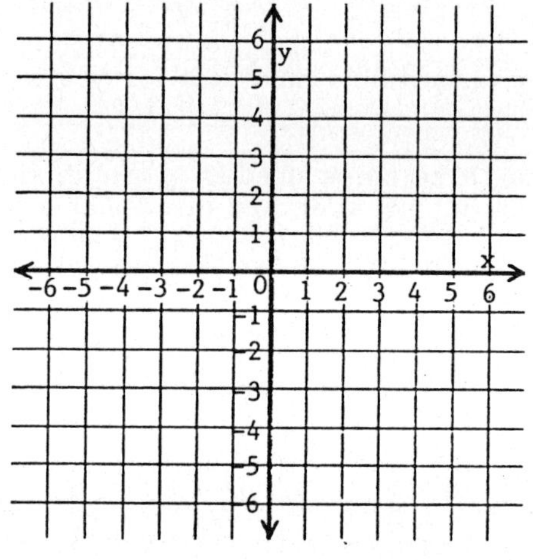

---

22. In each equation below, the numerical coefficient of x is a fraction. To avoid fractions for y, we substitute multiples of the denominator for x. That is, we substitute multiples of 3 at the left and multiples of 2 at the right. Complete the tables and graph each equation.

$y = -\frac{2}{3}x$ $\qquad$ $y = \frac{1}{2}x + 1$

x	y
6	
0	
-6	

x	y
4	
0	
-4	

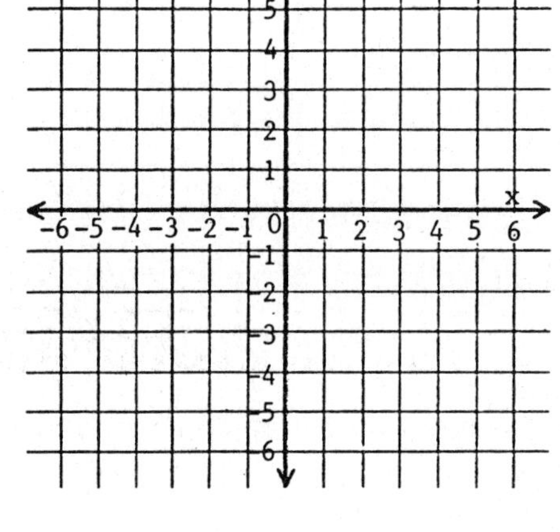

Answer to Frame 21:

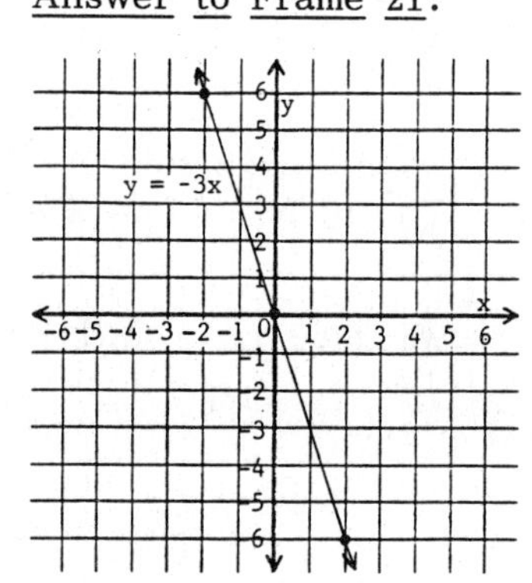

Answer to Frame 22:

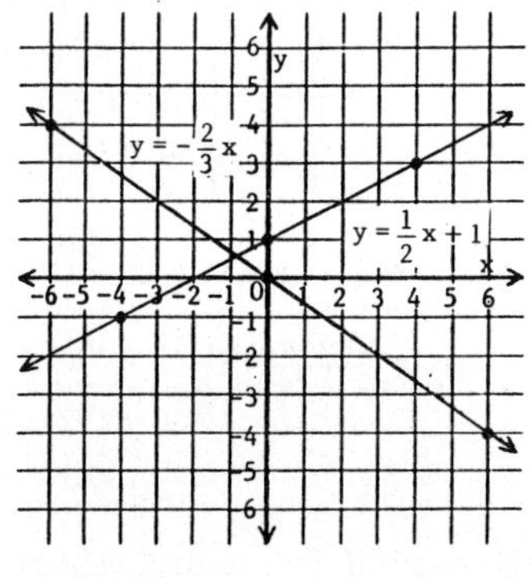

23. When graphing a linear equation, use three points that are far enough apart so that it is easy to draw the line accurately. To get the three points, substitute a negative value, 0, and a positive value for x. Pick your own three points to graph these.

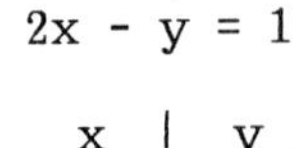

$2x - y = 1$

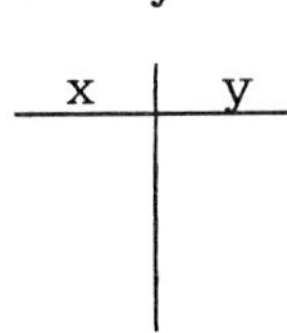

x	y

$y = -\frac{1}{2}x + 2$

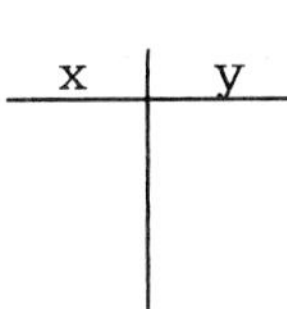

x	y

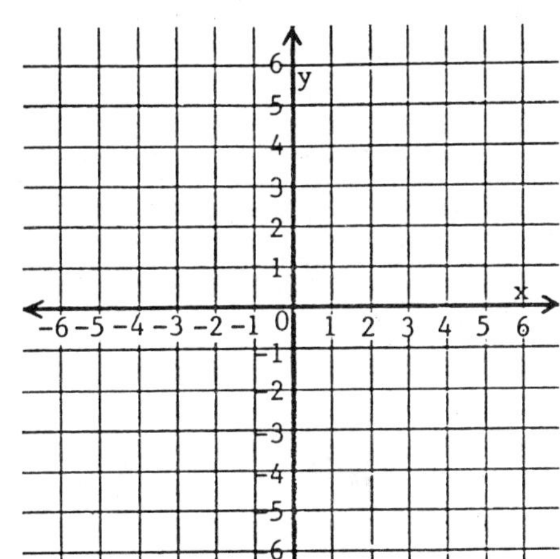

24. When graphing a linear equation, we frequently use the two intercepts with a third point as a check. The intercepts are the points where the line crosses the axes. An example is discussed below.

We graphed the linear equation below at the right.

$3x - 2y = 6$

The coordinates of the two intercepts are given.

The x-intercept is (2,0).

The y-intercept is (0,-3).

We can get the coordinates of the intercepts directly from the equation.

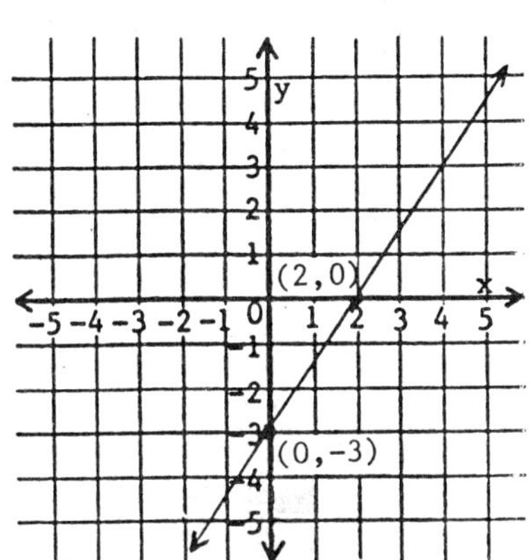

1) Finding the x-intercept: Since y is 0, we substitute 0 for y and solve for x.

$$3x - 2(0) = 6$$
$$3x = 6$$
$$x = 2$$

2) Finding the y-intercept: Since x is 0, we substitute 0 for x and solve for y.

$$3(0) - 2y = 6$$
$$-2y = 6$$
$$y = -3$$

Answer to Frame 23:

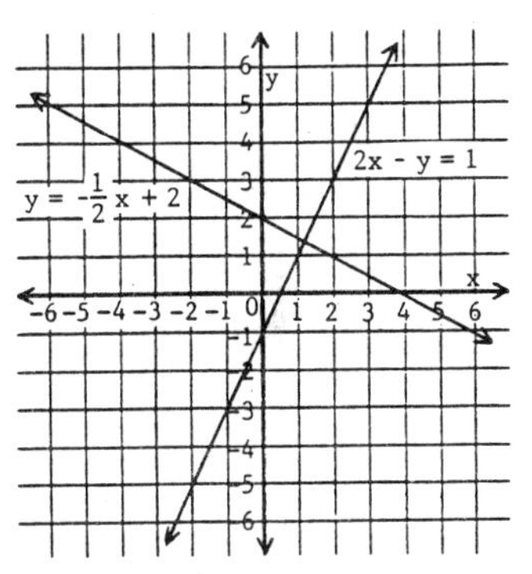

25. To find the coordinates of the intercepts, we do the following:

The x-intercept is (a,0). To find a, let y = 0 in the equation.
The y-intercept is (0,b). To find b, let x = 0 in the equation.

Find the coordinates of both intercepts for each equation.

a) $4y - 3x = 12$

x-intercept:

y-intercept:

b) $y = 2x - 5$

x-intercept:

y-intercept:

26. To graph $x + 2y = 4$ at the right, we plotted the two intercepts. Their coordinates are:

(0,2) (4,0)

We also plotted (-4,4) as a check.

a) Does (-4,4) lie on the line? ________

b) Is the graph probably correct? ________

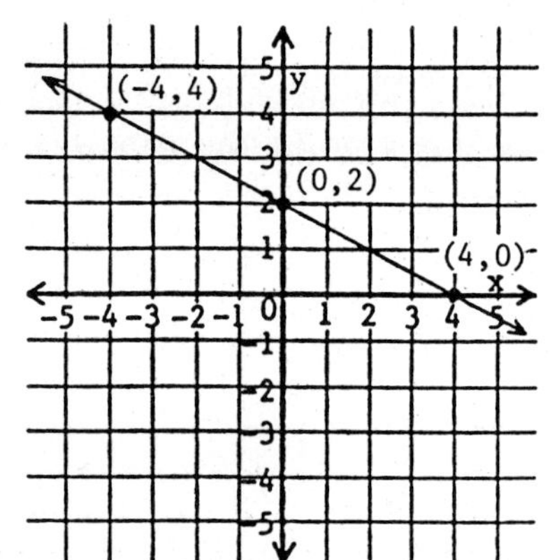

a) x-intercept (-4,0)
y-intercept (0,3)

b) x-intercept $\left(\frac{5}{2}, 0\right)$
y-intercept (0,-5)

27. Find the intercepts for each equation below and use them to graph the lines. Plot a third point as a check.

$4x - 2y = 8$ $2y = 5 - x$

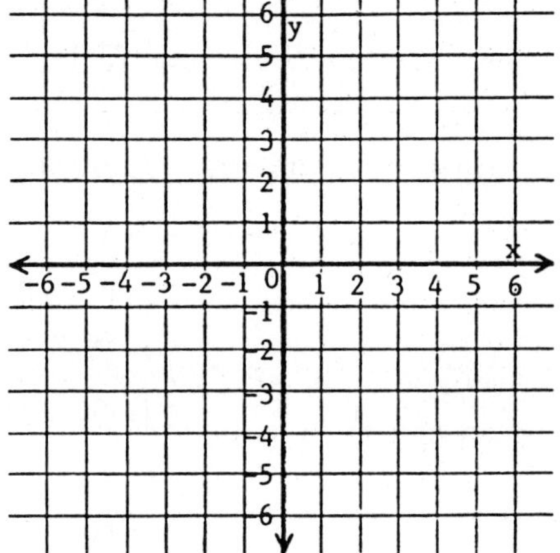

a) Yes

b) Yes

28. If we substitute 0 for x or y in the equation below, we get 0 for the other variable. Therefore, both intercepts are at the origin which is (0,0). To graph the equation, we need one more point and a third point as a check. Graph the equation at the right.

$y = -2x$

x	y
0	0

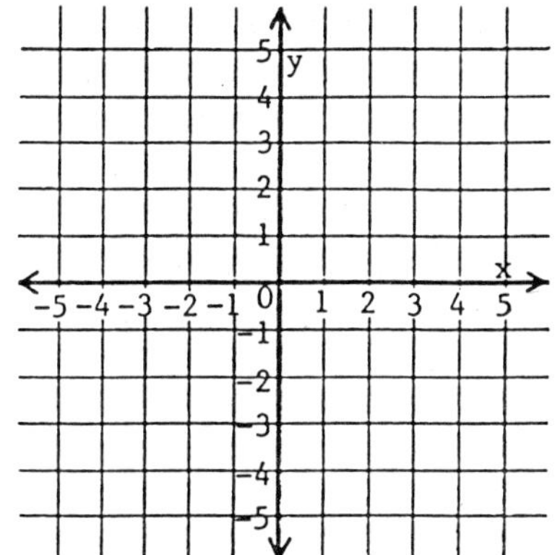

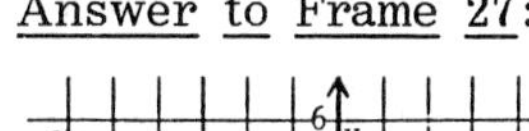

Answer to Frame 27:

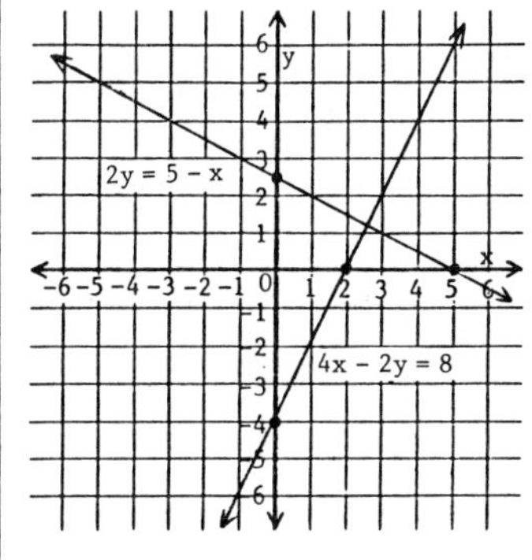

29. Two horizontal lines are shown at the right.

For line A, the y-coordinate is 2 for every value of x. Therefore, the equation of the line is y = 2.

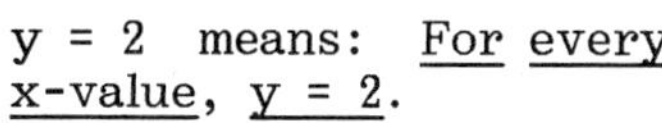

y = 2 means: For every x-value, y = 2.

For line B, the y-coordinate is -3 for every value of x. Therefore, the equation of the line is y = -3.

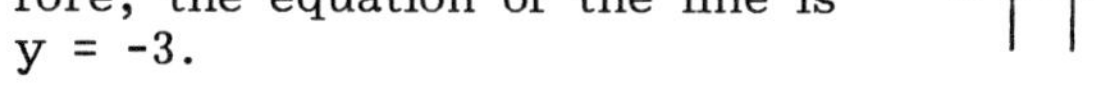

y = -3 means: For every x-value, y = ________.

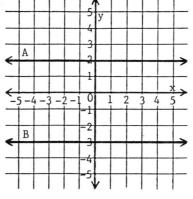

Answer to Frame 28:

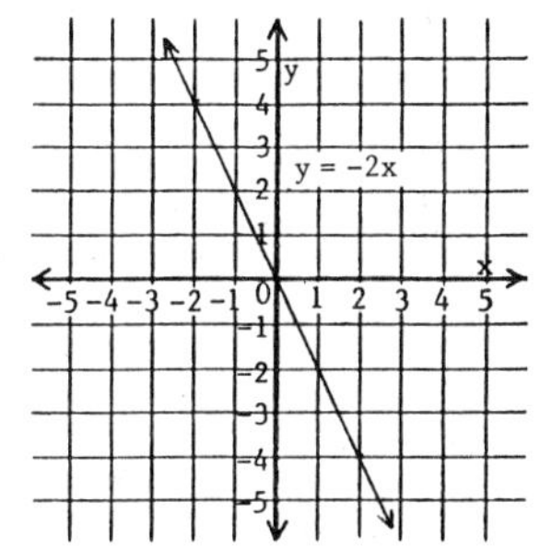

30. The x-axis is also a horizontal line on the coordinate system.

a) On the x-axis, for every x-value, y = ______.

b) Therefore, the equation of the x-axis is ______.

-3

31. Two vertical lines are shown at the right.

For line A, the x-coordinate is 4 for every value of y. Therefore, the equation of the line is x = 4.

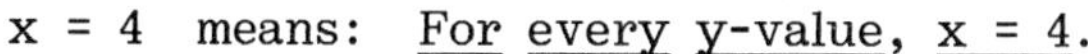

x = 4 means: For every y-value, x = 4.

For line B, the x-coordinate is -2 for every value of y. Therefore, the equation of the line is x = -2.

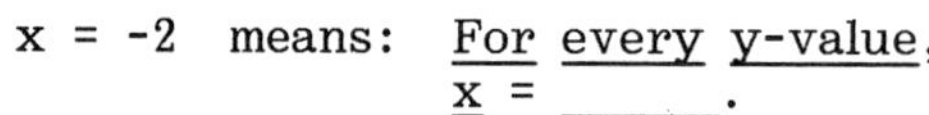

x = -2 means: For every y-value, x = ______.

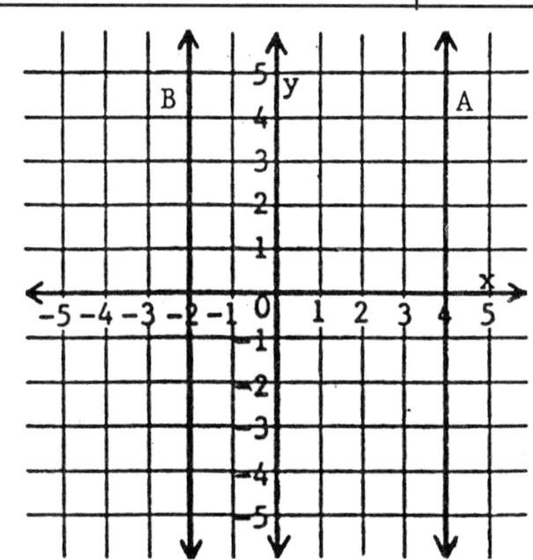

a) 0

b) y = 0

-2

32. The y-axis is also a vertical line on the coordinate system.

a) On the y-axis, for every y-value, x = _____.

b) Therefore, the equation of the y-axis is _________.

---

33. Four lines are drawn at the right. Write the equation of each line.

A: __________

B: __________

C: __________

D: __________

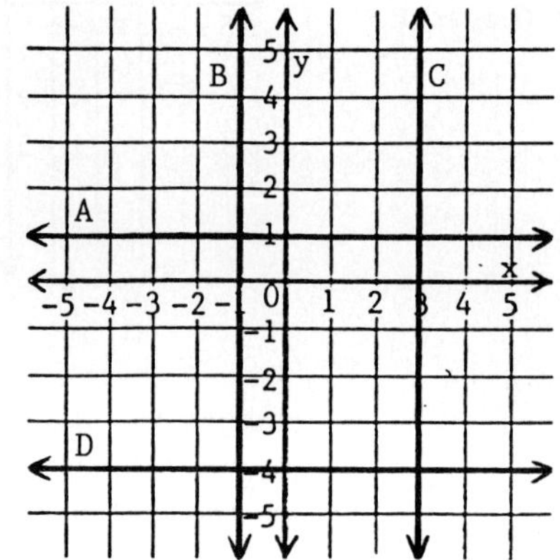

a) 0

b) x = 0

---

34. a) The graph of x = 0 is the __________ (x-axis, y-axis).

b) The graph of y = 0 is the __________ (x-axis, y-axis).

A: y = 1 C: x = 3

B: x = -1 D: y = -4

---

a) y-axis b) x-axis

SELF-TEST 24 (pages 333-346)

Decide whether the given ordered pair is a solution of the given equation.

1. (4,-3) for x - 2y = 8	2. (4,-4) for $y = \frac{1}{4}x - 5$

Complete the solution for each equation.

3. (4,) for x - y = 10	4. (,-5) for y = 4x - 3

5. The point (5,-3) lies in quadrant _____.

6. The point (-1,4) lies in quadrant _____.

7. The point (0,-2) lies on which axis? _____

8. The coordinates of the origin are __________.

Continued on following page.

SELF-TEST 24 - CONTINUED

Graph each equation.

9. $5x - 2y = 10$

10. $y = -\frac{1}{2}x$

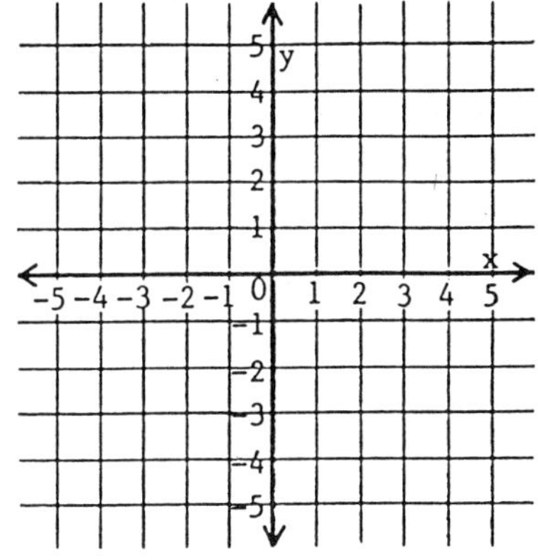

Write the equation of:

11. line A ____________

12. line B ____________

13. the x-axis ____________

14. the y-axis ____________

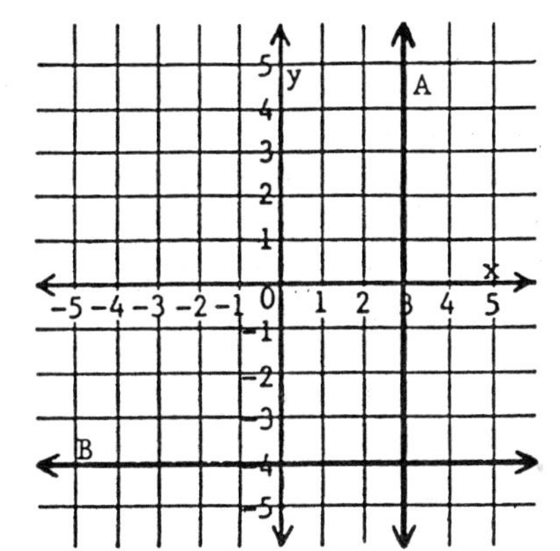

ANSWERS:

1. No
2. Yes
3. (4,-6)
4. $\left(-\frac{1}{2}, -5\right)$
5. 4
6. 2
7. y-axis
8. (0,0)
9-10.

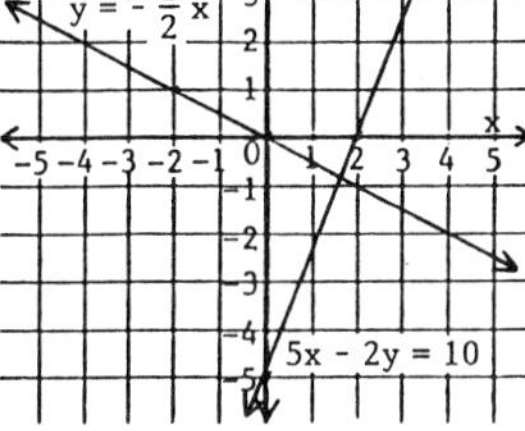

11. $x = 3$
12. $y = -4$
13. $y = 0$
14. $x = 0$

7-4 SLOPE OF A LINE

The slope of a graphed line is a measure of the steepness of the rise or fall of the line from left to right. We will discuss the slope of a line and the two-point formula for slope in this section.

35. Points P (1,1) and Q (4,5) are plotted on the line at the right. Δx and Δy are the changes in x and y from P to Q. The symbol Δ (pronounced delta) is used as an abbreviation for the phrase change in. That is:

Δx means change in x.

Δy means change in y.

The slope of a line is a ratio of the change in y to the change in x from one point to another point on the line. That is:

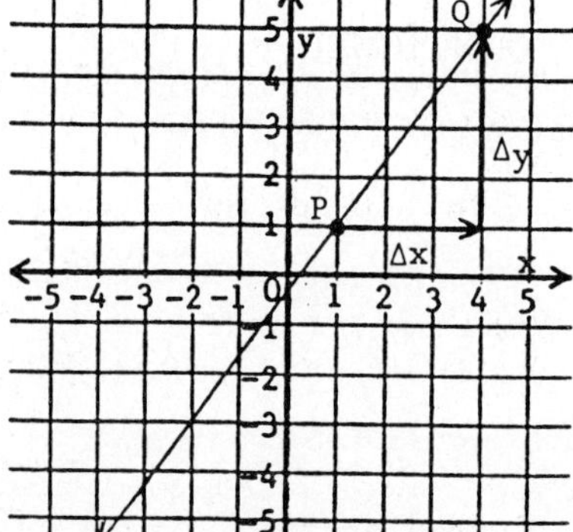

$$\text{Slope} = \frac{\Delta y}{\Delta x} = \frac{\text{increase or decrease in } y}{\text{increase in } x}$$

Let's use the changes from P to Q to compute the slope of the line on the graph.

Since x increases from 1 to 4, $\Delta x = 3$.

Since y increases from 1 to 5, $\Delta y = 4$.

Therefore, the slope $= \frac{\Delta y}{\Delta x} =$ ______

---

36. Let's use the changes from S (-4,2) to T (3,-2) to compute the slope of the line at the right.

$\frac{4}{3}$

Since x increases from -4 to 3, $\Delta x = 7$.

Since y decreases from 2 to -2, $\Delta y = -4$.

Therefore, the slope $= \frac{\Delta y}{\Delta x} =$ ______

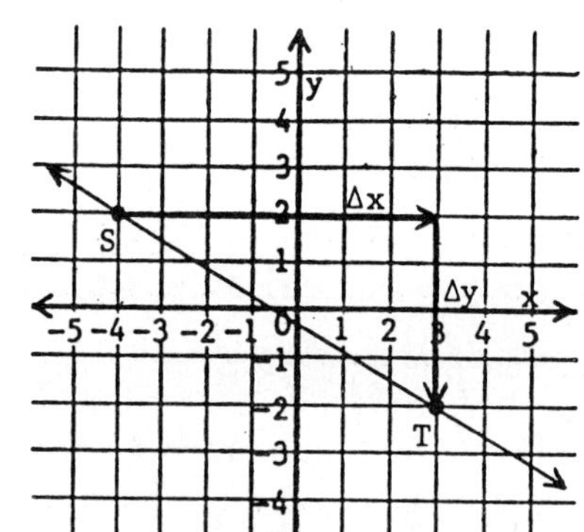
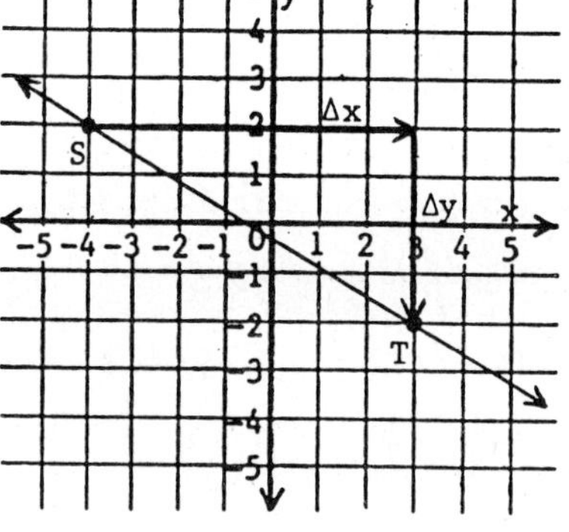

---

37. Slope is a ratio or fraction. When computing a slope, the ratio should always be reduced to lowest terms.

$\frac{-4}{7}$ or $-\frac{4}{7}$

a) If $\Delta x = 8$ and $\Delta y = -6$, the slope is _________.

b) If $\Delta x = 4$ and $\Delta y = 12$, the slope is _________.

---

a) $-\frac{3}{4}$

b) 3

38. No matter which pair of points we choose to compute the slope of a line, we always get the same value for the slope. As an example, we graphed the changes from A to B and from C to D on the line at the right.

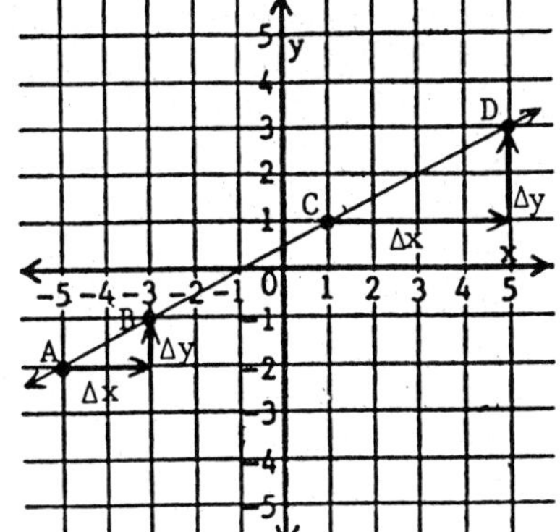

a) For A and B, $\Delta x = 2$ and $\Delta y = 1$. Therefore, the slope = ______.

b) For C and D, $\Delta x = 4$ and $\Delta y = 2$. Therefore, the slope = ______.

c) Did we get the same value for the slope with each pair? ______

---

39. The sign of the slope tells us whether a line rises or falls from left to right.

If its slope is positive, the line rises.

If its slope is negative, the line falls.

On the graph at the right, we have drawn four lines and labeled them #1, #2, #3, and #4.

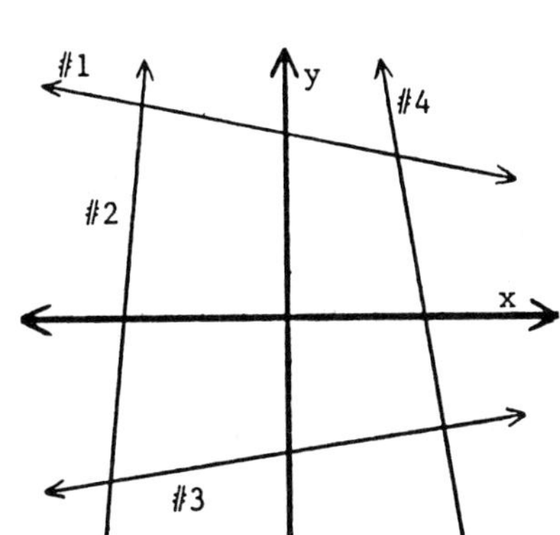

a) Which lines have a positive slope? 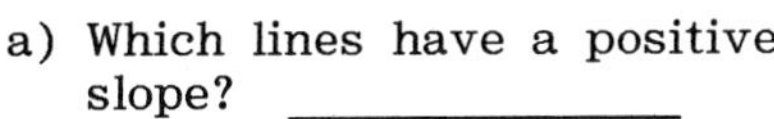

b) Which lines have a negative slope? 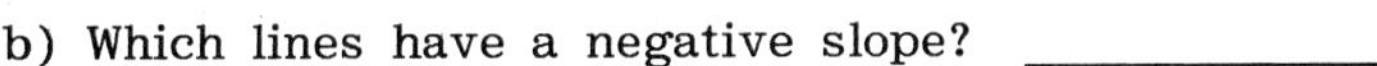

a) $\frac{1}{2}$

b) $\frac{1}{2}$ (from $\frac{2}{4}$)

c) Yes

---

40. The absolute value of the slope tells us how steep the the rise or fall of the line is.

We graphed three lines with positive slopes at the right. A is the steepest, then B, then C.

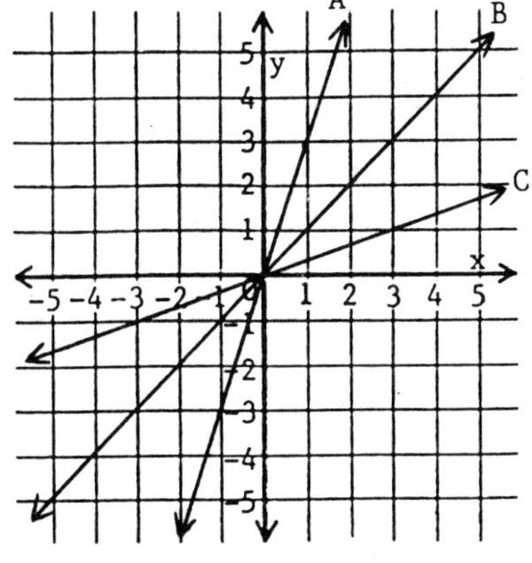

Line A has a slope of 3.

Line B has a slope of "1". (It forms a 45° angle with the x-axis.)

Line C has a slope of $\frac{1}{3}$.

Does the steepest line A have the slope with the largest absolute value? ______

a) lines #2 and #3

b) lines #1 and #4

Yes

41. We graphed three lines with negative slopes at the right. A is the steepest, then B, then C.

Line A has a slope of -4.

Line B has a slope of -1. (It forms a 45° angle with the x-axis.)

Line C has a slope of $-\frac{1}{4}$.

Does the steepest line A have the slope with the largest absolute value? ______

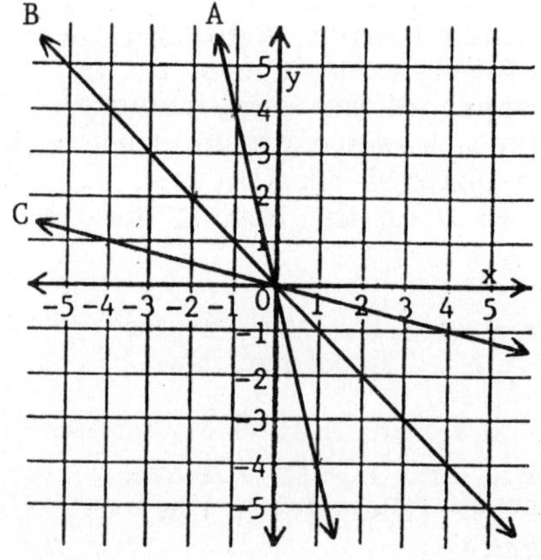

Yes

42. On the line at the right, $P_1(x_1,y_1)$ and $P_2(x_2,y_2)$ represent any two points. We can use their coordinates to find Δy and Δx. That is:

$$\Delta y = y_2 - y_1$$

$$\Delta x = x_2 - x_1$$

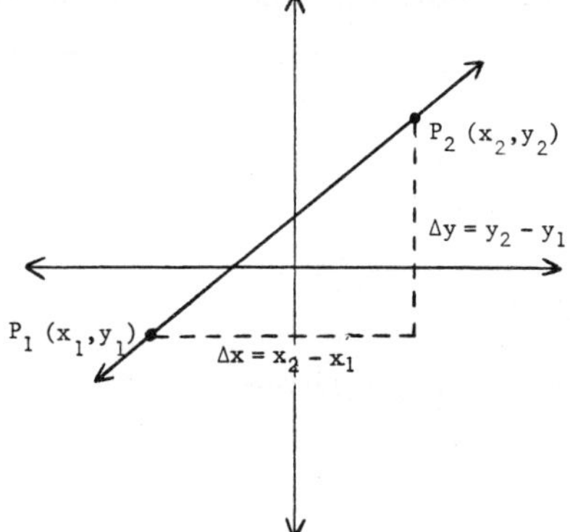

Using the letter m for slope, we can write the following formula for the slope of a line.

$$m = \frac{\Delta y}{\Delta x} = \frac{y_2 - y_1}{x_2 - x_1}$$

The line at the right passes through the points (2,-1) and (4,3). From the graph, you can see this fact:

$$m = \frac{\Delta y}{\Delta x} = \frac{4}{2} = 2$$

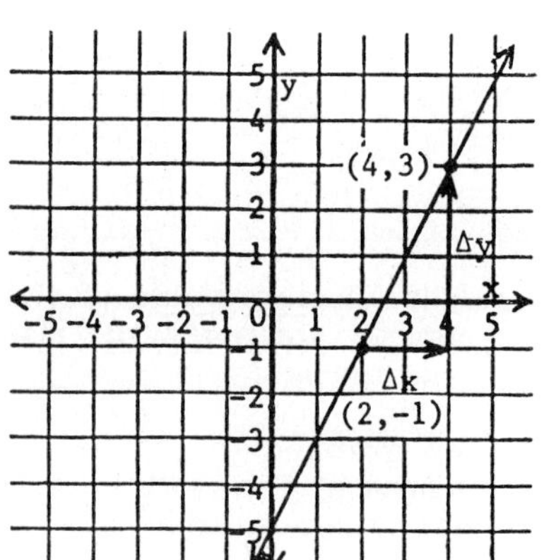

Let's use the two-point formula above to compute the slope. We can use either (2,-1) or (4,3) for (x_2,y_2).

Using (4,3) as (x_2,y_2), we get: $m = \frac{3 - (-1)}{4 - 2} = \frac{4}{2} = 2$

Using (2,-1) as (x_2,y_2), we get: $m = \frac{(-1) - 3}{2 - 4} = \frac{-4}{-2} = 2$

Did we get the correct value for the slope using both methods? ______

Yes

43. Let's use the same formula to find the slope of the line through (-4,0) and (0,2).

Note: When using the formula, it is helpful to sketch the two points as we have done to avoid gross errors.

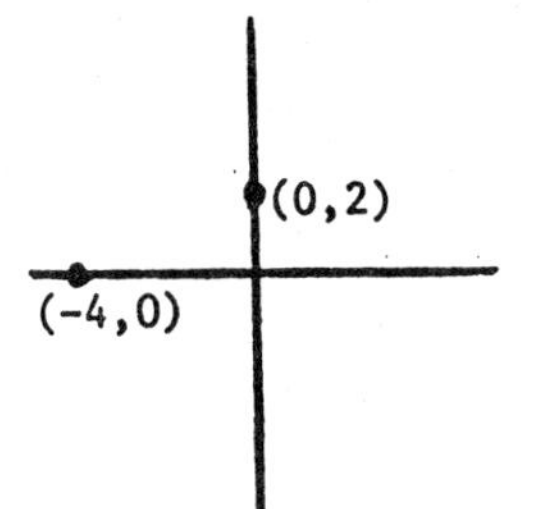

Using either (-4,0) or (0,2) as (x_2,y_2):

$$m = \frac{y_2 - y_1}{x_2 - x_1} = \_\_\_\_\_\_\_\_$$

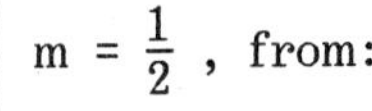

$m = \frac{1}{2}$, from:

$$\frac{2 - 0}{0 - (-4)} = \frac{2}{4}$$

or

$$\frac{0 - 2}{-4 - 0} = \frac{-2}{-4}$$

44. Use the formula to find the slope of the line through each pair of points below. (Sketch the points first.)

a) (5,-7) and (7,1)

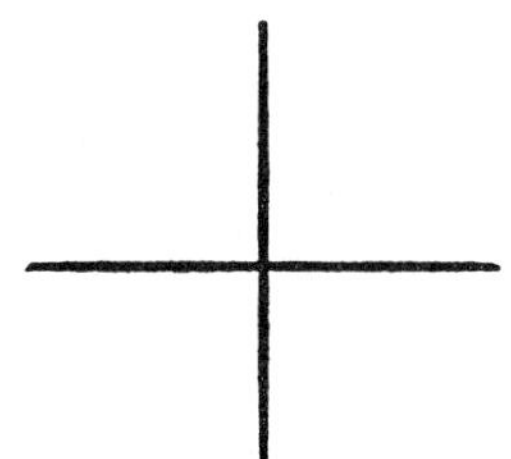

m = ________

b) (1,0) and (-5,4)

m = ________

a) m = 4

b) $m = -\frac{2}{3}$

45. Find the slope of the line through each pair of points. (Sketch the points first.)

a) (-8,0) and (0,20)

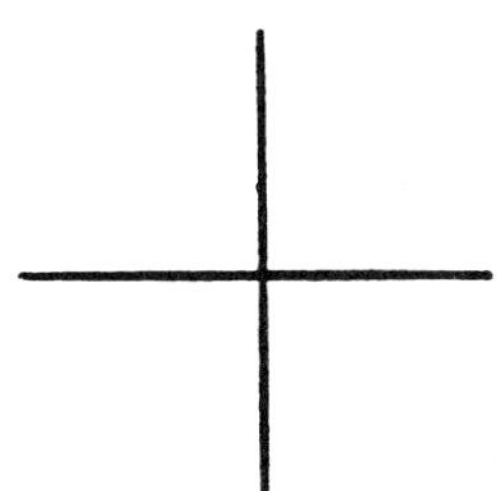

m = ________

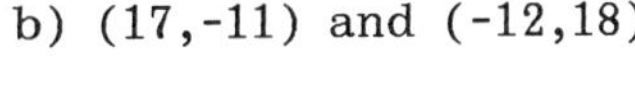

b) (17,-11) and (-12,18)

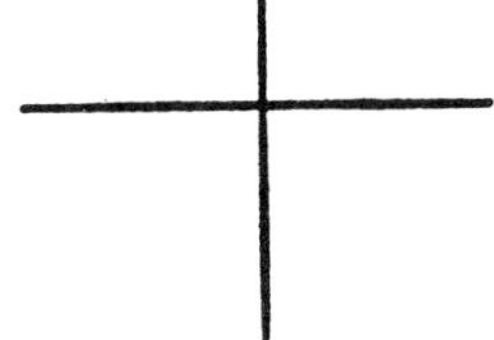

m = ________

---

a) $m = \frac{5}{2}$

b) m = -1

46. We know that the line at the right passes through (0,0) and (2,10). Let's use the two-point formula to find its slope.

$$m = \frac{y_2 - y_1}{x_2 - x_1} = \frac{10 - 0}{2 - 0} = \frac{10}{2} = 5$$

or

$$m = \frac{y_2 - y_1}{x_2 - x_1} = \frac{0 - 10}{0 - 2} = \underline{\hspace{2cm}}$$

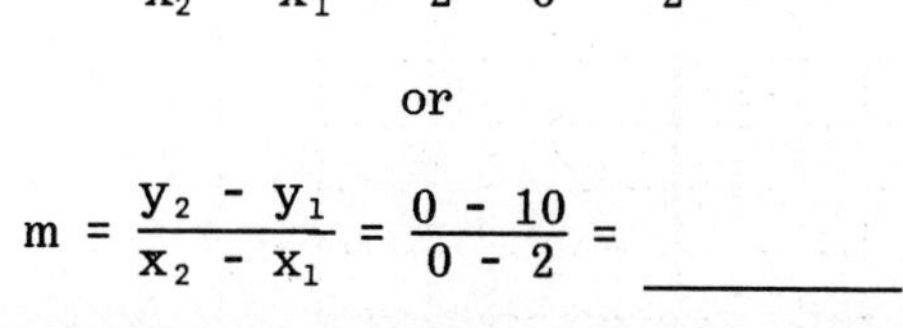

(2,10)

$\frac{-10}{-2} = 5$

47. Use the two-point formula to find the slope of a line passing through the origin and each point below. (Make a sketch.)

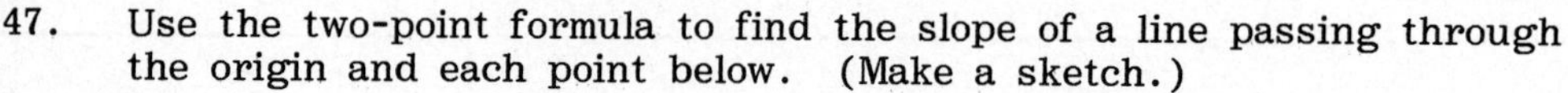

a) (-10,8)

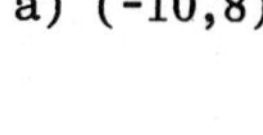

b) (-6,-7)

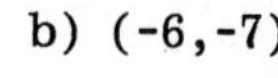

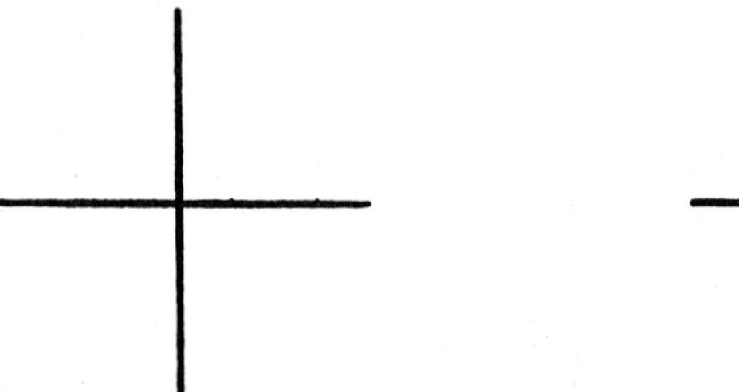

m = __________ m = __________

a) $m = -\frac{4}{5}$

b) $m = \frac{7}{6}$

48. The equation of the horizontal line at the right is y = 2. Since a horizontal line does not rise or fall, it seems that its slope should be 0. Let's use points C (1,2) and D (4,2) to confirm that fact.

$$m = \frac{y_2 - y_1}{x_2 - x_1} = \frac{2 - 2}{4 - 1} = \underline{\hspace{2cm}}$$

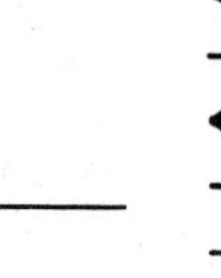

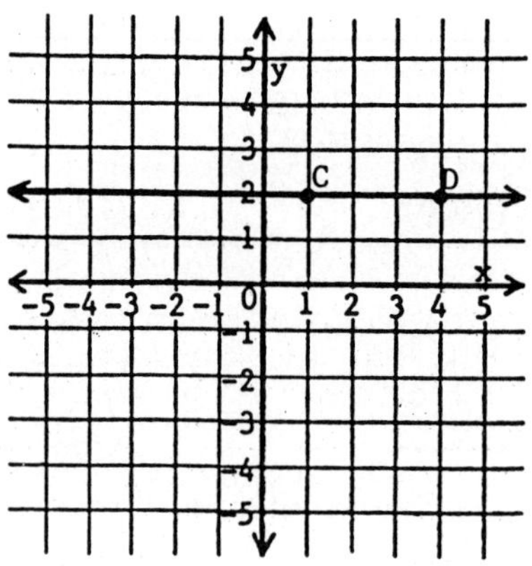

$\frac{0}{3} = 0$

49. The equation of the vertical line at the right is $x = 3$. Let's use P (3,2) and Q (3,4) to examine the slope of the line.

$$m = \frac{y_2 - y_1}{x_2 - x_1} = \frac{4 - 2}{3 - 3} = \frac{2}{0}$$

But division by 0 is undefined. Therefore, the slope of the line is undefined.

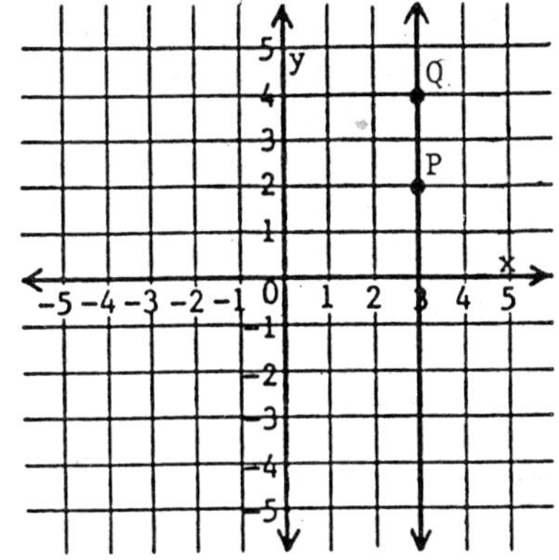

---

50. In the last two frames, we saw these facts:

1) The slope of any horizontal line is 0.

2) The slope of any vertical line is undefined.

What is the slope of each line below?

a) $y = 7$ m = ____ d) $x = 20$ m = ____

b) $x = -6$ m = ____ e) y-axis m = ____

c) $y = -30$ m = ____ f) x-axis m = ____

---

a) 0 c) 0 e) undefined
b) undefined d) undefined f) 0

7-5 SLOPE-INTERCEPT FORM

In this section, we will discuss the slope-intercept form of linear equations. We will show how linear equations in other forms can be rearranged to slope-intercept form. We will also show how the y-intercept and the slope can be used to graph a linear equation.

---

51. The following form of a linear equation is called slope-intercept form.

$$\boxed{y = mx + b}$$

where: m is the slope of the line.
b is the y-intercept of the line.

Continued on following page.

51. Continued

As an example, we graphed $y = 2x - 3$ below. The coordinates of points A and B are given.

1. In the equation, $m = 2$. To show that 2 is the slope of the line, we can use the changes in y and x from A to B.

$$m = \frac{\Delta y}{\Delta x} = \frac{4}{2} = 2$$

2. In the equation, $b = -3$. A is the y-intercept. Its coordinates are $(0,-3)$. -3 is the y-coordinate of A. Sometimes we say "-3 is the y-intercept".

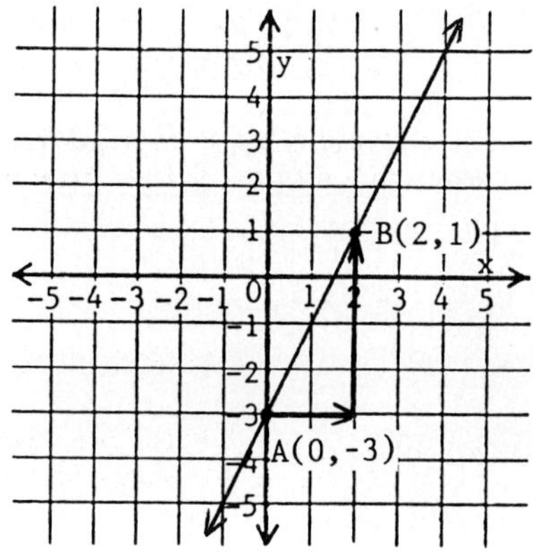

---

52. Following the examples, write the slope and y-intercept of the linear equations with the following slope-intercept forms.

Slope-intercept form	Slope	y-intercept
$y = x - 5$	1	(0,-5)
$y = -3x + 1$	-3	(0,1)
a) $y = \frac{3}{4}x + 8$	____	____
b) $y = -\frac{1}{2}x - \frac{5}{4}$	____	____

---

53. Following the example, write the slope-intercept form of the linear equations with the following slopes and y-intercepts.

	Slope	y-intercept	Slope-intercept form
	4	(0,-1)	$y = 4x - 1$
a)	$-\frac{5}{2}$	(0,4)	____
b)	-1	$(0,-\frac{1}{2})$	____

Answers to 52: a) $\frac{3}{4}$ (0,8) b) $-\frac{1}{2}$ $(0,-\frac{5}{4})$

---

54. To put the equation below in slope-intercept form, we solved for y. Notice how we wrote the x-term first on the right side. Put the other equation in slope-intercept form.

$$y - 2x = 7$$
$$y - 2x + 2x = 2x + 7$$
$$y = 2x + 7$$

$$x + y = 10$$

Answers to 53: a) $y = -\frac{5}{2}x + 4$ b) $y = -x - \frac{1}{2}$

Answer to 54: $y = -x + 10$

55. To get the additive inverse of a binomial, we replace each term with its additive inverse.

The additive inverse of $-9x + 5$ is $9x - 5$.

The additive inverse of $x - 1$ is $-x + 1$.

To put each equation below in slope-intercept form, we replaced each side with its additive inverse. Doing so is the same as multiplying both sides by -1.

$-y = -2x + 3$	$-y = x - 9$
$y = 2x - 3$	$y = -x + 9$

Notice how we used the principle above in the last step to put each equation below in slope-intercept form.

$3x - y = 10$	$x = 5 - y$
$-3x + 3x - y = -3x + 10$	$x + (-5) = -5 + 5 - y$
$-y = -3x + 10$	$x - 5 = -y$
$y = 3x - 10$	$y = -x + 5$

Put these in slope-intercept form.

a) $x - y = 4$ b) $8 - y = 7x$

---

a) $y = x - 4$

b) $y = -7x + 8$

56. To put the equation below in slope-intercept form, we divided $x + 6$ by 3. Put the other equation in slope-intercept form.

$3y = x + 6$

$y = \frac{x + 6}{3}$

$y = \frac{x}{3} + \frac{6}{3}$

$y = \frac{1}{3}x + 2$

$2y = -6x + 1$

$y = -3x + \frac{1}{2}$

57. Following the example, put the other equation in slope-intercept form.

$4x + 3y = 12$ | $x + 5y = 3$

$3y = -4x + 12$

$y = \frac{-4x + 12}{3}$

$y = -\frac{4}{3}x + 4$

$y = -\frac{1}{5}x + \frac{3}{5}$

58. Notice how we replaced each side with its additive inverse to get $4y = x - 7$ below. Put the other equation in slope-intercept form.

$x - 4y = 7$ | $5x - 3y = 15$

$-4y = 7 - x$

$4y = x - 7$

$y = \frac{x - 7}{4}$

$y = \frac{1}{4}x - \frac{7}{4}$

$y = \frac{5}{3}x - 5$

59. After putting an equation in slope-intercept form, we can easily identify its slope and y-intercept. For example:

Since $2x + y = 5$ is equivalent to $y = -2x + 5$:

Its slope is -2. Its y-intercept is (0,5).

Since $5x - 2y = 6$ is equivalent to $y = \frac{5}{2}x - 3$:

a) Its slope is ______. b) Its y-intercept is ______.

---

a) $\frac{5}{2}$ b) (0,-3)

60. The general slope-intercept form of linear equations is $y = mx + b$. However, the y-intercept of all lines through the origin is (0,0). Since $b = 0$ for all lines through the origin, their slope-intercept form is:

$$\boxed{y = mx}$$

If we know the slope of a line through the origin, we can write its equation. That is:

If $m = \frac{3}{4}$, $y = \frac{3}{4}x$ | If $m = -\frac{5}{2}$, $y = -\frac{5}{2}x$

We can find the slope of $y - 5x = 0$ by putting it in slope-intercept form. We get: $y = 5x$. Therefore, $m = 5$. Find the slope of each line below by putting it in slope-intercept form.

a) $4x + 3y = 0$ m = ______ b) $x - y = 0$ m = ______

---

a) $m = -\frac{4}{3}$ b) $m = 1$

61. If a linear equation is in slope-intercept form, we can use the y-intercept and the slope to graph the line. An example is discussed below.

For $y = \frac{3}{4}x + 2$, the slope is $\frac{3}{4}$ and the y-intercept is (0,2). We graphed the equation at the right. The steps are:

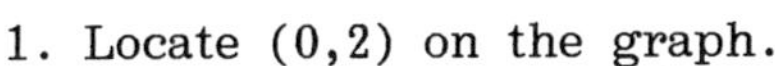

1. Locate (0,2) on the graph.
2. Move 4 units to the right for Δx and 3 units up for Δy to locate point P.
3. Draw the line through (0,2) and P.

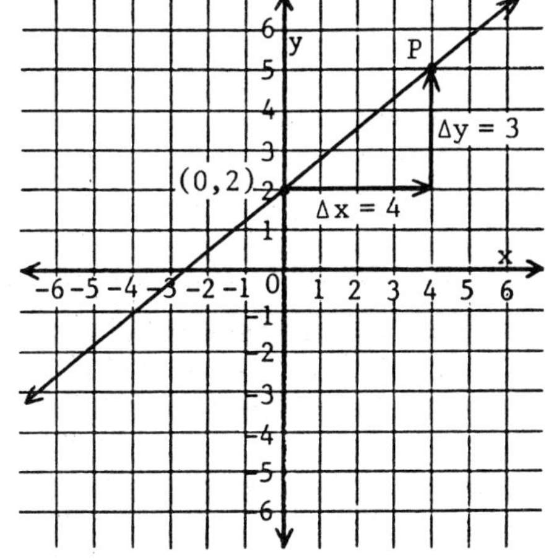

Use the same method to graph $y = -\frac{2}{3}x - 3$ on the same graph.

---

62. Sometimes it is more convenient to use an equivalent form for the slope. Below, for example, we used $\frac{-3}{3}$ instead of $\frac{-1}{1}$ for -1 so that P is farther away from (0,-3).

For $y = -x - 3$, the slope is -1 and the y-intercept is (0,-3). We graphed the equation at the right. The steps are:

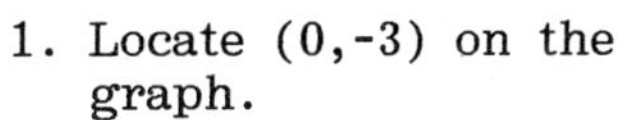

1. Locate (0,-3) on the graph.
2. Using $\frac{-3}{3}$ for -1, move 3 units to the right for Δx and 3 units down for Δy to locate point P.
3. Draw the line through (0,-3) and P.

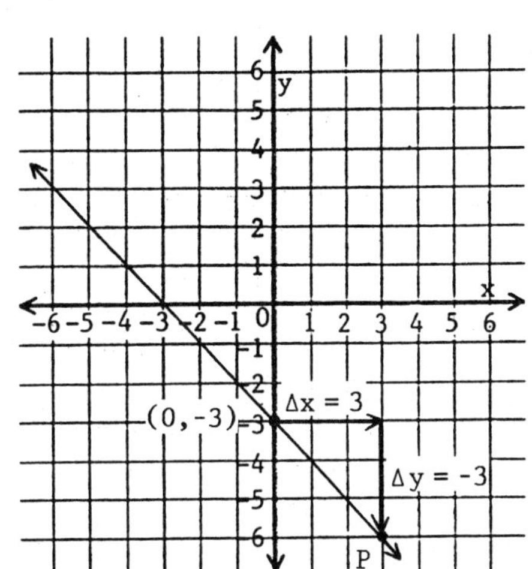

Use the same method to graph $y = 2x + 1$ on the same graph. Use $\frac{4}{2}$ for 2 so that P is farther away from (0,1).

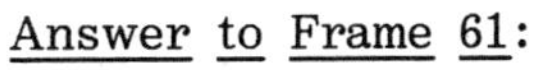

Answer to Frame 61:

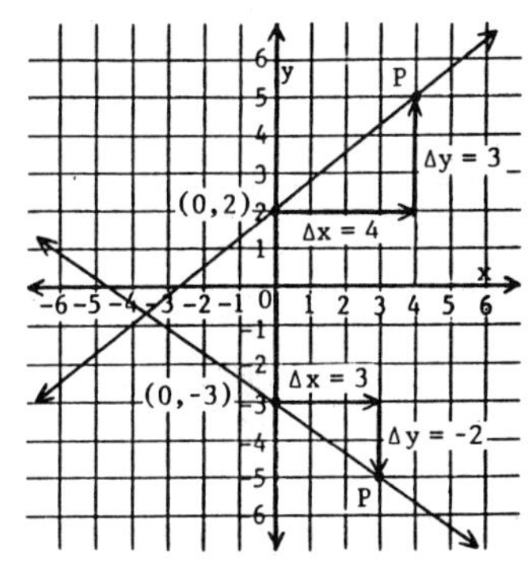

Answer to Frame 62:

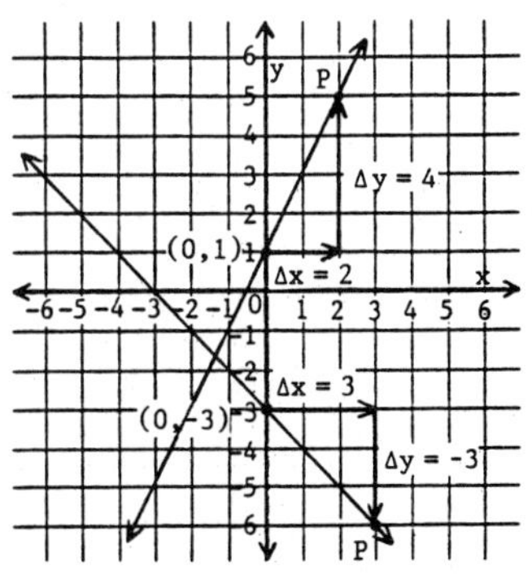

7-6 POINT-SLOPE FORM

In this section, we will discuss the point-slope form of linear equations.

---

63. For the line at the right, we know the slope m and the coordinates of point $P_1(x_1,y_1)$. Let $P(x,y)$ be any other point on the line. Then using the two-point formula for slope, we get:

$$\frac{y - y_1}{x - x_1} = m$$

Clearing the fraction, we get the point-slope form of a linear equation. It is:

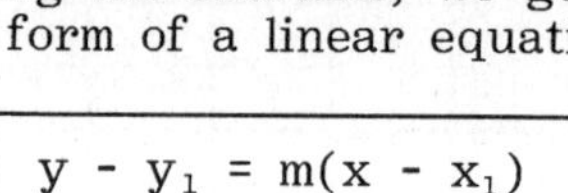

$$y - y_1 = m(x - x_1)$$

where: m is the slope of the line. $P_1(x_1,y_1)$ is a point whose coordinates are known.

Let's use the form above to get the equation of a line that passes through (4,5) and has a slope of 2.

$$y - y_1 = m(x - x_1)$$

$$y - 5 = 2(x - 4)$$

$$y - 5 = 2x - 8$$

Write the equation above in slope-intercept form. ______________

$y = 2x - 3$

64. We used the point-slope form to find the equation of one line below. We wrote it in slope-intercept form. Do the same for the other line.

Find the line with a slope of 4 that passes through (2,-3).

$$y - y_1 = m(x - x_1)$$

$$y - (-3) = 4(x - 2)$$

$$y + 3 = 4x - 8$$

$$y = 4x - 11$$

Find the line with a slope of $-\frac{1}{2}$ that passes through (4,-1)

$y = -\frac{1}{2}x + 1$

65. We found the equation of one line below. Find the equation of the other line.

Find the line with a slope of -1 that passes through (-2,4).

Find the line with a slope of $\frac{2}{3}$ that passes through (-6,-3).

$$y - y_1 = m(x - x_1)$$
$$y - 4 = -1[x - (-2)]$$
$$y - 4 = -1(x + 2)$$
$$y - 4 = -x - 2$$
$$y = -x + 2$$

$y = \frac{2}{3}x + 1$

66. When the coordinates of two points are given, we can use the slope formula to find the slope and then use the point-slope form with one of the points to find the equation of the line. An example is discussed.

Find the equation of the line through the points (-2,-4) and (2,2).

1. Use the slope formula to find the slope of the line.

$$m = \frac{y_2 - y_1}{x_2 - x_1} = \frac{2 - (-4)}{2 - (-2)} = \frac{6}{4} = \frac{3}{2}$$

2. Then use the point-slope formula with one of the points to find the equation. We used (2,2) below.

$$y - y_1 = m(x - x_1)$$
$$y - 2 = \frac{3}{2}(x - 2)$$
$$y - 2 = \frac{3}{2}x - 3$$
$$y = \frac{3}{2}x - 1$$

Using the same method, find the equations of the following lines.

a) through (-3,-3) and (1,5) b) through (-3,0) and (3,-2)

a) $y = 2x + 3$ b) $y = -\frac{1}{3}x - 1$

7-7 PARALLEL AND PERPENDICULAR LINES

In this section, we will show that the slopes of a pair of parallel or perpendicular lines have a special relationship.

---

67. Any two nonvertical lines that are parallel have the same slope. As an example, we graphed the following two equations at the right. The slope of each is $\frac{3}{2}$.

$$\text{A:} \quad y = \frac{3}{2}x + 1$$

$$\text{B:} \quad y = \frac{3}{2}x - 2$$

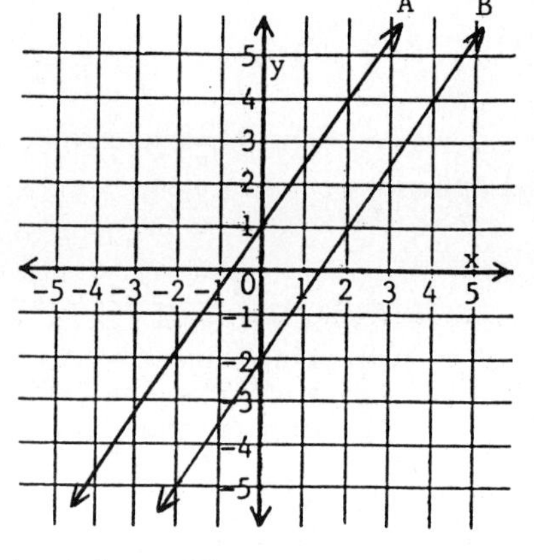

By writing them in slope-interrupt form, decide whether each pair of lines is parallel.

a) $3x - y = 5$
$\quad y - 3x = -1$

b) $4x + 5y = 20$
$\quad 5x - 4y = 20$

---

68. If two lines are perpendicular (meet at right angles), the product of their slopes is -1. As an example, we graphed the following two equations at the right. The product of their slopes is $\frac{1}{2}(-2) = -1$.

$$\text{A:} \quad y = \frac{1}{2}x - 1$$

$$\text{B:} \quad y = -2x + 1$$

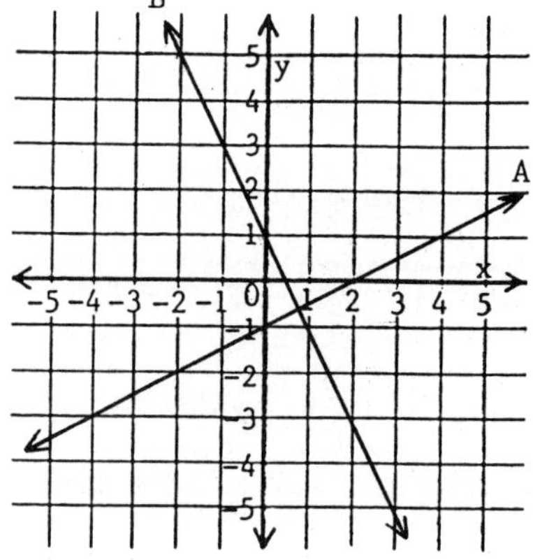

By writing them in slope-intercept form and then multiplying their slopes, decide whether each pair of lines is perpendicular.

a) 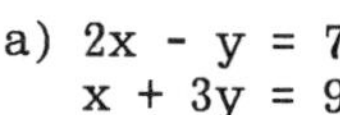
$2x - y = 7$
$x + 3y = 9$

b) 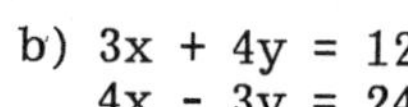
$3x + 4y = 12$
$4x - 3y = 24$

Answers to 67:

a) Parallel. Both slopes are 3.

b) Not parallel. The slopes are $-\frac{4}{5}$ and $\frac{5}{4}$.

Answers to 68:

a) Not perpendicular, since: $2\left(-\frac{1}{3}\right) = -\frac{2}{3}$

b) Perpendicular, since: $\left(-\frac{3}{4}\right)\left(\frac{4}{3}\right) = -1$

69. By computing the slopes, decide whether each pair of lines is parallel or perpendicular.

a) line through (-2,3) and (1,5)
line through (0,-1) and (2,-4)

b) line through (-2,1) and (1,-5)
line through (2,1) and (4,-3)

---

70. We can use the fact that parallel lines have the same slope to find the equation of a line. An example is given.

Find the equation of the line through point (5,3) that is parallel to 4x + 5y = 20.

1. Find the slope of 4x + 5y = 20.

$$4x + 5y = 20$$
$$5y = -4x + 20$$
$$y = -\frac{4}{5}x + 4$$

2. Using $m = -\frac{4}{5}$ with the point (5,3) write the equation of the line.

$$y - y_1 = m(x - x_1)$$
$$y - 3 = -\frac{4}{5}(x - 5)$$
$$y - 3 = -\frac{4}{5}x + 4$$
$$y = -\frac{4}{5}x + 7$$

Write the equation of the line through the given point and parallel to the given line.

a) (0,-3), 2x - y = 5

b) (-1,2), 3x + 2y = 6

a) Perpendicular, since:

$$\frac{2}{3}\left(-\frac{3}{2}\right) = -1$$

b) Parallel, since both slopes are -2.

71. We can use the fact that the product of the slopes of perpendicular lines is -1 to find the equation of a line. An example is given.

Find the slope-intercept form of the equation of the line through point (4,-1) that is perpendicular to $2x + y = 9$.

1. Find the slope of $2x + y = 9$.

$$2x + y = 9$$

$$y = -2x + 9$$

2. Using $m = \frac{1}{2}$ (the inverse of the reciprocal of -2) and the point (4,-1), write the equation of the line.

$$y - y_1 = m(x - x_1)$$

$$y - (-1) = \frac{1}{2}(x - 4)$$

$$y + 1 = \frac{1}{2}x - 2$$

$$y = \frac{1}{2}x - 3$$

Find the slope-intercept form of the equation of the line through the given point and perpendicular to the given line.

a) (3,2), $3x - y = 5$ b) (-2,-4), $2x + 3y = 12$

a) $y = 2x - 3$

b) $y = -\frac{3}{2}x + \frac{1}{2}$

---

a) $y = -\frac{1}{3}x + 3$

b) $y = \frac{3}{2}x - 1$

SELF-TEST 25 (pages 347-363)

Find the slope of:

1. line A ______ 2. line B ______

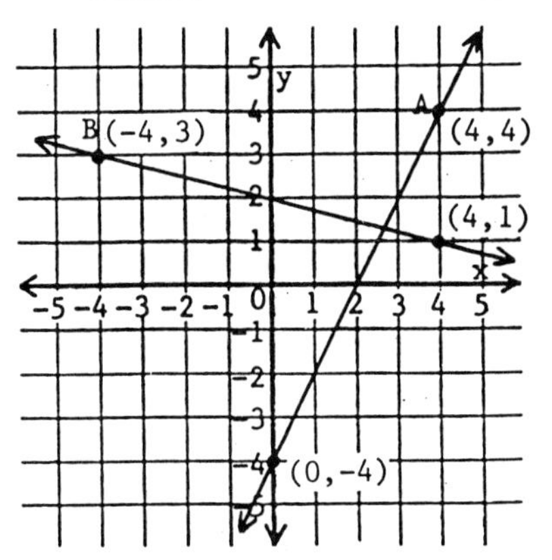

Find the slope of the line through each pair of points.

3. (-3,1) and (2,-4)

4. (0,0) and (-4,-6)

For each equation: (a) put it in slope-intercept form; (b) identify its slope; (c) identify the coordinates of its y-intercept.

5. $x - y = 5$

6. $3x + 4y = 8$

7. Find the slope-intercept form of the equation of the line with a slope of 2 that passes through (1,-3).

8. Find the slope-intercept form of the equation of the line with a slope of $-\frac{4}{3}$ that passes through (-3,5).

9. Decide whether this pair of lines is parallel or perpendicular.

$$2x + 4y = 7$$
$$y = -\frac{1}{2}x + 3$$

10. Write the equation of the line through (-2,5) that is perpendicular to $2x - y = 7$.

ANSWERS:

1. $m = 2$
2. $m = -\frac{1}{4}$
3. $m = -1$
4. $m = \frac{3}{2}$
5. a) $y = x - 5$ b) $m = 1$ c) (0,-5)
6. a) $y = -\frac{3}{4}x + 2$ b) $m = -\frac{3}{4}$ c) (0,2)
7. $y = 2x - 5$
8. $y = -\frac{4}{3}x + 1$
9. parallel
10. $y = -\frac{1}{2}x + 4$

7-8 GRAPHING LINEAR INEQUALITIES

The graph of any linear inequality is a half-plane. We will discuss the method for graphing linear inequalities in this section.

---

72. Three linear inequalities are shown below. They are the same as linear equations except that they contain a $>$, $<$, $\geq$, or $\leq$ instead of an = sign.

$y \geq x + 1$ $\qquad$ $y < -2x$ $\qquad$ $2x - y > 7$

A solution of a linear inequality is a pair of values that satisfies the inequality. For example:

(3,7) <u>is</u> a solution of $y \geq x + 1$, since:

$y \geq x + 1$

$7 \geq 3 + 1$

$7 \geq 4$ True

(5,3) <u>is</u> <u>not</u> a solution of $y \geq x + 1$, since:

$y \geq x + 1$

$3 \geq 5 + 1$

$3 \geq 6$ False

a) Is (-2,-5) a solution of $y < -2x$? ________

b) Is (4,3) a solution of $2x - y > 7$? ________

---

73. Any linear inequality has an infinite number of solutions. For example, all of the following are solutions of $y \geq x + 1$.

(3,5) $\qquad$ (0,2) $\qquad$ (-2,0) $\qquad$ (-5,-3)

The inequality $y \geq x + 1$ means both $y > x + 1$ and $y = x + 1$. To graph the inequality, we began by graphing the line $y = x + 1$ below. That line is a boundary that divides the coordinate system into two half-planes. All four solutions above lie in the half-plane <u>above</u> the line. Therefore, the graph of $y \geq x + 1$ includes the line for $y = x + 1$ and the half-plane above that line for $y > x + 1$. To show that on the graph, we shade the half-plane above the line as we have done below.

a) (1,5) is a solution of $y \geq x + 1$. Does it lie in the shaded half-plane? ________

b) (4,2) is not a solution of $y \geq x + 1$. Does it lie in the shaded half-plane? ________

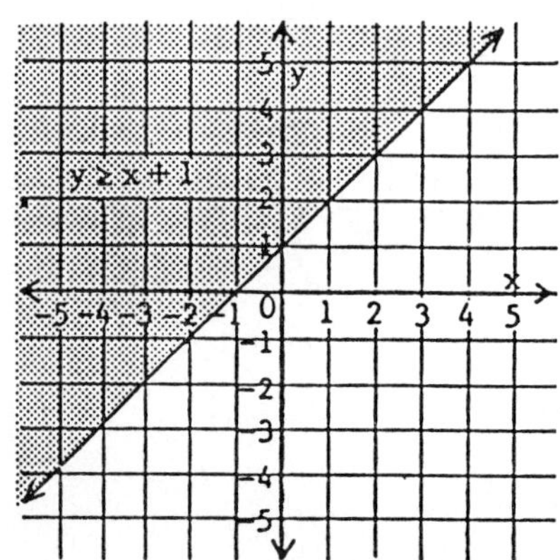

Answers to frame 72:

a) Yes, since $-5 < 4$ is true.

b) No, since $5 > 7$ is false.

74. All of the following are solutions of $y < -2x$.

$(-2,-5)$ $(-3,3)$ $(-4,-2)$ $(-5,1)$

Though the inequality does not include $y = -2x$, we still use that line as a boundary between the two half-planes. To show that the points on that line are not part of the graph, we used a dashed line below. Since all four solutions above lie in the half-plane <u>below</u> the line, the half-plane below the line is the graph of $y < -2x$. To show that on the graph, we shade the half-plane below the line as we have done below.

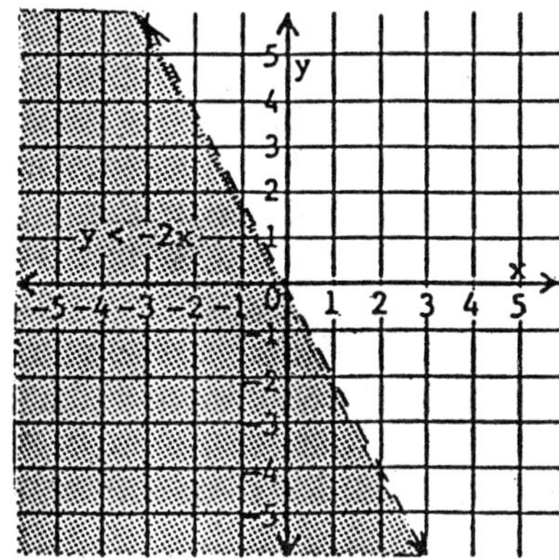

a) $(1,-5)$ is a solution of $y < -2x$. Does it lie in the shaded half-plane? __________

b) $(3,-2)$ is not a solution of $y < -2x$. Does it lie in the shaded half-plane? __________

a) Yes

b) No

---

75. The following can be used to determine whether to use a solid boundary line or a dashed boundary line.

Use a <u>solid line</u> if equality is included. The line is part of the graph. We use a solid line for these:

$y \geq 3x + 2$ $y \leq x$

Use a <u>dashed line</u> if equality is not included. The line is not part of the graph. We use a dashed line for these:

$y > -3x$ $y < x - 4$

We can use the following method to graph $3x + 2y < 6$.

1. Use the intercept method to graph the boundary $3x + 2y = 6$. The intercepts are $(0,3)$ and $(2,0)$. The boundary is dashed.

Continued on following page.

a) Yes

b) No

75. Continued

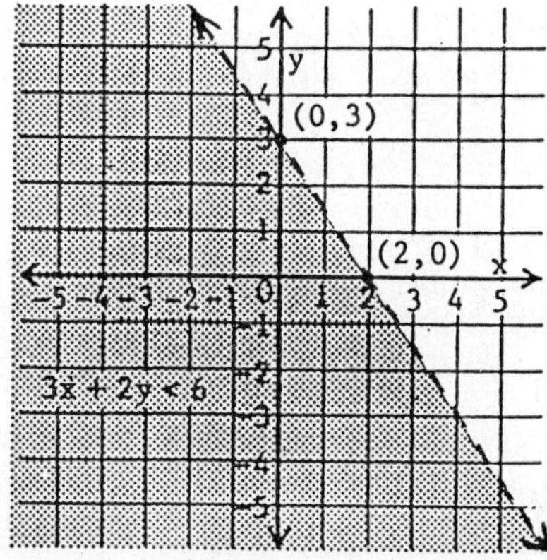

2. Then use a test point to determine the correct half-plane. Pick a point not on the line. We ordinarily use the origin (0,0).

$$3(0) + 2(0) < 6 \text{ is true.}$$

Since the origin is a solution, the half-plane containing (0,0) is correct. That is, the half-plane below the line is correct.

---

76. We can use the same method when the boundary is a line through the origin. As an example, we graphed $y - 3x \geq 0$ below. $y = 3x$ is the boundary. The boundary is solid.

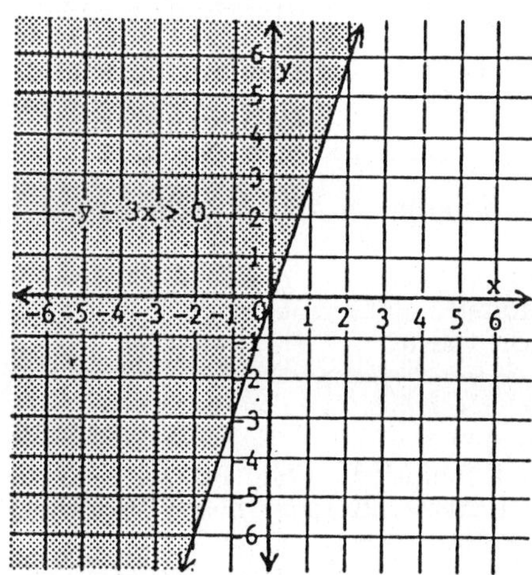

To determine the correct half-plane, we cannot use the origin because it is on the boundary line. We use some other point. If (1,0) is not on the line, it is easy to use.

$$0 - 3(1) > 0 \text{ is false.}$$

Since (1,0) is not a solution, the half-plane that does not contain (1,0) is correct. That is, the half-plane above the line is correct.

77. The following general method can be used to graph a linear inequality.

> 1. Graph the boundary. Use a solid line if the inequality contains $\geq$ or $\leq$. Use a dashed line if the inequality contains $>$ or $<$.
>
> 2. If the boundary line does not go through the origin, use the origin (0,0) as a test point. If (0,0) satisfies the inequality, shade the half-plane containing (0,0). If (0,0) does not satisfy the inequality, shade the other half-plane.
>
> 3. If the boundary line does go through the origin, use any other point as the test point. (1,0) is easy to use if it is not on the boundary line.

Using the method above, graph each inequality.

a) $3x + 4y < 12$

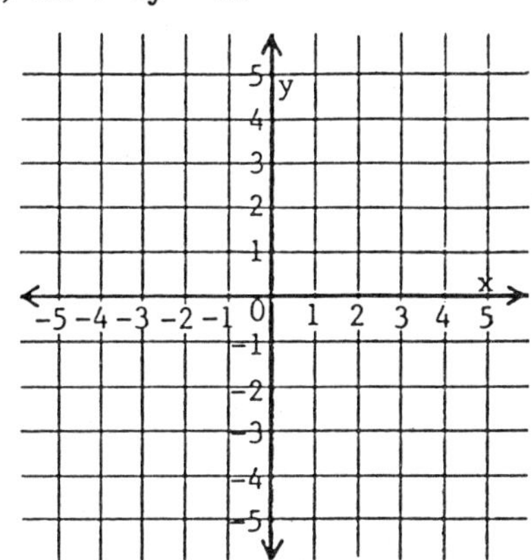

b) $y \leq 2x$

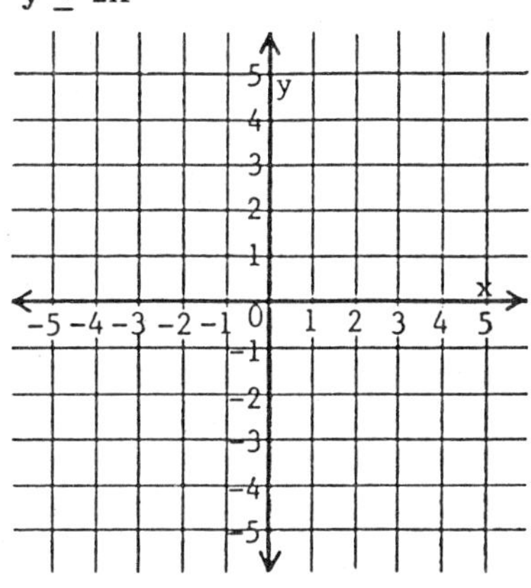

78. Using the same method, graph each inequality.

a) $2x - 5y \leq 10$

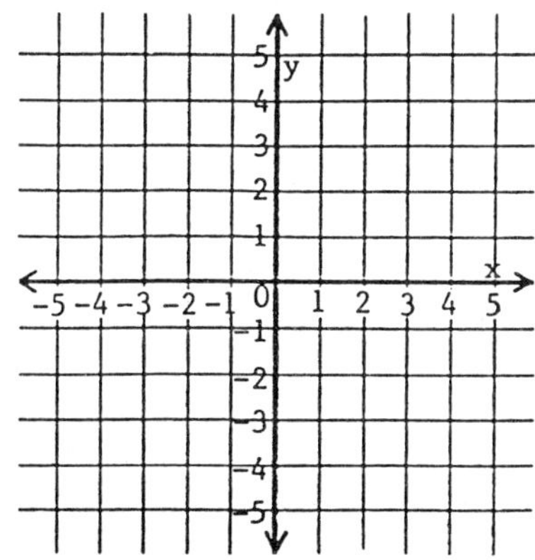

b) $y > -x + 2$

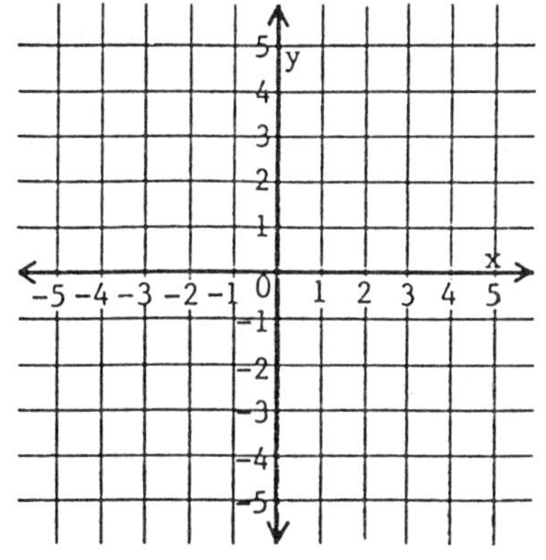

Answers to Frame 77:

a)

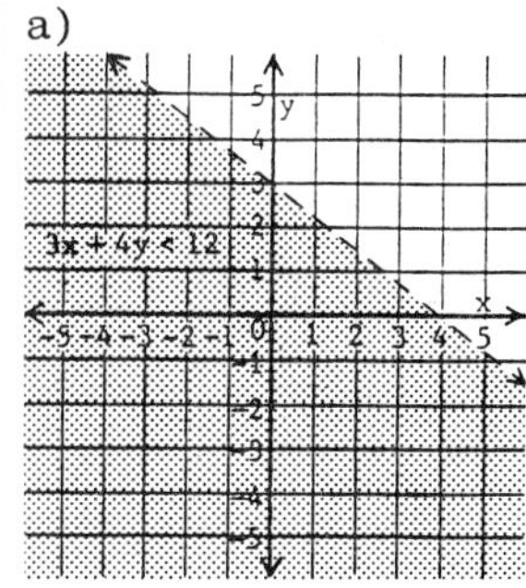

b)

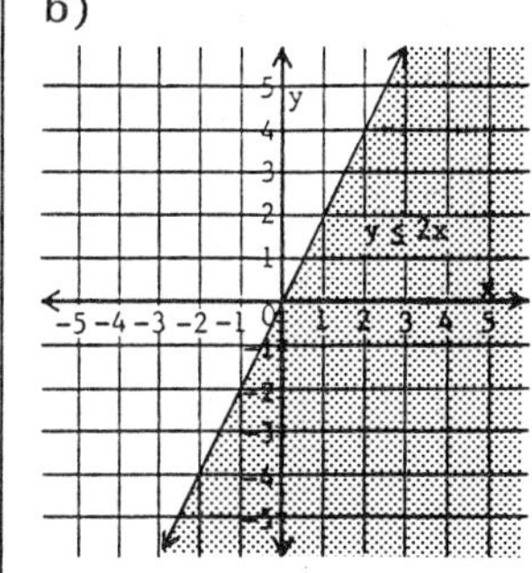

79. We graphed y > 2 below. The graph includes all points above the line y = 2. The line is dashed because y = 2 is not included. Graph $y \leq -1$ on the other graph.

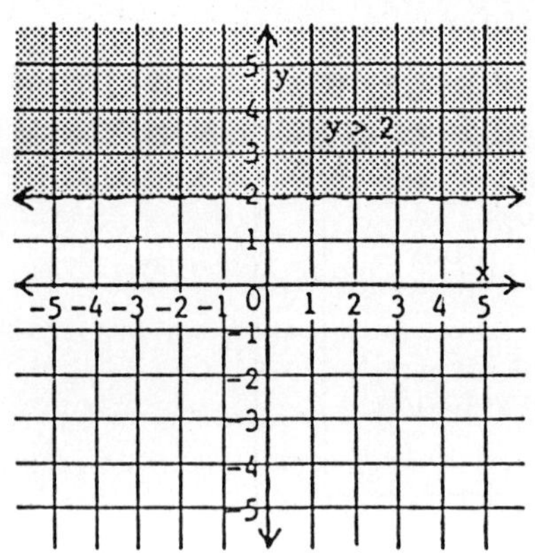

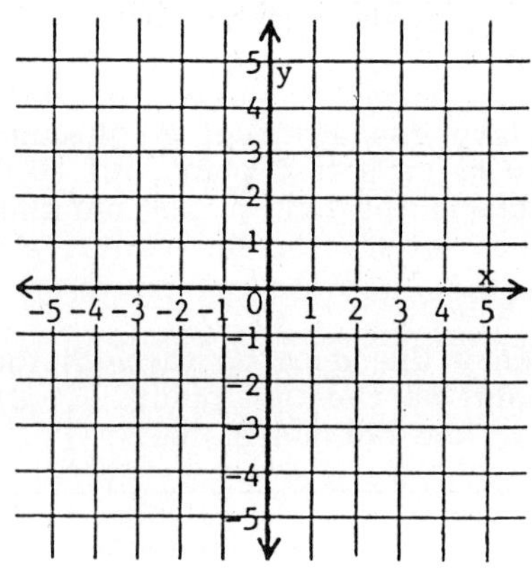

Answers to Frame 78:

a)

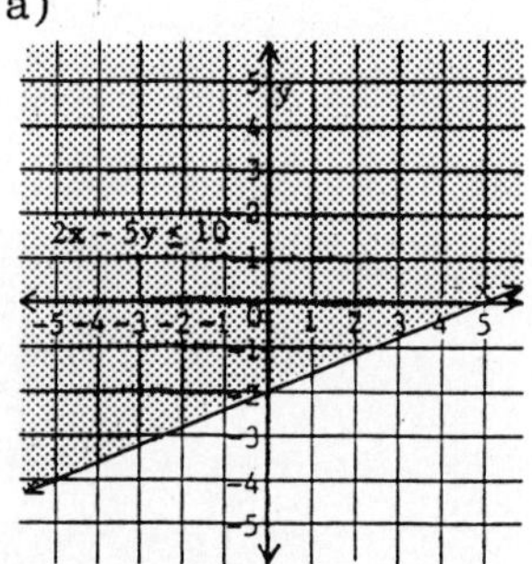

b)

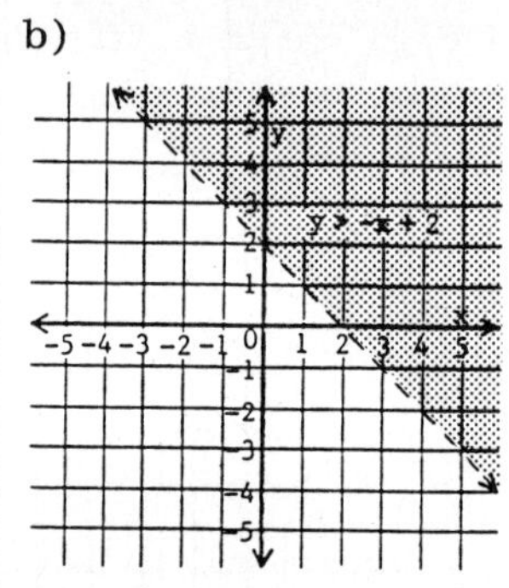

80. As we saw in an earlier chapter, $x \geq -2$ can be graphed on the number line. We get:

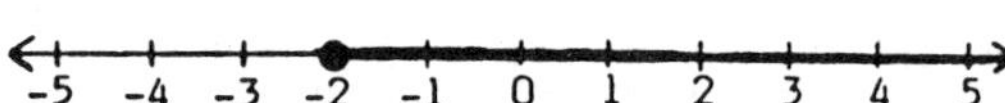

$x \geq -2$ can also be graphed on the coordinate system. We did so below. Its graph includes the line x = -2 and all points to the right of that line. Graph x < 3 on the other graph.

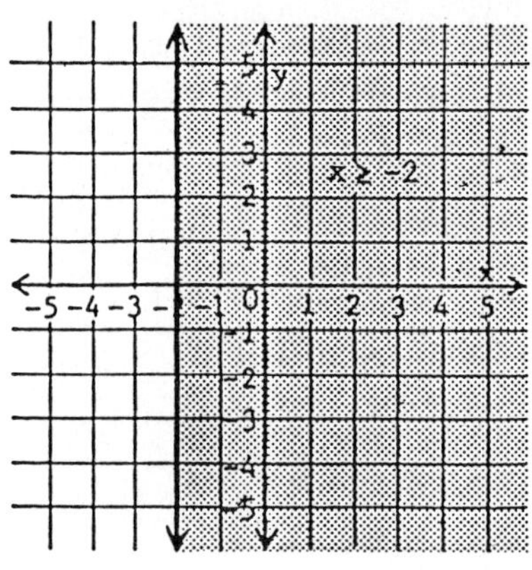

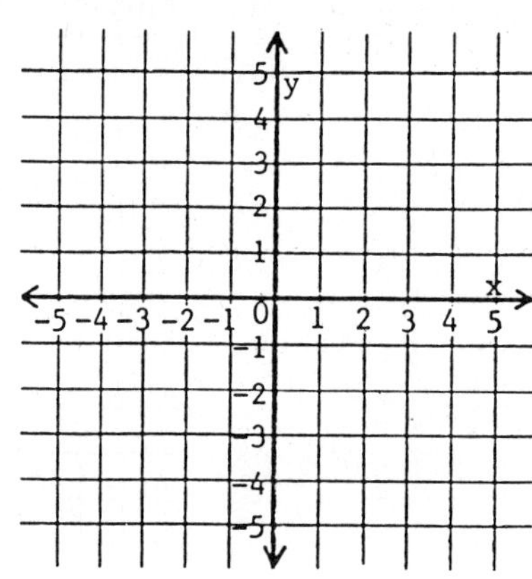

Answer to Frame 79:

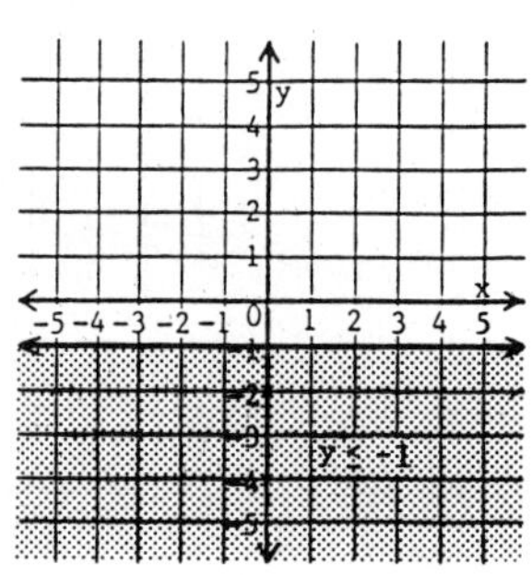

Answer to Frame 80:

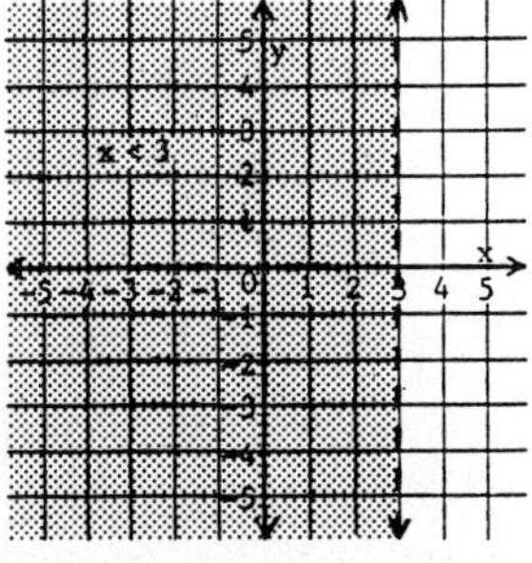

7-9 FUNCTIONS

In this section, we will define functions and discuss functional notation. The definition of functions is extended to formulas.

---

81. A relation is a set of ordered pairs. All of the following two-variable equations are relations including an infinite number of ordered pairs.

$$y = 5x$$
$$y = x^2 + 2x + 3$$
$$y = \pm\sqrt{x}$$
$$y = \sqrt{x - 4}$$
$$y = \frac{7}{x}$$

In each equation above, y is solved for. y is called the dependent variable because its value "depends" on the value substituted for x. x is called the independent variable. That is:

In $y = 3x - 1$: a) y is called the ______________ variable.

b) x is called the ______________ variable.

---

82. A function is a relation in which, for each value of the independent variable, there is only one value of the dependent variable.

$y = x^2 + 2x + 3$ is a function because for each value of x, there is only one value of y. That is:

if $x = 3$, $y = 18$

if $x = 0$, $y = 3$

if $x = -1$, $y = 2$

$y = \pm\sqrt{x}$ is not a function because for each positive value of x, there are two values of y. That is:

if $x = 4$, $y = 2$ and -2.

if $x = 9$, $y =$ ______ and ______.

Answers to 81: a) dependent b) independent

---

83. Since a relation is a function only if there is just one value of y for each value of x, we can use a vertical line test to determine if a graph represents a function. To show that fact, the two graphs below are discussed on the following page.

Answer to 82: 3 and -3

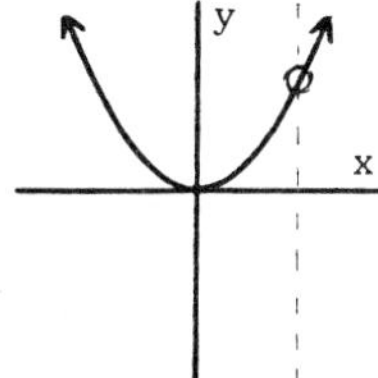

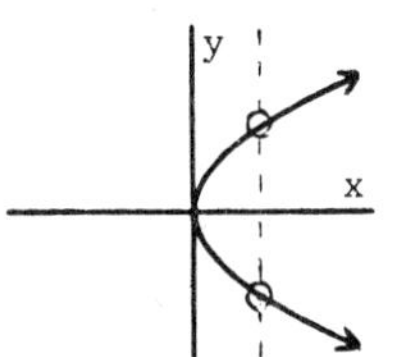

Continued on following page.

83. Continued

By mentally sliding the dotted vertical line along the horizontal axis, we can see these facts.

At the left, since the dotted line never intersects the graph more than once, there is only one y-value for each x-value. Therefore, the graph is a function.

At the right, since the dotted line usually intersects the graph twice, there is usually more than one y-value for each x-value. Therefore, the graph is not a function.

Which of the following represent functions? ____________

a)
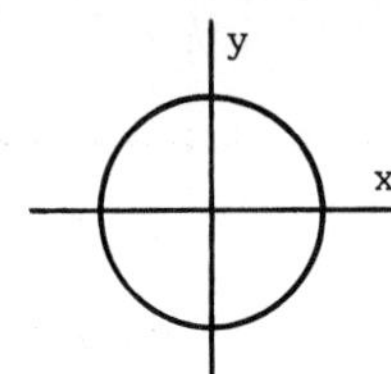

b)
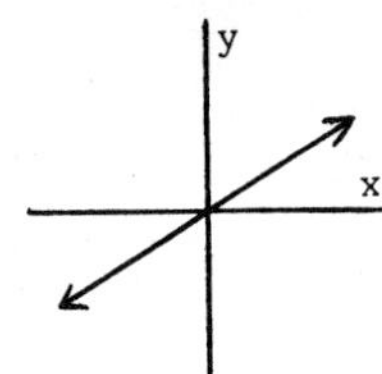

c)
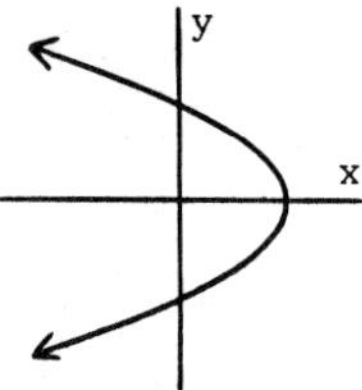

---

84. Which of the following represent functions? ____________

Only (b)

a)
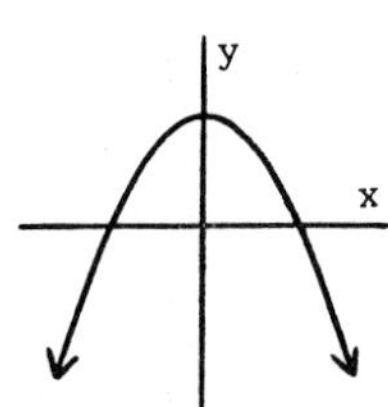

b)
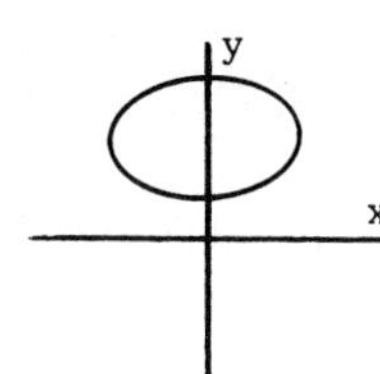

c)
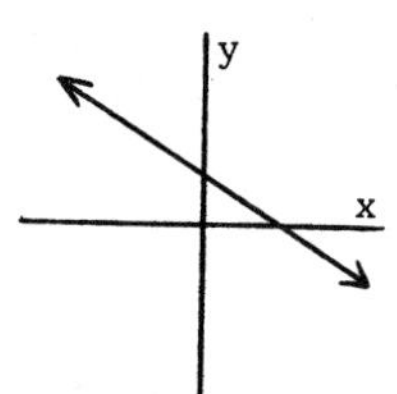

---

85. The functions below are not defined for one value of x because it leads to a division by 0 which is impossible.

Both (a) and (c)

$y = \frac{6}{x}$ is not defined for x = 0, since $y = \frac{6}{0}$

$y = \frac{5}{x - 3}$ is not defined for x = 3, since $y = \frac{5}{3 - 3} = \frac{5}{0}$

Each function below is not defined for one value of x. Name that x-value for each.

a) $y = \frac{9}{2x}$ b) $y = \frac{5}{x^2}$ c) $y = \frac{11}{x - 7}$

x = ______   x = ______   x = ______.

---

a) x = 0

b) x = 0

c) x = 7

86. The functions below are not defined for various values of x because those values lead to imaginary values of y. For example:

$y = \sqrt{x}$ is not defined for negative values of x (or $x < 0$) because they lead to imaginary values of y.

$y = \sqrt{x - 5}$ is not defined for values of x less than 5 (or $x < 5$) because they lead to imaginary values of y.

Each function below is not defined for various values of x. Name the x-values for each.

a) $y = \sqrt{2x}$ b) $y = \sqrt{x - 1}$ c) $y = \sqrt{x - 8}$

$x <$ ______ $x <$ ______ $x <$ ______

---

87. For a function, the set of first coordinates (or x) is called the domain; the set of second coordinates (or y) is called the range.

a) $x < 0$

b) $x < 1$

c) $x < 8$

For some functions, the domain includes all real numbers. Two functions of that type are:

$y = 10x$ $y = x^2 - 7x + 2$

For other functions, the domain is limited because the function is not defined for certain values of x. Two functions of that type are:

$y = \frac{7}{x - 6}$ (The domain is all real numbers except $x = 6$.)

$y = \sqrt{x - 9}$ (The domain is all real numbers except $x < 9$.)

Identify the domain for each function below.

a) $y = \frac{3}{5x}$ b) $y = 2x^2 - 1$ c) $y = \sqrt{x - 10}$

---

88. In a function, we say that y is a function of x. For example, y is a function of x in each equation below.

a) All real numbers except $x = 0$.

b) All real numbers.

c) All real numbers except $x < 10$.

$y = 3x$ $y = x^2 - 1$ $y = 2x^2 + x - 3$

For the phrase "function of x", we use the symbol f(x), which is read "f of x". The symbol f(x) does not mean "multiply f and x". Rather it is a general symbol for any function of x. Therefore, we can use it for each function above. We write:

$f(x) = 3x$

$f(x) = x^2 - 1$

$f(x) = 2x^2 + x - 3$

As you can see, f(x) is another way of stating y. Therefore:

Instead of $y = 10x$, we can write $f(x) = 10x$.

Instead of $y = x^2 - 5x - 9$, we can write ____________________.

89. We can use functional notation to evaluate a function for a particular value of x. For example, we used it below to evaluate $f(x) = 2x + 1$ for $x = 4$ and $x = -3$. To do so, we substituted 4 and -3 for x.

$f(x) = 2x + 1$	$f(x) = 2x + 1$
$f(4) = 2(4) + 1$	$f(-3) = 2(-3) + 1$
$f(4) = 9$	$f(-3) = -5$

The expressions f(4) and f(-3) are read "f of 4" and "f of -3".

f(4) means: find the value of the function when $x = 4$.

f(-3) means: find the value of the function when $x = -3$.

Given the same function $f(x) = 2x + 1$, find the following.

a) f(2) b) f(0) c) f(-2)

$f(x) = x^2 - 5x - 9$

---

90. Given the function $f(x) = 5x^2$, find these:

a) f(3) b) f(0) c) f(-4)

a) f(2) = 5
b) f(0) = 1
c) f(-2) = -3

---

91. a) Find f(-5) for this function. $f(x) = \frac{10}{x}$

b) Find f(29) for this function. $f(x) = \sqrt{x - 4}$

a) f(3) = 45
b) f(0) = 0
c) f(-4) = 80

---

92. Other letters besides f can be used to represent functions. For example, we can use f(x), h(x), A(x), P(x), and so on.

a) If $g(x) = x^2 - x$, find g(-4).

b) If $A(x) = x^2 + 5x - 10$, find A(0).

a) f(-5) = -2
b) f(29) = 5

---

93. Notice the steps used for the evaluation below.

If $f(x) = 2x - 1$, find $f(4) - f(-2)$

$f(x) = 2x - 1$

$f(4) = 2(4) - 1 = 8 - 1 = 7$

$f(-2) = 2(-2) - 1 = -4 - 1 = -5$

Therefore, $f(4) - f(-2) = 7 - (-5) = 7 + 5 = 12$

Continued on following page.

a) g(-4) = 20
b) A(0) = -10

93. Continued

For the same function $f(x) = 2x - 1$, do these:

a) Find $f(3) + f(-5)$.

b) Find $\frac{f(10) - f(6)}{4}$

---

94. We can also evaluate functions for algebraic expressions other than numbers. For example, we evaluated $f(x) = 3x + 5$ below for $x = 2a$ to get $f(2a)$ and $x = a + 1$ to get $f(a + 1)$. To do so, we substituted $2a$ and $a + 1$ for $\underline{x}$.

$$f(x) = 3x + 5$$
$$f(2a) = 3(2a) + 5$$
$$f(2a) = 6a + 5$$

$$f(x) = 3x + 5$$
$$f(a + 1) = 3(a + 1) + 5$$
$$f(a + 1) = 3a + 3 + 5$$
$$f(a + 1) = 3a + 8$$

Given the same function $f(x) = 3x + 5$, find these.

a) $f(5a)$

b) $f(a - 4)$

c) $f(2a + 7)$

Answers to 93: a) -6 b) 2

---

95. Given $H(x) = x^2 - 4$, we found $H(3a)$ and $H(a - 1)$ below.

$$H(x) = x^2 - 4$$
$$H(3a) = (3a)^2 - 4$$
$$H(3a) = 9a^2 - 4$$

$$H(x) = x^2 - 4$$
$$H(a - 1) = (a - 1)^2 - 4$$
$$H(a - 1) = a^2 - 2a + 1 - 4$$
$$H(a - 1) = a^2 - 2a - 3$$

Given the same function $H(x) = x^2 - 4$, find these.

a) $H(7a)$

b) $H(2b + 3)$

Answers to 94:

a) $f(5a) = 15a + 5$

b) $f(a - 4) = 3a - 7$

c) $f(2a+7) = 6a + 26$

Answers to 95:

a) $H(7a) = 49a^2 - 4$

b) $H(2b + 3) = 4b^2 + 12b + 5$

96. Given $P(x) = x + 2x^2$, we found $P(a + b)$ below.

$$P(x) = x + 2x^2$$
$$P(a + b) = (a + b) + 2(a + b)^2$$
$$P(a + b) = a + b + 2(a^2 + 2ab + b^2)$$
$$P(a + b) = a + b + 2a^2 + 4ab + 2b^2$$

Given $g(x) = x^2 - 5x$, find $g(a - b)$

$$g(x) = x^2 - 5x$$
$$g(a - b) =$$

---

97. Notice the steps used for the evaluation below.

Given $f(x) = x^2 - x$, find $\frac{f(x + h) - f(x)}{h}$

$$f(x) = x^2 - x$$
$$f(x + h) = (x + h)^2 - (x + h)$$
$$= x^2 + 2hx + h^2 - x - h$$

Therefore, $\frac{f(x + h) - f(x)}{h} = \frac{(x^2 + 2hx + h^2 - x - h) - (x^2 - x)}{h}$

$$= \frac{x^2 + 2hx + h^2 - x - h - x^2 + x}{h}$$
$$= \frac{2hx + h^2 - h}{h}$$
$$= \frac{\not{h}(2x + h - 1)}{\not{h}}$$
$$= 2x + h - 1$$

Use the same method for this evaluation.

Given $f(x) = 2x^2$, find $\frac{f(x + h) - f(x)}{h}$

Answer to 96: $g(a - b) = a^2 - 2ab + b^2 - 5a + 5b$

Answer to 97: $4x + 2h$

98. For $y = 2x + 1$, $f(x)$ is equivalent to y. Therefore, when graphing the function, we sometimes use $f(x)$ instead of y on the vertical axis as we have done below.

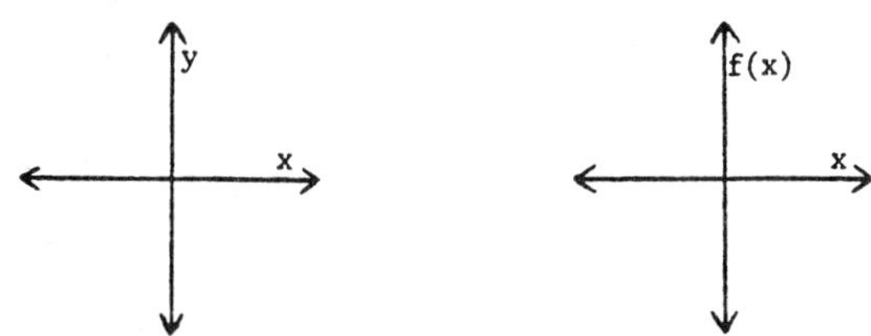

Using $f(x)$ on the vertical axis, we graphed the linear function $y = 2x + 1$ below. Graph the linear function $y = -x - 2$ on the same graph.

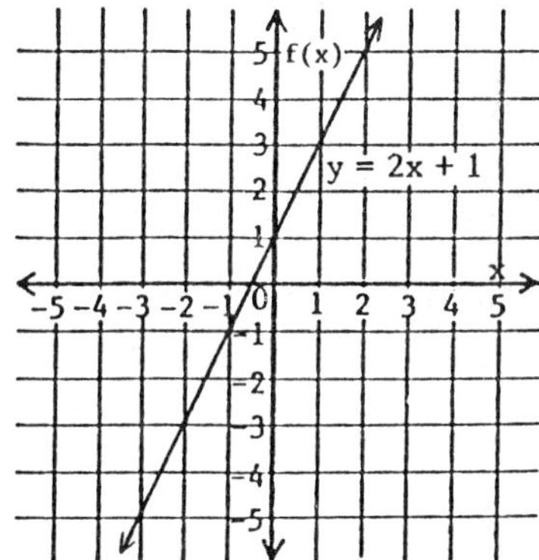

99. Formulas can also be functions. Three examples are given below.

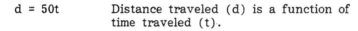

$d = 50t$	Distance traveled (d) is a function of time traveled (t).
$C = \pi d$	The circumference (c) of a circle is a function of the diameter (d) of the circle.
$A = s^2$	The area (A) of a square is a function of the side (s) of the square.

In a formula, letters other than x and y are the dependent and independent variables. That is:

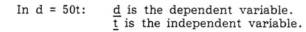

In $d = 50t$: d is the dependent variable. t is the independent variable.

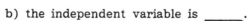

In $A = s^2$: a) the dependent variable is ______.

b) the independent variable is ______.

Answer to Frame 98:

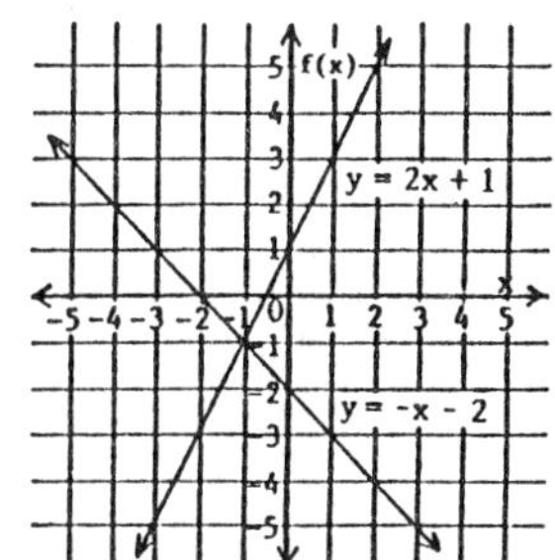

a) A

b) s

100. When a formula is a function, it can be stated in functional notation. That is:

For $d = 50t$, $d = f(t)$

For $C = \pi d$, $C = f(d)$

For $A = s^2$, $A = f(s)$

Usually we use the letter of the dependent variable instead of f for the functions above. That is:

Instead of $f(t) = 50t$, we write $d(t) = 50t$.

Instead of $f(d) = \pi d$, we write $C(d) = \pi d$.

Instead of $f(s) = s^2$, we write ____________.

---

101. We found d(2) and d(10) for $d(t) = 50t$ below.

$d(t) = 50t$	$d(t) = 50t$
$d(2) = 50(2)$	$d(10) = 50(10)$
$d(2) = 100$	$d(10) = 500$

a) Find C(4) for $C(d) = \pi d$.

b) Find A(5) for $A(s) = s^2$.

Answer to Frame 100: $A(s) = s^2$

---

102. When a function is a formula, the domain is not usually defined for negative values because they usually do not make sense. For example:

$d = 50t$ is not defined for negative values of t because negative periods of time do not make sense.

Therefore, when graphing a formula, usually only quadrant 1 is used. As an example, we graphed the linear formula $d = 50t$ below.

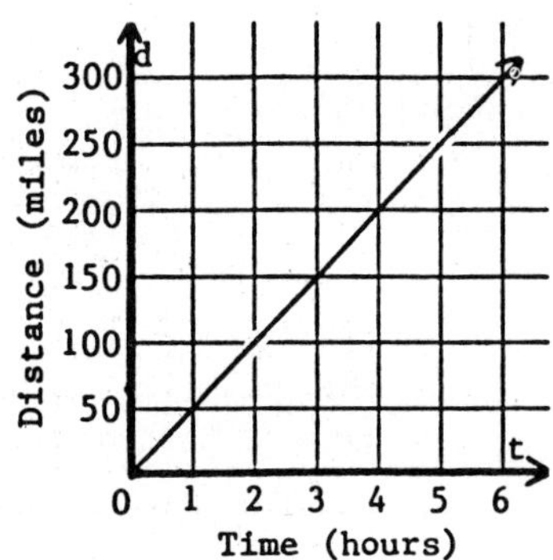

Given that $d(t) = 50t$, use the graph to complete these:

a) $d(1) =$ ________

b) $d(2) =$ ________

c) $d(4) =$ ________

d) $d(6) =$ ________

Answers to Frame 101: a) $C(4) = 4\pi$ b) $A(5) = 25$

Answers to Frame 102:

a) 50 c) 200

b) 100 d) 300

SELF-TEST 26 (pages 364-377)

(pages 364-377)

Graph each inequality.

1. $y < x - 2$

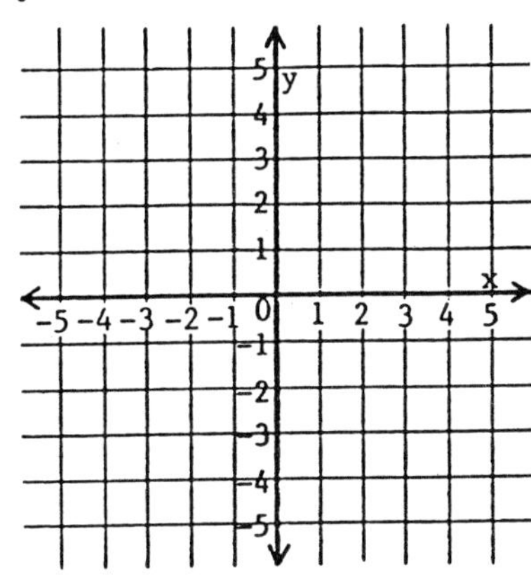

2. $2x + 3y \geq 6$

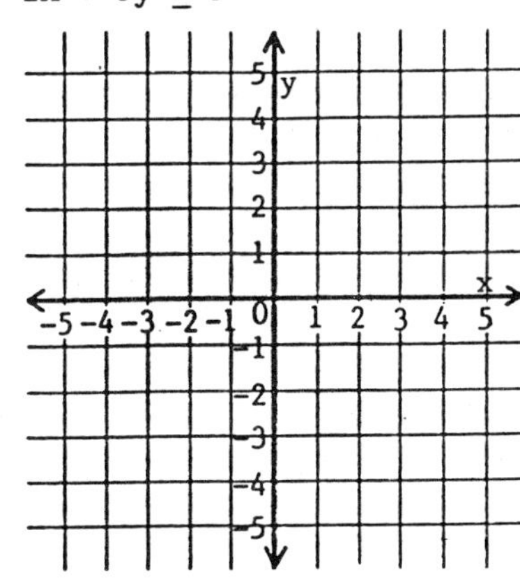

Identify the domain for each function below.

3. $y = \frac{5}{2x}$

4. $y = \sqrt{x - 8}$

5. $y = x^2 + 3x$

6. $y = \frac{10}{3x - 12}$

Evaluate each function for the given value.

7. If $f(x) = 3x + 5$, find $f(-4)$.

8. If $A(r) = \pi r^2$, find $A(10)$.

9. If $g(x) = x^2 - 3$, find $g(4) - g(-1)$.

10. If $H(x) = 5x + 1$, find $H(3a)$.

11. If $P(x) = x^2 - x$, find $P(a - b)$.

12. If $f(x) = 3x^2$, find $\frac{f(x + h) - f(x)}{h}$.

ANSWERS: 1.

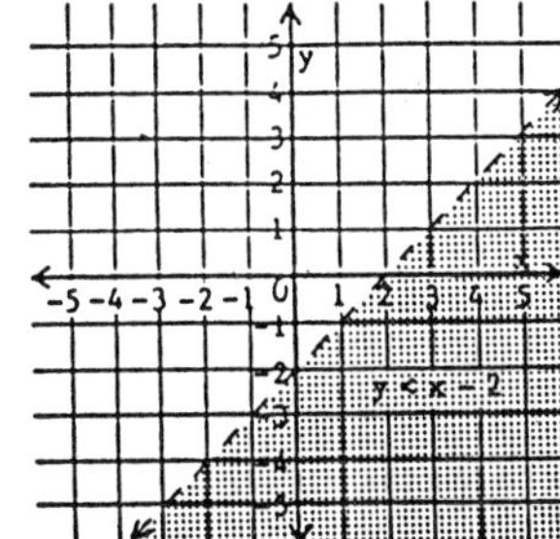

2.

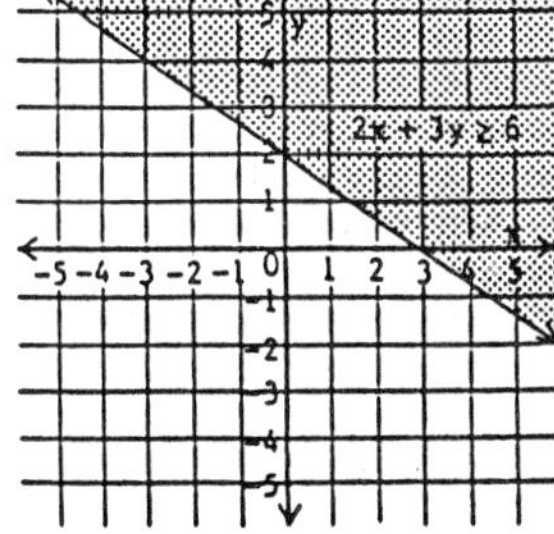

3. All real numbers except $x = 0$

4. All real numbers except $x < 8$

5. All real numbers

6. All real numbers except $x = 4$

7. -7

8. 100π

9. 15

10. $15a + 1$

11. $a^2 - 2ab + b^2 - a + b$

12. $6x + 3h$

7-10 DIRECT VARIATION

Any relationship of the form y = kx is called direct variation. We will discuss direct variation in this section.

---

103. If a car travels at a constant speed of 50 mph, the ratio of distance traveled (d) to time traveled (t) is a constant. The constant is 50. That is:

$$\frac{d}{t} = \frac{50}{1} = \frac{100}{2} = \frac{150}{3} = \frac{200}{4} = 50$$

The relationship above can be stated as an equation.

$$\frac{d}{t} = 50 \qquad \text{or} \qquad d = 50t$$

Any relationship of that type is called direct variation. Above, distance varies directly as time. When time increases, distance increases. When time decreases, distance decreases.

The general form for direct variation is given below.

y = kx	where k is called the constant of variation or the constant of proportionality.

Some examples of direct variation are:

y = 5x   d = 50t

a) In y = 5x, the constant of variation is ______.

b) In d = 50t, the constant of proportionality is ______.

---

104. The following language is used to state a direct variation.

For y = 4x, we say: y varies directly as x.
y is directly proportional to x.

a) For d = 100t, we say: ______ varies directly as ______.

b) For E = 6R, we say: ______ is directly proportional to ______.

Answers to 103: a) 5 b) 50

---

105. If we are given one specific pair of values in a direct variation, we can find k by substitution. For example, if y varies directly as x and y = 10 when x = 2, we get:

y = kx

10 = k(2)

$$k = \frac{10}{2} = 5$$

Since k = 5, the direct variation is y = 5x. Using that equation, we can find y for other values of x. For example, we found y when x = 3 below. Find y when x = 10.

y = 5x	y = 5x
y = 5(3)	
y = 15	______

Answers to 104: a) d ... t b) E ... R

106. Using the two steps from the last frame, we solved the problem below.

If E is directly proportional to R and E = 30 when R = 3, find E when R = 5.

$$E = kR \qquad 30 = k(3) \qquad k = \frac{30}{3} = 10$$

$$E = 10R \qquad E = 10(5) \qquad E = 50$$

Using the same steps, solve this one.

If d varies directly as t and d = 600 when t = 10, find d when t = 7.

Answer: y = 50

---

107. There is a proportion related to direct variation. To derive the proportion, we let (x_1,y_1) and (x_2,y_2) be two solutions of $y = kx$. We get:

$$y_1 = kx_1 \qquad \text{and} \qquad y_2 = kx_2$$

Dividing the first equation by the second, we get:

$$\frac{y_1}{y_2} = \frac{\not{k}x_1}{\not{k}x_2}$$

or

$$\boxed{\frac{y_1}{y_2} = \frac{x_1}{x_2}}$$: proportion related to direct variation

We can use the proportion above to solve a direct variation problem. An example is shown.

If y is directly proportional to x and y = 3 when x = 4, find y when x = 20. Using (4,3) and $(20,y_2)$ as the ordered pairs, we get:

$$\frac{y_1}{y_2} = \frac{x_1}{x_2}$$

$$\frac{3}{y_2} = \frac{4}{20}$$

$$20\not{y_2}\left(\frac{3}{\not{y_2}}\right) = \not{20}y_2\left(\frac{4}{\not{20}}\right)$$

$$60 = 4y_2$$

$$y_2 = 15$$

Continued on following page.

Answer: d = 420, from: d = 60t

107. Using the proportion method, solve this one:

If Q varies directly as P and Q = 5 when P = 2, find Q when P = 12.

---

108. We can use either of two methods to solve the problem below.

If T is directly proportional to S and T = 10 when S = 3, find T when S = 60.

Method 1: Find the constant of variation and then use the direct variation equation.

$$T = kS \qquad T = \frac{10}{3}S$$

$$10 = k(3) \qquad T = \frac{10}{\not{3}}(\not{60})^{20}$$

$$k = \frac{10}{3} \qquad T = 200$$

Method 2: Use the proportion related to direct variation.

$$\frac{T_1}{T_2} = \frac{S_1}{S_2}$$

$$\frac{10}{T_2} = \frac{3}{60}$$

$$60\not{T_2}\left(\frac{10}{\not{T_2}}\right) = \not{60}T_2\left(\frac{3}{\not{60}}\right)$$

$$600 = 3T_2$$

$$T_2 = 200$$

Did we get the same answer using both methods? ________

Answer to 107: $Q_2 = 30$, from:

$$\frac{5}{Q_2} = \frac{2}{12}$$

Answer to 108: Yes. We got 200.

109. Using either method, solve these:

a) When the resistance is constant, the amount of current (I) in an electric circuit varies directly as the applied voltage (E). If I = 30 amperes when E = 6 volts, find I when E = 24 volts.

b) The amount of stretch (s) in a spring is directly proportional to the force (F) applied to it. If s = 6 centimeters when F = 40 kilograms, find s when F = 90 kilograms.

a) 120 amperes b) 13.5 centimeters

7-11 INVERSE VARIATION

Any relationship of the form $y = \frac{k}{x}$ is called inverse variation. We will discuss inverse variation in this section.

110. Airplanes flying at different rates (or speeds) take different amounts of time to fly 500 miles. At 500 mph, it takes 1 hour. At 250 mph, it takes 2 hours. At 125 mph, it takes 4 hours. In each case, the product of the rate r and the time t is a constant. The constant is 500. That is:

$$rt = (500)(1) = (250)(2) = (125)(4) = 500$$

The relationship above can be stated as an equation.

$$rt = 500 \qquad \text{or} \qquad t = \frac{500}{r}$$

Any relationship of that type is called inverse variation. Above, time varies inversely as rate. When the rate increases, time decreases. When the rate decreases, time increases.

Continued on following page.

110. Continued

The general form for <u>inverse variation</u> is given below.

$$\boxed{y = \frac{k}{x}}$$, where <u>k</u> is called <u>the constant of variation</u> or <u>the variation constant</u>

Some examples of inverse variation are:

$y = \frac{12}{x}$ $t = \frac{20}{r}$ $V_1 = \frac{8.5}{P_1}$

a) In $y = \frac{12}{x}$, the constant of variation is ________.

b) In $V_1 = \frac{8.5}{P_1}$, the variation constant is ________.

---

111. The following language is used to state an inverse variation.

For $y = \frac{12}{x}$, we say: <u>y</u> varies <u>inversely</u> as <u>x</u>

or: <u>y</u> is <u>inversely proportional</u> to <u>x</u>.

a) In $t = \frac{20}{v}$, ______ varies inversely as ______.

b) In $V_1 = \frac{8.5}{P_1}$, ______ is inversely proportional to ______.

Answers to 110: a) 12 b) 8.5

---

112. If we are given one specific pair of values in an inverse variation, we can find <u>k</u> by substitution. For example, if <u>y</u> varies inversely as <u>x</u> and y = 50 when x = 4, we get:

$$y = \frac{k}{x}$$

$$50 = \frac{k}{4}$$

$$k = (50)(4) = 200$$

Since k = 200, the inverse variation is $y = \frac{200}{x}$. We can use that equation to find <u>y</u> for other values of <u>x</u>.

a) Find <u>y</u> when x = 5.

$$y = \frac{200}{x}$$

b) Find <u>y</u> when x = 50.

$$y = \frac{200}{x}$$

Answers to 111: a) t ... v b) V_1 ... P_1

Answers to 112: a) y = 40 b) y = 4

113. Using the two steps from the last frame, we solved the problem below.

If I is inversely proportional to R and I = 20 when R = 2, find I when R = 8.

$$I = \frac{k}{R} \qquad I = \frac{40}{R}$$

$$20 = \frac{k}{2} \qquad I = \frac{40}{8}$$

$$k = 40 \qquad I = 5$$

Using the same steps, solve this one.

If t varies inversely as v and t = 4 when v = 100, find t when v = 80.

---

t = 5, from:

$$t = \frac{400}{v}$$

114. There is a proportion related to inverse variation. To derive the proportion, we let (x_1,y_1) and (x_2,y_2) be two solutions of $y = \frac{k}{x}$. We get:

$$y_1 = \frac{k}{x_1} \quad \text{and} \quad y_2 = \frac{k}{x_2}$$

Dividing the first equation by the second, we get:

$$\frac{y_1}{y_2} = \frac{\frac{k}{x_1}}{\frac{k}{x_2}} = \frac{\cancel{k}}{x_1}\left(\frac{x_2}{\cancel{k}}\right) = \frac{x_2}{x_1}$$

or

$$\boxed{\frac{y_1}{y_2} = \frac{x_2}{x_1}}$$: proportion related to inverse variation

Note: In inverse variation, x_2 is the numerator and x_1 is the denominator.

We can use the proportion above to solve an inverse variation problem. An example is shown.

Continued on following page.

114. Continued

If y is inversely proportional to x and y = 12 when x = 2, find y when x = 3. Using (2,12) and $(3,y_2)$ as the ordered pairs, we get:

$$\frac{y_1}{y_2} = \frac{x_2}{x_1}$$

$$\frac{12}{y_2} = \frac{3}{2}$$

$$2y_2\left(\frac{12}{y_2}\right) = 2y_2\left(\frac{3}{2}\right)$$

$$24 = 3y_2$$

$$y_2 = 8$$

Using the proportion method, solve this one.

If T varies inversely as S and T = 4 when S = 100, find T when S = 5.

$T_2 = 80$, from:

$$\frac{4}{T_2} = \frac{5}{100}$$

115. We can use either of two methods to solve the problem below.

If m is inversely proportional to d and m = 12 when d = 40, find m when d = 10.

Method 1: Find the variation constant and then use the inverse variation equation.

$$m = \frac{k}{d} \qquad m = \frac{480}{d}$$

$$12 = \frac{k}{40} \qquad m = \frac{480}{10}$$

$$k = 480 \qquad m = 48$$

Method 2: Use the proportion related to inverse variation.

$$\frac{m_1}{m_2} = \frac{d_2}{d_1}$$

$$\frac{12}{m_2} = \frac{10}{40}$$

$$40m_2\left(\frac{12}{m_2}\right) = 40m_2\left(\frac{10}{40}\right)$$

$$480 = 10m_2$$

$$m_2 = 48$$

Did we get the same answer with both methods? ________

116. Using either method, solve these:

a) In an electric circuit with fixed power, the current (I) is inversely proportional to the voltage (E). If I = 8 amperes when E = 12 volts, find I when E = 10 volts.

b) The pressure (P) of a gas varies inversely as its volume (V). If P = 10 grams per square centimeter when V = 25 cubic centimeters, find P when V = 40 cubic centimeters.

Yes. We got 48.

a) I = 9.6 amperes   b) P = 6.25 grams per square centimeter

7-12 QUADRATIC VARIATION

In this section, we will discuss direct square variation and inverse square variation.

117. Any equation or formula of the form $y = kx^2$ is called direct square variation. Some examples are:

$y = 5x^2$ $P = 20I^2$ $A = .7854d^2$

In $y = kx^2$, k is called the constant of variation or the variation constant. The following language is used to state a direct square variation.

For $y = 5x^2$, we say: y varies directly as the square of x.

For $P = 20I^2$, we say: ____ varies directly as the square of ____.

P ... I

118. If we are given one specific pair of values in a direct square variation, we can find k by substitution. For example, if y varies directly as the square of x and y = 12 when x = 2, we get:

$$y = kx^2$$
$$12 = k(2)^2$$
$$12 = k(4)$$
$$k = 3$$

Since k = 3, the direct square variation is $y = 3x^2$. We can use that equation to find y for other values of x.

a) Find y when x = 4.

$y = 3x^2$

b) Find y when x = 10.

$y = 3x^2$

119. Using the two steps from the last frame, we solved the problem below.

If a varies directly as the square of m and a = 100 when m = 5, find a when m = 8.

$$a = km^2 \qquad a = 4m^2$$
$$100 = k(5)^2 \qquad a = 4(8)^2$$
$$100 = k(25) \qquad a = 4(64)$$
$$k = 4 \qquad a = 256$$

Using the same steps, solve this one.

If R varies directly as the square of V and R = 500 when V = 10, find R when V = 4.

a) $y = 48$

b) $y = 300$

---

120. There is a proportion related to $y = kx^2$. To derive the proportion, we let (x_1,y_1) and (x_2,y_2) be two solutions and get $y_1 = kx_1^2$, and $y_2 = kx_2^2$. Dividing, we get:

$$\frac{y_1}{y_2} = \frac{\cancel{k}x_1^2}{\cancel{k}x_2^2}$$

or

$$\boxed{\frac{y_1}{y_2} = \frac{x_1^2}{x_2^2}}$$: proportion related to direct square variation

We used the proportion above to solve the problem below.

If y varies directly as the square of x and y = 36 when x = 3, find y when x = 5. Using (3,36) and $(5,y_2)$ as the ordered pairs, we get:

$$\frac{y_1}{y_2} = \frac{x_1^2}{x_2^2}$$
$$\frac{36}{y_2} = \frac{3^2}{5^2}$$
$$\frac{36}{y_2} = \frac{9}{25}$$
$$25\cancel{y_2}\left(\frac{36}{\cancel{y_2}}\right) = \cancel{25}y_2\left(\frac{9}{\cancel{25}}\right)$$
$$900 = 9y_2$$
$$y_2 = 100$$

Continued on following page.

$R = 80$, from:

$R = 5V^2$

120. Continued

Use the proportion method to solve this one.

If C varies directly as the square of D and C = 50 when D = 5, find C when D = 7.

C = 98, from:

$$\frac{50}{C} = \frac{5^2}{7^2}$$

---

121. To solve a direct square variation problem, we can use either of two methods:

1. Find the variation constant and then use the direct square variation equation.
2. Use the proportion related to direct square variation.

Using either method, solve these:

a) The kinetic energy (E) of a moving object varies directly as the square of its velocity (v). If E = 400 ergs when v = 10 centimeters per second, find E when v = 8 centimeters per second.

b) The power (P) in an electric circuit varies directly as the square of the current (I). If P = 20 watts when I = 2 amperes, find P when I = 3 amperes.

---

a) E = 256 vigs

b) P = 45 watts

122. Any equation or formula of the form $\boxed{y = \frac{k}{x^2}}$ is called <u>inverse square variation</u>. Some examples are:

$$y = \frac{5}{x^2} \qquad F = \frac{65}{d^2} \qquad P = \frac{100}{t^2}$$

In $y = \frac{k}{x^2}$, k is called the <u>constant of variation</u> or the <u>variation constant</u>. The following language is used to state an inverse square variation.

For $y = \frac{5}{x^2}$, we say: y varies inversely as the square of x.

For $F = \frac{65}{d^2}$, we say: ______ varies inversely as the square of ______.

---

123. If we are given one specific pair of values in an inverse square variation, we can find k by substitution. For example, if y varies inversely as the square of x and $y = 12$ when $x = 2$, we get:

Answer (122): F ... d

$$y = \frac{k}{x^2}$$

$$12 = \frac{k}{2^2}$$

$$12 = \frac{k}{4}$$

$$k = 48$$

Since $k = 48$, the inverse square variation is $y = \frac{48}{x^2}$. We can use that equation to find y for other values of x.

a) Find y when $x = 4$. b) Find y when $x = 8$.

---

124. Using the two steps from the last frame, we solved the problem below.

Answer (123): a) $y = 3$ b) $y = \frac{3}{4}$ or 0.75

If H varies inversely as the square of P and H = 4 when P = 5, find H when P = 2.

$H = \frac{k}{P^2}$	$H = \frac{100}{P^2}$
$4 = \frac{k}{5^2}$	$H = \frac{100}{2^2}$
$4 = \frac{k}{25}$	$H = \frac{100}{4}$
$k = 100$	$H = 25$

Continued on following page.

124. Continued

Using the same steps, solve this one.

If m varies inversely as the square of s and m = 4 when s = 6, find m when s = 4.

---

125. There is also a proportion related to $y = \frac{k}{x^2}$. To derive the proportion, we let (x_1,y_1) and (x_2,y_2) be two solutions and get $y_1 = \frac{k}{x_1^2}$ and $y_2 = \frac{k}{x_2^2}$. Dividing, we get:

$$\frac{y_1}{y_2} = \frac{\frac{k}{x_1^2}}{\frac{k}{x_2^2}} = \frac{\cancel{k}}{x_1^2}\left(\frac{x_2^2}{\cancel{k}}\right) = \frac{x_2^2}{x_1^2}$$

or

$$\boxed{\frac{y_1}{y_2} = \frac{x_2^2}{x_1^2}}$$: proportion related to inverse square variation

Note: In inverse square variation, x_2^2 is the numerator and x_1^2 is the denominator.

We used the proportion above to solve the problem below.

If y varies inversely as the square of x and y = 80 when x = 3, find y when x = 4. Using (3,80) and $(4,y_2)$ as the ordered pairs, we get:

$$\frac{y_1}{y_2} = \frac{x_2^2}{x_1^2}$$

$$\frac{80}{y_2} = \frac{4^2}{3^2}$$

$$\frac{80}{y_2} = \frac{16}{9}$$

$$9\cancel{y_2}\left(\frac{80}{\cancel{y_2}}\right) = \cancel{9}y_2\left(\frac{16}{\cancel{9}}\right)$$

$$720 = 16y_2$$

$$y_2 = 45$$

Continued on following page.

m = 9, from:

$$m = \frac{144}{s^2}$$

125. Continued

Use the proportion method to solve this one.

If P varies inversely as the square of R and P = 8 when R = 10, find P when R = 5.

---

126. To solve an inverse square variation problem, we can also use either of two methods.

1. Find the variation constant and then use the inverse square variation equation.
2. Use the proportion related to inverse square variation.

Using either method, solve these:

a) The intensity (I) of a radio signal varies inversely as the square of its distance (d) from the transmitter. If I = 5 microvolts when d = 10 kilometers, find I when d = 5 kilometers.

b) The gravitational force (F) between two objects varies inversely as the square of the distance (d) between them. If F = 40 dynes when d = 10 centimeters, find F when d = 4 centimeters.

Answer to 125: P = 32, from:

$$\frac{8}{P} = \frac{5^2}{10^2}$$

Answers to 126:

a) I = 20 microvolts

b) F = 250 dynes

7-13 JOINT AND COMBINED VARIATION

In this section, we will discuss joint variation and combined variation. Both types of variation involve two or more independent variables.

---

127. Any equation or formula of the form $\boxed{y = kxz}$ is called joint variation. Some examples are:

$y = 4xz$ $\qquad$ $A = \frac{1}{2}bh$ $\qquad$ $V = 10hg^2$

When a formula is a joint variation, the units for the variables are frequently chosen so that $k = 1$. For example:

$E = IR$ $\qquad$ $F = ma$ $\qquad$ $P = I^2R$

The following language is used for a joint variation.

For $y = 4xz$, we say: y varies jointly as x and z.

For $P = I^2R$, we say: P varies jointly as R and the square of I.

For $E = IR$, we say: E varies jointly as ____ and ____.

---

128. To solve the problem below, we found the variation constant and then used $y = 4xz$ to complete the solution.

Answer to 127: I and R

y varies jointly as x and z. If $y = 80$ when $x = 2$ and $z = 10$, find y when $x = 3$ and $z = 6$.

$y = kxz$	$y = 4xz$
$80 = k(2)(10)$	$y = 4(3)(6)$
$80 = 20k$	$y = 12(6)$
$k = 4$	$y = 72$

Use the same method to solve this problem.

P varies jointly as Q and R. If $P = 40$ when $Q = 2$ and $R = 4$, find P when $Q = 5$ and $R = 10$.

Answer to 128: $P = 250$

129. We used the same two-step method to solve one problem below. Solve the other problem.

b varies jointly as c and the square of d. If $b = 200$ when $c = 4$ and $d = 5$, find b when $c = 2$ and $d = 10$.

$b = kcd^2$	$b = 2cd^2$
$200 = k(4)(5^2)$	$b = 2(2)(10^2)$
$200 = k(4)(25)$	$b = 2(2)(100)$
$200 = 100k$	$b = 400$
$k = 2$	

m varies jointly as s and the square of t. If $m = 90$ when $s = 2$ and $t = 3$, find m when $s = 8$ and $t = 5$.

$m = 1,000$

130. Combined variation includes both direct and inverse variation. Some examples and the language used to describe them are given below.

$y = \frac{kx}{z}$ — y varies directly as x and inversely as z.

$p = \frac{kst}{v}$ — p varies jointly as s and t and inversely as v.

For the formulas below, the variation constant k equals "1".

$F = \frac{mv^2}{r}$ — F varies jointly as m and the square of v and inversely as r.

$F = \frac{m_1m_2}{rd^2}$ — F varies jointly as m_1 and m_2 and inversely as r and the square of d.

Write an equation or formula for each of these.

a) y varies directly as x and inversely as b and the square of z. ____________

b) P_1 varies jointly as P_2 and V_2 and inversely as V_1, with $k = 1$. ____________

---

a) $y = \frac{kx}{bz^2}$

b) $P_1 = \frac{P_2 V_2}{V_1}$

131. We used a two-step process to solve the problem below. Solve the other problem.

y varies jointly as x and z and inversely as w. If y = 15 when x = 2, z = 5, and w = 4, find y when x = 3, z = 10, and w = 9.

$$y = \frac{kxz}{w} \qquad y = \frac{6xz}{w}$$

$$15 = \frac{k(2)(5)}{4} \qquad y = \frac{6(3)(10)}{9}$$

$$15 = \frac{10k}{4} \qquad y = \frac{180}{9}$$

$$4(15) = \cancel{4}\,\frac{10k}{\cancel{4}} \qquad y = 20$$

$$60 = 10k$$

$$k = 6$$

b varies directly as c and inversely as the square of d. If b = 10 when c = 20 and d = 4, find b when c = 45 and d = 6.

b = 10

---

132. Solve each problem:

a) The wind force (F) on a vertical surface varies jointly as the area (A) of of the surface and the square of the wind velocity (v). When v = 20 mph and A = 1 sq ft, F = 2 lbs. Find F when v = 30 mph and A = 2 sq ft.

b) The volume (V) of a gas varies directly as the temperature (T) and inversely as the pressure (P). If V = 288 cm^3 when T = 32° and P = 10 kg/cm^2, find V when T = 38° and P = 15 kg/cm^2.

---

a) F = 9 lbs, from:

$$F = .005\ Av^2$$

b) V = 228 cm^2, from:

$$V = \frac{90T}{P}$$

SELF-TEST 27 (pages 378-394)

1. If y varies directly as x and $y = 30$ when $x = 5$, find y when $x = 7$.

2. If d is inversely proportional to b and $d = 20$ when $b = 3$, find d when $b = 12$.

3. If distance d is directly proportional to time t and $d = 28$ m when $t = 8$ sec, find d when $t = 20$ sec.

4. If current I in an electric circuit varies inversely as the resistance R and $I = 3$ amperes when $R = 10$ ohms, find I when $R = 12$ ohms.

5. If y varies directly as the square of x and $y = 200$ when $x = 5$, find y when $x = 9$.

6. If p varies inversely as the square of q, and $p = 50$ when $q = 2$, find p when $q = 5$.

7. The intensity I of light varies inversely as the square of the distance d from the light source. If $I = 80\ W/m^2$ when $d = 4m$, find I when $d = 10m$.

8. The pressure P in a liquid varies jointly with the depth h and density D of the liquid. If $P = 180\ gr/cm^2$ when $h = 1.25$ cm and $D = 144\ gr/cm^3$, find P when $h = 1.45$ cm and $D = 120\ gr/cm^3$.

ANSWERS:

1. $y = 42$
2. $d = 5$
3. $d = 70$ m
4. $I = 2.5$ amperes
5. $y = 648$
6. $p = 8$
7. $I = 12.8\ W/m^2$
8. $P = 174\ gr/cm^2$

SUPPLEMENTARY PROBLEMS - CHAPTER 7

Assignment 24

Decide whether the given ordered pair is a solution of the given equation.

1. (2,10) for y = 5x
2. (-1,-4) for y = 3x - 1
3. (-2,12) for 2x + y = 9
4. (0,-3) for x - y = 3
5. (-10,-5) for $y = -\frac{1}{2}x$
6. (-6,-5) for $y = -\frac{2}{3}x - 1$

Complete the solutions below for y = 3x + 6.

7. (-1,)
8. (,-6)
9. (0,)
10. (,0)

Complete the solutions below for 4x + y = 8.

11. (3,)
12. (,4)
13. (0,)
14. (,0)

15. Write the coordinates of each point.

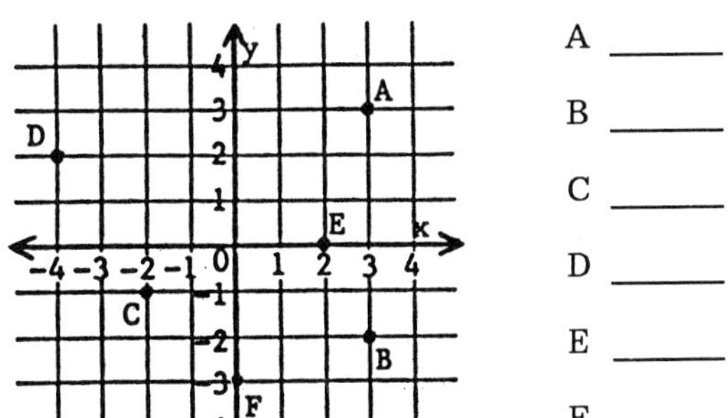

A ______
B ______
C ______
D ______
E ______
F ______

16. Estimate the coordinates of each point.

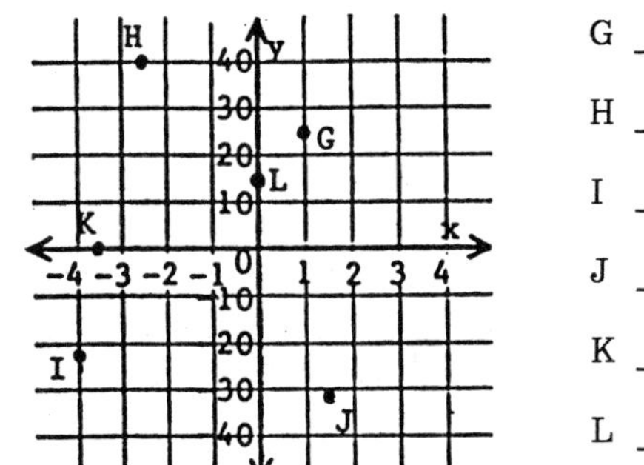

G ______
H ______
I ______
J ______
K ______
L ______

17. State the number of the quadrant in which each point lies.
a) (-8,1) b) (20,-5) c) (2,13) d) (-1,-3) e) (-32,50)

18. Which of the following points lie on the x-axis?
a) (0,-7) b) (-7,0) c) (0,0) d) (30,0) e) (0,1)

19. Which of the following points lie on the y-axis?
a) (0,4) b) (1,-1) c) (-3,0) d) (0,0) e) (0,-50)

20. What name is given to the point (0,0)?

Graph each equation.

21. $y = x$

22. $y = -2x - 1$

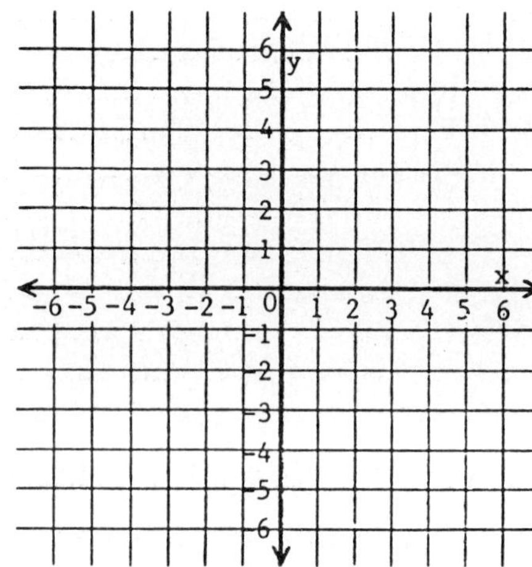

Graph each equation.

23. $x + y = 2$

24. $y = \frac{1}{2}x - 3$

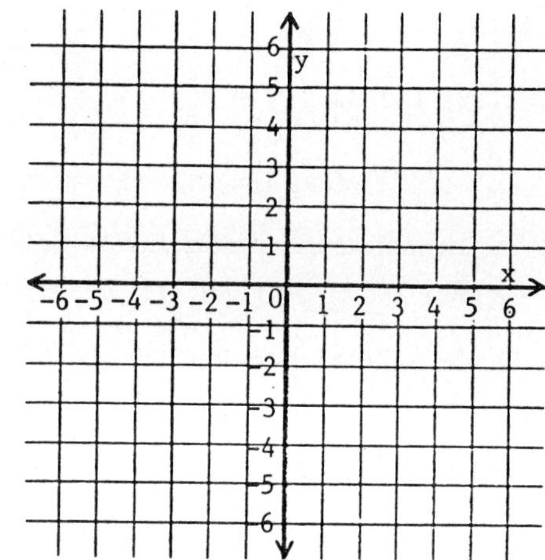

Find the coordinates of the x-intercept of each equation.

25. $y = 2x + 6$ 26. $3x + 5y = 15$ 27. $4x - 3y = 2$ 28. $2y = 5x + 8$

Find the coordinates of the y-intercept of each equation.

29. $y = 3x - 9$ 30. $2x + 5y = 10$ 31. $3y = 4x + 1$ 32. $6x - 2y = 5$

Use the intercept method to graph each equation.

33. $y = x + 2$

34. $x + 3y = 3$

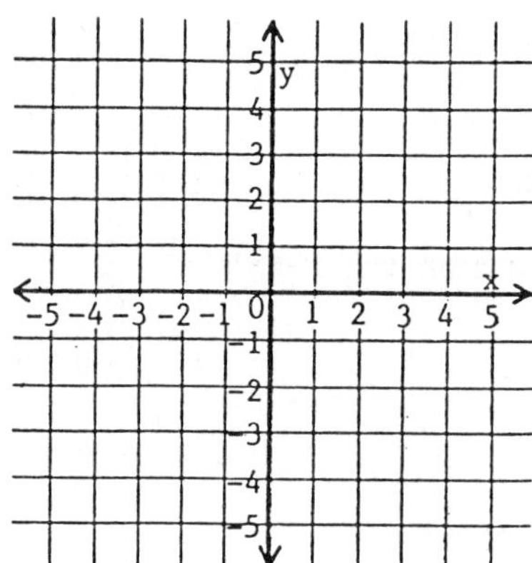

35. $3x - 2y = 6$

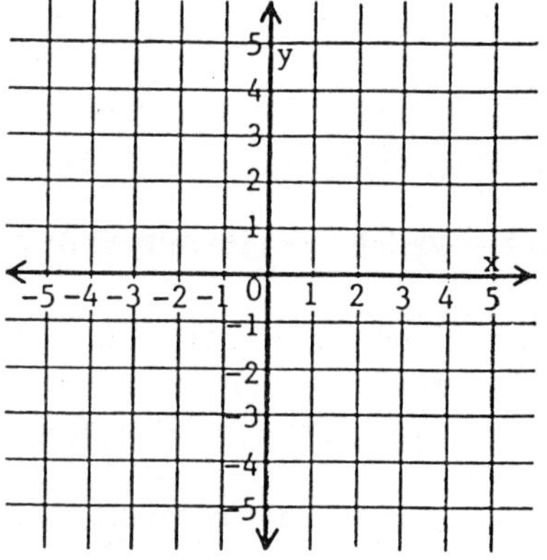

Write the equation of each line.

36. line A: 

37. line B: ____________

38. line C: ____________

39. line D: ____________

40. the x-axis: ____________

41. the y-axis: ____________

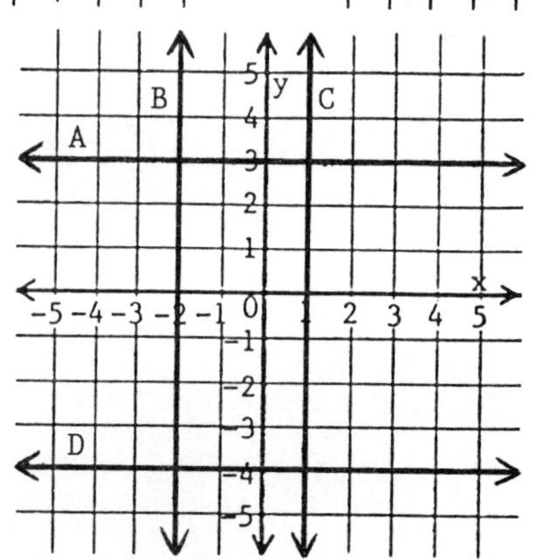

Assignment 25

Find the slope of each line graphed below.

1. Line A

2. Line B

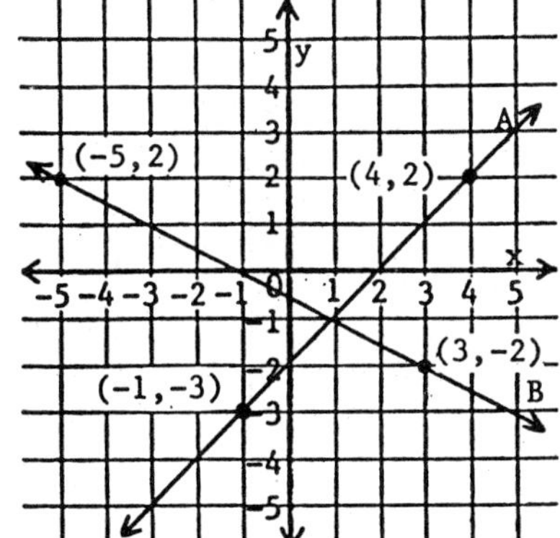

3. Line C

4. Line D

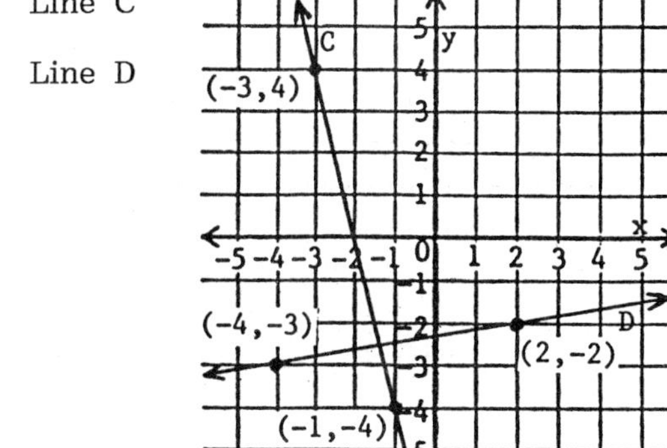

Five lines have these slopes: Line A: 4 Line B: -1 Line C: -5 Line D: $\frac{1}{2}$ Line E: -3

5. Which lines rise from left to right?

6. Which line has the steepest rise?

7. Which lines fall from left to right?

8. Which line has the steepest fall?

Find the slope of the line through each pair of points.

9. (2,3) and (-1,-3)

10. (6,8) and (10,2)

11. (-4,4) and (0,8)

12. (-4,2) and (2,-4)

13. (-3,-4) and (5,-2)

14. (0,6) and (10,0)

15. (-20,10) and (30,35)

16. (0,0) and (-6,-12)

17. (0,0) and (2,-8)

What is the slope of: 18. Any horizontal line? 19. Any vertical line?

For each equation, identify the slope and the coordinates of the y-intercept.

20. $y = 5x + 2$

21. $y = x - 3$

22. $y = -\frac{2}{5}x + \frac{3}{2}$

23. $y = \frac{1}{2}x$

Write the slope-intercept form of the linear equation whose:

24. Slope is -1 and y-intercept is (0,2)

25. Slope is $\frac{1}{2}$ and y-intercept is (0,-5)

26. Slope is $\frac{1}{3}$ and y-intercept is (0,-1)

27. Slope is $\frac{8}{3}$ and y-intercept is (0,0)

Put each equation in slope-intercept form.

28. $y = 6 - 4x$

29. $y - 3x = 2$

30. $x - y = 5$

31. $4y = x - 8$

32. $3x + 6y = 10$

33. $x - 5y = 0$

Using the given point and slope, graph each line below.

34. through (-2,-6) with a slope of $\frac{4}{3}$

35. through (-3,4) with a slope of $-\frac{2}{5}$

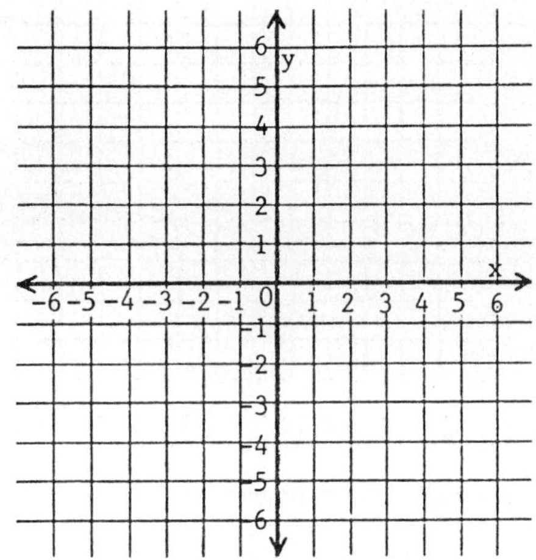

Using the y-intercept and the slope, graph each line below.

36. $y = \frac{2}{3}x + 2$

37. $y = -2x - 2$

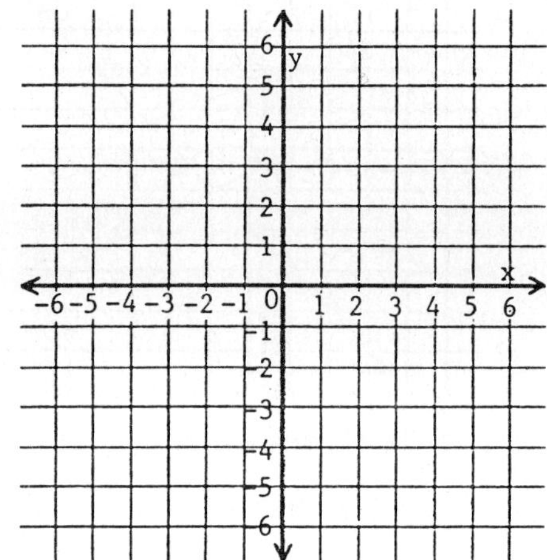

Find the slope-intercept form of the equation of the line whose:

38. Slope is 3 and passes through (2,0)

39. Slope is -1 and passes through (3,-1)

40. Slope is $\frac{1}{3}$ and passes through (-6,3)

41. Slope is -4 and passes through (1,2)

42. Slope is $-\frac{1}{2}$ and passes through (-4,5)

43. Slope is 5 and passes through (0,0)

Find the slope-intercept form of the equation of the line through each pair of points.

44. (-1,-2) and (3,1)

45. (0,1) and (2,-5)

46. (2,4) and (6,2)

Decide whether each pair of lines is parallel or perpendicular.

47. $2x = y$ and $2y - 4x = 5$

48. $4x + 5y = 20$ and $5x - 4y = 20$

49. $2y - x = 7$ and $y + 2x = 3$

50. $3x + y = 0$ and $6x + 2y = 9$

51. the line through (1,2) and (4,4) and the line through (-4,-3) and (2,1)

52. the line through (-2,-3) and (1,-2) and the line through (3,4) and (5,-2)

Write the slope-intercept form of the line through the given point and parallel to the given line.

53. (0,2); $y = 3x$

54. (-1,-3); $x + y = 7$

55. (-4,2); $2x - 4y = 3$

Write the slope-intercept form of the line through the given point and perpendicular to the given line.

56. (4,1); $y = -2x$

57. (0,-3); $x + 4y = 3$

58. (6,-2); $3x - 4y = 5$

Assignment 26

Graph each inequality.

1. $y < x$

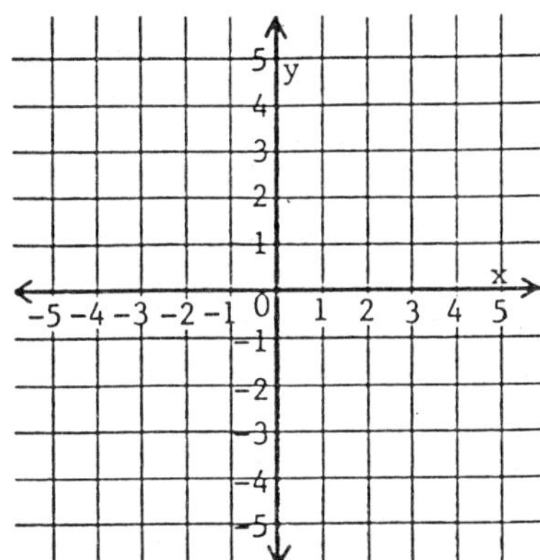

2. $y \geq x + 3$

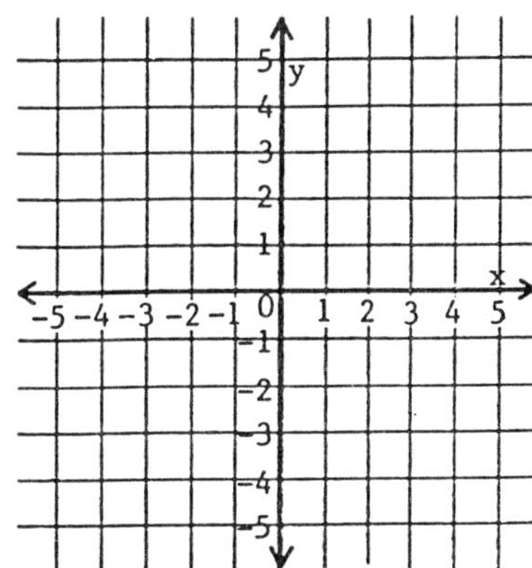

3. $y > 1 - x$

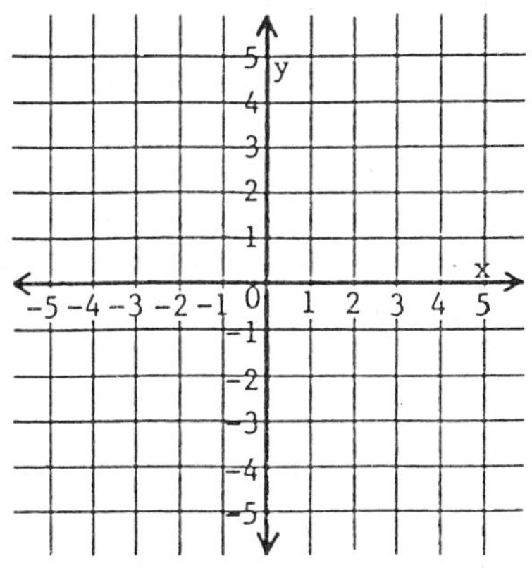

4. $x - 2y \leq 2$

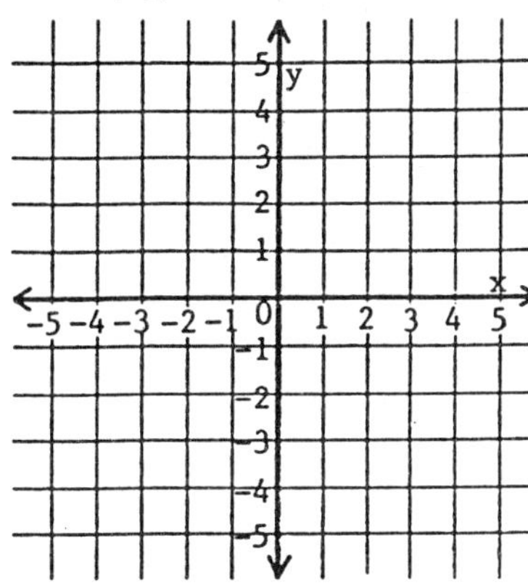

5. $x + 2y > 4$

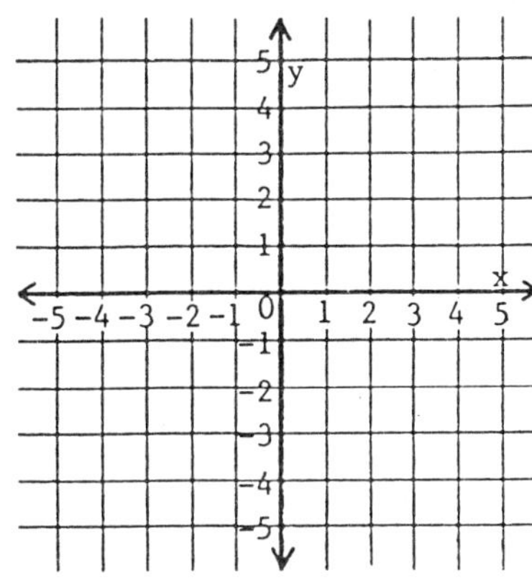

6. $3x - 5y \geq 15$

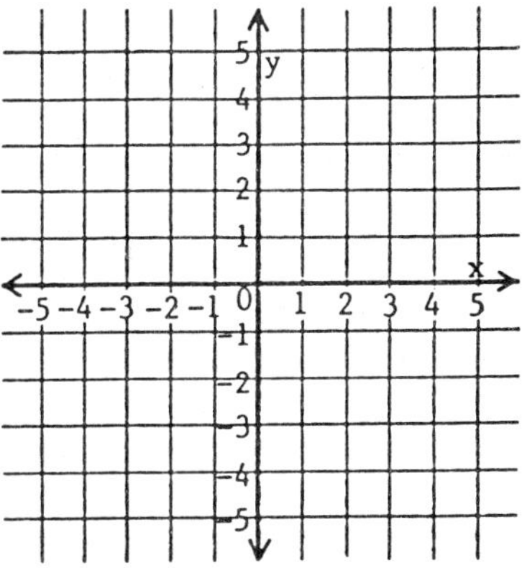

7. $5x + 2y < 10$

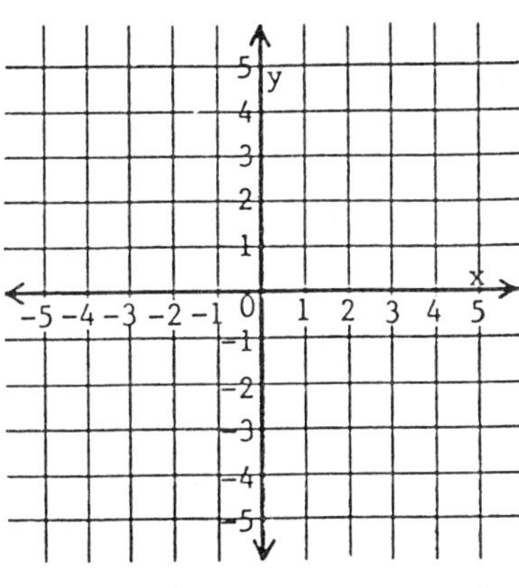

8. $y < -2$

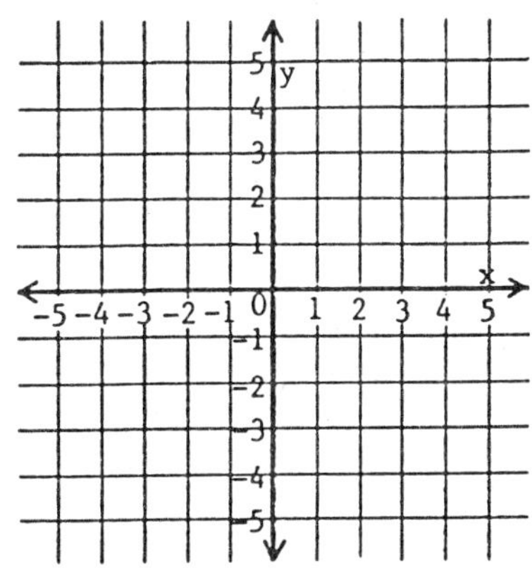

9. $x \geq 1$

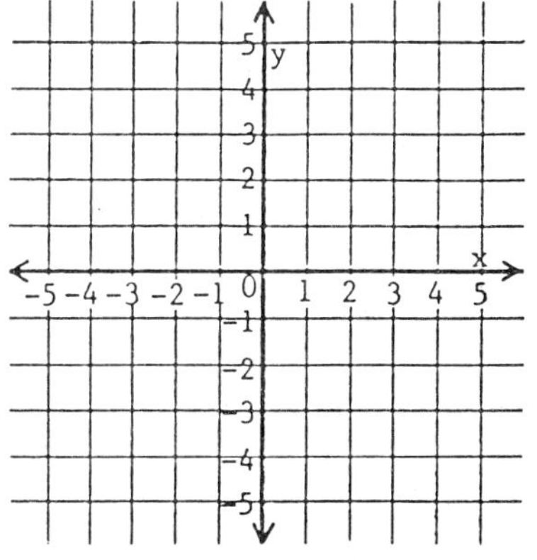

Identify the domain of each function.

10. $y = \frac{5}{x}$ 11. $y = 5x$ 12. $y = \frac{6}{x - 5}$ 13. $y = \frac{9}{2x - 4}$

14. $y = \sqrt{3x}$ 15. $y = x^2 - 3$ 16. $y = \sqrt{x - 9}$ 17. $y = \sqrt{2x - 6}$

18. Which of the following are functions? Use the vertical line test.

a)
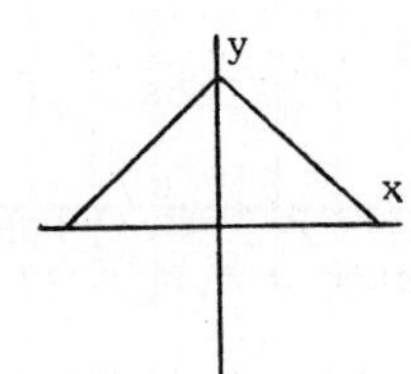

b)
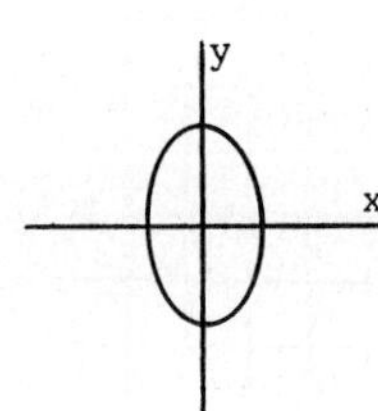

c)
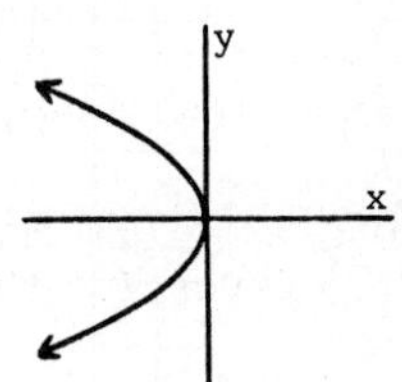

d)
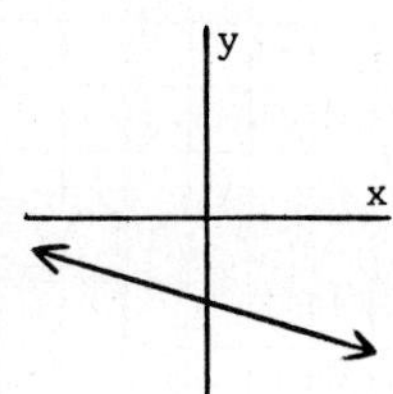

Evaluate each function for the given value.

19. If $f(x) = 3x - 2$, find $f(5)$.

20. If $f(x) = 1 - 2x$, find $f(-3)$.

21. If $g(x) = 4x^2$, find $g(-1)$.

22. If $P(x) = \frac{10}{x - 1}$, find $P(3)$.

23. If $f(x) = \sqrt{x - 1}$, find $f(10)$.

24. If $H(x) = x^2 - 2x$, find $H(-4)$.

25. If $h(x) = x^2 - 3x + 5$, find $h(3)$.

26. If $f(x) = 5x$, find $f(2) + f(-1)$.

27. If $A(x) = 3x + 1$, find $A(4) - A(-3)$.

28. If $g(x) = x + 3$, find $\frac{g(7) + g(3)}{4}$

29. If $f(x) = x + 7$, find $f(2b)$.

30. If $P(x) = 2x + 7$, find $P(a - 3)$.

31. If $g(x) = 3x - 1$, find $g(2a - 5)$.

32. If $h(x) = x^2 - 9$, find $h(a + 2)$.

33. If $H(x) = 2x + x^2$, find $H(a + b)$.

34. If $f(x) = x^2 - 3x$, find $f(a - b)$.

35. If $f(x) = 3x^2$, find $\frac{f(x + h) - f(x)}{h}$.

36. If $f(x) = x^2 - 2x$, find $\frac{f(x + h) - f(x)}{h}$.

37. If $d(t) = 100t$, find $d(4)$.

38. If $C(d) = \pi d$, find $C(7)$.

39. If $A(s) = s^2$, find $A(12)$.

40. If $A(r) = \pi r^2$, find $A(5)$.

Assignment 27

1. If $\underline{y}$ varies directly as $\underline{x}$ and $y = 350$ when $x = 50$, find $\underline{y}$ when $x = 15$.

2. If H is directly proportional to P and $H = 450$ when $P = 15$, find H when $P = 25$.

3. The distance $\underline{d}$ traveled by a car varies directly as time $\underline{t}$. If $d = 170$ kilometers when $t = 2$ hours, find $\underline{d}$ when $t = 7$ hours.

4. In an electric circuit, the current I is directly proportional to the applied voltage E. If $I = 4$ amperes when $E = 20$ volts, find I when $E = 45$ volts.

5. If $\underline{y}$ varies inversely as $\underline{x}$ and $y = 5$ when $x = 20$, find $\underline{y}$ when $x = 10$.

6. If $\underline{p}$ is inversely proportional to $\underline{q}$ and $p = 100$ when $q = 4$, find $\underline{p}$ when $q = 80$.

7. The volume V of a gas is inversely proportional to its pressure P. If V = 25 liters when P = 500 gr/cm^2, find V when P = 625 gr/cm^2.

8. The time t needed to fill a tank varies inversely as the rate r of pumping. If t = 40 min when r = 650 kL, find t when r = 800 kL.

9. If y varies directly as the square of x and y = 240 when x = 4, find y when x = 3.

10. If G varies directly as the square of H and G = 216 when H = 6, find G when H = 10.

11. If an object is dropped, the distance d that it falls varies directly as the square of the time t that it falls. If d = 64 ft when t = 2 sec, find d when t = 5 sec.

12. When the brakes are applied, the stopping distance d of a car varies directly as the square of the speed r. If d = 125 ft when r = 50 mph, find d when r = 70 mph.

13. If y varies inversely as the square of x and y = 100 when x = 8, find y when x = 20.

14. If S varies inversely as the square of T and S = 2 when T = 40, find S when T = 4.

15. The intensity I of a sound varies inversely as the square of the distance d from the source. If I = 12 microwatts when d = 10m, find I when d = 20m.

16. The gravitational force F between two objects varies inversely as the square of the distance d between them. If F = 1000 dynes when d = 2 cm, find F when d = 20 cm.

17. If y varies jointly as x and the square of z and y = 200 when x = 5 and z = 2, find y when x = 2 and z = 5.

18. If b varies directly as c and inversely as d and b = 12 when c = 40 and d = 10, find b when c = 80 and d = 5.

19. The simple interest I earned in a given time varies jointly as the principle P and the interest rate r. If I = $185 when P = $1,000 and r = 9.25%, find I when P = $3,000 and r = 7.75% or .0775.

20. The volume V of a gas varies directly as the temperature T and inversely as the pressure P. If V = 304 cm^3 when T = 38° and P = 10 kg/cm^2, find V when T = 44° and P = 16 kg/cm^2.

8 Systems of Equations

In this chapter, we define a system of two equations and solve systems of that type by the graphing method, the addition method, and the substitution method. We will define a system of three equations and solve systems of that type by the addition method. Various word problems involving systems of both types are included. We will also discuss the determinant method for solving systems of both types.

8-1 THE GRAPHING METHOD

In this section, we will define a system of two linear equations and discuss the graphing method for solving systems of that type.

---

1. Two linear equations with the same variables form a system of two linear equations. An example is given below.

$$x + y = 7$$

$$2x - y = 8$$

A solution of a system of two linear equations is an ordered pair that satisfies both equations. To show that (5,2) or $x = 5$, $y = 2$ is a solution of the system above, we substituted 5 for x and 2 for y in each equation on the following page.

Continued on following page.

1. Continued

$$x + y = 7$$
$$5 + 2 = 7$$
$$7 = 7$$

$$2x - y = 8$$
$$2(5) - 2 = 8$$
$$10 - 2 = 8$$
$$8 = 8$$

a) Is (10,30) a solution of this system? ________

$$y = 3x$$
$$4x + y = 70$$

b) Is (4,-5) a solution of this system? ________

$$3x - y = 17$$
$$5x + 4y = 9$$

a) Yes b) No

2. We graphed both equations in the system below.

$$x + y = 5$$
$$x - y = 1$$

The coordinates of any point on line A satisfy $x + y = 5$. The coordinates of any point on line B satisfy $x - y = 1$. The only point that lies on both lines is the point of intersection (3,2). Since it is the only point whose coordinates satisfy both equations, (3,2) is the only solution of the system.

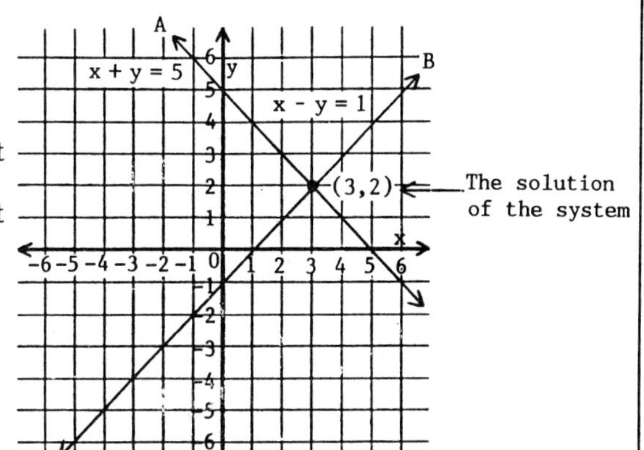

Check the solution (3,2) in each equation below.

$$x + y = 5$$

$$x - y = 1$$

$$x + y = 5$$
$$3 + 2 = 5$$
$$5 = 5$$

$$x - y = 1$$
$$3 - 2 = 1$$
$$1 = 1$$

3. Let's use the graphing method to solve this system.

$$y = 2x$$
$$y - x = 2$$

a) Graph the equations.

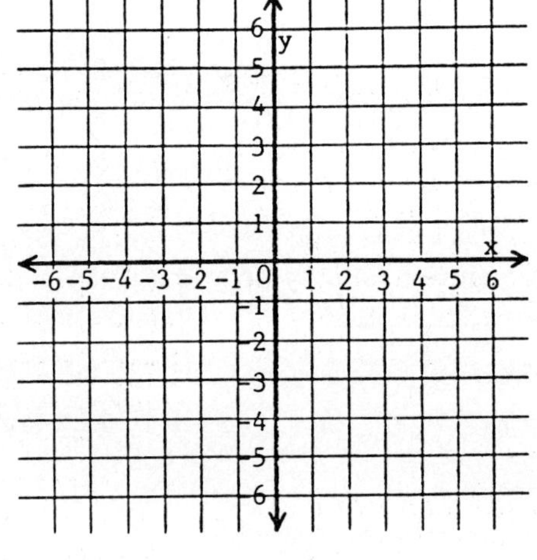

b) The solution of the system is: ________

---

4. Solve this system graphically.

$$x + y = 0$$
$$2x - y = 3$$

a) Graph each equation.

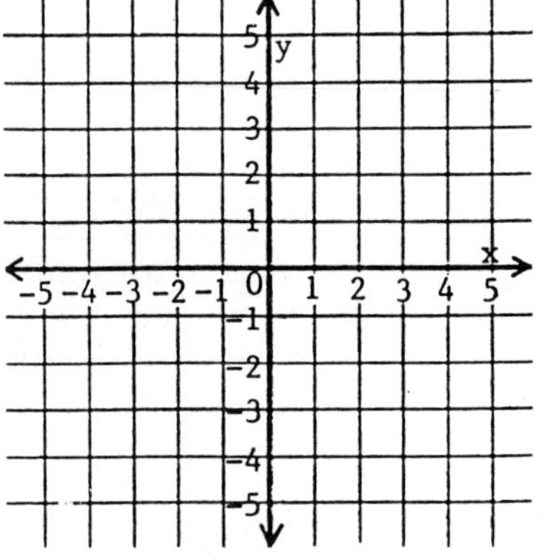

b) The solution of the system is: ________

Answers to 3:

a)

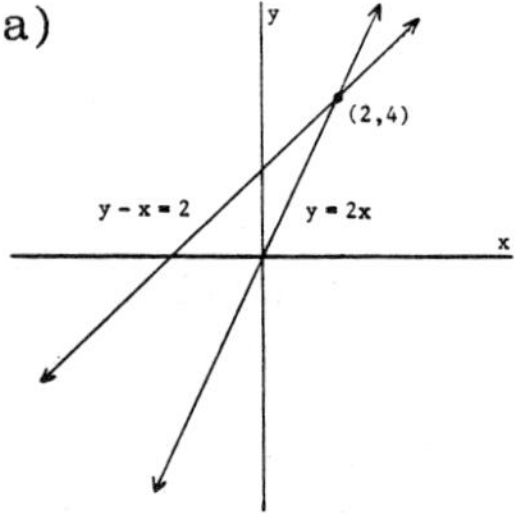

b) (2,4)

---

5. A system has one solution if its graph is two lines that intersect at one point. A system can have no solution or an infinite number of solutions. An example of each type is discussed below.

SYSTEM WITH NO SOLUTION

The system below is graphed at the right. Notice that the lines are parallel. They have the same slope, $m = 3$.

$$y = 3x + 2$$
$$y = 3x - 1$$

Since the parallel lines do not intersect, the equations have no common solution. Therefore, the system has no solution. A system with no solution is called an inconsistent system.

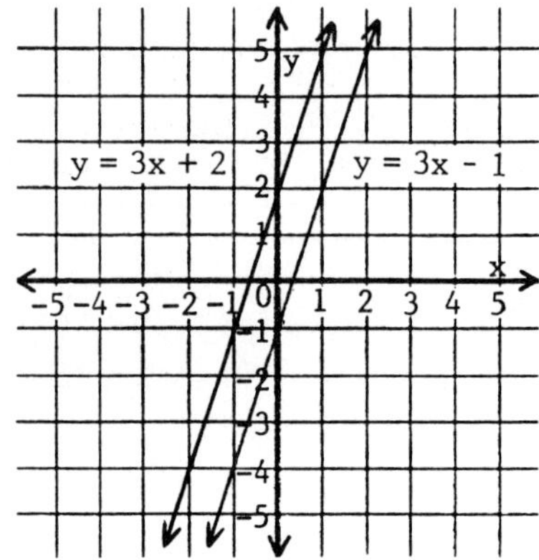

Answers to 4:

a)

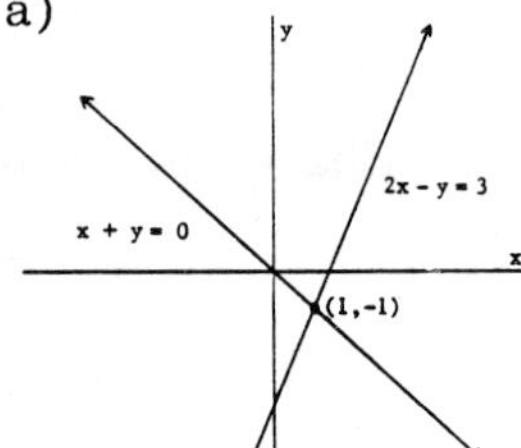

b) (1,-1)

Continued on following page.

5. Continued

SYSTEM WITH AN INFINITE NUMBER OF SOLUTIONS

In the system below, the bottom equation can be obtained by multiplying the top equation by 2. The system is graphed at the right. Notice that each equation has the same graph. Equations with the same graph are called dependent equations.

$$x + y = 2$$
$$2x + 2y = 4$$

Since the lines are identical, the equations have an infinite number of common solutions. Therefore, the system has an infinite number of solutions.

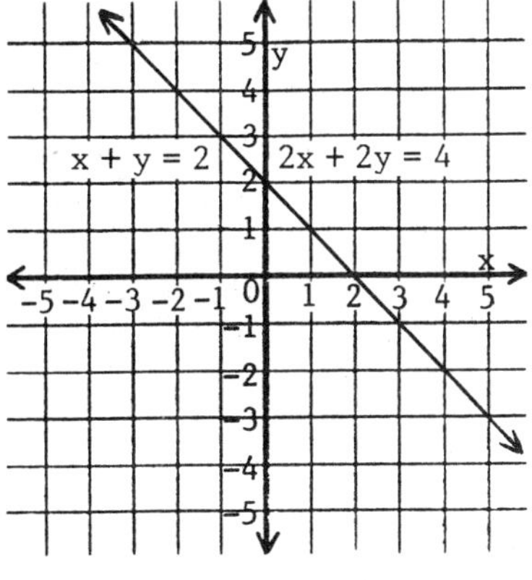

8-2 THE ADDITION METHOD

Since the graphing method is not efficient or accurate, an algebraic method is ordinarily used to solve systems of equations. In this section, we will discuss an algebraic method called the addition method.

---

6. The addition method for solving systems is based on the following property of addition.

If	A = B
and	C = D
then	A + C = B + D

The property above can be used to add the two equations in a system. In the addition method for solving systems, we add the equations in order to eliminate a variable. For example, we eliminated y by adding the equations below. Add the other two equations.

$$\begin{array}{r} x + 2y = 3 \\ \underline{x - 2y = 5} \\ 2x \quad = 8 \end{array} \qquad \begin{array}{r} 3x + y = 7 \\ \underline{-3x - 5y = 1} \end{array}$$

$-4y = 8$

7. Let's use the addition method to solve the system at the right. The two steps are described.

$$x + 2y = 5$$
$$3x - 2y = 7$$

1. Finding the value of x. By adding the two equations, we can eliminate y and solve for x.

$$\begin{array}{rl} x + 2y &= 5 \\ \underline{3x - 2y} &\underline{= 7} \\ 4x \qquad &= 12 \\ x &= 3 \end{array}$$

2. Finding the value of y. We can now find the corresponding value of y by substituting 3 for x in either of the original equations.

$$\begin{array}{rl} x + 2y &= 5 \\ 3 + 2y &= 5 \\ 2y &= 2 \\ y &= 1 \end{array} \qquad \begin{array}{rl} 3x - 2y &= 7 \\ 3(3) - 2y &= 7 \\ 9 - 2y &= 7 \\ -2y &= -2 \\ y &= 1 \end{array}$$

The solution is (3,1) or $x = 3$, $y = 1$. The solution checks.

Use the addition method to solve each system below.

a) $2x + y = 10$
$2x - y = 6$

b) $x - 4y = 16$
$x + 4y = 0$

---

a) (4,2) b) (8,-2)

8. If we add the equations in the system at the right, we get $3x + y = 11$. Neither variable is eliminated.

$$x + 2y = 7$$
$$2x - y = 4$$

However, if the $-y$ in the bottom equation were a $-2y$, we could eliminate y by adding. To get a $-2y$ at the right, we multiplied both sides of the bottom equation by 2. Then we added and solved for x.

$$\begin{array}{rll} x + 2y &= 7 & \\ \underline{4x - 2y} &\underline{= 8} & \text{(Multiplied by 2)} \\ 5x \qquad &= 15 & \\ x &= 3 & \end{array}$$

Substituting 3 for x in one of the original equations, we found the corresponding value of y.

$$\begin{array}{rl} x + 2y &= 7 \\ 3 + 2y &= 7 \\ 2y &= 4 \\ y &= 2 \end{array}$$

Continued on following page.

8. Continued

We get (3,2) as a solution. Check it in each original equation below.

$x + 2y = 7$ $\qquad$ $2x - y = 4$

---

9. Let's solve the system at the right by eliminating $\underline{x}$. To do so, we can multiply either equation by -1. Let's multiply the bottom equation by -1.

$$\begin{aligned} x + y &= 5 \\ x + 3y &= 19 \end{aligned}$$

a) Write the new system obtained if the bottom equation is multiplied by -1.

b) Solve the system.

Answers to 8:

$$\begin{aligned} 3 + 2(2) &= 7 \\ 3 + 4 &= 7 \\ 7 &= 7 \end{aligned}$$

$$\begin{aligned} 2(3) - 2 &= 4 \\ 6 - 2 &= 4 \\ 4 &= 4 \end{aligned}$$

---

10. By eliminating either letter, solve each system below.

a) $$\begin{aligned} x + 4y &= 10 \\ 3x - 2y &= 16 \end{aligned}$$

b) $$\begin{aligned} 2x + y &= 0 \\ 2x + 3y &= 8 \end{aligned}$$

Answers to 9:

a) $$\begin{aligned} x + y &= 5 \\ -x - 3y &= -19 \end{aligned}$$

b) (-2,7)

11. Sometimes we have to multiply both equations by a number before adding the equations. An example is discussed below.

a) (6,1)

b) (-2,4)

We could eliminate y in the system at the right if the 3y were a 15y and the 5y were a -15y.

$$5x + 3y = 2$$
$$3x + 5y = -2$$

To get a 15y in the top equation, we multiplied it by 5 at the right. To get a -15y in the bottom equation, we multiplied it by -3 at the right. Then we added to eliminate y and solve for x.

$$25x + 15y = 10 \quad \text{(Multiplied by 5)}$$
$$-9x - 15y = 6 \quad \text{(Multiplied by -3)}$$
$$16x = 16$$
$$x = 1$$

Substituting "1" for x in the original top equation, we found the corresponding value of y at the right.

$$5x + 3y = 2$$
$$5(1) + 3y = 2$$
$$5 + 3y = 2$$
$$3y = -3$$
$$y = -1$$

We got (1,-1) as the solution. Check it in each original equation.

$5x + 3y = 2$ $\qquad$ $3x + 5y = -2$

---

12. Let's solve the system at the right by eliminating x. We can do so if we get a 10x in the top equation and a -10x in the bottom equation.

$$5x - 2y = 0$$
$$2x - 3y = -11$$

a) Write the new system obtained if the top equation is multiplied by 2 and the bottom equation is multiplied by -5.

b) Solve the system.

$$5(1) + 3(-1) = 2$$
$$5 + (-3) = 2$$
$$2 = 2$$

$$3(1) + 5(-1) = -2$$
$$3 + (-5) = -2$$
$$-2 = -2$$

---

a) $10x - 4y = 0$
$-10x + 15y = 55$

b) (2,5)

13. By eliminating either letter, solve each system below.

a) $2x + 5y = 9$
$3x - 2y = 4$

b) $3x + 4y = 6$
$2x + 3y = 5$

---

14. In the system below, both equations are <u>in standard form</u> $Ax + By = C$. That is, the variables are on the left side and the constant is on the right side.

$$\begin{aligned} 3x + y &= 7 \\ x - 5y &= 9 \end{aligned}$$

In the systems below, $y = x - 8$ and $x = 2y$ are not in standard form.

$$\begin{aligned} y &= x - 8 \\ 5x + y &= 4 \end{aligned} \qquad \begin{aligned} 3x - y &= 4 \\ x &= 2y \end{aligned}$$

When an equation in a system is not in standard form, we must convert it to standard form before using the addition method. We converted $y = x - 8$ to standard form below. <u>Notice how we lined up the variables on the left side</u>. Write the other system with $x = 2y$ in standard form.

$$\begin{aligned} -x + y &= -8 \\ 5x + y &= 4 \end{aligned}$$

a) (2,1) b) (-2,3)

---

15. In the system at the right, $y = 3x - 8$ is not in the standard form $Ax + By = C$.

$$\begin{aligned} x - y &= 2 \\ y &= 3x - 8 \end{aligned}$$

To solve the system, we begin by converting $y = 3x - 8$ to standard form.

$$\begin{aligned} -3x + y &= -3x + 3x - 8 \\ -3x + y &= -8 \end{aligned}$$

Then we can use the addition method to solve the standard system at the right.

$$\begin{aligned} x - y &= 2 \\ -3x + y &= -8 \end{aligned}$$

a) Solve the standard system.

b) Check your solution in each original equation.

$x - y = 2$ $\qquad$ $y = 3x - 8$

$$\begin{aligned} 3x - y &= 4 \\ x - 2y &= 0 \end{aligned}$$

16. The solution for a system is written as an ordered pair. Where the variables are different than x and y, they are written in alphabetical order in the ordered pair. For example:

$c = 5,\ d = -1$ is written $(5,-1)$

$s = -3,\ t = 4$ is written $(-3,4)$

Following the steps in the last frame, solve each system.

a) $2p + q = 30$
   $p = 2q$

b) $2a = 7 - 3b$
   $3a - 2b = 4$

a) (3,1)

b) $3 - 1 = 2$
   $2 = 2$

   $1 = 3(3) - 8$
   $1 = 9 - 8$
   $1 = 1$

---

17. Earlier we saw that a system of equations could have no solution or an infinite number of solutions. Let's see what happens with the addition method for these two special cases.

SYSTEM WITH NO SOLUTION

If we graphed the equations at the right, the lines would be parallel. Therefore, the system has no solution. It is called an inconsistent system.

$$\begin{aligned} y - 2x &= 5 \\ y - 2x &= 3 \end{aligned}$$

We tried the addition method at the right by multiplying the bottom equation by -1. We got the false equation 0 = 2.

$$\begin{aligned} y - 2x &= 5 \\ -y + 2x &= -3 \\ \hline 0 &= 2 \end{aligned}$$

If you try the addition method and get a false equation like 0 = 2, the system has no solution.

SYSTEM WITH AN INFINITE NUMBER OF SOLUTIONS

If we graphed the system at the right, the lines would be identical. Therefore, the system has an infinite number of solutions. The equations are called dependent equations.

$$\begin{aligned} x - 2y &= 3 \\ 2x - 4y &= 6 \end{aligned}$$

We tried the addition method at the right by multiplying the top equation by -2. We got the equation 0 = 0.

$$\begin{aligned} -2x + 4y &= -6 \\ 2x - 4y &= 6 \\ \hline 0 &= 0 \end{aligned}$$

If you try the addition method and get the equation 0 = 0, the system has an infinite number of solutions.

a) (12,6)

b) (2,1)

8-3 THE SUBSTITUTION METHOD

The substitution method is a second algebraic method for solving systems of equations. We will discuss the substitution method in this section.

---

18. Instead of adding equations to eliminate a variable, we can substitute to eliminate a variable. An example is discussed below.

    In the system at the right, y is solved for in the bottom equation.

    $$\begin{aligned} 3x - 2y &= 4 \\ y &= x + 2 \end{aligned}$$

    We can eliminate y by substituting $x + 2$ for y in the top equation. We substituted, simplified, and solved for x at the right.

    $$\begin{aligned} 3x - 2y &= 4 \\ 3x - 2(x + 2) &= 4 \\ 3x - (2x + 4) &= 4 \\ 3x - 2x - 4 &= 4 \\ x &= 8 \end{aligned}$$

    Substituting 8 for x in the bottom equation, we can solve for y. We get:

    $$\begin{aligned} y &= x + 2 \\ y &= 8 + 2 \\ y &= 10 \end{aligned}$$

    Check (8,10) in each original equation below.

    $3x - 2y = 4$ $\qquad\qquad$ $y = x + 2$

---

19. Use the substitution method to solve each system.

    a) $\begin{aligned} y &= 3x \\ 5x - y &= 4 \end{aligned}$ $\qquad$ b) $\begin{aligned} 4x + y &= 7 \\ y &= x - 8 \end{aligned}$

Answers to 18:

$$\begin{aligned} 3(8) - 2(10) &= 4 \\ 24 - 20 &= 4 \\ 4 &= 4 \end{aligned}$$

$$\begin{aligned} 10 &= 8 + 2 \\ 10 &= 10 \end{aligned}$$

---

20. Neither variable is solved for in the system below.

    $$\begin{aligned} x - 3y &= 0 \\ 2x + y &= 4 \end{aligned}$$

    To use the substitution method, we must solve for one variable in one equation.

    If we solve for x in the top equation or y in the bottom equation, we get non-fractional solutions.

    $\begin{aligned} x - 3y &= 0 \\ x &= 3y \end{aligned}$ $\qquad\qquad$ $\begin{aligned} 2x + y &= 4 \\ y &= 4 - 2x \end{aligned}$

Continued on following page.

Answers to 19:

a) (2,6)

b) (3,-5)

20. Continued

If we solve for y in the top equation or x in the bottom equation, we get fractional solutions.

$$x - 3y = 0 \qquad 2x + y = 4$$
$$x = 3y \qquad 2x = 4 - y$$
$$y = \frac{x}{3} \qquad x = \frac{4 - y}{2}$$

Since it is easier to substitute non-fractional solutions, we would use either $x = 3y$ or $y = 4 - 2x$ for the substitution. Notice that we got the non-fractional solutions by solving for a variable whose coefficient is "1".

Which variable in each system would you solve for to get a non-fraction solution?

a) $4x - 5y = 0$
$x - 2y = 8$

b) $3x + y = 11$
$2x - 3y = 4$

c) $2p - 5q = 7$
$4p + q = 9$

---

21. Let's use the substitution method to solve this system.

$$a + 4b = 6$$
$$5a - 3b = 7$$

a) Solve for a in the top equation to get a non-fractional solution.

b) Substitute that solution for a in the bottom equation and then find the value of b.

c) Substitute in one of the original equations to find the value of a.

d) The solution is: ____________

Answers to 20:

a) x in the bottom equation

b) y in the top equation

c) q in the bottom equation

---

22. Let's use the substitution method to solve this system. Since no variable has "1" as a coefficient, we cannot avoid a fractional solution.

$$2x - 3y = 9$$
$$4x + 3y = 9$$

1. Solve for x in the top equation.

$$2x = 3y + 9$$
$$x = \frac{3y + 9}{2}$$

2. Substitute that solution for x in the bottom equation and then find the value of y.

$$\overset{2}{\cancel{4}}\left(\frac{3y + 9}{\cancel{2}}\right) + 3y = 9$$
$$6y + 18 + 3y = 9$$
$$9y + 18 = 9$$
$$9y = -9$$
$$y = -1$$

Answers to 21:

a) $a = 6 - 4b$

b) $5(6-4b)-3b = 7$
$30-20b - 3b = 7$
$30 - 23b = 7$
$-23b = -23$
$b = 1$

c) $a = 2$

d) (2,1)

Continued on following page.

22. Continued

3. Substitute -1 for y in the top equation and find the value of x.

$$\begin{aligned} 2x - 3y &= 9 \\ 2x - 3(-1) &= 9 \\ 2x + 3 &= 9 \\ 2x &= 6 \\ x &= 3 \end{aligned}$$

The solution of the system is __________.

---

23. When deciding whether to use the addition or the substitution method, the following suggestions can be used.

(3,-1)

1. When a variable is solved for in one equation, the substitution method is probably better.
2. When neither variable is solved for but one has a coefficient of "1" (so that the solution for it is non-fractional), either method can be used.
3. Otherwise, use the addition method.
4. When in doubt, use the addition method.

Identify the method you would use to solve each of these.

a) $3x - y = 10$
   $x + 3y = 10$

b) $x + y = 6$
   $x = 2y$

c) $2p + 5q = 9$
   $3p - 2q = 4$

---

24. To solve the system at the right, we begin by clearing the fractions in each equation. We did so below.

$$\frac{x}{2} - \frac{y}{3} = \frac{5}{6}$$
$$\frac{x}{5} - \frac{y}{4} = \frac{1}{10}$$

a) Either method

b) The substitution method

c) The addition method

For the top equation, the LCD is 6. We get:

$$6\left(\frac{x}{2} - \frac{y}{3}\right) = 6\left(\frac{5}{6}\right)$$
$$\overset{3}{\cancel{6}}\left(\frac{x}{\cancel{2}}\right) - \overset{2}{\cancel{6}}\left(\frac{y}{\cancel{3}}\right) = \overset{1}{\cancel{6}}\left(\frac{5}{\cancel{6}}\right)$$
$$3x - 2y = 5$$

For the bottom equation, the LCD is 20. We get:

$$20\left(\frac{x}{5} - \frac{y}{4}\right) = 20\left(\frac{1}{10}\right)$$
$$\overset{4}{\cancel{20}}\left(\frac{x}{5}\right) - \overset{5}{\cancel{20}}\left(\frac{y}{4}\right) = \overset{2}{\cancel{20}}\left(\frac{1}{10}\right)$$
$$4x - 5y = 2$$

The resulting system with non-fractional equations is shown at the right.

$$\begin{aligned} 3x - 2y &= 5 \\ 4x - 5y &= 2 \end{aligned}$$

a) Which method would you use to solve the system? __________________

b) Solve the system.

25. To solve the system at the right, we begin by clearing the decimals in each equation. We did so below.

$$.1x + .3y = 9$$
$$.05x + .12y = 3.9$$

a) Addition Method

b) (3,2)

For the top equation, we multiply both sides by 10.

$$10(.1x + .3y) = 10(9)$$
$$10(.1x) + 10(.3y) = 90$$
$$x + 3y = 90$$

For the bottom equation, we multiply both sides by 100.

$$100(.05x + .12y) = 100(3.9)$$
$$100(.05x) + 100(.12y) = 390$$
$$5x + 12y = 390$$

The resulting system with non-decimal equations is shown at the right.

$$x + 3y = 90$$
$$5x + 12y = 390$$

a) Which method would you use to solve the system? 

b) Solve the system.

a) Either method   b) (30,20)

SELF-TEST 28 (pages 402-414)

1. Use the graphing method to solve this system of equations.

$$x - 2y = 6$$
$$2x + y = 2$$

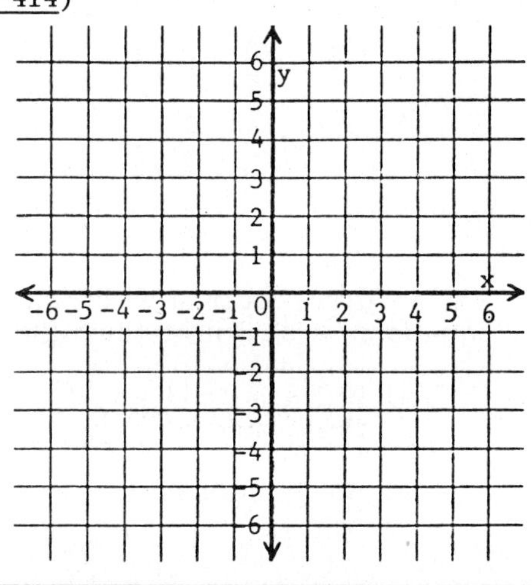

Continued on following page.

SELF-TEST - Continued

2. Use the addition method to solve this system.

$$2x - 3y = 7$$
$$5x + 2y = 8$$

3. Use the substitution method to solve this system.

$$8x - 3y = 1$$
$$y = 2 - 2x$$

Use any method to solve each system.

4. $\frac{x}{2} + \frac{y}{3} = 6$

$\frac{x}{4} - 1 = \frac{y}{6}$

5. $1.5x + 2y = 6.5$
$x - 2.5 = 0.5y$

ANSWERS: 
1. (2,-2)

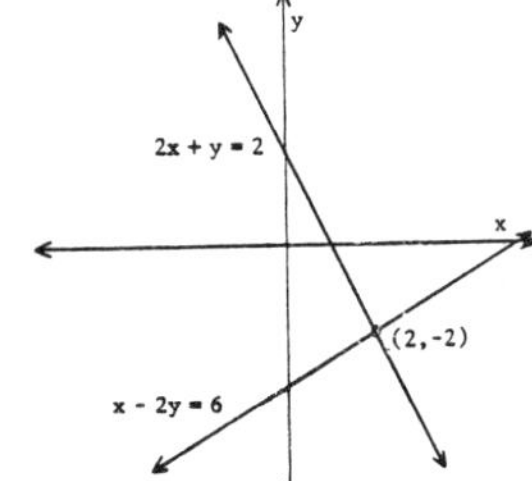

2. (2,-1)

3. $(\frac{1}{2}, 1)$

4. (8,6)

5. (3,1)

8-4 WORD PROBLEMS

Systems of equations can be used to solve various word problems. We will discuss problems of that type in this section.

26. The following problem can be solved by setting up and solving a system of equations.

> The sum of two numbers is 19. Three times the first number minus four times the second number is 8. Find the two numbers.

Using x for the first number and y for the second number, we can set up two equations.

The sum of two numbers is 19.

$$x + y = 19$$

Three times the first number minus four times the second number is 8.

$$3x - 4y = 8$$

Therefore, the problem can be solved by solving the system at the right.

$$\begin{aligned} x + y &= 19 \\ 3x - 4y &= 8 \end{aligned}$$

Multiplying the top equation by 4, we get this system.

$$\begin{aligned} 4x + 4y &= 76 \\ 3x - 4y &= 8 \end{aligned}$$

Adding the two equations, we can find the value of x.

$$\begin{aligned} 7x &= 84 \\ x &= 12 \end{aligned}$$

Substituting 12 for x in the top equation, we can find the value of y.

$$\begin{aligned} 12 + y &= 19 \\ y &= 7 \end{aligned}$$

Therefore, the first number is _____ and the second number is _____.

---

first is 12
second is 7

27. The geometry problem below can be solved by means of a system of equations.

> The perimeter of a rectangle is 260 cm. If its length is 30 centimeters longer than its width, what are its length and width?

Using L for length and W for width, we can set up two equations.

The perimeter of a rectangle is 260 cm.

$$2L + 2W = 260$$

The length is 30 cm longer than the width.

$$L = W + 30$$

Therefore, we can solve the problem by solving the system at the right.

$$\begin{aligned} 2L + 2W &= 260 \\ L &= W + 30 \end{aligned}$$

Substituting W + 30 for L in the top equation, we can solve for W.

$$\begin{aligned} 2(W + 30) + 2W &= 260 \\ 2W + 60 + 2W &= 260 \\ 4W &= 200 \\ W &= 50 \end{aligned}$$

Substituting 50 for W in the bottom equation, we can solve for L.

$$\begin{aligned} L &= 50 + 30 \\ L &= 80 \end{aligned}$$

Therefore, its length is _____ cm and its width is _____ cm.

28. Following the examples in the last two frames, solve these:

a) The difference between two numbers is 4. Twice the larger plus three times the smaller is 58. Find the two numbers.

b) The perimeter of a rectangular lot is 216 meters. If the length is twice the width, find the length and width.

length is 80 cm
width is 50 cm

---

29. We can use a system of equations to solve this problem.

A woman invests a total of $4,200 in two accounts. Part is invested at 10% and the rest at 8%. In one year, the investments earn $372 in simple interest. Find the amount invested at each rate.

Letting x equal the amount invested at 10% and y equal the amount invested at 8%, we can set up the following two equations.

$$x + y = 4{,}200$$

$$10\%x + 8\%y = 372$$

Therefore, we can solve the problem by solving the system at the right.

$$\begin{aligned} x + y &= 4{,}200 \\ .1x + .08y &= 372 \end{aligned}$$

Multiplying the bottom equation by 100, we can clear the decimals.

$$\begin{aligned} x + y &= 4200 \\ 10x + 8y &= 37{,}200 \end{aligned}$$

Multiplying the top equation by -8 and adding, we can eliminate y and solve for x.

$$\begin{aligned} -8x - 8y &= -33{,}600 \\ 10x + 8y &= 37{,}200 \\ \hline 2x &= 3{,}600 \\ x &= 1{,}800 \end{aligned}$$

Substituting 1800 for x in the top equation, we can solve for y.

$$\begin{aligned} 1{,}800 + y &= 4{,}200 \\ y &= 2{,}400 \end{aligned}$$

Therefore, __________ is invested at 10% and __________ is invested at 8%.

a) larger is 14
smaller is 10

b) length is 72 m
width is 36 m

$1,800 at 10%
$2,400 at 8%

30. We can also use a system of equations to solve this problem.

It takes a boat 2 hours to go 14 miles downstream and $3\frac{1}{2}$ hours to return. Find the speed of the current and the speed of the boat in still water.

Letting x be the speed of the boat in still water and y be the speed of the current, we organized the data in the chart below.

	Distance	Rate	Time
Downstream	14	$x + y$	2
Upstream	14	$x - y$	$3\frac{1}{2}$

Using the relationship $r = \frac{d}{t}$ from $d = rt$, we can set up the following two equations.

$$x + y = \frac{14}{2}$$

$$x + y = 7$$

$$x - y = \frac{14}{3\frac{1}{2}}$$

$$x - y = \frac{14}{\frac{7}{2}}$$

$$x - y = \overset{2}{\cancel{14}}\left(\frac{2}{\cancel{7}}\right)$$

$$x - y = 4$$

Therefore, we can solve the problem by solving the system at the right.

$$\begin{aligned} x + y &= 7 \\ x - y &= 4 \\ \hline 2x &= 11 \\ x &= 5\tfrac{1}{2} \\ x + y &= 7 \\ 5\tfrac{1}{2} + y &= 7 \\ y &= 1\tfrac{1}{2} \end{aligned}$$

Therefore, the speed of the boat in still water is ________ mph and the speed of the current is ________ mph.

speed in still water = $5\frac{1}{2}$ mph

speed of current = $1\frac{1}{2}$ mph

31. Following the examples in the last frame, solve these.

a) A man invests a total of $3,600 in two accounts. Part is invested at 9% and the rest at 7%. In one year, the investments earn $296 in simple interest. Find the amount invested at each rate.

b) A plane took 2 hours to fly 1080 km against a headwind. The return trip with the wind took $1\frac{4}{5}$ hours. Find the speed of the plane in still air and the speed of the wind.

---

32. A system of equations can be used to solve this problem.

Tickets to the school play cost $3 for adults and $2 for students. On Friday night, 490 tickets were sold for total receipts of $1,205. How many tickets of each kind were sold?

Letting x equal the number of adult tickets and y equal the number of student tickets, we can set up two equations.

The total number of tickets sold was 490.

$$x + y = 490$$

At $3 for adults and $2 for students total receipts were $1,205.

$$3x + 2y = 1205$$

Therefore, we can solve the problem by solving the system at the right.

$$x + y = 490$$
$$3x + 2y = 1205$$

Multiplying the top equation by -2 and adding, we can solve for x. Then we can substitute to solve for y.

$$-2x - 2y = -980$$
$$3x + 2y = 1205$$
$$x = 225$$

$$x + y = 490$$
$$225 + y = 490$$
$$y = 265$$

Therefore, there were _______ adult tickets sold and _________ student tickets sold.

Answers to frame 31:

a) $2,200 at 9%
$1,400 at 7%

b) speed in still air = 570 km/h

speed of wind = 30 km/h

Answers to frame 32:

225 adult tickets
265 student tickets

33. Here is a mixture problem that can be solved with a system of equations.

> Solution A is 30% acid and solution B is 60% acid. How many liters of each solution should be mixed to get 60 liters of a solution that is 40% acid?

Letting x represent the number of liters of the 30% acid and y represent the number of liters of the 60% acid, we can summarize the information in the table below.

Liters of solution	Percent	Liters of acid
x	30%	.3x
y	60%	.6y
60	40%	.4(60) = 24

To get the 60 liters of the new solution, x liters of the first solution and y liters of the second solution must be mixed. Therefore:

$$x + y = 60$$

The amount of acid in the new solution must be 40% of 60 liters, which is .4(60) = 24 liters. This amount will be made up by the acid in the two original solutions. These amounts of acid are 30% of x or .3x and 60% of y or .6y. Therefore:

$$30\%x + 60\%y = 40\%(60)$$

$$.3x + .6y = 24$$

or

$$3x + 6y = 240 \quad \text{(Multiplied by 10)}$$

Therefore, the problem can be solved by solving the system at the right.

$$x + y = 60$$
$$3x + 6y = 240$$

Multiplying the top equation by -3, we can use the addition method.

$$-3x - 3y = -180$$
$$\underline{3x + 6y = 240}$$
$$3y = 60$$
$$y = 20$$

$$x + y = 60$$
$$x + 20 = 60$$
$$x = 40$$

Therefore, ________ liters of solution A and ________ liters of solution B should be used.

---

40 liters of A
20 liters of B

34. Following the examples in the last two frames, solve these:

a) A bank teller receives a savings deposit in \$5 and \$10 bills. There were a total of 30 bills worth \$235. How many bills of each type were deposited?

b) Solution A is 40% alcohol and solution B is 80% alcohol. How much of each is needed to make 100 milliliters of a solution that is 55% alcohol?

a) 17 \$10 bills
13 \$5 bills

b) 62.5 milliliters of A
37.5 milliliters of B

8-5 SYSTEMS OF THREE EQUATIONS

In this section, we will show how the addition method can be used to solve systems of three equations.

35. Three linear equations with the same variables form a system of three equations. An example is given below.

$$\begin{aligned} x + y + z &= 4 \\ x - 2y - z &= 1 \\ 2x - y - 2z &= -1 \end{aligned}$$

A solution of a system of three linear equations is an ordered triple that satisfies all three equations. To show that (2,-1,3) is a solution of the system above, we substituted 2 for $\underline{x}$, -1 for $\underline{y}$, and 3 for $\underline{z}$ in each equation.

$$\begin{aligned} x + y + z &= 4 \\ 2 + (-1) + 3 &= 4 \\ 4 &= 4 \end{aligned}$$

$$\begin{aligned} x - 2y - z &= 1 \\ 2 - 2(-1) - 3 &= 1 \\ 2 + 2 - 3 &= 1 \\ 1 &= 1 \end{aligned}$$

$$\begin{aligned} 2x - y - 2z &= -1 \\ 2(2) - (-1) - 2(3) &= -1 \\ 4 + 1 - 6 &= -1 \\ -1 &= -1 \end{aligned}$$

Is (2,4,3) or (4,2,-1) a solution of the system below? ____________

$$\begin{aligned} x - y + z &= 1 \\ 2x - 3y + 4z &= -2 \\ 3x - 2y - z &= 9 \end{aligned}$$

36. For convenience, we numbered the equations in the system below.

(4,2,-1) is a solution

$$\begin{aligned} &(1) \quad 4x - y + z = 6 \\ &(2) \quad -3x + 2y - z = -3 \\ &(3) \quad 2x + y + 2z = 3 \end{aligned}$$

The three steps needed to solve the system by the addition method are discussed below.

1. Eliminate one variable to get a system of two equations with two variables.

a) Add a pair of equations to eliminate one variable. We added (1) and (2) below to eliminate z.

$$\begin{array}{lrcl} (1) & 4x - y + z & = & 6 \\ (2) & -3x + 2y - z & = & -3 \\ \hline (4) & x + y & = & 3 \end{array}$$

b) Add a different pair of equations to eliminate the same variable. We added (2) and (3) below to eliminate z. To do so, we had to multiply equation (2) by 2.

$$\begin{array}{lrcl} (2) & -3x + 2y - z & = & -3 \\ (3) & 2x + y + 2z & = & 3 \end{array} \qquad \begin{array}{lrcl} & -6x + 4y - 2z & = & -6 \\ & 2x + y + 2z & = & 3 \\ \hline (5) & -4x + 5y & = & -3 \end{array}$$

2. Solve the system of two equations to find the values of the two variables x and y.

a) We multiplied the top equation by 4 below to eliminate x and solve for y.

$$\begin{array}{lrcl} (4) & x + y & = & 3 \\ (5) & -4x + 5y & = & -3 \end{array} \qquad \begin{array}{rcl} 4x + 4y & = & 12 \\ -4x + 5y & = & -3 \\ \hline 9y & = & 9 \\ y & = & 1 \end{array}$$

b) We substituted "1" for y in the top equation to solve for x.

$$\begin{aligned} x + y &= 3 \\ x + 1 &= 3 \\ x &= 2 \end{aligned}$$

3. Substitute 2 for x and "1" for y in one of the three original equations to solve for z. We substituted in equation (1).

$$\begin{aligned} 4x - y + z &= 6 \\ 4(2) - 1 + z &= 6 \\ 8 - 1 + z &= 6 \\ 7 + z &= 6 \\ z &= -1 \end{aligned}$$

The solution is (2,1,-1). We checked it in the three original equations below.

$$\begin{aligned} 4x - y + z &= 6 \\ 4(2) - 1 + (-1) &= 6 \\ 8 - 1 - 1 &= 6 \\ 6 &= 6 \end{aligned} \qquad \begin{aligned} -3x + 2y - z &= -3 \\ -3(2) + 2(1) - (-1) &= -3 \\ -6 + 2 + 1 &= -3 \\ -3 &= -3 \end{aligned} \qquad \begin{aligned} 2x + y + 2z &= 3 \\ 2(2) + 1 + 2(-1) &= 3 \\ 4 + 1 + (-2) &= 3 \\ 3 &= 3 \end{aligned}$$

37. Let's use the addition method to solve the system below.

$$(1)\quad x + y + z = 6$$

$$(2)\quad 2x - y + 3z = 9$$

$$(3)\quad -x + 2y + 2z = 9$$

a) Write the system of two equations obtained if y is eliminated by adding (1) and (2) and then adding (2) and (3).

b) Solve the system of two equations to find the values of x and z.

c) Substitute those values for x and z in one of the three original equations to find the corresponding value of y.

d) The solution of the system is: ____________

a) $3x + 4z = 15$
$3x + 8z = 27$

b) $x = 1$, $z = 3$

c) $y = 2$

d) (1,2,3)

38. In the system below, the variable $\underline{z}$ does not appear in equation (1). Therefore, the system is easier to solve.

$$(1) \qquad x - y = 4$$

$$(2) \qquad x - 2y + z = 5$$

$$(3) \qquad 2x + 3y - 2z = 3$$

a) Let's eliminate $\underline{z}$ to get a system of two equations. One of the equations is (1) above. Find the other equation by adding (2) and (3) to eliminate $\underline{z}$.

b) Solve the system of two equations to find the values of $\underline{x}$ and $\underline{y}$.

c) Substitute those values for $\underline{x}$ and $\underline{y}$ in one of the three original equations to find the corresponding value of $\underline{z}$.

d) The solution of the system is: ____________

---

a) $x - y = 4$
$4x - y = 13$

b) $x = 3,\ y = -1$

c) $z = 0$

d) $(3,-1,0)$

8-6 WORD PROBLEMS

In this section, we will solve some word problems that involve a system of three equations.

39. We can use a system of three equations to solve this problem.

> The sum of three numbers is 6. The first number minus the second plus the third is 8. The third minus the first is 2 more than the second. Find the three numbers.

Letting the three numbers be $\underline{x}$, $\underline{y}$, and $\underline{z}$, we can set up three equations.

The sum of three numbers is 6.

$$x + y + z = 6$$

The first number minus the second plus the third is 8.

$$x - y + z = 8$$

The third minus the first is 2 more than the second.

$$z - x = y + 2$$

Therefore, we have the system of three equations below. Notice that we rewrote the system to put the third equation in the standard form Ax + By + Cz = D.

$$\begin{array}{lll} (1)\ x + y + z = 6 & & (1)\ x + y + z = 6 \\ (2)\ x - y + z = 8 & \text{or} & (2)\ x - y + z = 8 \\ (3)\ z - x = y + 2 & & (3)\ -x - y + z = 2 \end{array}$$

To solve the system, we'll begin by eliminating $\underline{y}$. To do so, we added (1) and (2) and then (1) and (3) below.

$$\begin{array}{ll} (1)\ x + y + z = 6 & (1)\ x + y + z = 6 \\ (2)\ \underline{x - y + z = 8} & (3)\ \underline{-x - y + z = 2} \\ (4)\ 2x \quad + 2z = 14 & (5)\ \qquad 2z = 8 \end{array}$$

Notice that we also eliminated $\underline{x}$ when adding (1) and (3). Therefore, solving the system below for $\underline{x}$ and $\underline{z}$ is simplified.

$$\begin{array}{ll} (4) & 2x + 2z = 14 \\ (5) & \qquad 2z = 8 \end{array}$$

$z = 4$ (Solving 2z = 8)

$2x + 2(4) = 14$ (Substituting in 2x + 2z = 14)

$$\begin{aligned} 2x + 8 &= 14 \\ 2x &= 6 \\ x &= 3 \end{aligned}$$

Substituting 4 for $\underline{z}$ and 3 for $\underline{x}$ in equation (1), we have solved for $\underline{y}$ below.

$$\begin{aligned} x + y + z &= 6 \\ 3 + y + 4 &= 6 \\ y &= -1 \end{aligned}$$

The solution of the system is (3,-1,4) or x = 3, y = -1, z = 4. Therefore, the first number is ______, the second number is ______, and the third number is ______.

40. We can use a system of three equations to solve this problem.

In triangle ABC, angle A is 20° more than angle B. Angle C is twice angle B. Find the size of each angle.

Letting the three angles be A, B, and C, we can set up three equations.

We know that the sum of the angles in a triangle is 180°.

$$A + B + C = 180$$

Angle A is 20° more than angle B.

$$A = B + 20$$

Angle C is twice angle B.

$$C = 2B$$

Therefore, we have the system of three equations below. Notice that we rewrote the system to put the two bottom equations in standard form.

$$\begin{aligned} A + B + C &= 180 \\ A &= B + 20 \\ C &= 2B \end{aligned} \qquad \text{or} \qquad \begin{aligned} A + B + C &= 180 \\ A - B &= 20 \\ 2B - C &= 0 \end{aligned}$$

If we solve the system, we get: A = 60, B = 40, and C = 80. Therefore, angle A contains ______°, angle B contains ______°, and angle C contains ______°.

first = 3
second = -1
third = 4

41. Following the examples in the last two frames, solve these.

a) The sum of three numbers is 20. The first minus the second equals the third. The second is 1 more than twice the third. Find the three numbers.

b) In triangle ABC, angle B is three times angle C. Angle A is 40° more than angle B. Find the size of each angle.

A = 60°
B = 40°
C = 80°

42. We can also use a system of three equations to solve this problem.

John has a total score of 261 on three tests. The sum of the scores on his first and second tests is 97 more than his third score. His second score is 5 more than his first score. Find the scores on the three tests.

Letting x, y, and z equal the three test scores, we can set up three equations.

The sum of the three scores is 261.

$$x + y = z = 261$$

The sum of the scores on his first and second tests is 97 more than his third score.

$$x + y = z + 97$$

His second score is 5 more than his first score.

$$y = x + 5$$

Therefore, we have the system of three equations below. We rewrote the system to put the bottom two equations in standard form.

$$\begin{array}{rcl} x + y + z = 261 & & x + y + z = 261 \\ x + y = z + 97 & \text{or} & x + y - z = 97 \\ y = x + 5 & & -x + y \quad = 5 \end{array}$$

If we solve the system, we get x = 87, y = 92, and z = 82. Therefore, his first score was ______, his second score was ______, and his third score was ______.

a) first = 10
second = 7
third = 3

b) A = 100°
B = 60°
C = 20°

---

43. Following the example in the last frame, solve these.

a) Three basketball players score a total of 63 points in one game. Player A scores 9 points less than player B. Together, players A and B score 23 more points than player C. How many points did each player score?

b) Three production lines can produce 248 cars in one week. Lines A and B together can produce 163. Lines B and C together can produce 98 more than line A. How many can each line produce alone?

first = 87

second = 92

third = 82

---

a) A = 17 points
B = 26 points
C = 20 points

b) A = 75 cars
B = 88 cars
C = 85 cars

SELF-TEST 29 (pages 415-428)

Solve each problem.

1. The perimeter of a rectangular lot is 480 yd. The length is 50 yd longer than the width. Find the length and width.

2. A car radiator holds 12 liters. We want to add pure antifreeze to a solution that is 4% antifreeze to fill the radiator with a solution that is 20% antifreeze. How many liters of each should be used?

3. Solve this system.

$$\begin{aligned} x + y + z &= 7 \\ 2x + 2y - z &= 5 \\ x - y + z &= 9 \end{aligned}$$

4. Solve this problem.

   To meet a monthly quota, a car dealer must sell 80 new cars. He must sell 10 more small cars than intermediates and 5 more intermediates than large cars. How many of each type must he sell?

ANSWERS:

1. L = 145 yd, W = 95 yd
2. 2 liters of pure antifreeze, 10 liters of 4% solution.
3. $x = 5$, $y = -1$, $z = 3$
4. 35 small cars
   25 intermediate cars
   20 large cars

8-7 EVALUATING DETERMINANTS

In this section, we will discuss the methods for evaluating second-order and third-order determinants.

---

44. A determinant is a square array of numbers enclosed by two vertical lines. Two examples are shown.

$$\begin{vmatrix} 2 & 1 \\ 7 & 5 \end{vmatrix} \qquad \begin{vmatrix} -3 & 8 \\ 6 & -4 \end{vmatrix}$$

The general form for the above determinants is shown below. Since there are 2 rows and 2 columns, the determinant is called a second-order or two-by-two determinant. Each number in a determinant is called an element.

Column 1, Column 2

Row 1, Row 2

$$\begin{vmatrix} a_1 & b_1 \\ a_2 & b_2 \end{vmatrix}$$

Any second-order determinant equals one number. To find that number, we use the definition below.

$$\begin{vmatrix} a_1 & b_1 \\ a_2 & b_2 \end{vmatrix} = a_1b_2 - a_2b_1$$

That is, a determinant equals the difference of the products of the diagonals. Using the definition above, we get.

$$\begin{vmatrix} 4 & 2 \\ 3 & 5 \end{vmatrix} = (4)(5) - (3)(2) = 20 - 6 = 14$$

$$\begin{vmatrix} 3 & -5 \\ 2 & 4 \end{vmatrix} = (3)(4) - (2)(-5) = 12 - (-10) = \_\_\_\_$$

---

45. Using the definition in the last frame, find the value of each determinant.

Answer to frame 44: 22

a) $\begin{vmatrix} 5 & -2 \\ 3 & 4 \end{vmatrix} =$

b) $\begin{vmatrix} -7 & 3 \\ 2 & 2 \end{vmatrix} =$

c) $\begin{vmatrix} 6 & -5 \\ -5 & -6 \end{vmatrix} =$

Answers to frame 45:

a) 26, from 20 - (-6)

b) -20, from -14 - 6

c) -61, from -36 - 25

46. When one element in a determinant is 0, the evaluation is simplified. Evaluate these.

a) $\begin{vmatrix} 4 & -1 \\ 0 & 2 \end{vmatrix} =$ ______________________

b) $\begin{vmatrix} 1 & -3 \\ -2 & 0 \end{vmatrix} =$ ______________________

---

a) 8

b) -6

---

47. The general form for a <u>third-order</u> or <u>three-by-three</u> determinant is shown below. Notice that there are 3 rows and 3 columns.

	Column 1	Column 2	Column 3
Row 1	a_1	b_1	c_1
Row 2	a_2	b_2	c_2
Row 3	a_3	b_3	c_3

An example of a third-order determinant is given below.

$$\begin{vmatrix} 3 & 0 & -1 \\ -2 & 6 & 8 \\ 4 & -7 & 2 \end{vmatrix}$$

Each element in a third-order determinant has a <u>minor</u> which is a second-order determinant. We find the minor for an element by crossing out its row and column.

For -2: 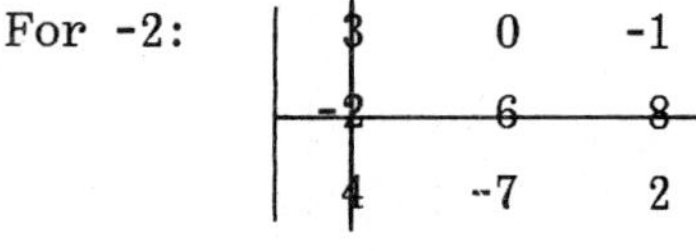 the minor is $\begin{vmatrix} 0 & -1 \\ -7 & 2 \end{vmatrix}$

For 6: 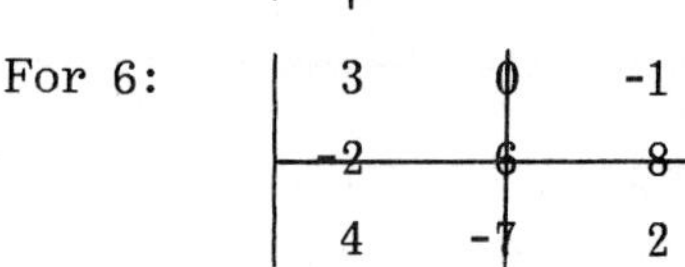 the minor is $\begin{vmatrix} 3 & -1 \\ 4 & 2 \end{vmatrix}$

For 2: $\begin{vmatrix} 3 & 0 & -1 \\ -2 & 6 & 8 \\ 4 & -7 & 2 \end{vmatrix}$ the minor is:

$\begin{vmatrix} 3 & 0 \\ -2 & 6 \end{vmatrix}$

48. Any third-order determinant equals one number. To find that number, we use the definition below.

$$\begin{vmatrix} a_1 & b_1 & c_1 \\ a_2 & b_2 & c_2 \\ a_3 & b_3 & c_3 \end{vmatrix} = a_1b_2c_3 + b_1c_2a_3 + c_1a_2b_3 - a_3b_2c_1 - b_3c_2a_1 - c_3a_2b_1$$

By rearranging terms and factoring, we get:

$$\begin{vmatrix} a_1 & b_1 & c_1 \\ a_2 & b_2 & c_2 \\ a_3 & b_3 & c_3 \end{vmatrix} = a_1(b_2c_3 - b_3c_2) - a_2(b_1c_3 - b_3c_1) + a_3(b_1c_2 - b_2c_1)$$

The expressions in parentheses can be expressed as second-order determinants which are the minors of a_1, a_2, and a_3.

$$\begin{vmatrix} a_1 & b_1 & c_1 \\ a_2 & b_2 & c_2 \\ a_3 & b_3 & c_3 \end{vmatrix} = a_1\begin{vmatrix} b_2 & c_2 \\ b_3 & c_3 \end{vmatrix} - a_2\begin{vmatrix} b_1 & c_1 \\ b_3 & c_3 \end{vmatrix} + a_3\begin{vmatrix} b_1 & c_1 \\ b_2 & c_2 \end{vmatrix}$$

First column (points to a_1, a_2, a_3); Minor of a_1; Minor of a_2; Minor of a_3

The expression above is called <u>the expansion of the determinant by minors</u> of the elements of the first column.

We evaluated the determinant below by expanding by minors of the elements of the first column.

$$\begin{vmatrix} 1 & -2 & 3 \\ 4 & -3 & 1 \\ 2 & 1 & -2 \end{vmatrix} = 1\begin{vmatrix} -3 & 1 \\ 1 & -2 \end{vmatrix} - 4\begin{vmatrix} -2 & 3 \\ 1 & -2 \end{vmatrix} + 2\begin{vmatrix} -2 & 3 \\ -3 & 1 \end{vmatrix}$$

$$= 1(6 - 1) - 4(4 - 3) + 2[-2 - (-9)]$$

$$= 1(5) - 4(1) + 2(7)$$

$$= 5 - 4 + 14$$

$$= 15$$

Use the same method to evaluate this determinant.

$$\begin{vmatrix} 2 & 1 & -3 \\ 1 & -2 & 2 \\ 3 & -1 & -2 \end{vmatrix} =$$

49. Be careful of the signs when one or more of the elements in the first column are negative. An example is shown.

$$\begin{vmatrix} -2 & 0 & 1 \\ -1 & -2 & 2 \\ 1 & -1 & -3 \end{vmatrix} = -2\begin{vmatrix} -2 & 2 \\ -1 & -3 \end{vmatrix} - (-1)\begin{vmatrix} 0 & 1 \\ -1 & -3 \end{vmatrix} + 1\begin{vmatrix} 0 & 1 \\ -2 & 2 \end{vmatrix}$$

$$= -2[6 - (-2)] + 1[0 - (-1)] + 1[0 - (-2)]$$
$$= -2(8) + 1(1) + 1(2)$$
$$= -16 + 1 + 2$$
$$= -13$$

Use the same method to evaluate this determinant.

$$\begin{vmatrix} 3 & 2 & -1 \\ -2 & 4 & -3 \\ -1 & 0 & 1 \end{vmatrix} =$$

5, from:

12 + 5 - 12

---

50. When one or more elements in the first column is 0, the evaluation is simplified. An example is shown.

$$\begin{vmatrix} 0 & -1 & 0 \\ 2 & 4 & -2 \\ 3 & 1 & 5 \end{vmatrix} = 0\begin{vmatrix} 4 & -2 \\ 1 & 5 \end{vmatrix} - 2\begin{vmatrix} -1 & 0 \\ 1 & 5 \end{vmatrix} + 3\begin{vmatrix} -1 & 0 \\ 4 & -2 \end{vmatrix}$$

$$= 0[20 - (-2)] - 2(-5 - 0) + 3(2 - 0)$$
$$= 0(22) - 2(-5) + 3(2)$$
$$= 0 + 10 + 6$$
$$= 16$$

Use the same method to evaluate this determinant.

$$\begin{vmatrix} 3 & -1 & -5 \\ 0 & 4 & 3 \\ 0 & 2 & 1 \end{vmatrix} =$$

18, from:

12 + 4 + 2

51. Up to this point, we have evaluated third-order determinants by expanding by minors of the first column. However, when factoring in frame 49, we could have factored out the elements of any column or row. Therefore, a third-order determinant can be evaluated by expanding by minors of any row or column. The correct signs for the terms of any expansion are given in the figure below.

$$\begin{vmatrix} + & - & + \\ - & + & - \\ + & - & + \end{vmatrix}$$

We evaluated the determinant below by expanding by minors of the <u>second row</u>.

$$\begin{vmatrix} 3 & -2 & -1 \\ 1 & 3 & -3 \\ 2 & -1 & 2 \end{vmatrix} = -1\begin{vmatrix} -2 & -1 \\ -1 & 2 \end{vmatrix} + 3\begin{vmatrix} 3 & -1 \\ 2 & 2 \end{vmatrix} - (-3)\begin{vmatrix} 3 & -2 \\ 2 & -1 \end{vmatrix}$$

$$= -1(-4 - 1) + 3[6 - (-2)] + 3[-3 - (-4)]$$

$$= -1(-5) + 3(8) + 3(1)$$

$$= 5 + 24 + 3$$

$$= 32$$

Evaluate the same determinant by expanding by minors of the <u>third column</u>.

$$\begin{vmatrix} 3 & -2 & -1 \\ 1 & 3 & -3 \\ 2 & -1 & 2 \end{vmatrix} =$$

-6, from:

-6 + 0 + 0

---

52. a) Evaluate this determinant by expanding by minors of the <u>first row</u>.

$$\begin{vmatrix} 1 & -2 & -3 \\ 4 & -1 & 1 \\ 2 & 0 & 2 \end{vmatrix} =$$

b) Evaluate this determinant by expanding by minors of the <u>second column</u>.

$$\begin{vmatrix} 1 & 4 & 1 \\ 2 & -1 & -2 \\ 3 & -2 & 1 \end{vmatrix} =$$

32, from:

7 + 3 + 22

53. As we saw earlier, it is easier to evaluate a determinant if we expand by minors of a row or column that contains one or more 0's. Let's do that with the determinants below.

a) Evaluate this determinant by expanding by minors of the second column.

$$\begin{vmatrix} 2 & 1 & 4 \\ -3 & 0 & 2 \\ -2 & 1 & 5 \end{vmatrix} =$$

b) Evaluate this determinant by expanding by minors of the first row.

$$\begin{vmatrix} 0 & 2 & 0 \\ 3 & -1 & 1 \\ 1 & -2 & 2 \end{vmatrix} =$$

a) 4, from:

$-2 + 12 - 6$

b) -38, from:

$-32 + 2 - 8$

a) -5, from:

$11 + 0 - 16$

b) -10, from:

$0 - 10 + 0$

8-8 THE DETERMINANT METHOD - CRAMER'S RULE

In this section, we will discuss the determinant method for solving systems of two and three equations. The determinant method is called Cramer's rule.

54. The standard form for a system of two equations is given below.

$$a_1x + b_1y = c_1$$
$$a_2x + b_2y = c_2$$

The a's and b's are coefficients and the c's are constants. We can use the a's, b's, and c's to set up determinants to solve the system. The three determinants used are D, D_x, and D_y. They are shown below.

$$D = \begin{vmatrix} a_1 & b_1 \\ a_2 & b_2 \end{vmatrix} \qquad D_x = \begin{vmatrix} \overset{\downarrow}{c_1} & b_1 \\ c_2 & b_2 \end{vmatrix} \qquad D_y = \begin{vmatrix} a_1 & \overset{\downarrow}{c_1} \\ a_2 & c_2 \end{vmatrix}$$

Continued on following page.

54. Continued

Determinant D is formed by the coefficients of the x's and y's. Notice these points about D_x and D_y.

1) To get D_x from D, we replace the a's (the coefficients of the x's) with the c's (see the arrow).
2) To get D_y from D, we replace the b's (the coefficients of the y's) with the c's (see the arrow).

Using the determinants above, we can set up the following solutions for x and y. They are known as Cramer's rule.

$$x = \frac{D_x}{D} = \frac{\begin{vmatrix} c_1 & b_1 \\ c_2 & b_2 \end{vmatrix}}{\begin{vmatrix} a_1 & b_1 \\ a_2 & b_2 \end{vmatrix}} \qquad y = \frac{D_y}{D} = \frac{\begin{vmatrix} a_1 & c_1 \\ a_2 & c_2 \end{vmatrix}}{\begin{vmatrix} a_1 & b_1 \\ a_2 & b_2 \end{vmatrix}}$$

55. Let's use Cramer's rule to solve this system.

$$5x - 2y = 19$$
$$7x + 3y = 15$$

First we compute D, D_x, and D_y.

$$D = \begin{vmatrix} 5 & -2 \\ 7 & 3 \end{vmatrix} = 15 - (-14) = 15 + 14 = 29$$

$$D_x = \begin{vmatrix} 19 & -2 \\ 15 & 3 \end{vmatrix} = 57 - (-30) = 57 + 30 = 87$$

$$D_y = \begin{vmatrix} 5 & 19 \\ 7 & 15 \end{vmatrix} = 75 - 133 = -58$$

Then using D, D_x, and D_y, we can solve for x and y. We get:

$$x = \frac{D_x}{D} = \frac{87}{29} = 3 \qquad y = \frac{D_y}{D} = \frac{-58}{29} = -2$$

Show that (3,-2) satisfies each equation in the system below.

$5x - 2y = 19$ $\qquad$ $7x + 3y = 15$

$$\begin{aligned} 5(3) - 2(-2) &= 19 \\ 15 + 4 &= 19 \\ 19 &= 19 \end{aligned} \qquad \begin{aligned} 7(3) + 3(-2) &= 15 \\ 21 + (-6) &= 15 \\ 15 &= 15 \end{aligned}$$

56. Let's use the determinant method to solve this system.

$$x + y = 7$$
$$x - y = 3$$

To do so, compute D, D_x, and D_y and then use them to find x and y.

$D = \begin{vmatrix} & \\ & \end{vmatrix} =$ ____________

$D_x = \begin{vmatrix} & \\ & \end{vmatrix} =$ ____________

$D_y = \begin{vmatrix} & \\ & \end{vmatrix} =$ ____________

$x = \frac{D_x}{D} =$ ________ $y = \frac{D_y}{D} =$ ________

$D = \begin{vmatrix} 1 & 1 \\ 1 & -1 \end{vmatrix}$
$= -1 - 1 = -2$

$D_x = \begin{vmatrix} 7 & 1 \\ 3 & -1 \end{vmatrix}$
$= -7 - 3 = -10$

$D_y = \begin{vmatrix} 1 & 7 \\ 1 & 3 \end{vmatrix}$
$= 3 - 7 = -4$

$x = \frac{-10}{-2} = 5$

$y = \frac{-4}{-2} = 2$

57. Let's use the determinant method to solve this system.

$$4a + 3b = 1$$
$$3a - 2b = 22$$

Compute D, D_a, and D_b and then use them to find a and b.

D = ____________

$D_a =$ ____________

$D_b =$ ____________

$a = \frac{D_a}{D} =$ ________ $b = \frac{D_b}{D} =$ ________

$D = \begin{vmatrix} 4 & 3 \\ 3 & -2 \end{vmatrix} = -17$

$D_a = \begin{vmatrix} 1 & 3 \\ 22 & -2 \end{vmatrix} = -68$

$D_b = \begin{vmatrix} 4 & 1 \\ 3 & 22 \end{vmatrix} = 85$

$a = \frac{-68}{-17} = 4$

$b = \frac{85}{-17} = -5$

58. The system below is not in standard form. We must put it in standard form before we use the determinant method.

$$4y = 20 - x$$
$$3x - 8y = 0$$

Converting the system to standard form, we get:

$$x + 4y = 20$$
$$3x - 8y = 0$$

Continued on following page.

58. Continued

Now compute D, D_x, and D_y and then use them to find x and y.

D =

D_x =

D_y =

$x = \frac{D_x}{D} =$ ________ $y = \frac{D_y}{D} =$ ________

---

59. Cramer's rule can be extended to solving systems of three equations. To do so, we start with the standard form of a system of three equations.

$$a_1x + b_1y + c_1z = d_1$$
$$a_2x + b_2y + c_2z = d_2$$
$$a_3x + b_3y + c_3z = d_3$$

The a's, b's, and c's are coefficients and the d's are constants. We can use the a's, b's, c's, and d's to set up determinants to solve the system. The four determinants used are D, D_x, D_y, D_z. They are shown below.

$$D = \begin{vmatrix} a_1 & b_1 & c_1 \\ a_2 & b_2 & c_2 \\ a_3 & b_3 & c_3 \end{vmatrix} \qquad D_x = \begin{vmatrix} \downarrow \\ d_1 & b_1 & c_1 \\ d_2 & b_2 & c_2 \\ d_3 & b_3 & c_3 \end{vmatrix}$$

$$D_y = \begin{vmatrix} & \downarrow & \\ a_1 & d_1 & c_1 \\ a_2 & d_2 & c_2 \\ a_3 & d_3 & c_3 \end{vmatrix} \qquad D_z = \begin{vmatrix} & & \downarrow \\ a_1 & b_1 & d_1 \\ a_2 & b_2 & d_2 \\ a_3 & b_3 & d_3 \end{vmatrix}$$

Determinant D is formed by the coefficients of the x's, y's, and z's. Notice these points about D_x, D_y, and D_z.

1) To get D_x from D, we replace the a's (the coefficients of the x's) with the d's (see the arrow).
2) To get D_y from D, we replace the b's (the coefficients of the y's) with the d's (see the arrow).
3) To get D_z from D, we replace the c's (the coefficients of the z's) with the d's (see the arrow).

Using the determinants above, we can set up the following solutions for x, y, and z.

$$x = \frac{D_x}{D} \qquad y = \frac{D_y}{D} \qquad z = \frac{D_z}{D}$$

D = -20
D_x = -160
D_y = -60
x = 8
y = 3

60. Let's use Cramer's rule to solve this system.

$$x - 2y + 3z = 6$$
$$2x - y - z = -3$$
$$x + y + z = 6$$

First we compute D, D_x, D_y, and D_z. To evaluate each determinant, we expanded by minors of the first column.

$$D = \begin{vmatrix} 1 & -2 & 3 \\ 2 & -1 & -1 \\ 1 & 1 & 1 \end{vmatrix} = 1\begin{vmatrix} -1 & -1 \\ 1 & 1 \end{vmatrix} - 2\begin{vmatrix} -2 & 3 \\ 1 & 1 \end{vmatrix} + 1\begin{vmatrix} -2 & 3 \\ -1 & -1 \end{vmatrix}$$
$$= 1(-1 + 1) -2(-2 - 3) + 1(2 + 3) = 15$$

$$D_x = \begin{vmatrix} 6 & -2 & 3 \\ -3 & -1 & -1 \\ 6 & 1 & 1 \end{vmatrix} = 6\begin{vmatrix} -1 & -1 \\ 1 & 1 \end{vmatrix} - (-3)\begin{vmatrix} -2 & 3 \\ 1 & 1 \end{vmatrix} + 6\begin{vmatrix} -2 & 3 \\ -1 & -1 \end{vmatrix}$$
$$= 6(-1 + 1) + 3(-2 - 3) + 6(2 + 3) = 15$$

$$D_y = \begin{vmatrix} 1 & 6 & 3 \\ 2 & -3 & -1 \\ 1 & 6 & 1 \end{vmatrix} = 1\begin{vmatrix} -3 & -1 \\ 6 & 1 \end{vmatrix} - 2\begin{vmatrix} 6 & 3 \\ 6 & 1 \end{vmatrix} + 1\begin{vmatrix} 6 & 3 \\ -3 & -1 \end{vmatrix}$$
$$= 1(-3 + 6) - 2(6 - 18) + 1(-6 + 9) = 30$$

$$D_z = \begin{vmatrix} 1 & -2 & 6 \\ 2 & -1 & -3 \\ 1 & 1 & 6 \end{vmatrix} = 1\begin{vmatrix} -1 & -3 \\ 1 & 6 \end{vmatrix} - 2\begin{vmatrix} -2 & 6 \\ 1 & 6 \end{vmatrix} + 1\begin{vmatrix} -2 & 6 \\ -1 & -3 \end{vmatrix}$$
$$= 1(-6 + 3) - 2(-12 - 6) + 1(6 + 6) = 45$$

Then using the values for D, D_x, D_y, and D_z, we can solve for x, y, and z.

$$x = \frac{D_x}{D} = \frac{15}{15} = 1 \qquad y = \frac{D_y}{D} = \frac{30}{15} = 2 \qquad z = \frac{D_z}{D} = \frac{45}{15} = 3$$

The solution is (1,2,3). Check the solution in each original equation below.

$$x - 2y + 3z = 6 \qquad 2x - y - z = -3 \qquad x + y + z = 6$$

$$1 - 2(2) + 3(3) = 6$$
$$1 - 4 + 9 = 6$$
$$6 = 6$$

$$2(1) - 2 - 3 = -3$$
$$2 - 2 - 3 = -3$$
$$-3 = -3$$

$$1 + 2 + 3 = 6$$
$$6 = 6$$

61. Let's use the determinant method to solve this system.

$$\begin{aligned} 2x - y + z &= 2 \\ x + 3y + 2z &= 7 \\ 5x + y - z &= -9 \end{aligned}$$

Compute D, D_x, D_y, and D_z and then use them to find x, y, and z.

$D = \left| \quad \right| =$

$D_x = \left| \quad \right| =$

$D_y = \left| \quad \right| =$

$D_z = \left| \quad \right| =$

$x = \frac{D_x}{D} =$ ______ $y = \frac{D_y}{D} =$ ______ $z = \frac{D_z}{D} =$ ______

$$D = \begin{vmatrix} 2 & -1 & 1 \\ 1 & 3 & 2 \\ 5 & 1 & -1 \end{vmatrix} = -35$$

$$D_x = \begin{vmatrix} 2 & -1 & 1 \\ 7 & 3 & 2 \\ -9 & 1 & -1 \end{vmatrix} = 35$$

$$D_y = \begin{vmatrix} 2 & 2 & 1 \\ 1 & 7 & 2 \\ 5 & -9 & -1 \end{vmatrix} = 0$$

$$D_z = \begin{vmatrix} 2 & -1 & 2 \\ 1 & 3 & 7 \\ 5 & 1 & -9 \end{vmatrix} = -140$$

x = -1, y = 0, z = 4

62. Let's use the determinant method to solve this system. Notice that we rewrote the system in standard form at the right.

$$\begin{aligned} x + 2y &= -1 \\ 2x - z &= 5 \\ 2y + z &= -3 \end{aligned} \qquad \begin{aligned} x + 2y &= -1 \\ 2x - z &= 5 \\ 2y + z &= -3 \end{aligned}$$

Compute D, D_x, D_y, and D_z and then use them to find x, y, and z.

$D = \left| \quad \right| =$

$D_x = \left| \quad \right| =$

$D_y = \left| \quad \right| =$

$D_z = \left| \quad \right| =$

$x = \frac{D_x}{D} =$ ______ $y = \frac{D_y}{D} =$ ______ $z = \frac{D_z}{D} =$

$$D = \begin{vmatrix} 1 & 2 & 0 \\ 2 & 0 & -1 \\ 0 & 2 & 1 \end{vmatrix} = -2 \qquad D_x = \begin{vmatrix} -1 & 2 & 0 \\ 5 & 0 & -1 \\ -3 & 2 & 1 \end{vmatrix} = -6 \qquad D_y = \begin{vmatrix} 1 & -1 & 0 \\ 2 & 5 & -1 \\ 0 & -3 & 1 \end{vmatrix} = 4 \qquad D_z = \begin{vmatrix} 1 & 2 & -1 \\ 2 & 0 & 5 \\ 0 & 2 & -3 \end{vmatrix} = -2$$

x = 3, y = -2, z = 1

SELF-TEST 30 (pages 429-440)

Evaluate each determinant.

1. $\begin{vmatrix} 1 & -3 & 2 \\ -1 & 4 & -3 \\ 2 & -1 & 1 \end{vmatrix}$

2. $\begin{vmatrix} 5 & 0 & 2 \\ 1 & -3 & 0 \\ 0 & 4 & -2 \end{vmatrix}$

3. Use the determinant method to solve this system.

$$\begin{aligned} 3x - 2y &= 18 \\ 2x + y &= 5 \end{aligned}$$

4. Use the determinant method to solve this system.

$$\begin{aligned} x + 2y - z &= 7 \\ 4x + y + 2z &= 9 \\ 2x - 3y + z &= 1 \end{aligned}$$

ANSWERS:

1. 2
2. 38
3. x = 4, y = -3
4. x = 3, y = 1, z = -2

SUPPLEMENTARY PROBLEMS - CHAPTER 8

Assignment 28

Use the graphing method to solve each system.

1. $x + y = 4$
   $x - y = 2$

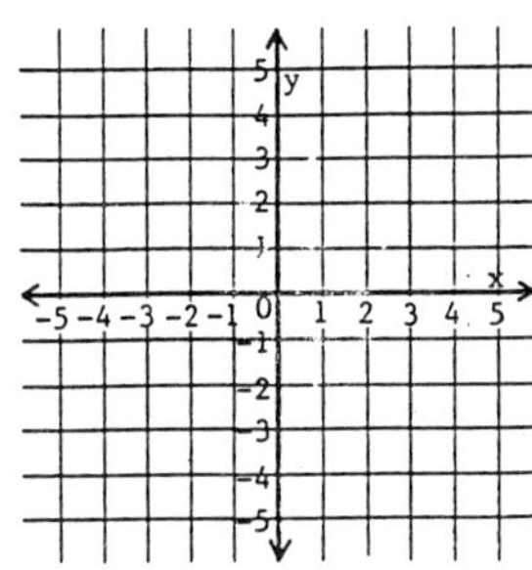

2. $2x - y = -4$
   $x + y = 1$

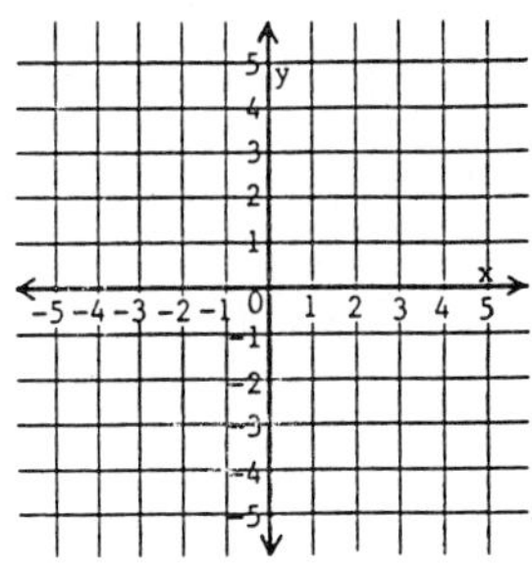

Use the addition method to solve each system.

3. $5x - 4y = 23$, $3x + 4y = 1$
4. $x + 9y = 65$, $4x - 9y = 35$
5. $3x + 2y = 28$, $2x - y = 7$
6. $5x + 2y = 3$, $3x + 6y = 21$
7. $6x - y = 19$, $2x - y = 7$
8. $7x + 3y = 26$, $9x - 2y = 10$
9. $5x + 7y = 32$, $4x + 5y = 25$
10. $11x - 6y = 8$, $12x - 5y = 1$
11. $2x + y = 13$, $y = x - 2$
12. $2x = 6 - 5y$, $2y = 4x - 36$
13. $.3x - .1y = 1.1$, $.2x + .1y = .9$
14. $.21a + .08b = .46$, $.17a + .04b = .36$

Use the substitution method to solve each system.

15. $x = 3y$, $2x + y = 14$
16. $3x - 8y = 12$, $x = 4y$
17. $5x - y = 8$, $y = 2x - 5$
18. $y = 2x - 3$, $7x - 3y = 8$
19. $2x = 5y$, $3y - x = 2$
20. $4y - 5x = 2$, $4x = 3y$
21. $3x - 2y = 40$, $4x - 3y = 50$
22. $7p + 3q = 12$, $5p - 14 = 6q$

Use any method to solve each system.

23. $7x - 3y = 34$, $5x + 3y = 14$
24. $x + 4y = 5$, $x + 5y = 7$
25. $x + 2y = 15$, $x = 4y$
26. $5a - 4b = 17$, $2a - 5b = 17$
27. $\frac{x}{2} + \frac{y}{3} = 4$, $\frac{3x}{4} + \frac{y}{6} = 4$
28. $2x + \frac{y}{5} = 20$, $5x - y = 5$
29. $.1x + .2y = 3$, $.5x - .4y = 8$
30. $2.4x + 3.2 = y$, $3y - 12 = 4x$

Assignment 29

Solve each problem.

1. The sum of two numbers is 214 and their difference is 68. Find the two numbers.
2. The sum of two numbers is 70. The first number is four times the second number. Find the two numbers.
3. The perimeter of a rectangular room is 60 ft. If the width is 6 ft less than the length, find the length and width.
4. We want to construct a rectangle whose perimeter is 42 cm and whose length is twice the width. Find the length and width.

5. A woman invested a total of \$7,500, part at 7% and the rest at 8%. If the simple interest for the two investments for one year was \$560, how much was invested at each rate?

6. A man invested a certain amount of money at 6% and \$2,000 more than that at 8%. If the simple interest for the two investments for one year is \$510, find the amount invested at each rate.

7. A jet averages 484 mph with the wind and 416 mph into the wind. Find the speed of the wind and the speed of the jet in still air.

8. It takes a boat $2\frac{1}{2}$ hours to go 15 miles downstream and 5 hours to return. Find the speed of the current and the speed of the boat in still water.

9. One day a store sold 50 T-shirts. White ones cost \$7.95 and blue ones cost \$9.95. If the total receipts were \$453.50, how many of each color were sold?

10. A taxi charges a flat rate plus a certain amount per mile. If a 12-mile trip costs \$18 and a 6-mile trip costs \$10.50, find the flat rate and the amount per mile.

11. How much of a 10% solution of alcohol and a 50% solution of alcohol should be mixed to make 200 milliliters of a 35% solution?

12. A clerk wants to mix candy worth \$4.80 per pound with candy worth \$2.40 per pound to get 40 pounds of a mixture worth \$3.30 per pound. How many pounds of the \$4.80 candy and the \$2.40 candy should he mix?

Use the addition method to solve each system.

13. $\begin{aligned} 2x - y + z &= 9 \\ x - 3y - z &= 4 \\ 3x + y + z &= 10 \end{aligned}$

14. $\begin{aligned} x + y - 2z &= 8 \\ 2x - y + 2z &= 4 \\ x + y &= 2 \end{aligned}$

15. $\begin{aligned} x - y + z &= -6 \\ 2x + y + z &= 2 \\ x + 2y - z &= 7 \end{aligned}$

16. $\begin{aligned} x + y + z &= 5 \\ x + 2y - z &= 13 \\ x - y - 2z &= 3 \end{aligned}$

17. $\begin{aligned} x + y + 2z &= 4 \\ x - 2y + z &= 5 \\ y + z &= 1 \end{aligned}$

18. $\begin{aligned} x + y &= z + 9 \\ y &= x + 2 \\ x &= 8 - z \end{aligned}$

Solve each problem.

19. The sum of three numbers is 7. The first number minus three times the second plus twice the third is 0. The first number plus the second minus the third is -3. Find the three numbers.

20. The sum of three numbers is 3. The first number minus the second plus the third is 7. The difference between the third number and twice the second is 3 less than the first. Find the three numbers.

21. In triangle ABC, angle A is 40° less than angle C. Angle A is three times angle B. Find the size of each angle.

22. In triangle ABC, the sum of angles A and C is 20° more than angle B. Angle B is 25° more than angle C. Find the size of each angle.

23. Joan has a total score of 268 on three tests. The sum of the scores on her first and third tests is 98 more than her second test. Her first score is 5 more than her second score. Find the scores on the three tests.

24. Three masons, A, B, and C, can lay 650 blocks in one hour. A and B together can lay 425 blocks. B and C together can lay 445 blocks. How many blocks could each mason lay by himself?

Assignment 30

Evaluate each determinant.

1. $\begin{vmatrix} 4 & 2 \\ 3 & 6 \end{vmatrix}$

2. $\begin{vmatrix} 3 & 10 \\ 2 & 5 \end{vmatrix}$

3. $\begin{vmatrix} -1 & 1 \\ 4 & 2 \end{vmatrix}$

4. $\begin{vmatrix} 5 & 4 \\ -3 & 2 \end{vmatrix}$

5. $\begin{vmatrix} -2 & 1 \\ -3 & 4 \end{vmatrix}$

6. $\begin{vmatrix} -6 & -1 \\ -10 & -5 \end{vmatrix}$

7. $\begin{vmatrix} 4 & 9 \\ -6 & 0 \end{vmatrix}$

8. $\begin{vmatrix} -3 & 0 \\ 8 & 1 \end{vmatrix}$

9. $\begin{vmatrix} 2 & 1 & -1 \\ 3 & 1 & 4 \\ 1 & 2 & 1 \end{vmatrix}$

10. $\begin{vmatrix} -1 & 2 & 2 \\ -2 & 3 & -1 \\ 2 & 1 & -1 \end{vmatrix}$

11. $\begin{vmatrix} 4 & -2 & 1 \\ -1 & 6 & 2 \\ -1 & 3 & 5 \end{vmatrix}$

12. $\begin{vmatrix} -2 & 5 & 1 \\ 0 & 3 & 4 \\ -1 & 2 & 1 \end{vmatrix}$

13. $\begin{vmatrix} 4 & 1 & -1 \\ 2 & 0 & 3 \\ -2 & -1 & 5 \end{vmatrix}$

14. $\begin{vmatrix} 6 & 2 & 0 \\ -3 & -1 & -4 \\ 2 & 1 & 0 \end{vmatrix}$

Use the determinant method to solve each system.

15. $\begin{aligned} 4x + 5y &= 10 \\ 3x + 2y &= 11 \end{aligned}$

16. $\begin{aligned} 3x - 2y &= 6 \\ 4x - 3y &= 10 \end{aligned}$

17. $\begin{aligned} 5x + 8y &= 6 \\ 2x - 3y &= 21 \end{aligned}$

18. $\begin{aligned} 9a - 1 &= 5b \\ 27 - 6b &= 8a \end{aligned}$

19. $\begin{aligned} x + y + z &= 3 \\ 2x + y - z &= 0 \\ x + 2y + z &= 6 \end{aligned}$

20. $\begin{aligned} x + 2y + z &= 8 \\ 2x - y - z &= 6 \\ x - 2y + z &= 0 \end{aligned}$

21. $\begin{aligned} 2x + y + z &= 5 \\ x - 3y - z &= 2 \\ 4x - y - z &= 1 \end{aligned}$

22. $\begin{aligned} x + y - z &= -4 \\ 2x - y + 2z &= 15 \\ x - 2y &= 13 \end{aligned}$

23. $\begin{aligned} x - y &= 7 \\ 2y - z &= 3 \\ 2x - 5y &= 8 \end{aligned}$

24. $\begin{aligned} p + r &= q - 4 \\ r &= p + 1 \\ p &= q - 3 \end{aligned}$

9 Second-Degree Equations and Their Graphs

In this chapter, we will discuss second-degree equations whose graphs are either parabolas, circles, ellipses, or hyperbolas. The relationship between the equations and some properties of the graphed figures is emphasized. The distance formula, nonlinear systems of equations, and the inverses of functions are also discussed.

9-1 GRAPHING QUADRATIC FUNCTIONS

In this section, we will graph quadratic functions of the forms $f(x) = ax^2$, $f(x) = a(x - h)^2$, and $f(x) = a(x - h)^2 + k$. The graphs of all quadratic functions are cup-shaped curves called <u>parabolas</u>.

---

1. The general form of a <u>quadratic function</u> is given below.

$$y = ax^2 + bx + c \quad \text{or} \quad f(x) = ax^2 + bx + c \qquad (a \neq 0)$$

To graph a quadratic function, we substitute values for $\underline{x}$ and compute the corresponding values of $\underline{y}$ or $f(x)$. (In this section, we will use $\underline{y}$ and $f(x)$ interchangeably.)

The simplest quadratic function is $y = x^2$. We graphed it on the following page. To do so, we made up a solution-table, plotted the points, and then drew a smooth curve through the plotted points. Notice in the solution-table that we used 0, some positive values, and some negative values for $\underline{x}$.

Continued on following page.

1. Continued

$y = x^2$

x	y
-3	9
-2	4
-1	1
0	0
1	1
2	4
3	9

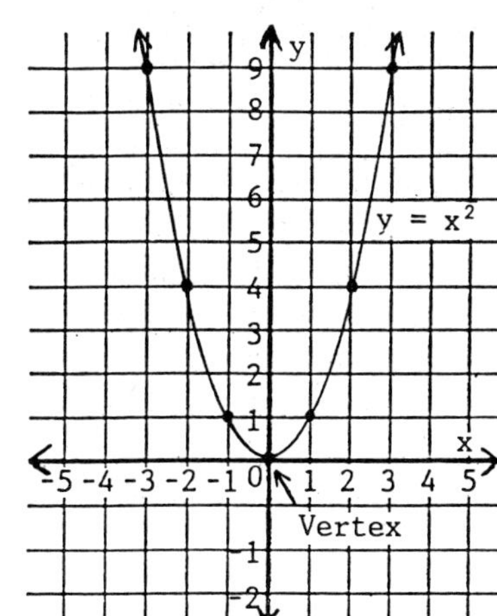

The graph of any quadratic function is a cup-shaped curve like the one above. Curves of that type are called parabolas. Note:

1. The point (0,0) is called the vertex of the parabola.

2. The y-axis is the axis of symmetry of the parabola. That is, if the parabola were folded on that axis, the two halves of the curve would match.

---

2. When graphing a quadratic function, plot enough points so that the outline of the parabola is clear. Then draw a smooth curve through the plotted points. Graph $y = \frac{1}{2}x^2$ below.

$y = \frac{1}{2}x^2$

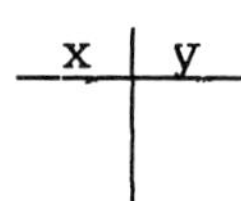

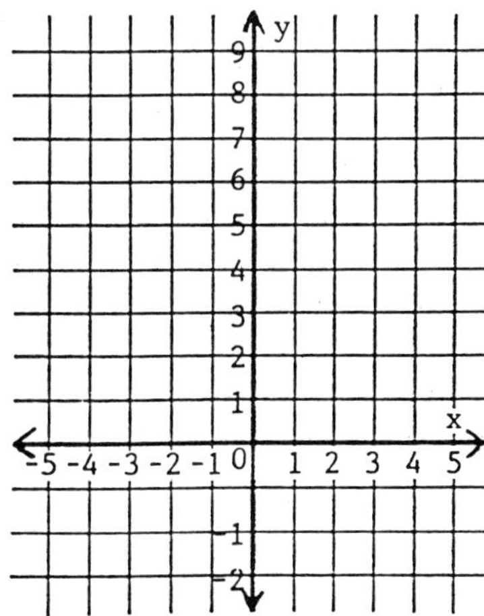

Answer To Frame 2:

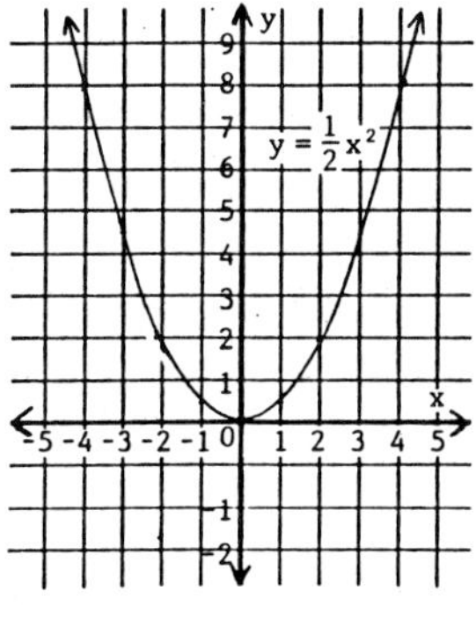

3. Use the same method to graph $y = 2x^2$ below.

$y = 2x^2$

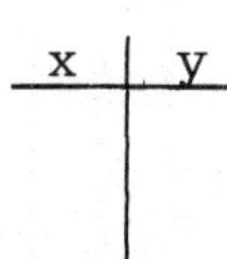

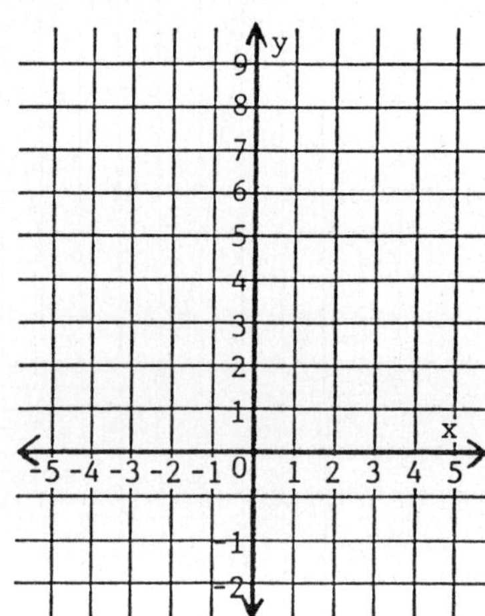

4. The graphs of $y = \frac{1}{2}x^2$, $y = x^2$, and $y = 2x^2$ are shown on the same axes below.

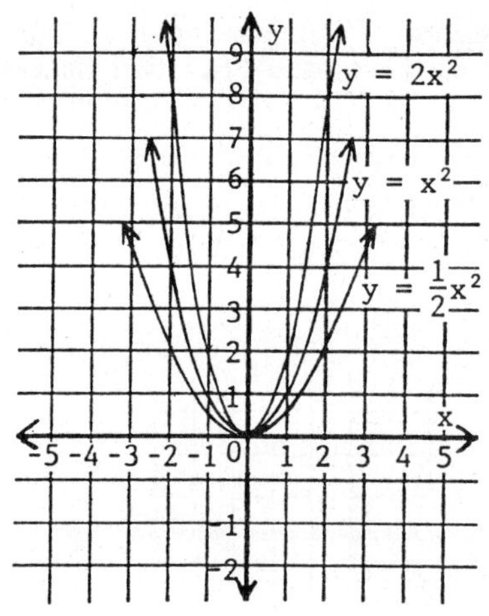

Note: 1. The graph of $y = \frac{1}{2}x^2$ is a wider parabola than the graph of $y = x^2$.

The graph of $y = 2x^2$ is a narrower parabola than the graph of $y = x^2$.

2. The vertex of each parabola is at the origin.

3. The axis of symmetry of each parabola is the y-axis.

Answer To Frame 3:

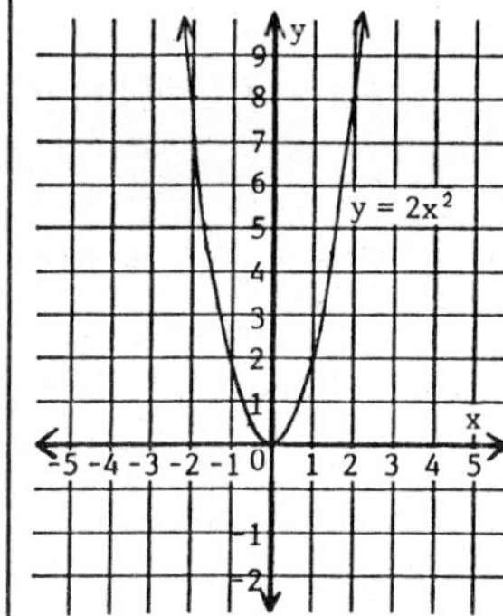

5. We graphed $y = -x^2$ at the left below. Notice that the parabola opens downward. However, the vertex is still at the origin and the axis of symmetry is still the y-axis. Graph $y = -2x^2$ at the right below.

$y = -x^2$

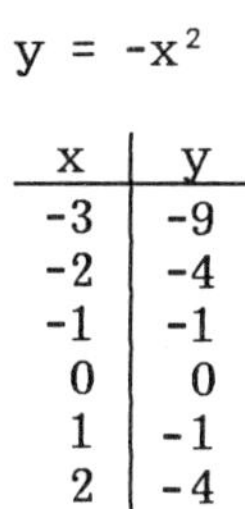

x	y
-3	-9
-2	-4
-1	-1
0	0
1	-1
2	-4
3	-9

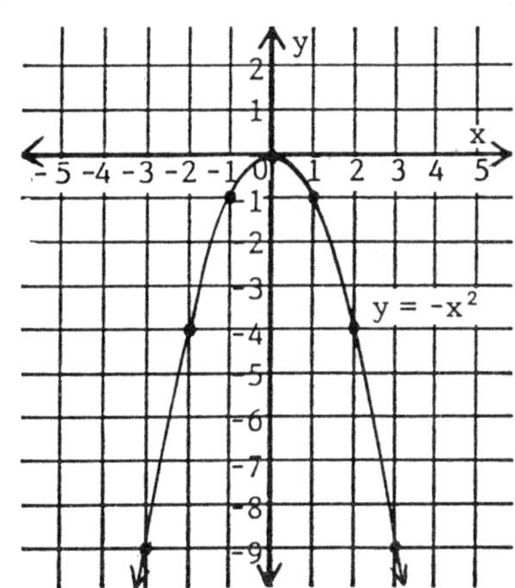

$y = -2x^2$

x	y

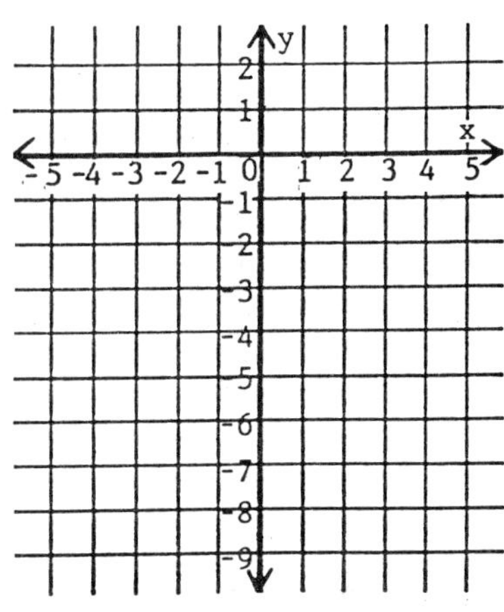

6. When graphing functions of the form $y = ax^2$, the parabola opens upward if $\underline{a}$ is positive and downward if $\underline{a}$ is negative.

Answer either upward or downward for these:

a) The graph of $y = -\frac{1}{2}x^2$ opens ______________.

b) The graph of $y = 3x^2$ opens ______________.

c) The graph of $y = -5x^2$ opens ______________.

Answer To Frame 5:

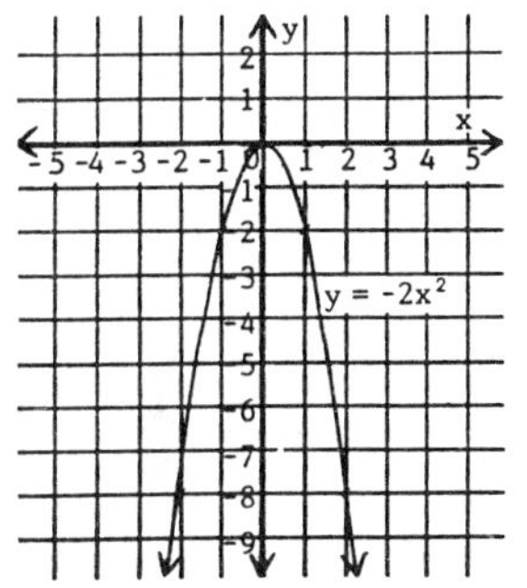

7. We graphed the quadratic function $f(x) = (x - 2)^2$ below.

$f(x) = (x - 2)^2$

x	f(x)
-1	9
0	4
2	0
3	1
4	4
5	9

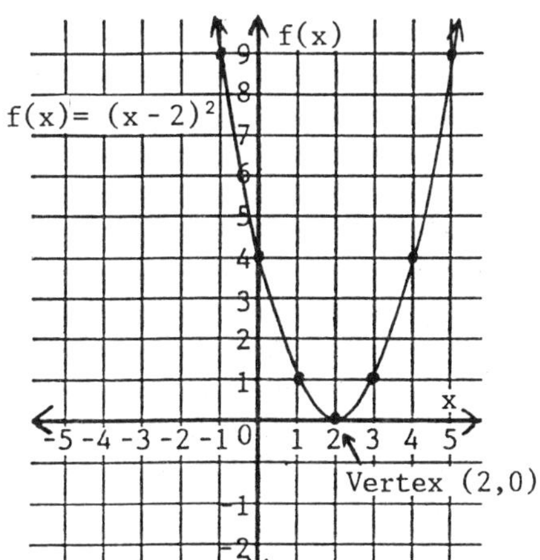

The parabola above is just like the graph of $f(x) = x^2$ except that it is moved 2 units to the right.

Note: 1. The vertex is the point (2,0).
2. The axis of symmetry is the line $x = 2$.

a) downward

b) upward

c) downward

8. We graphed the quadratic function $f(x) = 2(x + 2)^2$ below.

$f(x) = 2(x + 2)^2$

x	f(x)
-4	8
-3	2
-2	0
-1	2
0	8

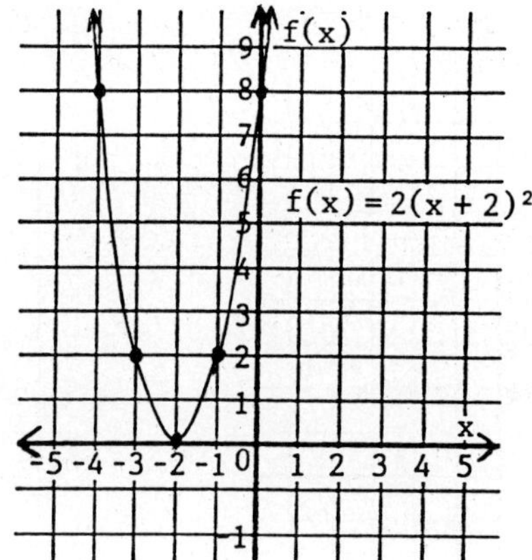

The parabola above is just like the graph of $f(x) = 2x^2$ except that it is moved 2 units to the left.

a) The vertex is the point ________.

b) The axis of symmetry is the line ________.

---

9. We graphed the quadratic function $f(x) = -\frac{1}{2}(x - 3)^2$ below.

a) (-2,0)

b) x = -2

$f(x) = -\frac{1}{2}(x - 3)^2$

x	f(x)
-1	-8
0	$-4\frac{1}{2}$
1	-2
2	$-\frac{1}{2}$
3	0
4	$-\frac{1}{2}$
5	-2
6	$-4\frac{1}{2}$
7	-8

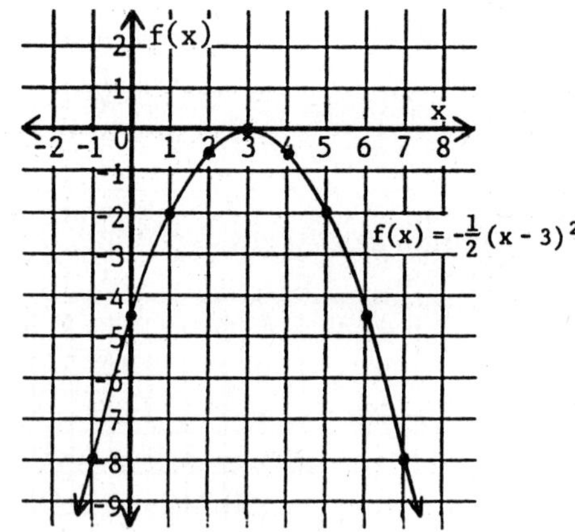

The parabola above opens downward. It is just like the graph of $f(x) = -\frac{1}{2}x^2$ except that it is shifted 3 units to the right.

a) The vertex is the point ________.

b) The axis of symmetry is the line ________.

---

a) (3,0)

b) x = 3

10. When graphing quadratic functions of the form $f(x) = a(x - h)^2$, the parabola is the same as $f(x) = ax^2$ except that it is shifted to the right or left. If $\underline{h}$ is positive, it is shifted to the right. If $\underline{h}$ is negative, it is shifted to the left. The vertex is (h,0) and the axis of symmetry is the line x = h.

For $f(x) = \frac{1}{2}(x - 1)^2$, h = 1. Therefore:

The parabola is shifted 1 unit to the right.
The vertex is the point (1,0).
The axis of symmetry is the line x = 1.

For $f(x) = -2(x + 4)^2$ or $-2[x - (-4)]^2$, h = -4. Therefore:

a) The parabola is shifted ______ units to the __________.

b) The vertex is the point __________.

c) The axis of symmetry is the line __________.

a) 4 units to the left

b) (-4,0)

c) x = -4

11. Graph each quadratic function below.

a) $f(x) = \frac{1}{2}(x - 2)^2$

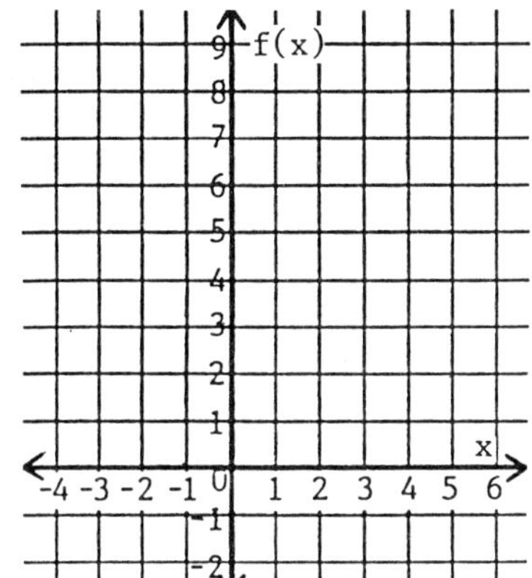

b) $f(x) = -2(x + 3)^2$

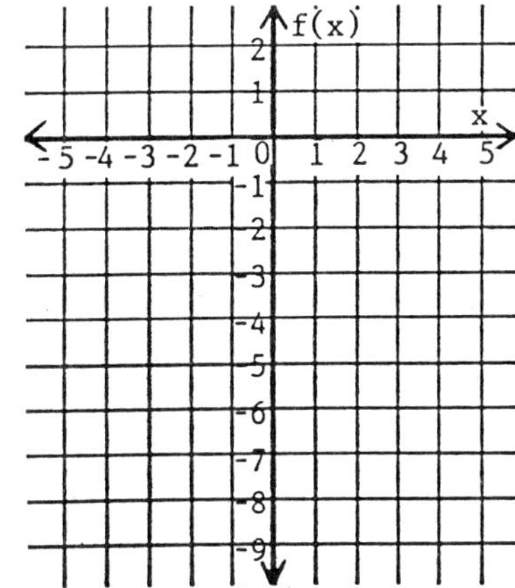

Answers To Frame 11:

a)

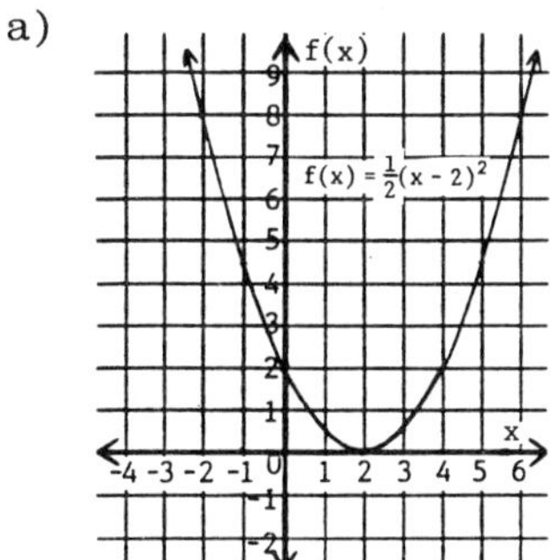

b)

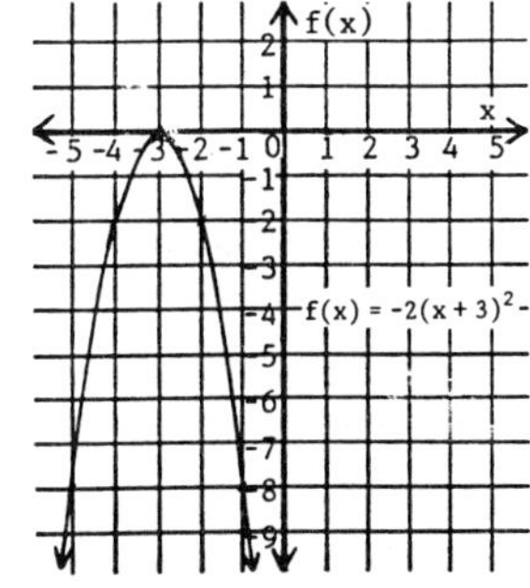

12. We graphed the quadratic function $f(x) = (x - 1)^2 + 2$ below.

$f(x) = (x - 1)^2 + 2$

x	f(x)
-2	11
-1	6
0	3
1	2
2	3
3	6
4	11

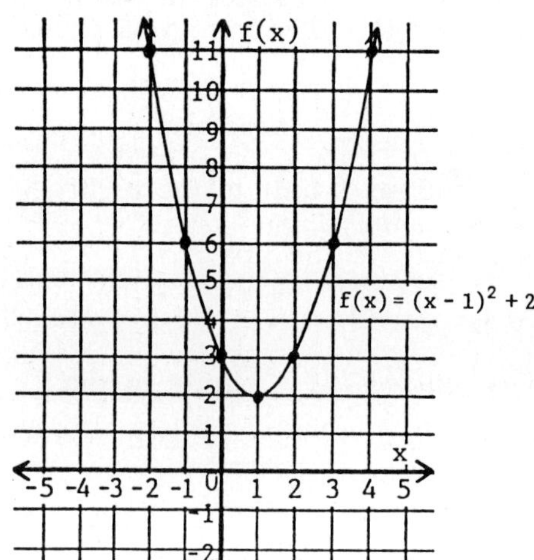

The parabola above is just like the graph of $f(x) = (x - 1)^2$ except that it is moved 2 units up because 2 is added to each function value.

Note: 1. The vertex is the point (1,2).

2. The axis of symmetry is the line $x = 1$.

---

13. We graphed the quadratic function $f(x) = \frac{1}{2}(x + 3)^2 - 2$ below.

$f(x) = \frac{1}{2}(x + 3)^2 - 2$

x	f(x)
-7	6
-6	$2\frac{1}{2}$
-5	0
-4	$-1\frac{1}{2}$
-3	-2
-2	$-1\frac{1}{2}$
-1	0
0	$2\frac{1}{2}$
1	6

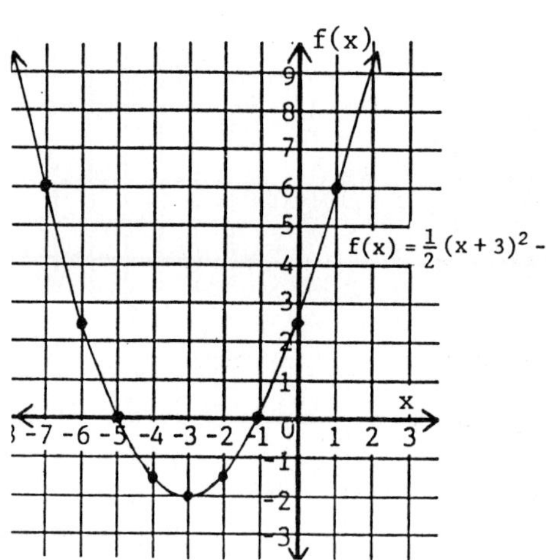

The parabola above is just like the graph of $f(x) = \frac{1}{2}(x + 3)^2$ except that it is moved 2 units down because -2 is added to each function value.

a) The vertex is at the point ________.

b) The axis of symmetry is the line ________.

---

a) (-3,-2)

b) $x = -3$

14. We graphed the quadratic function $f(x) = -2(x + 1)^2 + 3$ below.

$f(x) = -2(x + 1)^2 + 3$

x	f(x)
-3	-5
-2	1
$-1\frac{1}{2}$	$2\frac{1}{2}$
-1	3
$-\frac{1}{2}$	$2\frac{1}{2}$
0	1
1	-5

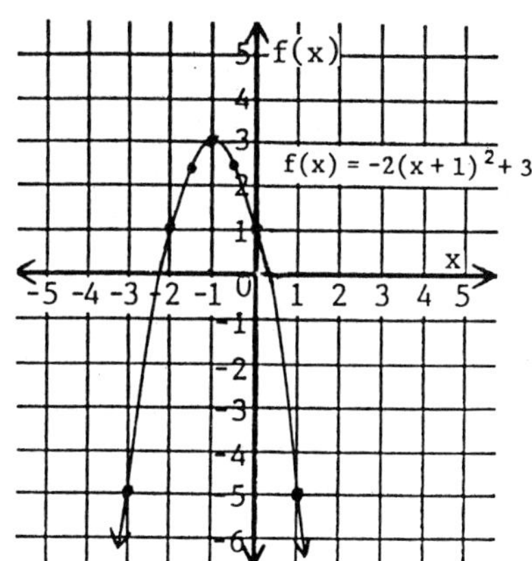

The parabola above opens downward. It is just like the graph of $f(x) = -2(x + 1)^2$ except that it is moved up 3 units because 3 is added to each function value.

a) The vertex is at the point ___________.

b) The axis of symmetry is the line _________.

---

a) (-1,3)

b) x = -1

15. When graphing quadratic functions of the form $f(x) = a(x - h)^2 + k$, the parabola is the same as $f(x) = a(x - h)^2$ except that it is shifted up or down. If $\underline{k}$ is positive, it is shifted up. If $\underline{k}$ is negative, it is shifted down. The vertex is at (h,k) and the axis of symmetry is the line x = h.

For $f(x) = 3(x - 2)^2 + 1$, k = 1. Therefore:

The parabola is shifted 1 unit up.
The vertex is the point (2,1).
The axis of symmetry is the line x = 2.

For $f(x) = -\frac{1}{2}(x + 1)^2 - 4$ or $-\frac{1}{2}[x - (-1)]^2 - 4$, k = -4. Therefore:

a) The parabola is shifted _____ units ___________.

b) The vertex is the point ___________.

c) The axis of symmetry is the line _________.

---

a) 4 units down

b) (-1,-4)

c) x = -1

16. Graph each quadratic function below.

a) $f(x) = -\frac{1}{2}(x - 2)^2 - 1$

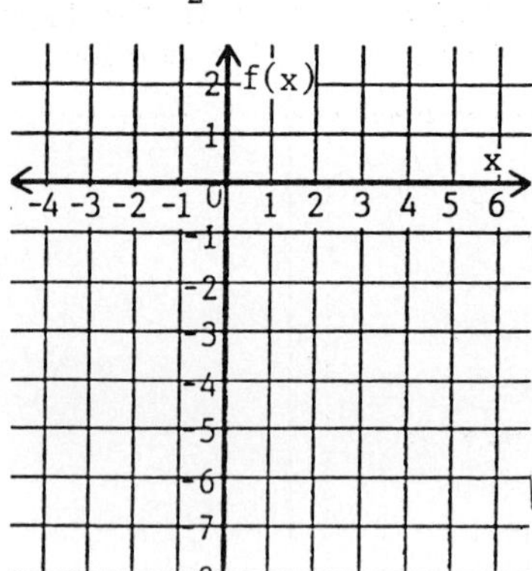

b) $f(x) = 2(x + 3)^2 - 3$

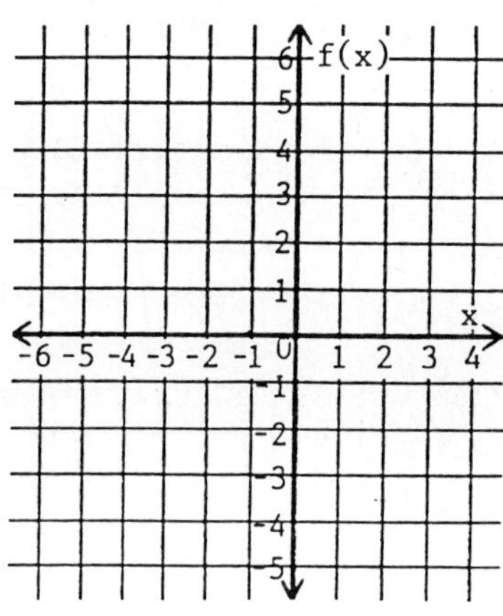

Answers To Frame 16:

a)

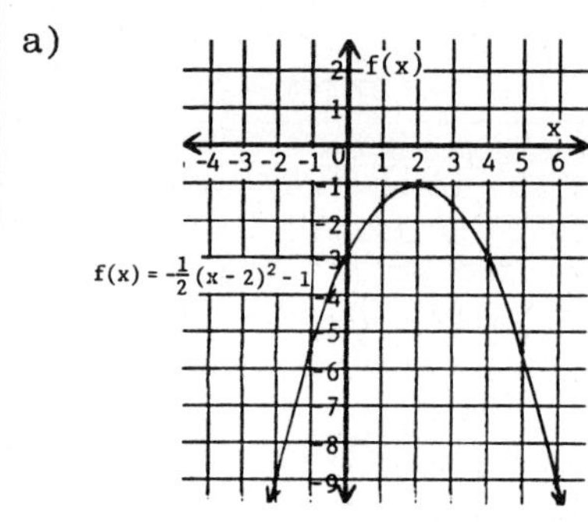

b)

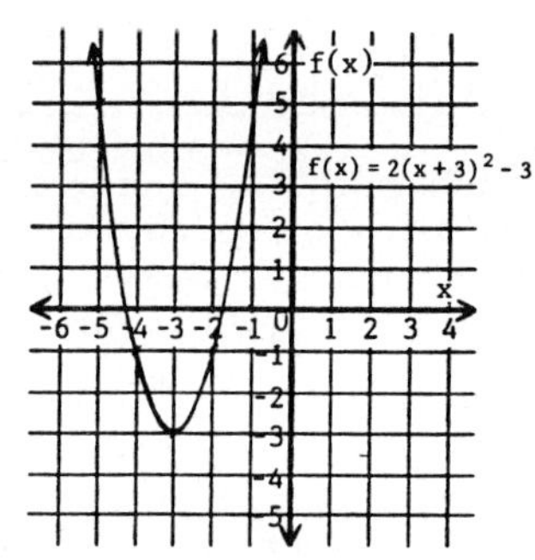

17. The function $f(x) = x^2 - 2$ can be written in the form $f(x) = (x - h)^2 + k$ by substituting 0 for h. That is:

$$f(x) = x^2 - 2 \quad \text{can be written} \quad f(x) = (x - 0)^2 - 2$$

We graphed $f(x) = x^2 - 2$ below. Notice that the vertex is (0,2), the axis of symmetry is the y-axis, and the parabola is shifted 2 units down. Graph $f(x) = -2x^2 + 3$ in the space provided.

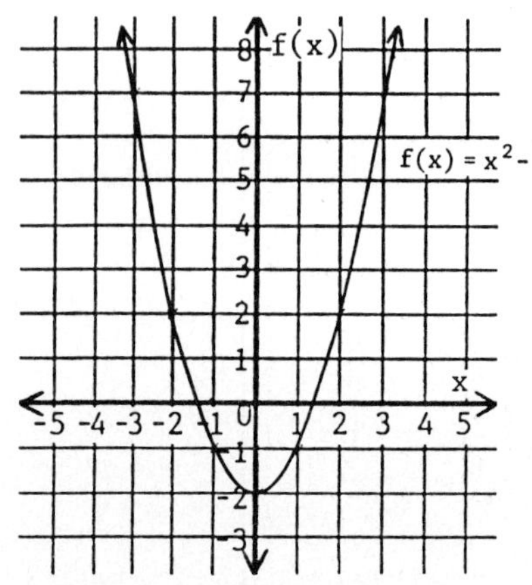

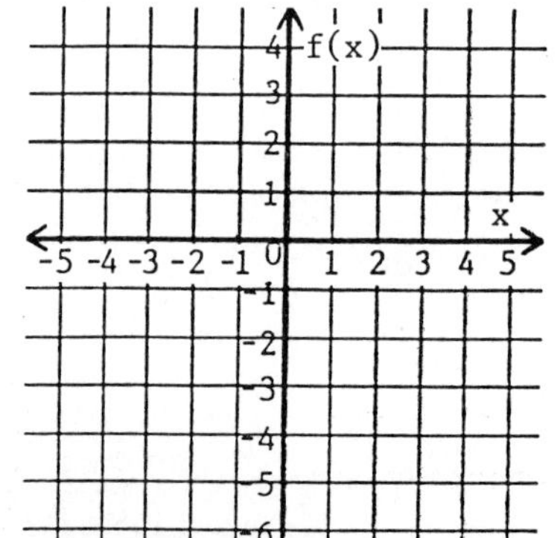

Answer To Frame 17:

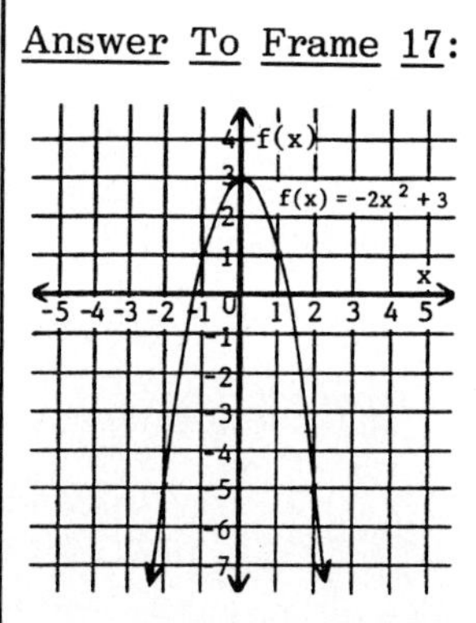

9-2 GRAPHING FUNCTIONS OF THE FORM: $f(x) = ax^2 + bx + c$

To graph a quadratic function in the form $f(x) = ax^2 + bx + c$, we can convert it to the form $f(x) = a(x - h)^2 + k$ by completing the square. We will discuss the method in this section.

---

18. To graph $f(x) = x^2 - 4x + 1$, we can convert it to the form $f(x) = a(x - h)^2 + k$ by completing the square. We did so below.

$$f(x) = x^2 - 4x + 1$$
$$f(x) = (x^2 - 4x) + 1$$

We must complete the square within the parentheses. Since half of -4 is -2 and $(-2)^2 = 4$, we add 4 - 4 or 0 inside the parentheses. We get:

$$f(x) = (x^2 - 4x + 4 - 4) + 1$$
$$f(x) = (x^2 - 4x + 4) - 4 + 1$$
$$f(x) = (x - 2)^2 - 3$$

Therefore, the parabola opens upward, has its vertex at (2,-3) and has x = 2 as its axis of symmetry. The graph is shown below.

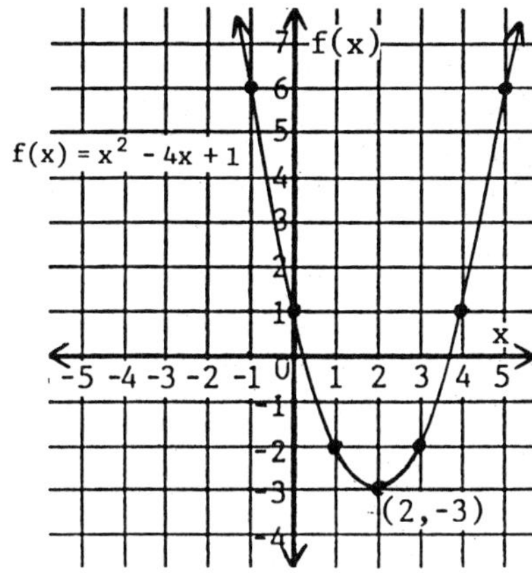

---

19. Let's use the method in the last frame to graph $f(x) = x^2 + 6x + 10$.

a) Convert the function to the form $f(x) = a(x + h)^2 + k$.

b) The vertex is ___________.

c) The axis of symmetry is ___________.

d) Graph the function.

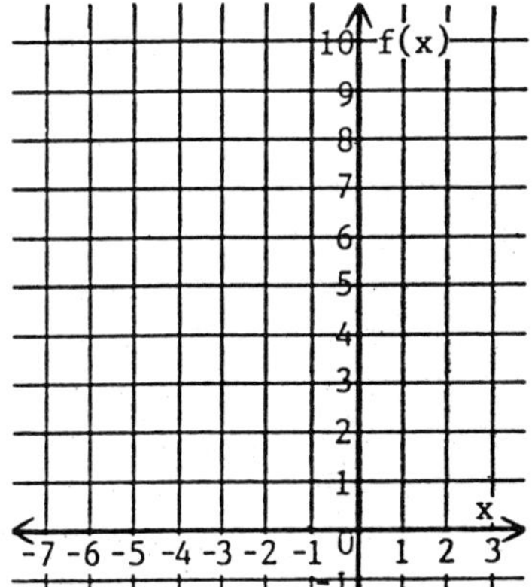

20. To graph $f(x) = 2x^2 + 4x - 1$, we complete the square on $2x^2 + 4x$. We begin by factoring out the 2 so that the coefficient of x^2 is "1".

$$f(x) = 2x^2 + 4x - 1$$
$$f(x) = 2(x^2 + 2x) - 1$$

Since half of 2 is "1" and $1^2 = 1$, we add $1 - 1$ inside the parentheses.

$$f(x) = 2(x^2 + 2x + 1 - 1) - 1$$

To remove the -1 from the parentheses, we must multiply it by 2 because of the distributive principle.

$$f(x) = 2(x^2 + 2x + 1) + 2(-1) - 1$$
$$f(x) = 2(x + 1)^2 - 3$$

Therefore, the parabola opens upward, has its vertex at (-1,-3), and has $x = -1$ as its axis of symmetry. The graph is shown below.

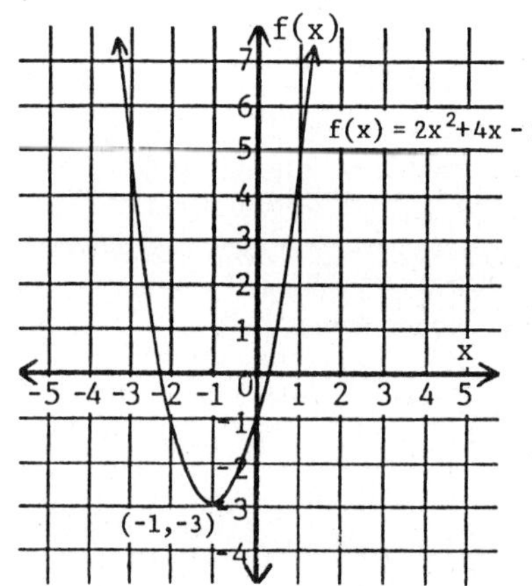

a) $f(x) = (x+3)^2 + 1$

b) (-3,1)

c) $x = -3$

d)

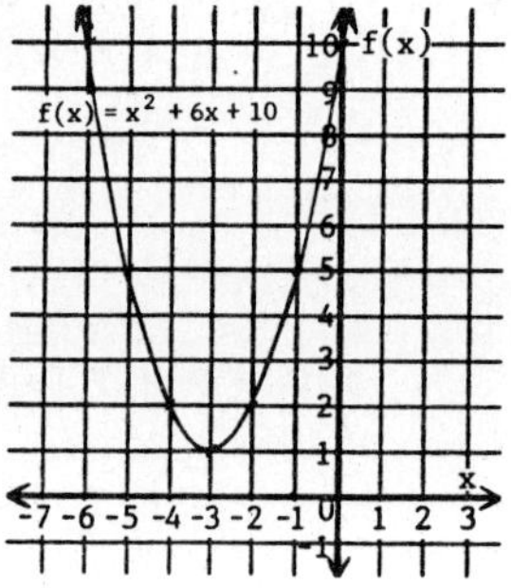

21. Let's use the method in the last frame to graph $f(x) = 2x^2 - 12x + 19$.

a) Convert the function to the form $f(x) = a(x - h)^2 + k$.

b) The vertex is ___________.

c) The axis of symmetry is ___________.

d) Graph the function.

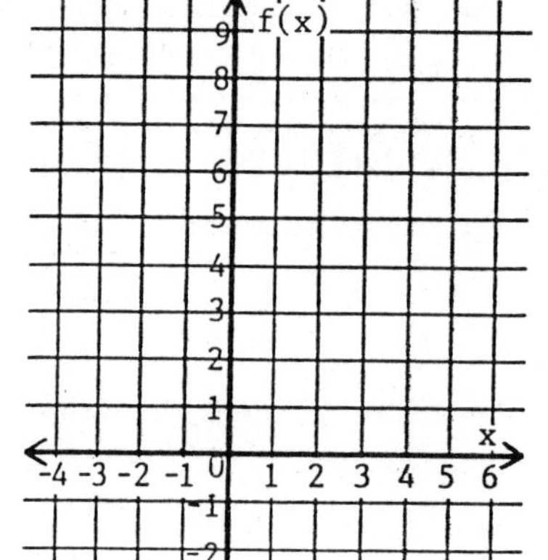

22. To graph $f(x) = -3x^2 + 12x - 8$, we complete the square on $-3x^2 + 12x$. We begin by factoring out the -3 so that the coefficient of x^2 is "1".

$$f(x) = -3x^2 + 12x - 8$$
$$f(x) = -3(x^2 - 4x) - 8$$

Since half of -4 is -2 and $(-2)^2 = 4$, we add 4 - 4 inside the parentheses.

$$f(x) = -3(x^2 - 4x + 4 - 4) - 8$$

To remove the -4 from the parentheses, we must multiply it by -3 because of the distributive principle.

$$f(x) = -3(x^2 - 4x + 4) - 3(-4) - 8$$
$$f(x) = -3(x - 2)^2 + 4$$

Therefore, the parabola opens downward, has its vertex at (2,4), and has x = 2 as its axis of symmetry. The graph is shown below.

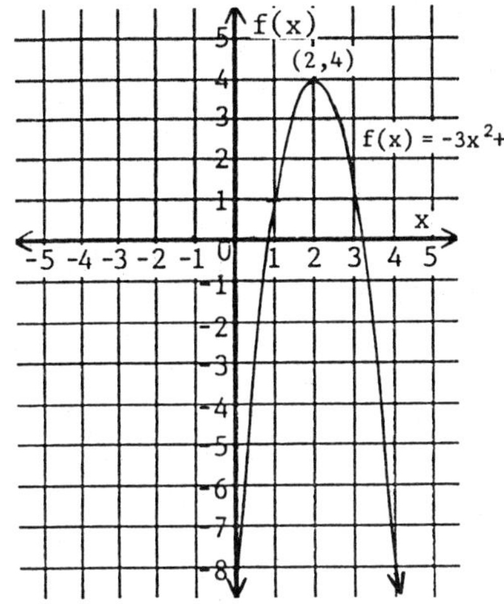

a) $f(x) = 2(x - 3)^2 + 1$

b) (3,1)

c) x = 3

d)

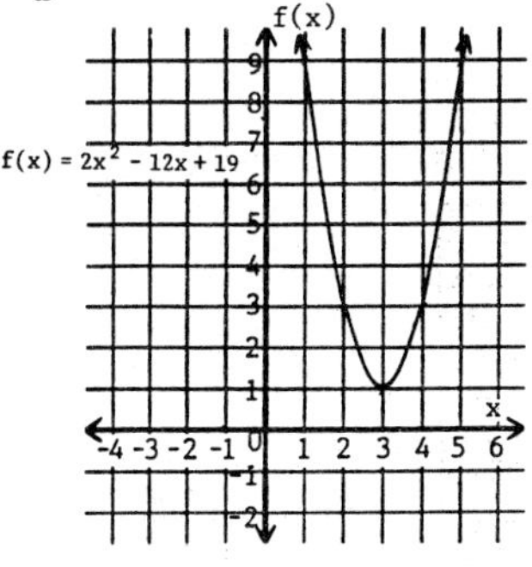

---

23. Let's use the method in the last frame to graph $f(x) = -2x^2 - 4x - 3$.

a) Convert the function to the form $f(x) = a(x - h)^2 + k$.

b) The vertex is ____________.

c) The axis of symmetry is ____________.

d) Graph the function.

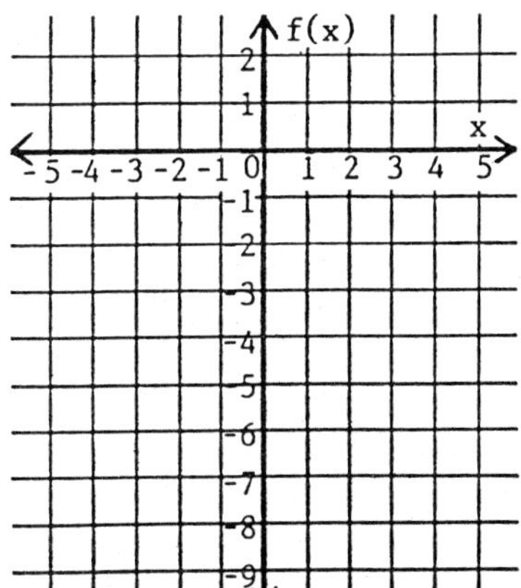

Answers To Frame 23:

a) $f(x) = -2(x + 1)^2 - 1$

b) (-1,-1)

c) $x = -1$

d) 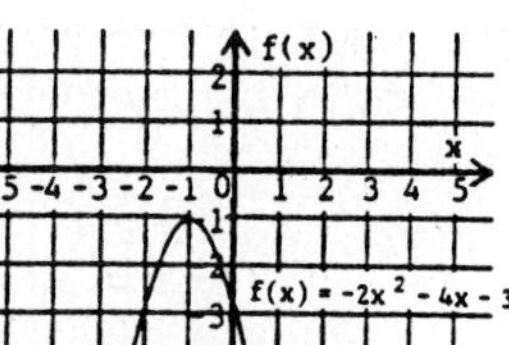

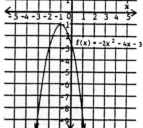

SELF-TEST 31 (pages 444-456)

State whether each parabola opens "upward" or "downward".

1. $y = -3x^2 + 1$ __________

2. $y = \frac{1}{2}(x - 1)^2 - 3$ __________

For $f(x) = 2(x + 3)^2 - 4$:

3. the vertex is __________.

4. the axis of symmetry is __________.

Graph each function.

5. $f(x) = x^2 + 2$

6. $f(x) = -\frac{1}{2}(x - 2)^2$

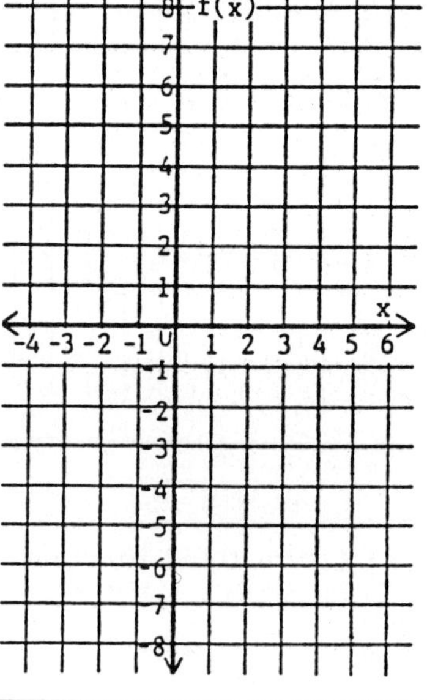

Given $f(x) = 2x^2 + 4x - 1$

7. Convert the function to the form $f(x) = a(x - h)^2 + k$.

8. Graph the function.

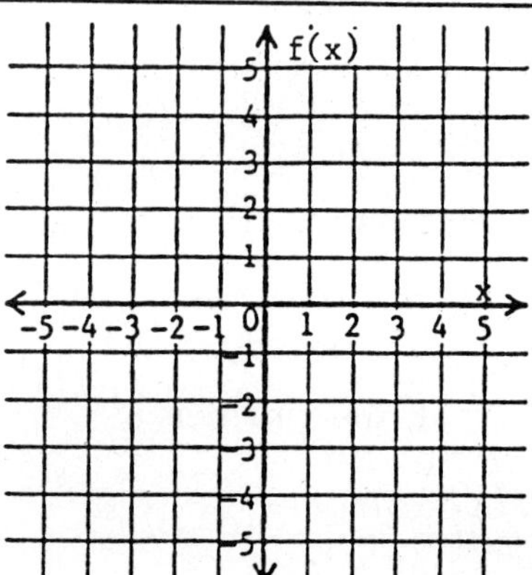

ANSWERS TO SELF-TEST 31:

1. downward
2. upward
3. (-3,-4)
4. x = -3

5-6.

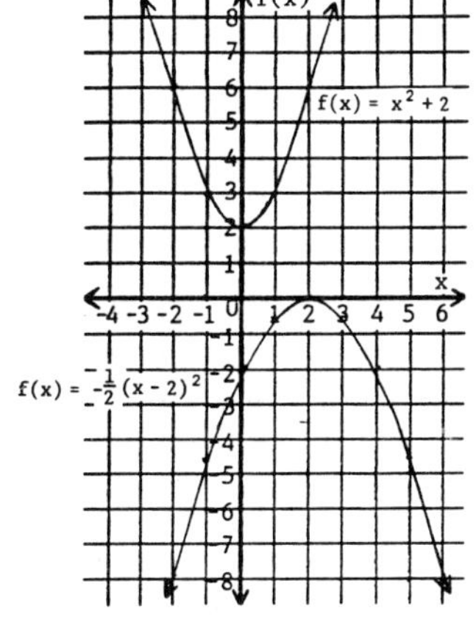

7. $f(x) = 2(x + 1)^2 - 3$

8.

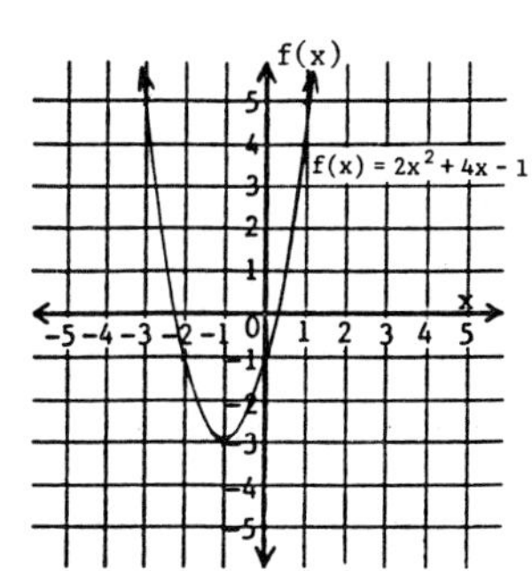

9-3 INTERCEPTS

The points at which a parabola crosses the x-axis are called the x-intercepts. We will discuss x-intercepts in this section.

24. The graph of $f(x) = x^2 - x - 6$ is shown at the right. The points at which the parabola crosses the x-axis are called the x-intercepts. From the graph, you can see that the x-intercepts are (-2,0) and (3,0).

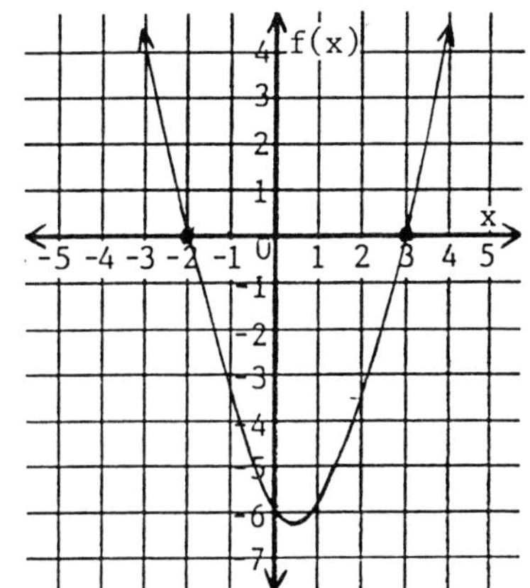

Since f(x) is always 0 on the x-axis, we can find the x-intercepts from the equation by setting f(x) = 0. We get:

$$f(x) = x^2 - x - 6$$
$$0 = x^2 - x - 6$$
$$0 = (x + 2)(x - 3)$$

x + 2 = 0	x - 3 = 0
x = -2	x = 3

Therefore, the x-intercepts are (-2,0) and (3,0). We sometimes refer to the x-coordinates as the intercepts. We say that the x-intercepts are -2 and 3.

Find the x-intercepts for the function below.

$$f(x) = x^2 + 3x - 10$$

25. The graph of $f(x) = x^2 - 4x$ is shown at the right. To find the x-intercepts, we let $f(x) = 0$ and factored below.

$$f(x) = x^2 - 4x$$
$$0 = x^2 - 4x$$
$$0 = x(x - 4)$$
$$x = 0 \text{ and } 4$$

Therefore, the x-intercepts are (0,0) and (4,0).

Find the x-intercepts for these.

a) $f(x) = x^2 - x$ b) $f(x) = x^2 + 5x$

(-5,0) and (2,0)

---

26. The graph of $f(x) = x^2 + 4x + 1$ is shown at the right. We have to use the quadratic formula to find the intercepts.

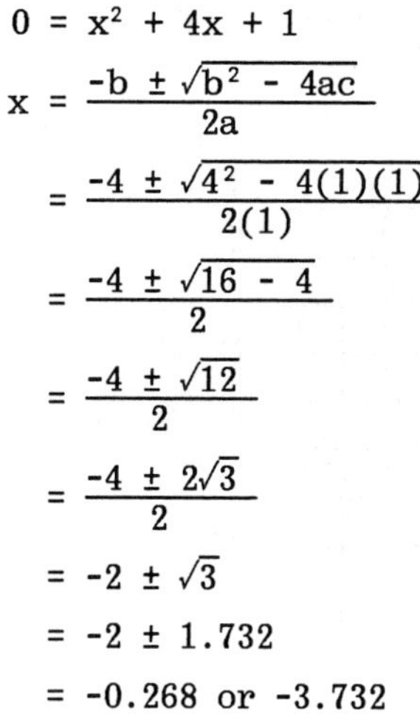

$$0 = x^2 + 4x + 1$$
$$x = \frac{-b \pm \sqrt{b^2 - 4ac}}{2a}$$
$$= \frac{-4 \pm \sqrt{4^2 - 4(1)(1)}}{2(1)}$$
$$= \frac{-4 \pm \sqrt{16 - 4}}{2}$$
$$= \frac{-4 \pm \sqrt{12}}{2}$$
$$= \frac{-4 \pm 2\sqrt{3}}{2}$$
$$= -2 \pm \sqrt{3}$$
$$= -2 \pm 1.732$$
$$= -0.268 \text{ or } -3.732$$

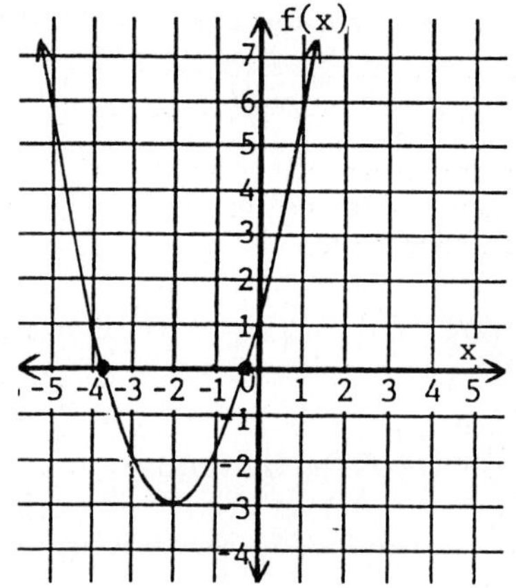

Therefore, the x-intercepts are (-3.732,0) and (-0.268,0), or approximately (-3.7,0) and (-0.3,0).

Use the quadratic formula to find the x-intercepts for this function.

$$f(x) = x^2 - 2x - 1$$

a) (0,0) and (1,0)

b) (0,0) and (-5,0)

27. The graph of $f(x) = x^2 + 4x + 4$ is shown at the right. There is only one x-intercept. It is at (-2,0). We can see what that means algebraically by using the factoring method.

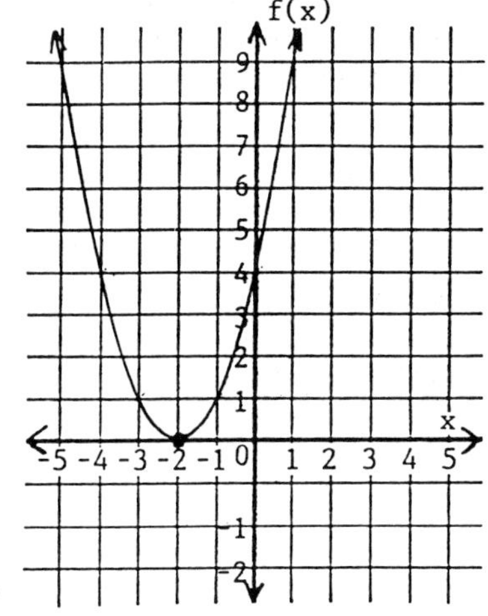

$$0 = x^2 + 4x + 4$$

$$0 = (x + 2)(x + 2)$$

$x + 2 = 0$	$x + 2 = 0$
$x = -2$	$x = -2$

Since $x^2 + 4x + 4$ is a perfect square, we get -2 as a double root.

Find the one x-intercept for this function.

$$f(x) = 4x^2 - 12x + 9$$

$(1 + \sqrt{2}, 0)$ and $(1 - \sqrt{2}, 0)$

or

(2.414,0) and (-0.414,0)

---

28. We graphed $f(x) = x^2 - 4x + 5$ at the right. There is no x-intercept. We can see what that means algebraically by using the quadratic formula.

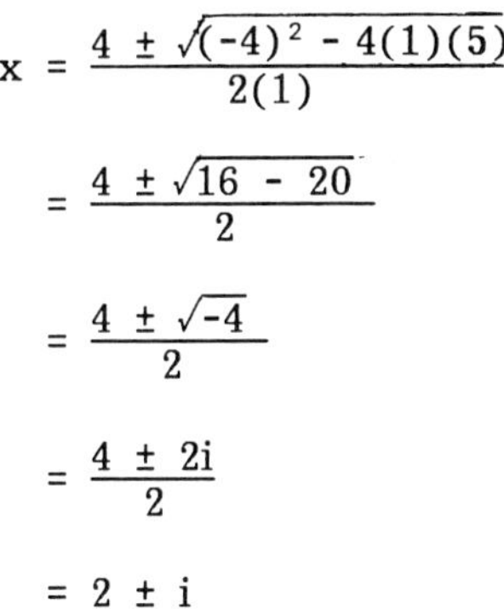

$$x = \frac{4 \pm \sqrt{(-4)^2 - 4(1)(5)}}{2(1)}$$

$$= \frac{4 \pm \sqrt{16 - 20}}{2}$$

$$= \frac{4 \pm \sqrt{-4}}{2}$$

$$= \frac{4 \pm 2i}{2}$$

$$= 2 \pm i$$

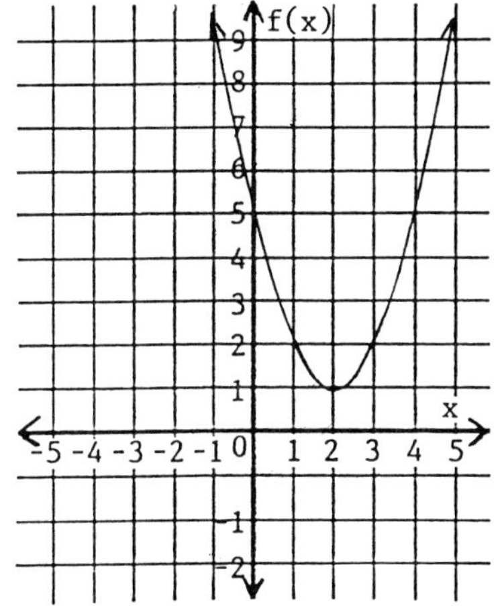

When there is no x-intercept, we get complex numbers as the solutions for x.

Find the x-intercepts for this function.

$$f(x) = x^2 - 6x + 10$$

$\left(\frac{3}{2}, 0\right)$

29. To find the x-intercepts of the function below, we began by squaring the binomial and simplifying to put it in the form $f(x) = ax^2 + bx + c$. Then we let $f(x) = 0$ and used the quadratic formula. Find the x-intercepts of the other function.

$$f(x) = 2(x - 1)^2 - 4$$
$$= 2(x^2 - 2x + 1) - 4$$
$$= 2x^2 - 4x + 2 - 4$$
$$= 2x^2 - 4x - 2$$

$$0 = 2x^2 - 4x - 2$$
$$0 = x^2 - 2x - 1$$

$$x = \frac{2 \pm \sqrt{(-2)^2 - 4(1)(-1)}}{2(1)}$$
$$= \frac{2 \pm \sqrt{4 + 4}}{2}$$
$$= \frac{2 + \sqrt{8}}{2}$$
$$= \frac{2 \pm 2\sqrt{2}}{2}$$
$$= 1 \pm \sqrt{2}$$
$$= 1 \pm 1.414$$
$$= 2.414 \text{ and } -0.414$$

The x-intercepts are (2.414,0) and (-0.414,0) or approximately (2.4,0) and (-0.4,0).

$$f(x) = (x + 2)^2 - 1$$

There are none, since:

$$x = 3 \pm i$$

---

30. In $f(x) = 3(x - 1)^2$, $k = 0$. Therefore, we can find its x-intercepts in two different ways. We did so below. At the left, we squared the binomial and then used the factoring method. At the right, we used the square-root method.

$$f(x) = 3(x - 1)^2$$
$$= 3(x^2 - 2x + 1)$$
$$= 3x^2 - 6x + 3$$

$$0 = 3x^2 - 6x + 3$$
$$0 = x^2 - 2x + 1$$
$$0 = (x - 1)(x - 1)$$
$$x = 1 \text{ and } 1$$

$$f(x) = 3(x - 1)^2$$
$$0 = 3(x - 1)^2$$
$$\frac{0}{3} = (x - 1)^2$$
$$0 = (x - 1)^2$$
$$0 = x - 1$$
$$x = 1$$

$f(x) = 3(x - 1)^2$ has one x-intercept at (1,0). That intercept is also the vertex of the parabola.

(-1,0) and (-3,0)

Continued on following page.

30. Continued

Find the x-intercept for these.

a) $f(x) = (x - 5)^2$

b) $f(x) = 4(x + 3)^2$

---

a) (5,0) b) (-3,0)

31. We used the square-root method to find the x-intercepts below.

$f(x) = x^2 - 4$	$f(x) = x^2 + 4$
$0 = x^2 - 4$	$0 = x^2 + 4$
$x^2 = 4$	$x^2 = -4$
$x = \pm 2$	$x = \pm 2i$

$f(x) = x^2 - 4$ has x-intercepts at (-2,0) and (2,0).
$f(x) = x^2 + 4$ does not have x-intercepts.

Find the x-intercept of these.

a) $f(x) = \frac{1}{2}x^2$

b) $f(x) = 3x^2 + 3$

---

a) (0,0)

b) There are none, since:

$x = \pm i$

9-4 MAXIMUM AND MINIMUM VALUES

For a quadratic function, the value of $f(x)$ at the vertex is either a maximum value or a minimum value. We will discuss maximum and minimum values in this section.

---

32. We graphed $f(x) = 2(x - 3)^2 - 4$ at the right. The vertex is the point $(3,-4)$. The parabola opens upward because $a > 0$.

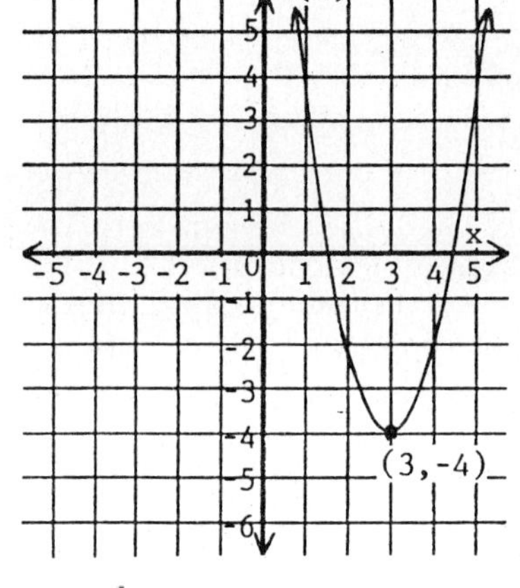

When a parabola opens upward, the function value $f(x)$ at the vertex is a <u>minimum</u> value. That is, the value of $f(x)$ at the vertex is <u>less than</u> the value of $f(x)$ at any other point. The minimum function value is <u>k</u>. For $f(x) = 2(x - 3)^2 - 4$, the minimum function value is -4.

Give the minimum function value for each of these.

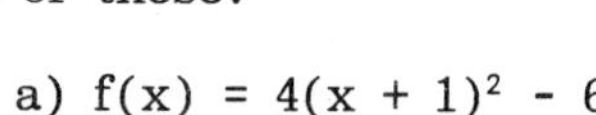

a) $f(x) = 4(x + 1)^2 - 6$ b) $f(x) = \frac{1}{2}(x - 4)^2 + 5$

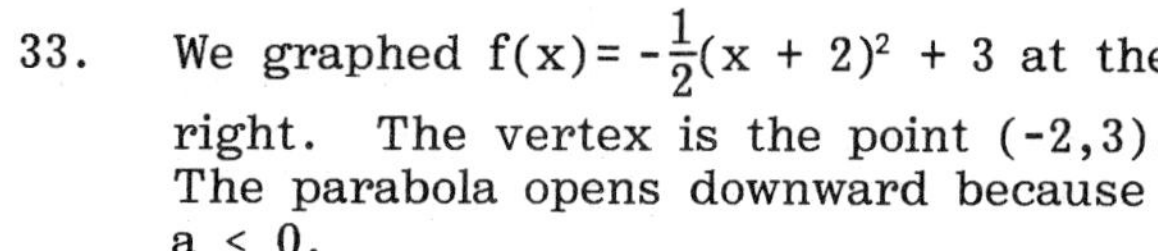

33. We graphed $f(x) = -\frac{1}{2}(x + 2)^2 + 3$ at the right. The vertex is the point $(-2,3)$. The parabola opens downward because $a < 0$.

When a parabola opens downward, the function value $f(x)$ at the vertex is a <u>maximum</u> value. That is, the value of $f(x)$ at the vertex is <u>greater than</u> the value of $f(x)$ at any other point. The maximum function value is <u>k</u>. For $f(x) = -\frac{1}{2}(x + 2)^2 + 3$, the maximum function value is 3.

Give the maximum function value of each of these.

a) $f(x) = -3(x - 1)^2 + 6$ b) $f(x) = -\frac{1}{2}(x + 5)^2 - 1$

a) -6 b) 5

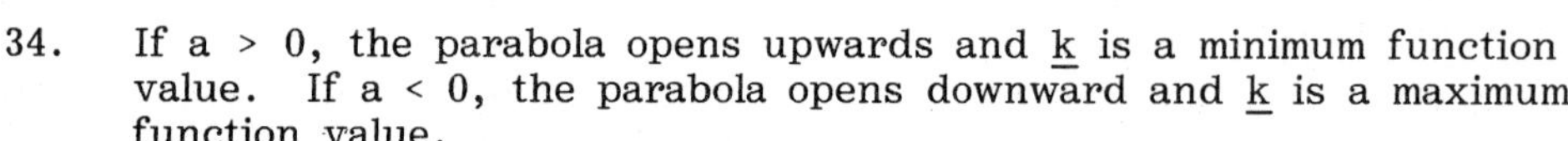

34. If $a > 0$, the parabola opens upwards and <u>k</u> is a minimum function value. If $a < 0$, the parabola opens downward and <u>k</u> is a maximum function value.

Give the maximum or minimum function value for these.

a) $f(x) = 2(x + 3)^2 - 3$ c) $f(x) = -\frac{1}{2}(x - 1)^2 - 4$

b) $f(x) = -2(x - 3)^2 + 5$ d) $f(x) = \frac{1}{2}(x + 1)^2 + 7$

a) 6 b) -1

35. When k is 0, the maximum or minimum function value is 0. For example:

For $f(x) = 2x^2$, the minimum function value is 0.

For $f(x) = -3(x + 1)^2$, the maximum function value is 0.

Give the maximum or minimum function value for these.

a) $f(x) = -x^2$

b) $f(x) = (x + 4)^2 - 1$

c) $f(x) = 4(x - 3)^2$

d) $f(x) = -(x - 3)^2 + 2$

a) Minimum: -3

b) Maximum: 5

c) Maximum: -4

d) Minimum: 7

---

36. To find the maximum or minimum value for $f(x) = -3x^2 - 6x + 2$, we must complete the square to find k.

$$f(x) = -3x^2 - 6x + 2$$
$$= -3(x^2 + 2x) + 2$$
$$= -3(x^2 + 2x + 1 - 1) + 2$$
$$= -3(x^2 + 2x + 1) - 3(-1) + 2$$
$$= -3(x + 1)^2 + 5$$

The maximum function value is 5.

Find the maximum or minimum function value for these.

a) $f(x) = x^2 + 6x + 7$

b) $f(x) = -2x^2 + 4x - 5$

a) Maximum: 0

b) Minimum: -1

c) Minimum: 0

d) Maximum: 2

---

37. Some word problems can be solved by finding a maximum value. An example is shown.

Find the maximum area of a rectangle if the sum of the length and width is 40 meters. What length and width give that area?

Letting L equal the length and 40 - L equal the width, we can set up a quadratic function. We get:

$$A = LW$$
$$= L(40 - L)$$
$$= 40L - L^2$$
$$= -L^2 + 40L$$

Continued on following page.

a) Minimum: -2

$f(x) = (x + 3)^2 - 2$

b) Maximum: -3

$f(x) = -2(x - 1)^2 - 3$

37. Continued

Then we can complete the square to find the maximum area and the value of L at which it occurs.

$$A = -(L^2 - 40L + 400 - 400)$$
$$= -(L^2 - 40L + 400) - (-400)$$
$$= -(L - 20)^2 + 400$$

The maximum area is 400. Since the vertex is (20,400), the maximum area occurs when L = 20 and W = 20. That is, the maximum area occurs when the rectangle is a square.

Using the same method, solve this problem.

Of all pairs of numbers whose sum is 24, find the pair with the maximum product. Let x and 24 - x equal the number.

---

38. If a model rocket is fired vertically upward at an initial velocity v_0, its distance s (in feet) above the ground at time t (in seconds) is given by the formula below.

$$s = v_0 t - 16t^2$$

We can use the formula to solve maximum height problems. An example is shown. Solve the other problem.

Find the maximum height the rocket reaches and the time needed to reach that height if v_0 = 160 ft/sec.

$$s = v_0 t - 16t^2$$
$$= 160t - 16t^2$$
$$= -16t^2 + 160t$$
$$= -16(t^2 - 10t)$$
$$= -16(t^2 - 10t + 25 - 25)$$
$$= -16(t^2 - 10t + 25) - 16(-25)$$
$$= -16(t - 5)^2 + 400$$

Since the vertex is (5,400), the rocket reaches a maximum height of 400 ft in 5 sec.

Continued on following page.

The pair is 12 and 12.

Their product is 144.

38. Continued

Find the maximum height the rocket reaches and the time needed to reach that height if v_o = 96 ft/sec.

It reaches 144 ft in 3 sec.

9-5 HORIZONTAL PARABOLAS

The graph of any equation of the form $x = ay^2 + by + c$ is a horizontal parabola. We will discuss parabolas of that type in this section.

39. Up to this point, we have discussed parabolas whose equations have the form $y = ax^2 + bx + c$. The parabolas have been vertical. If we interchange x and y, we get an equation of the following form.

$$x = ay^2 + by + c \qquad (a \neq 0)$$

The graph of any equation of the form above is a horizontal parabola. Since the parabola is horizontal, the equation is not a function. As an example, we graphed $x = y^2$ below. Notice that the vertex is (0,0) and the axis of symmetry is the x-axis.

$x = y^2$

x	y
9	-3
4	-2
1	-1
0	0
1	1
4	2
9	3

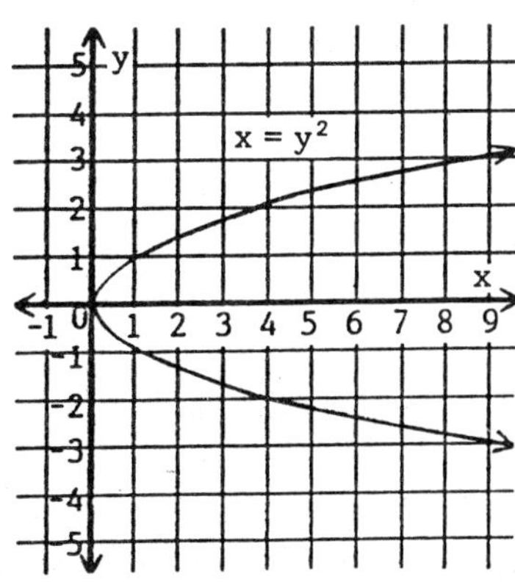

Continued on following page.

39. Continued

Graph $x = 2y^2$ below. It is easier to find pairs of values by substituting for y. Use the y-values provided in the table. Notice that the parabola is narrower than the graph of $x = y^2$.

$x = 2y^2$

x	y
	-2
	-1
	0
	1
	2

Answer To Frame 39:

x	y
8	-2
2	-1
0	0
2	1
8	2

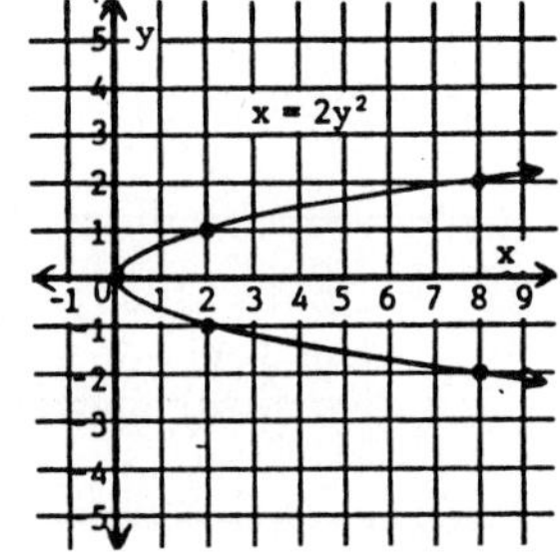

40. To graph $x = (y - 2)^2 - 3$, it is easier to substitute values of y. We did so below and then graphed the equation. Notice that the vertex is (-3,2) and the axis of symmetry is the line y = 2.

$x = (y - 2)^2 - 3$

x	y
1	0
-2	1
-3	2
-2	3
1	4

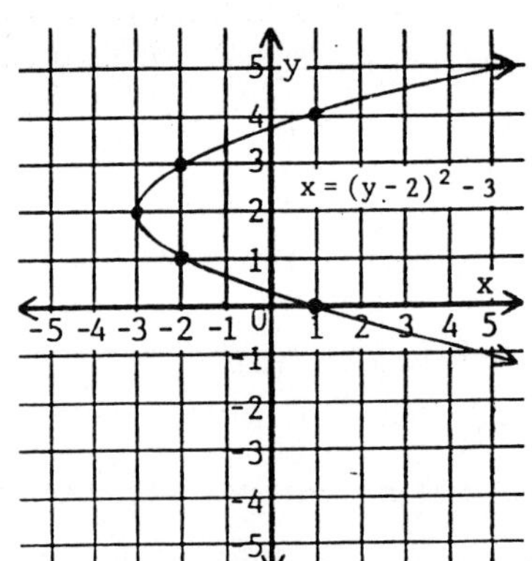

Graph $x = (y + 1)^2 + 2$ below. Substitute values for y. Notice that the vertex is (2,-1) and the axis of symmetry is the line y = -1.

$x = (y + 1)^2 + 2$

x	y

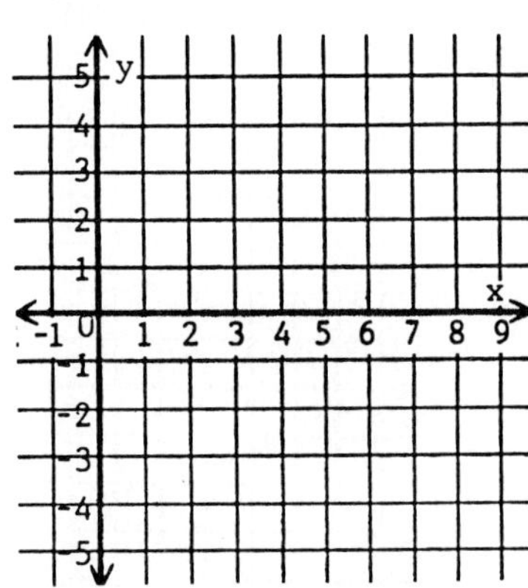

41. To graph $x = -2y^2 + 12y - 17$, we can complete the square on y to put it in the form $x = a(y - k)^2 + h$. We did so below.

$$\begin{aligned} x &= -2y^2 + 12y - 17 \\ &= -2(y^2 - 6y) - 17 \\ &= -2(y^2 - 6y + 9 - 9) - 17 \\ &= -2(y^2 - 6y + 9) - 2(-9) - 17 \\ &= -2(y - 3)^2 + 1 \end{aligned}$$

We graphed the equation below. Since a is negative, the parabola opens to the left.

$x = -2(y - 3)^2 + 1$

x	y
-7	1
-1	2
1	3
-1	4
-7	5

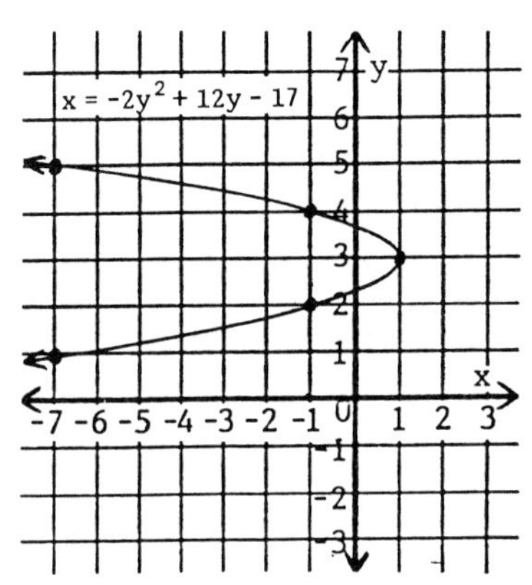

Use the graph for these:

a) The vertex is the point ________.

b) The axis of symmetry is the line ________.

Answer To Frame 40:

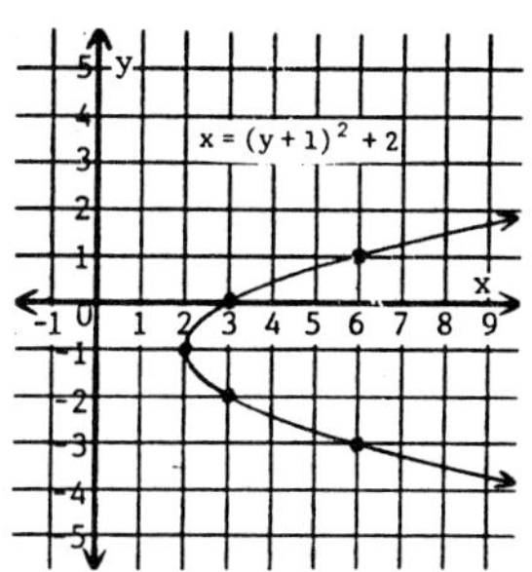

a) (1,3) b) y = 3

SELF-TEST 32 (pages 457-468)

Find the x-intercepts.

1. $f(x) = x^2 + 2x - 8$	2. $f(x) = (x + 5)^2$	3. $f(x) = x^2 + 9$

Give the maximum or minimum function value.

4. $f(x) = -3(x - 1)^2 + 4$

5. $f(x) = 2x^2 + 5$

6. $f(x) = (x + 5)^2 - 2$

7. $f(x) = -\frac{1}{2}x^2$

8. Find the maximum or minimum function value.

$$f(x) = 2x^2 + 8x + 5$$

9. Graph: $x = y^2 - 1$

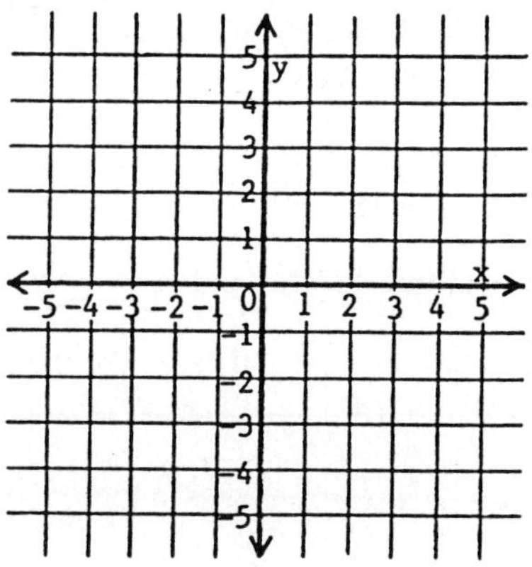

ANSWERS:

1. (-4,0) and (2,0)	4. Maximum: 4	7. Maximum: 0
2. (-5,0)	5. Minimum: 5	8. Minimum: -3
3. None	6. Minimum: -2	

9.

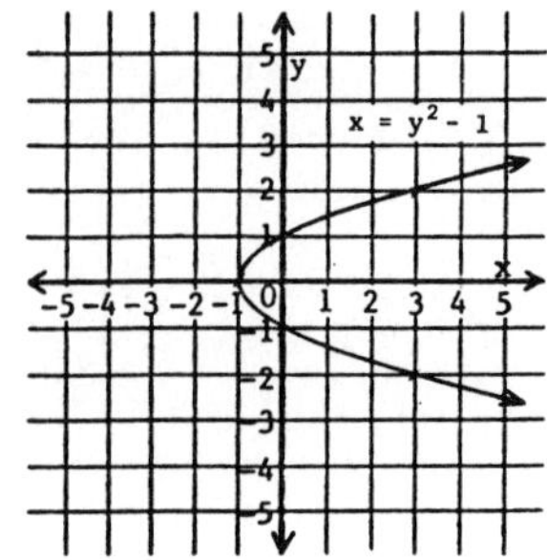

9-6 THE DISTANCE FORMULA

In this section, we will discuss a formula that can be used to find the distance between any two points on a graph.

42. If two points are on a vertical line, they have the same x-value. $d = |y_2 - y_1|$ can be used to find the distance between them.

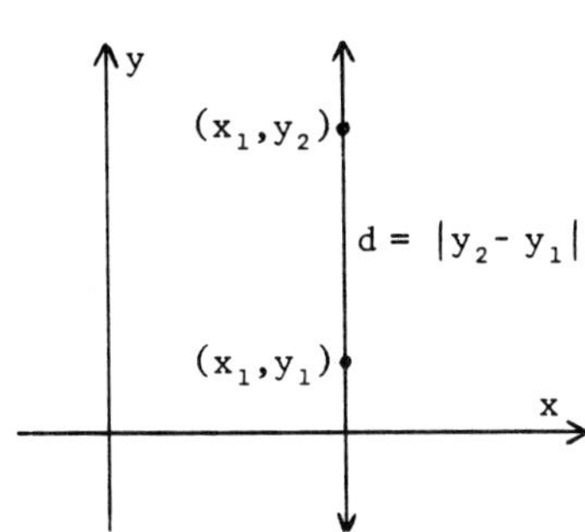

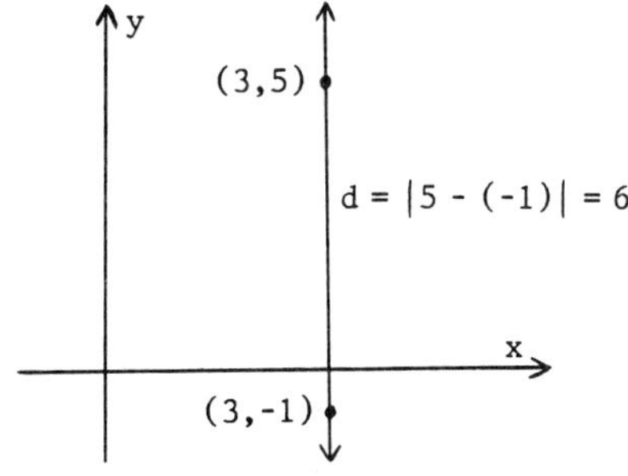

In $d = |y_2 - y_1|$, we use absolute value so that we can subtract either way. For example, we can then subtract either way at the right and get $d = 6$.

$$d = |5 - (-1)| = |6| = 6$$
$$d = |(-1) - 5| = |-6| = 6$$

Each pair of points below lies on a vertical line. Find the distance between them.

a) (-3,10) and (-3,2) b) (5,1) and (5,-9)

---

a) 8 b) 10

43. If two points are on a horizontal line, they have the same y-value. $d = |x_2 - x_1|$ can be used to find the distance between them.

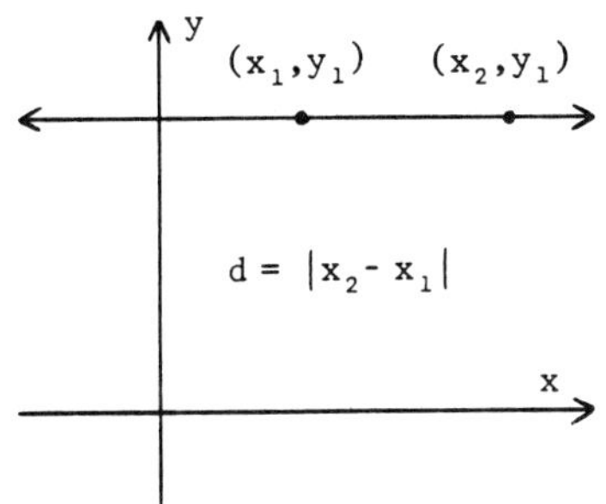

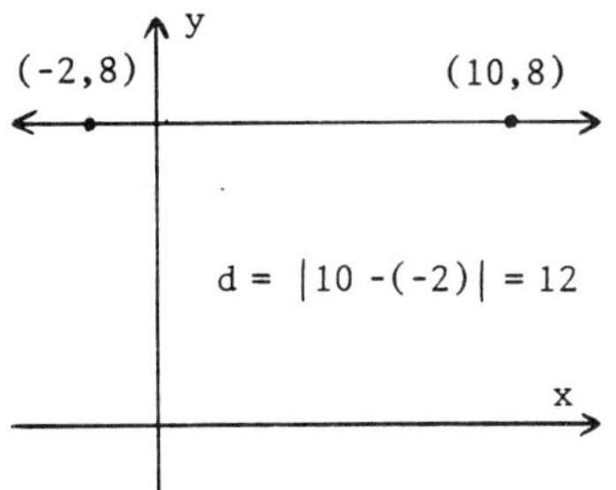

In $d = |x_2 - x_1|$, we use absolute value so that we can subtract either way. For example, we can then subtract either way and get $d = 12$.

$$d = |10 - (-2)| = |12| = 12$$
$$d = |(-2) - 10| = |-12| = 12$$

Continued on following page.

43. Continued

Each pair of points below lies on a horizontal line. Find the distance between them.

a) (1,-5) and (8,-5) b) (-6,3) and (0,3)

---

44. In the figure below, the points (x_1,y_1) and (x_2,y_2) do not lie on either a vertical or horizontal line. We completed a right triangle. The coordinates of the third corner are (x_2,y_1).

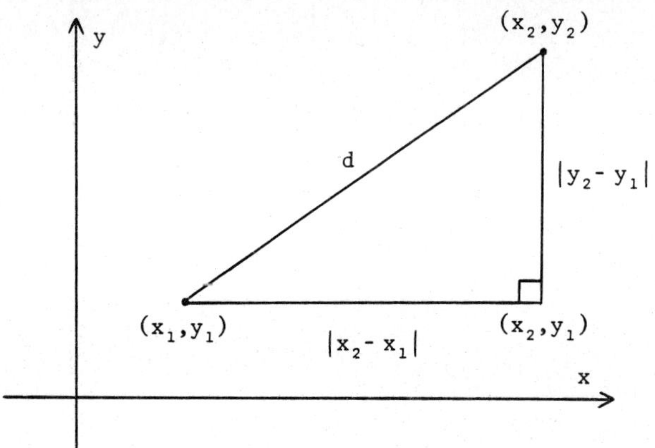

The lengths of the legs of the right triangle are $|x_2 - x_1|$ and $|y_2 - y_1|$. The length d of the hypotenuse is the distance between the two points. Using the Pythagorean Theorem, we get:

$$d^2 = |x_2 - x_1|^2 + |y_2 - y_1|^2$$

And since squares are always positive, we can drop the absolute value signs. We get:

$$d^2 = (x_2 - x_1)^2 + (y_2 - y_1)^2$$

By taking the principle square root of each side, we get the distance formula.

> The Distance Formula: distance between (x_1,y_1) and (x_2,y_2)
>
> $$d = \sqrt{(x_2 - x_1)^2 + (y_2 - y_1)^2}$$

Let's use the distance formula to find the distance between (2,3) and (5,-1). Substituting, we get:

$$d = \sqrt{(5 - 2)^2 + (-1 - 3)^2} = \sqrt{3^2 + (-4)^2} = \sqrt{25} = 5$$

Find the distance between each pair of points.

a) (-2,3) and (4,-5) b) (6,4) and (1,-8)

Answers (43): a) 7 b) 6

45. To find the distance between (-2,3) and (5,-1) below, we used the square root table.

$$d = \sqrt{[5 - (-2)]^2 + (-1 - 3)^2} = \sqrt{7^2 + (-4)^2} = \sqrt{65} = 8.062$$

Find the distance between each pair of points.

a) (4,-7) and (-1,-5) b) (0,6) and 3,-3)

a) 10	b) 13

---

46. The formula can also be used to find the distance between two points on a vertical or horizontal line. An example is shown. Use the formula to find the distance between the other two points.

(4,3) and (4,8)

$$d = \sqrt{(4 - 4)^2 + (8 - 3)^2}$$
$$= \sqrt{0^2 + 5^2}$$
$$= \sqrt{25}$$
$$= 5$$

(-1,-3) and (6,-3)

a) $\sqrt{29} = 5.385$

b) $\sqrt{90} = 9.487$

7

9-7 CIRCLES

In this section, we will discuss the equations of circles and graph some circles.

---

47. A circle is the set of all points in a plane which are a fixed distance from a fixed point. The fixed point is called the center and the fixed distance is called the radius.

For the circle at the right, the center is (a,b), the radius is r, and (x,y) is any point on the circle. Using the distance formula with (a,b) and (x,y), we get:

$$\sqrt{(x - a)^2 + (y - b)^2} = r$$

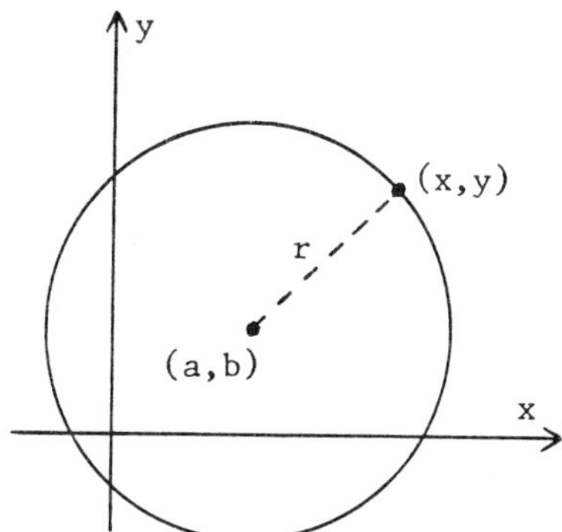

Continued on following page.

47. Continued

Squaring both sides, we get the following equation for a circle.

<u>Equation Of A Circle</u>: with center (a,b) and radius <u>r</u>
$(x - a)^2 + (y - b)^2 = r^2$ (Standard form)

Given the equation of a circle, we can identify the center and radius.

For $(x - 3)^2 + (y - 1)^2 = 16$: the center is (3,1)
the radius is 4, since $\sqrt{16} = 4$

For $(x - 2)^2 + (y - 5)^2 = 49$: a) the center is ________.

b) the radius is ________.

a) (2,5)

b) 7

48. We graphed the circle at the left below. Its center is (2,3) and its radius is 4. Graph the other circle.

$(x - 2)^2 + (y - 3)^2 = 16$

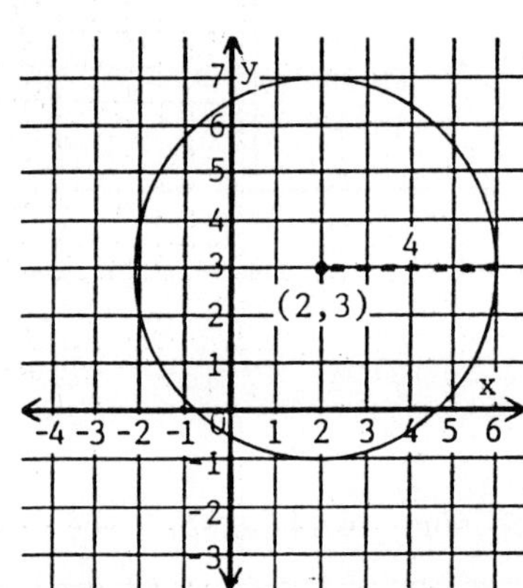

$(x - 1)^2 + (y - 2)^2 = 9$

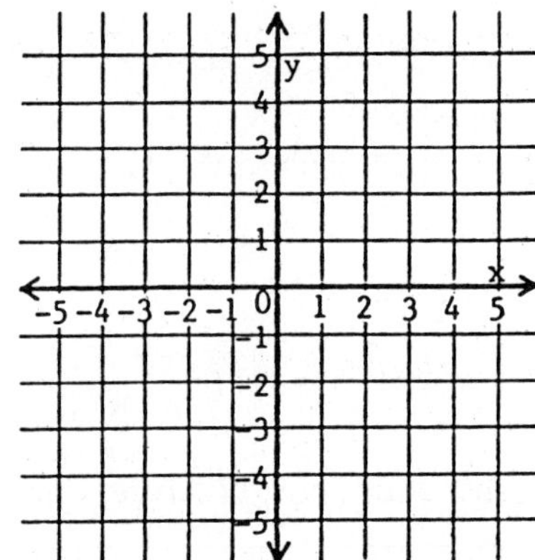

Answer To Frame 48:

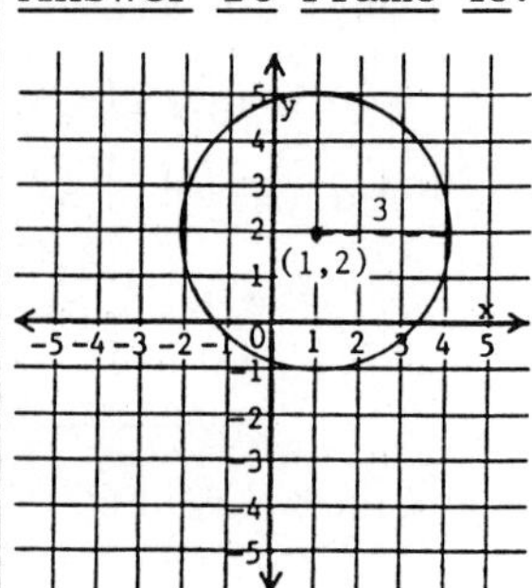

49. In the standard form of the equation of a circle, both binomials are subtractions. To find the center of the circle at the left below, we had to put it in standard form first. Find the center and radius of the other circle.

$(x - 5)^2 + (y + 2)^2 = 36$

$(x - 5)^2 + [y - (-2)]^2 = 36$

The center is (5,-2).

The radius is 6.

$(x + 1)^2 + (y - 7)^2 = 4$

a) The center is ________.

b) The radius is ________.

a) (-1,7)

b) 2

50. We graphed one circle below. Its center is (-2,1) and its radius is 3. Graph the other circle.

$(x + 2)^2 + (y - 1)^2 = 9$ $(x + 1)^2 + (y + 2)^2 = 16$

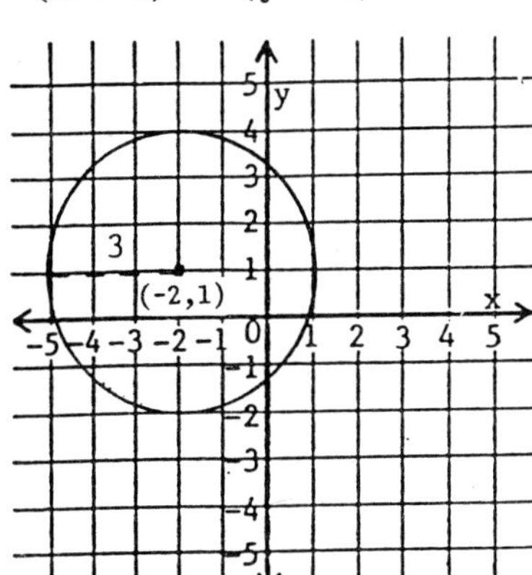

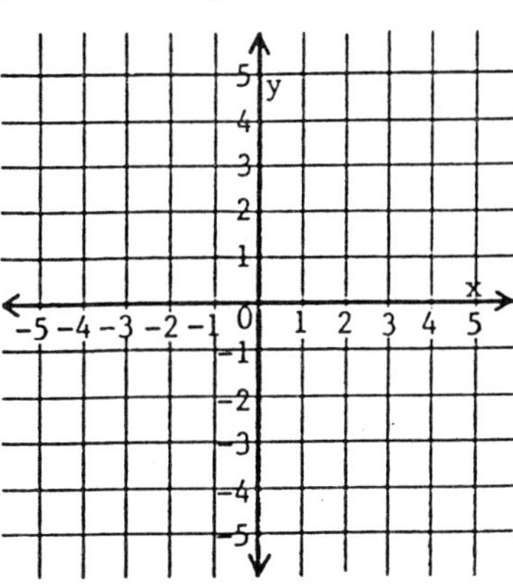

51. When the center of a circle is at the origin, the center (a,b) is (0,0). Therefore, circles of that type have a simpler equation.

> <u>Equation of circle with center at origin and radius r</u>
>
> $$x^2 + y^2 = r^2$$

For $x^2 + y^2 = 64$: the center is (0,0).
the radius is 8.

For $x^2 + y^2 = 81$: a) the center is ________.

b) the radius is ________.

Answer To Frame 50:

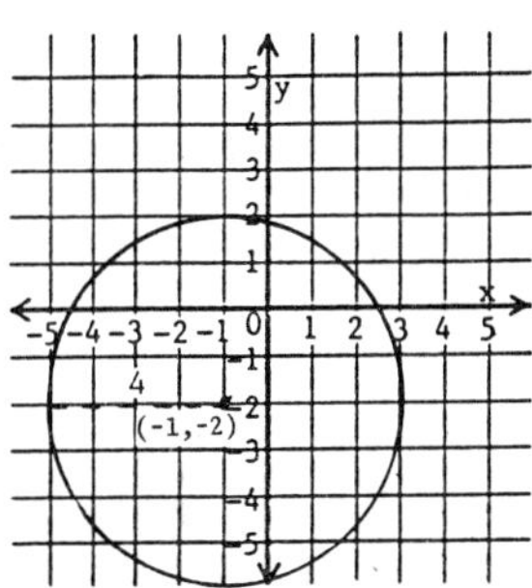

52. We graphed one circle below. Its center is (0,0) and its radius is 3. Graph the other circle.

$x^2 + y^2 = 9$ $x^2 + y^2 = 25$

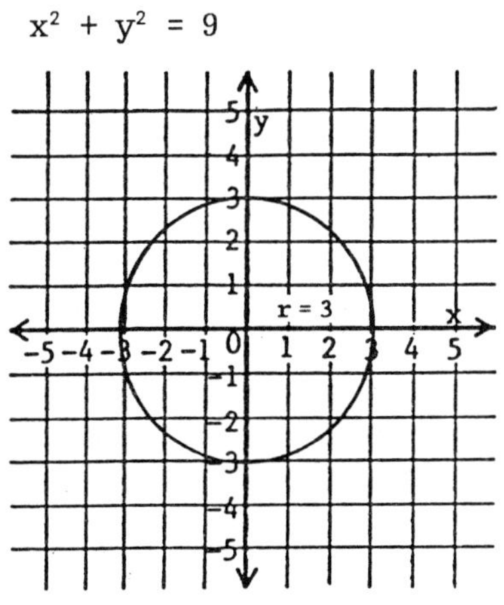

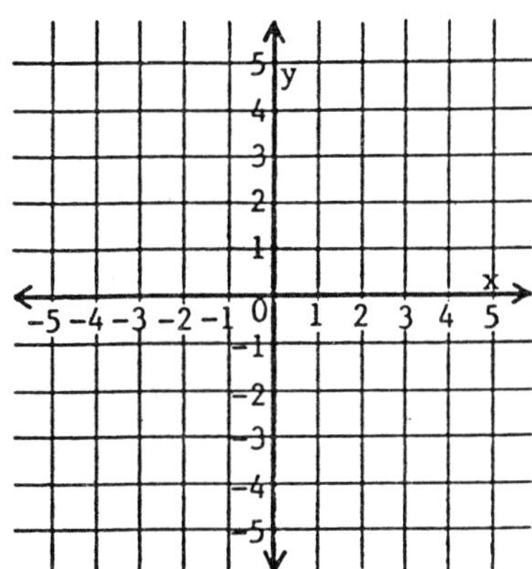

a) (0,0)

b) 9

53. To find the equation of the circle with center (4,-5) and radius $\sqrt{3}$, we can substitute those values into the standard form of the equation.

$$(x - a)^2 + (y - b)^2 = r^2$$

$$(x - 4)^2 + [y - (-5)]^2 = (\sqrt{3})^2$$

$$(x - 4)^2 + (y + 5)^2 = 3$$

Find the equation of each of the following circles.

a) center (1,6) and radius 5 ________________

b) center (-7,2) and radius 10 ________________

c) center (0,0) and radius $\sqrt{7}$ ________________

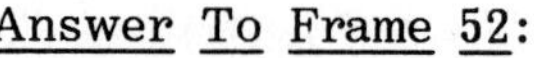
Answer To Frame 52:

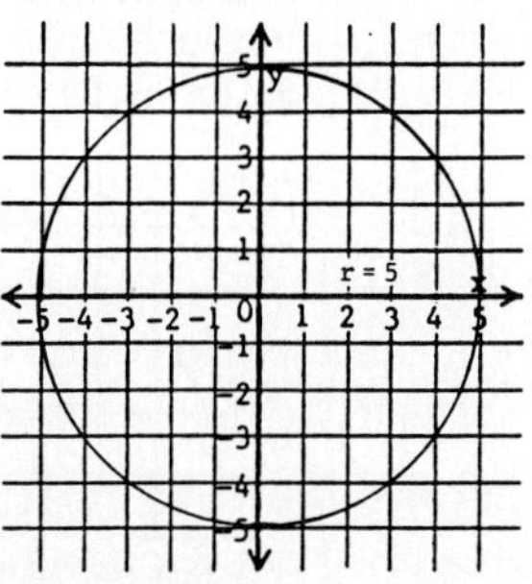

---

54. For the equation of a circle below, we squared the binomials and simplified.

$$(x - 2)^2 + (y - 3)^2 = 16$$

$$x^2 - 4x + 4 + y^2 - 6y + 9 = 16$$

$$x^2 + y^2 - 4x - 6y - 3 = 0$$

We can get back to the standard form of the equation by completing the square on x and y. We get:

$$(x^2 - 4x) + (y^2 - 6y) - 3 = 0$$

$$(x^2 - 4x + 4 - 4) + (y^2 - 6y + 9 - 9) - 3 = 0$$

$$(x^2 - 4x + 4) + (y^2 - 6y + 9) - 4 - 9 - 3 = 0$$

$$(x - 2)^2 + (y - 3)^2 = 16$$

Therefore, when an equation contains both x^2 and y^2 terms, it may be an equation of a circle. By completing the squares on x and y, show that the equation below is the equation of a circle. Identify the center and radius.

$$x^2 + y^2 - 10x - 2y + 17 = 0$$

a) $(x-1)^2+(y-6)^2=25$

b) $(x+7)^2+(y-2)^2=100$

c) $x^2 + y^2 = 7$

$(x - 5)^2 + (y - 1)^2 = 9$

The center is (5,1).

The radius is 3.

55. By completing the squares on x and y, show that the equation below is the equation of a circle. Identify the center and radius.

$$x^2 + y^2 + 12x - 8y + 27 = 0$$

$(x+6)^2 + (y-4)^2 = 25$

The center is (-6,4).

The radius is 5.

9-8 ELLIPSES

The orbits of satellites and the paths of planets around the sun are oval-shaped curves called ellipses. In this section, we will discuss the equations of ellipses with centers at the origin and we will graph some ellipses.

56. An ellipse is the set of all points in a plane such that the sum of the distances from two fixed points (called foci) is constant. For the ellipse below, the foci are at (c,0) and (-c,0), the x-intercepts are at (a,0) and (-a,0), and the y-intercepts are at (0,b) and (0,-b).

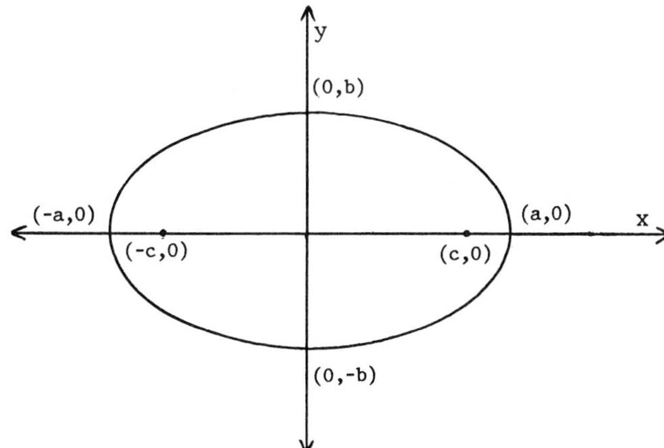

The standard form of the equation of an ellipse is given below.

Equation Of An Ellipse: with its center at the origin
$\frac{x^2}{a^2} + \frac{y^2}{b^2} = 1$ $a,b > 0$ (Standard Form)

Continued on following page.

56. Continued

As you can see, the x-intercepts and y-intercepts are given by a^2 and b^2 in the equation.

For $\frac{x^2}{4} + \frac{y^2}{25} = 1$: the x-intercepts are (2,0) and (-2,0).
the y-intercepts are (0,5) and (0,-5).

For $\frac{x^2}{16} + \frac{y^2}{9} = 1$: a) the x-intercepts are ________ and ________.

b) the y-intercepts are ________ and ________.

---

a) (4,0) and (-4,0)

b) (0,3) and (0,-3)

57. We graphed the ellipse below by plotting the intercepts and then sketching an ellipse through them.

$$\frac{x^2}{16} + \frac{y^2}{4} = 1$$

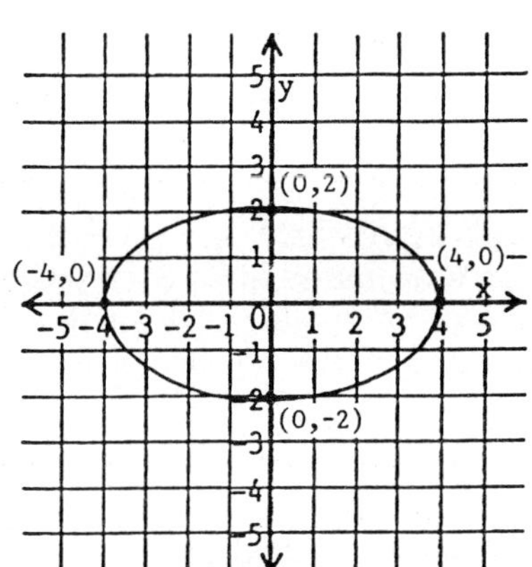

Use the same method to graph each ellipse below.

a) $\frac{x^2}{4} + \frac{y^2}{9} = 1$

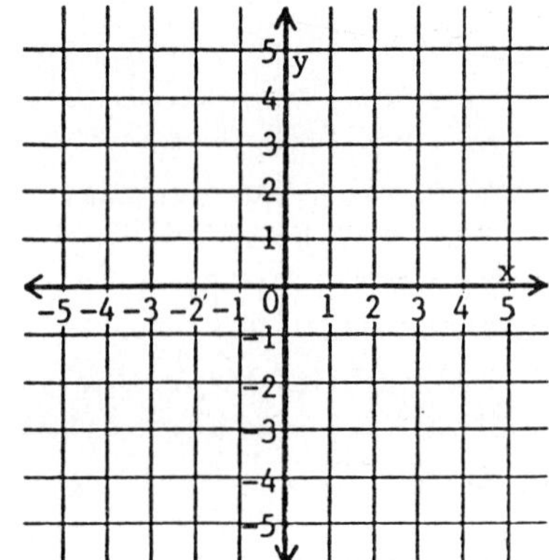

b) $\frac{x^2}{25} + \frac{y^2}{9} = 1$

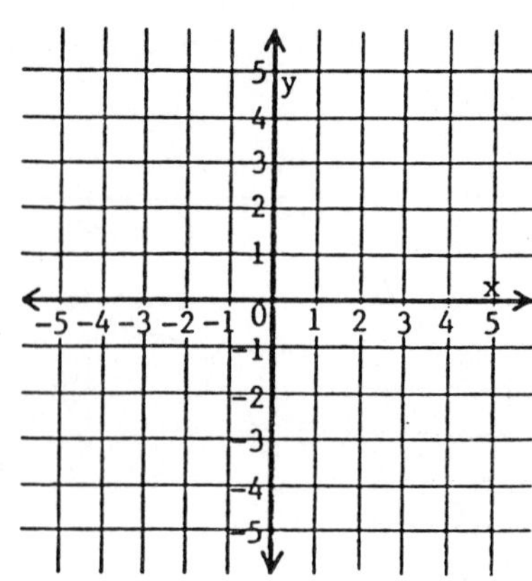

Answers To Frame 57:

a)

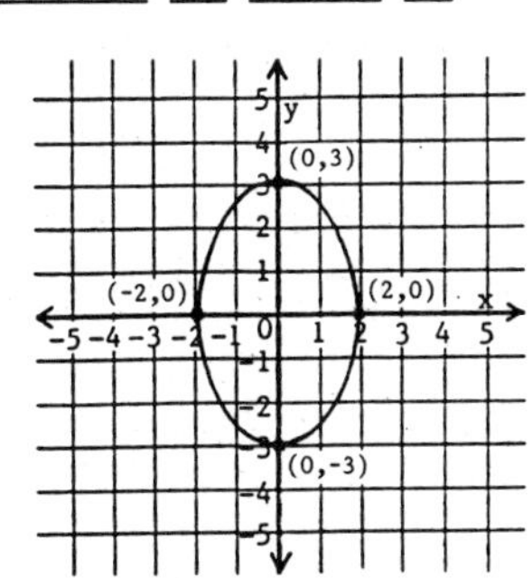

b)

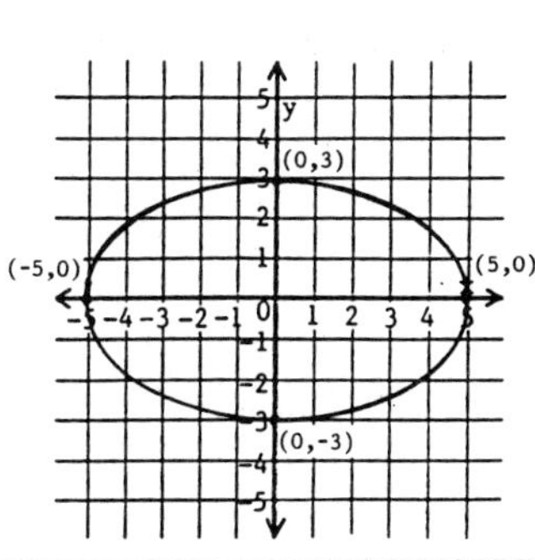

SELF-TEST 33 (pages 469-477)

Find the distance between each pair of points.

1. (-4,7) and (1,7)

2. (-3,2) and (2,-5)

Identify the center and radius of each circle.

3. $(x + 4)^2 + (y - 3)^2 = 25$

Center: ________

Radius: ________

4. $x^2 + y^2 = 36$

Center: ________

Radius: ________

5. Find the center and radius of this circle.

$$x^2 + y^2 - 4x + 10y + 25 = 0$$

6. Find the intercepts of this ellipse.

$$\frac{x^2}{36} + \frac{y^2}{9} = 1$$

x-intercepts: ________ and ________

y-intercepts: ________ and ________

7. Graph each circle.

$(x - 1)^2 + (y + 2)^2 = 9$

$x^2 + y^2 = 4$

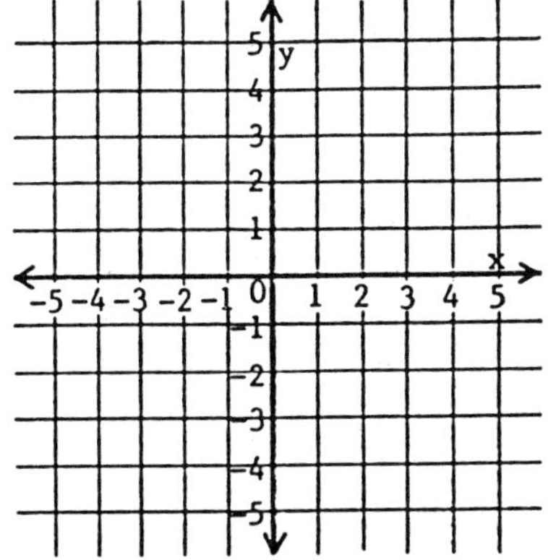

8. Graph this ellipse.

$$\frac{x^2}{9} + \frac{y^2}{16} = 1$$

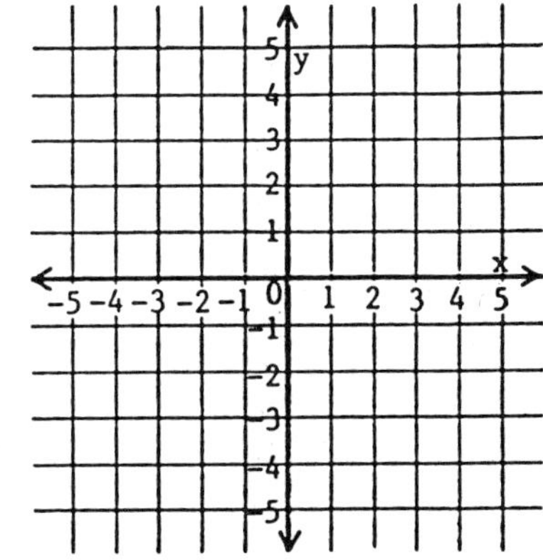

ANSWERS TO SELF-TEST 33:

1. 5
2. $\sqrt{74} = 8.602$
3. (-4,3), r = 5
4. (0,0), r = 6
5. (2,-5), r = 2
6. (6,0) and (-6,0)
   (0,3) and (0,-3)

7.

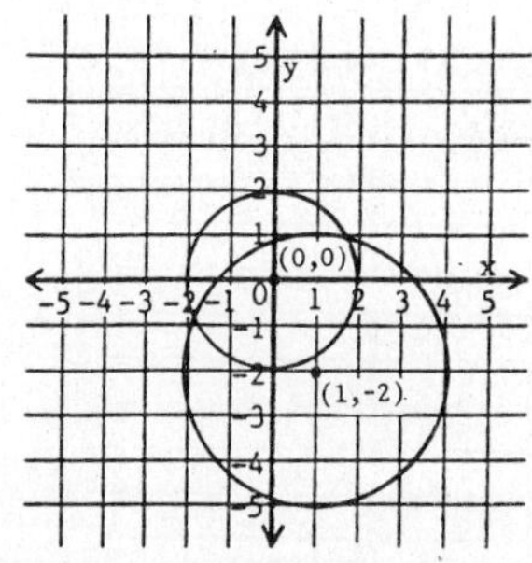

8.

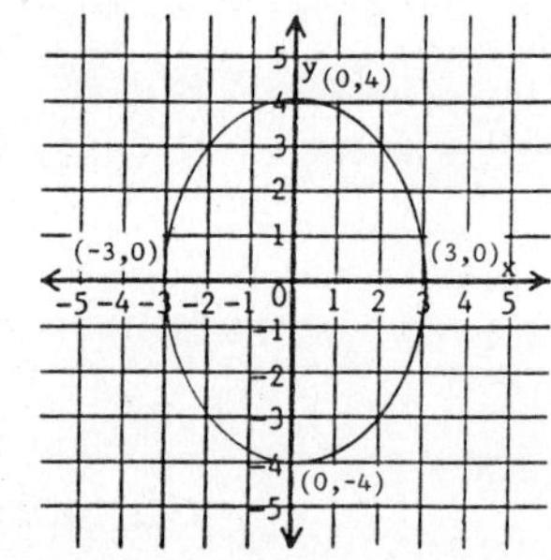

9-9 HYPERBOLAS

In this section, we will discuss the equations of hyperbolas and graph some hyperbolas.

58. A hyperbola is a figure with two curved parts that are mirror images of each other. The standard forms of the equations of hyperbolas that cross either the x-axis or the y-axis are given below.

Equations Of Hyperbolas: with intercepts		
Intercepts on x-axis	$\frac{x^2}{a^2} - \frac{y^2}{b^2} = 1$	a,b > 0
Intercepts of y-axis	$\frac{y^2}{b^2} - \frac{x^2}{a^2} = 1$	a,b > 0

Sketches of hyperbolas with intercepts on either the x-axis or the y-axis are shown below.

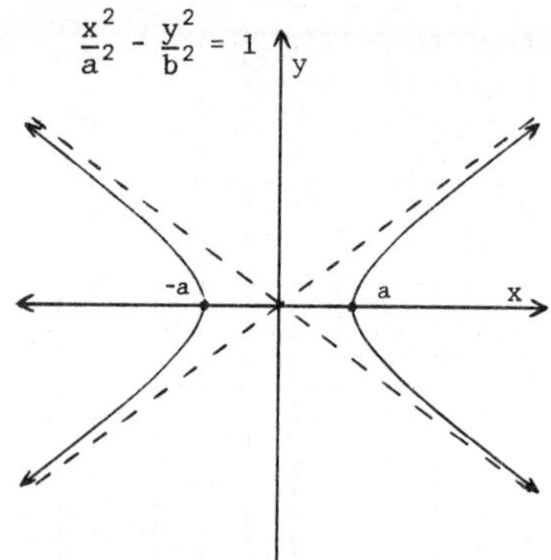

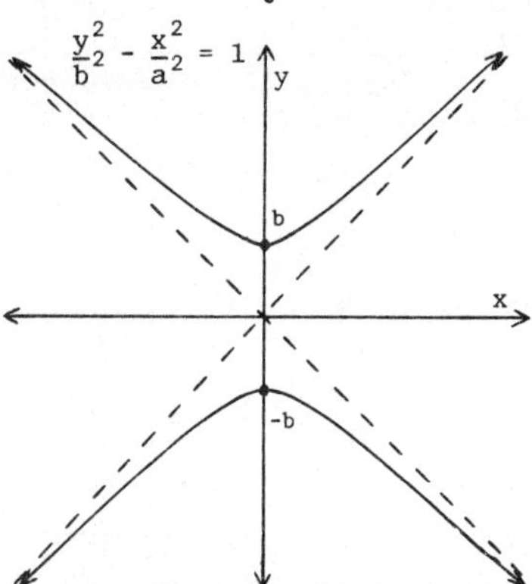

Asymptotes: The dashed lines through the origin are called asymptotes. As the curve gets farther away from the origin, it gets closer and closer to the asymptotes.

Intercepts: The hyperbola on the left crosses the x-axis at a and -a. Therefore, the x-intercepts are (a,0) and (-a,0).

The hyperbola on the right crosses the y-axis at b and -b. Therefore, the y-intercepts are (0,b) and (0,-b).

59. The positive term in the equation of a hyperbola tells us whether the intercepts are on the x-axis or the y-axis. For example:

In $\frac{x^2}{25} - \frac{y^2}{36} = 1$, the x^2 term is positive. Therefore, the intercepts are on the x-axis. They are (5,0) and (-5,0).

In $\frac{y^2}{4} - \frac{x^2}{81} = 1$, the y^2 term is positive. Therefore, the intercepts are on the y-axis. They are (0,2) and (0,-2).

Write the intercepts for each hyperbola.

a) $\frac{y^2}{9} - \frac{x^2}{49} = 1$ b) $\frac{x^2}{64} - \frac{y^2}{4} = 1$

_________ and _________ _________ and _________

---

60. The equations of the asymptotes are the same for both types of hyperbolas. The equations are:

$$y = \frac{b}{a}x \quad \text{and} \quad y = -\frac{b}{a}x$$

Let's find the equations of the asymptotes for these:

For $\frac{x^2}{25} - \frac{y^2}{36}$, a = 5 and b = 6.

The asymptotes are: $y = \frac{6}{5}x$ and $y = -\frac{6}{5}x$

For $\frac{y^2}{9} - \frac{x^2}{36}$, a = 6 and b = 3.

The asymptotes are: _________ and _________

a) (0,3) and (0,-3)

b) (8,0) and (-8,0)

---

61. We graphed the following hyperbola at the right. Since the x^2 term is positive, its intercepts are on the x-axis.

$$\frac{x^2}{4} - \frac{y^2}{9} = 1$$

To do so, we used three steps:

1. Sketch the asymptotes. Since a = 2 and b = 3, their equations are:

$$y = \frac{3}{2}x \quad \text{and} \quad y = -\frac{3}{2}x$$

2. Plot the x-intercepts. Since a = 2, the x-intercepts are (2,0) and (-2,0).

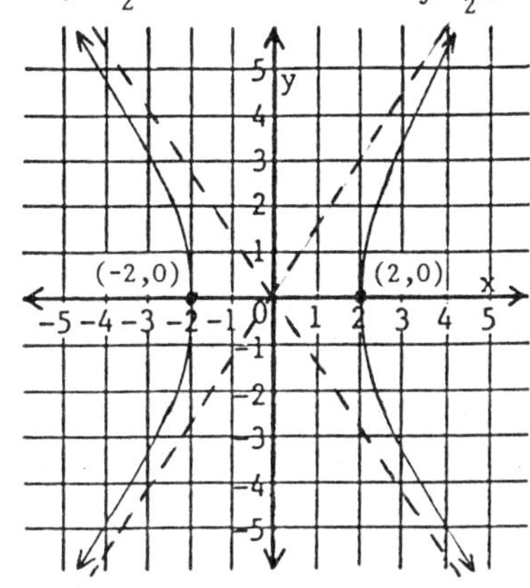

3. Sketch the graph. Draw a curve through each intercept and have it approach closer and closer to the asymptotes.

$y = \frac{1}{2}x$ and $y = -\frac{1}{2}x$

Continued on following page.

61. Continued

Let's use the same steps to graph this hyperbola.

$$\frac{x^2}{9} - \frac{y^2}{4} = 1$$

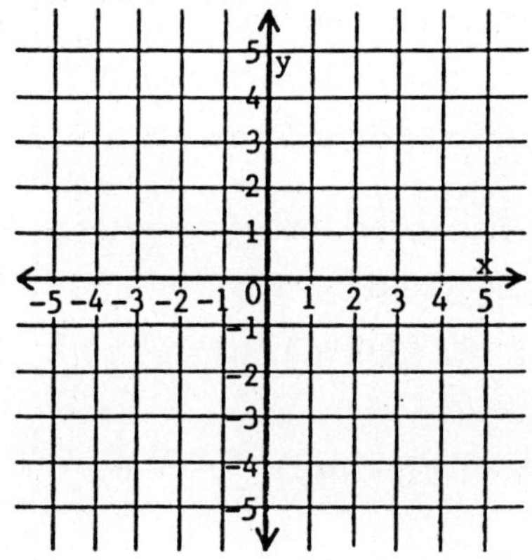

a) The equations of the asymptotes are:

__________ and __________

b) The coordinates of the x-intercepts are:

__________ and __________

c) Sketch the graph.

---

62. We can use a, -a, b, and -b to draw a rectangle that can be used to draw the asymptotes. For example, we drew a rectangle through x = 3 and -3 and y = 4 and -4 below. The asymptotes are the diagonals of the rectangle. Draw a rectangle and the asymptotes for the other hyperbola.

$$\frac{x^2}{9} - \frac{y^2}{16} = 1$$

a = 3, b = 4

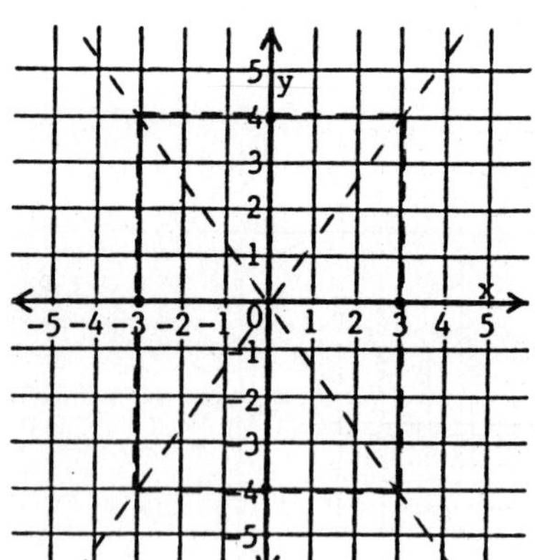

$$\frac{y^2}{4} - \frac{x^2}{16} = 1$$

a = 4, b = 2

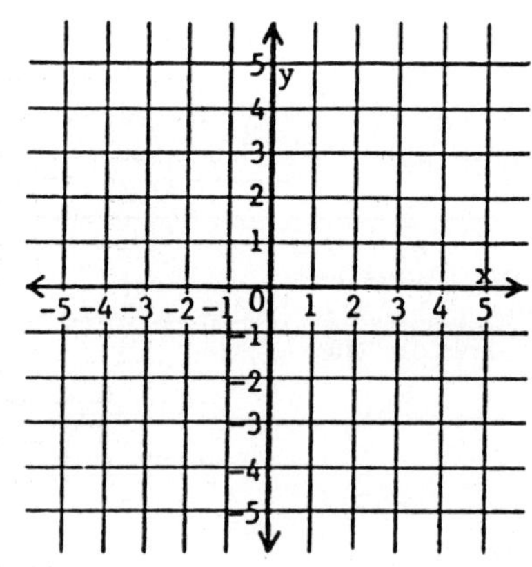

a) $y = \frac{2}{3}x$ and

$y = -\frac{2}{3}x$

b) (3,0) and (-3,0)

c)

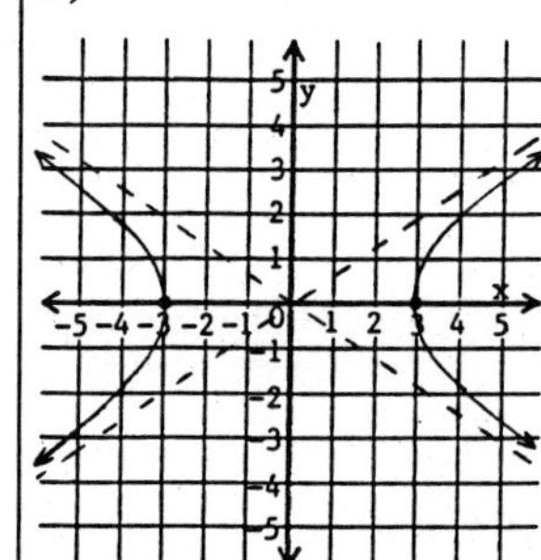

Answer To Frame 62:

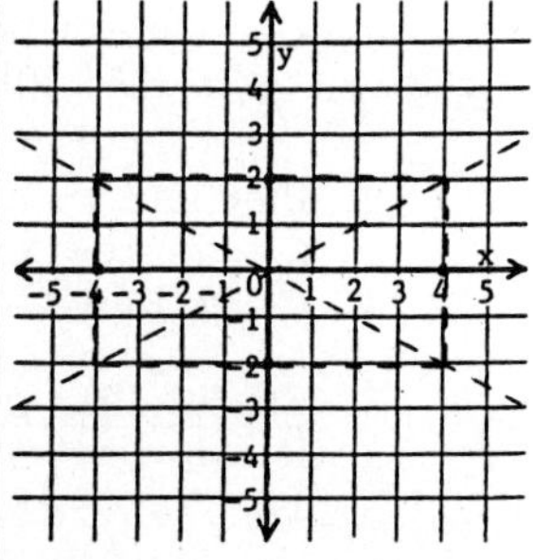

63. We graphed the following hyperbola at the right. Since the y^2 term is positive, the intercepts are on the y-axis.

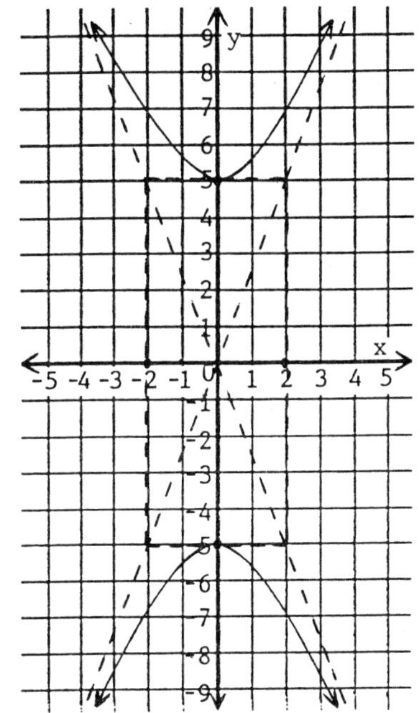

$$\frac{y^2}{25} - \frac{x^2}{4} = 1$$

To do so, we used these steps:

1. We used x = 2 and -2 and y = 5 and -5 to draw a rectangle. Then we drew the asymptotes through the corners.

   Note: The equations of the asymptotes are
   $$y = \frac{5}{2}x \text{ and } y = -\frac{5}{2}x.$$

2. Then we drew the curves through the y-intercepts at (0,5) and (0,-5). Notice that the curves touch the top and bottom of the rectangle.

Use the same steps to graph the hyperbola below.

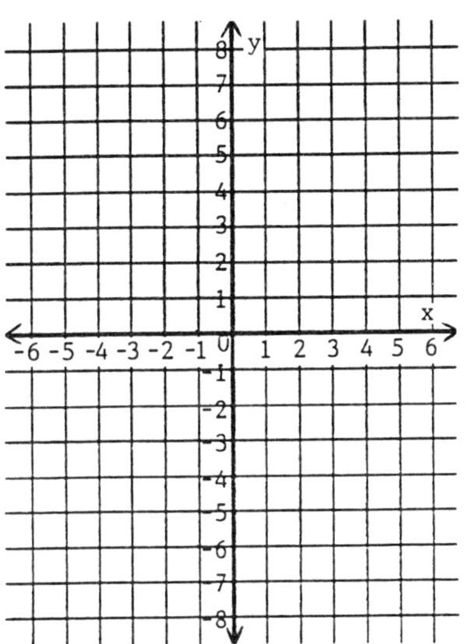

$$\frac{y^2}{16} - \frac{x^2}{9} = 1$$

$a = 3$, $b = 4$

---

64. There are also hyperbolas that have the axes as their asymptotes. Hyperbolas of that type have no intercepts. The standard form of their equation is given below.

Equation Of Hyperbolas:	with no intercepts
$xy = k$	$(k = \emptyset)$

Continued on following page.

Answer To Frame 63:

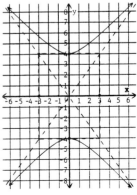

64. Continued

If k is positive, the two branches of the hyperbola are in quadrants 1 and 3. If k is negative, the two branches of the hyperbola are in quadrants 2 and 4. As examples, we graphed $xy = 8$ and $xy = -8$. To do so, we plotted the points in the tables.

$xy = 8$

x	y
-8	-1
-4	-2
-2	-4
-1	-8
1	8
2	4
4	2
8	1

$xy = -8$

x	y
-8	1
-4	2
-2	4
-1	8
1	-8
2	-4
4	-2
8	-1

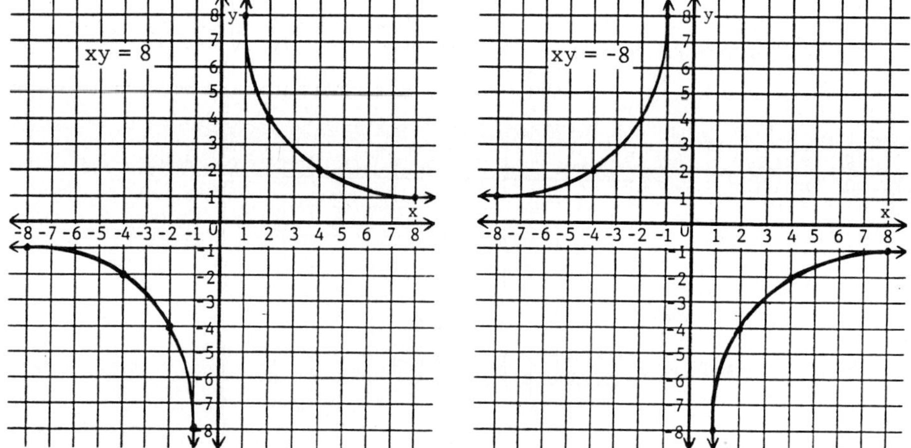

Using the same method, graph the hyperbolas below. Make up your own solution-table.

$xy = 6$

$xy = -6$

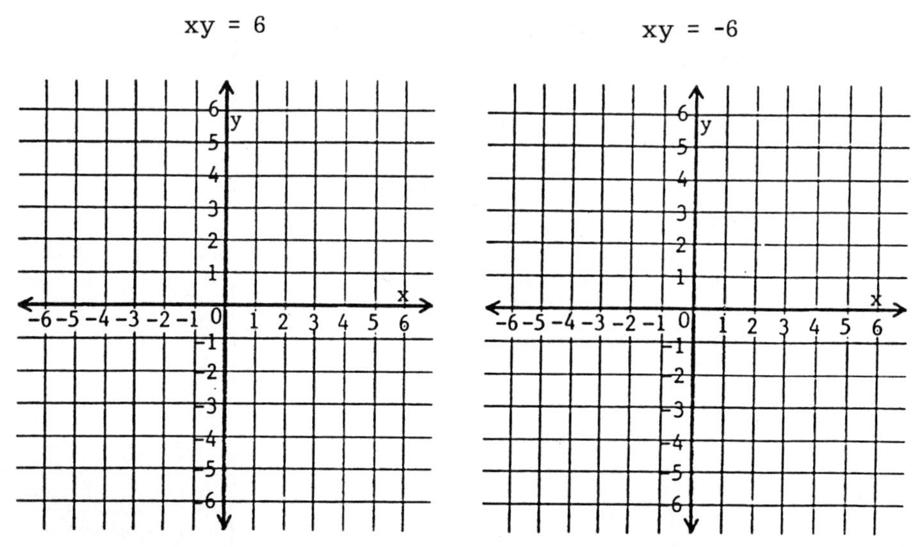

Answers To Frame 64:

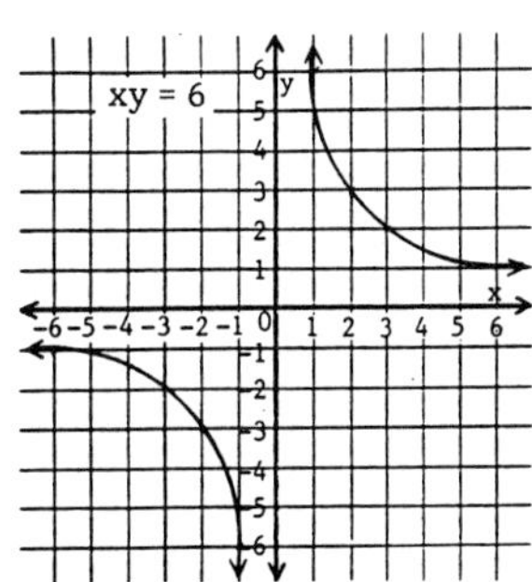

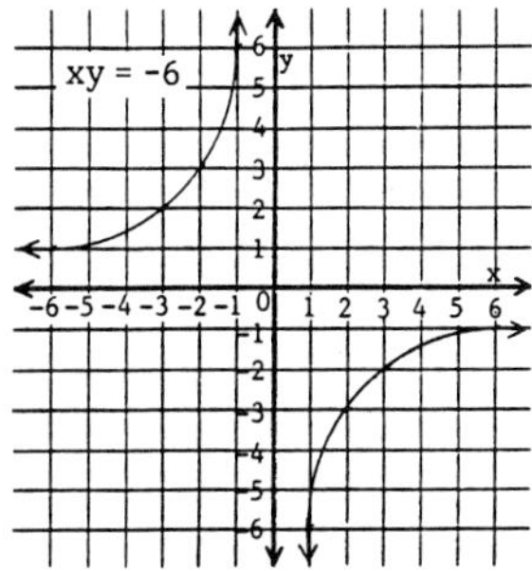

9-10 NONLINEAR SYSTEMS OF EQUATIONS

A system of equations containing at least one nonlinear equation is called a nonlinear system of equations. We will discuss the methods used to solve systems of that type in this section.

65. When a system contains one linear and one second-degree equation, it can have two, one, or zero real number solutions. The three possibilities are shown in the graphs below.

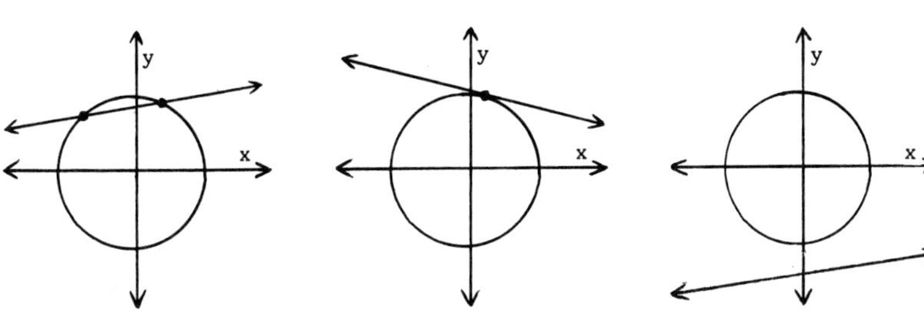

Figure 1 Figure 2 Figure 3

In Figure 1, the line and circle intersect at two points. Therefore, there are two real number solutions.

In Figure 2, the line and circle intersect at one point. Therefore, there is one real number solution.

In Figure 3, the line and circle do not intersect. Therefore, there are ________ real number solutions.

zero

66. When a system contains one linear and one second-degree equation, we use the substitution method to solve it. An example is discussed below.

The system below contains a circle and a line. It is graphed at the right. The two points of intersection are (-5,0) and (4,3).

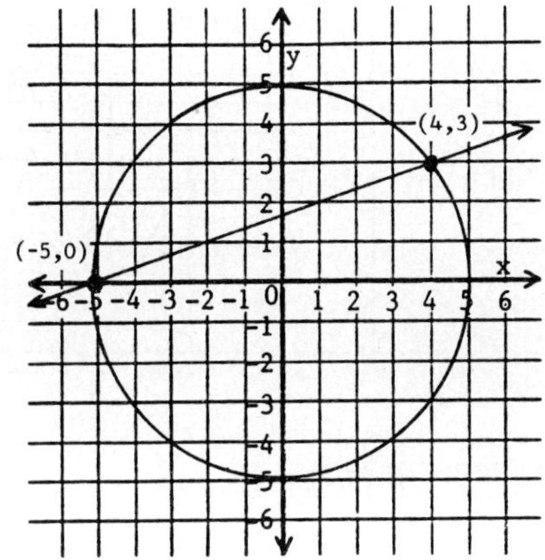

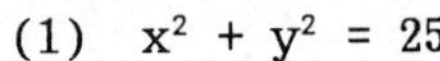

(1) $x^2 + y^2 = 25$

(2) $x + 5 = 3y$

To solve it algebraically, we begin by solving for $\underline{x}$ in the linear equation (2).

(3) $x = 3y - 5$

We then substitute $3y - 5$ for $\underline{x}$ in the nonlinear equation (1) and solve for $\underline{y}$.

$$(3y - 5)^2 + y^2 = 25$$

$$9y^2 - 30y + 25 + y^2 = 25$$

$$10y^2 - 30y = 0$$

$$10y(y - 3) = 0$$

$$y = 0 \text{ and } 3$$

Substituting 0 and 3 for $\underline{y}$ in equation (3), we can find the values of $\underline{x}$.

$x = 3y - 5$	$x = 3y - 5$
$x = 3(0) - 5$	$x = 3(3) - 5$
$x = -5$	$x = 4$

Therefore, the two solutions are (-5,0) and (4,3). The solutions check. Do they correspond to the points of intersection on the graph? ________

Yes

67. Following the method in the last frame, solve each system.

a) $x^2 + y^2 = 34$
$x = y - 2$

b) $4x^2 + 9y^2 = 36$
$3y - 6 = 2x$

68. The system below contains a parabola and a line. It is graphed at the right. The one point of intersection is (2,1).

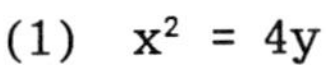

(1) $x^2 = 4y$

(2) $x - y = 1$

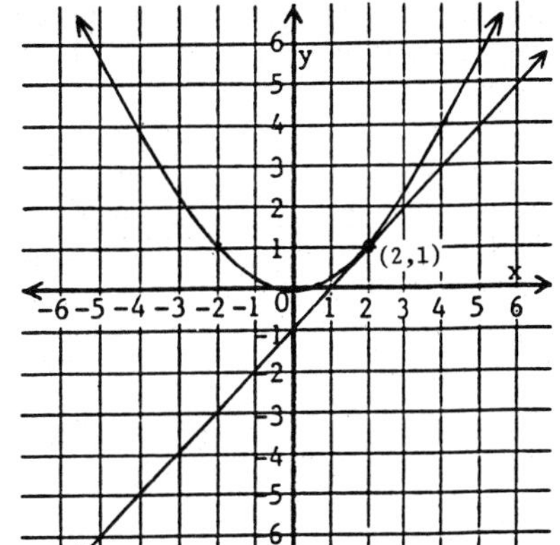

Solving for $\underline{x}$ in the linear equation (2), we get:

(3) $x = y + 1$

Substituting $y + 1$ for $\underline{x}$ in the nonlinear equation (1) and solving for $\underline{y}$, we get:

$$(y + 1)^2 = 4y$$
$$y^2 + 2y + 1 = 4y$$
$$y^2 - 2y + 1 = 0$$
$$(y - 1)(y - 1) = 0$$
$$y = 1$$

Substituting "1" for $\underline{y}$ in equation (3), we can find the value of $\underline{x}$.

$$x = y + 1$$
$$x = 1 + 1$$
$$x = 2$$

Therefore, the one solution is (2,1). The solution checks. It also corresponds to the one solution on the graph.

a) (3,5) and (-5,-3)

b) (0,2) and (-3,0)

---

69. Following the method in the last frame, solve each system.

a) $x^2 = 2y$
$2y + 1 = 2x$

b) $y = x^2$
$y + 6 = 5x$

70. Each system below contains a hyperbola and a line. We solved one system. Solve the other one.

(1)	$xy = 12$	$xy = 8$
(2)	$y + 2 = 2x$	$y + 4x = 12$

Solving for $\underline{y}$ in equation (2), we get:

(3) $y = 2x - 2$

Substituting $2x - 2$ for $\underline{y}$ in equation (1) and solving for $\underline{x}$, we get:

$$x(2x - 2) = 12$$
$$2x^2 - 2x = 12$$
$$2x^2 - 2x - 12 = 0$$
$$x^2 - x - 6 = 0$$
$$(x - 3)(x + 2) = 0$$
$$x = 3 \text{ and } -2$$

Substituting 3 and -2 for $\underline{x}$ in equation (3), we can solve for $\underline{y}$.

$y = 2x - 2$	$y = 2x - 2$
$y = 2(3) - 2$	$y = 2(-2) - 2$
$y = 4$	$y = -6$

The two solutions are (3,4) and (-2,-6). The solutions check.

a) $(1,\frac{1}{2})$

b) (2,4) and (3,9)

---

71. When a system contains two second-degree equations, it can have four, three, two, one, or zero real number solutions. The five possibilities are shown in the graphs below.

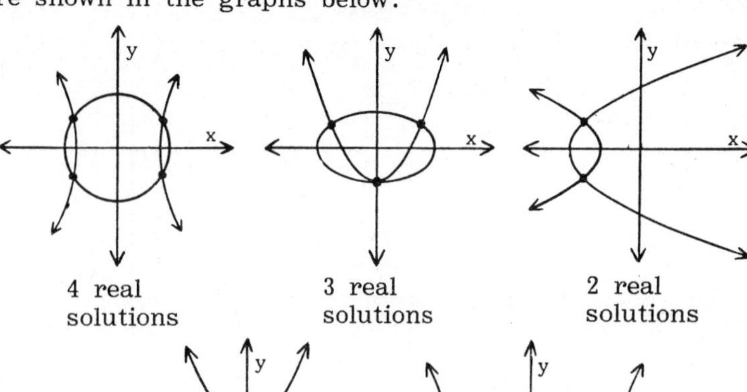

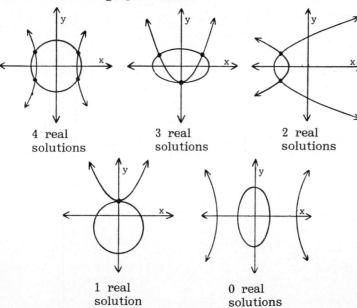

4 real solutions

3 real solutions

2 real solutions

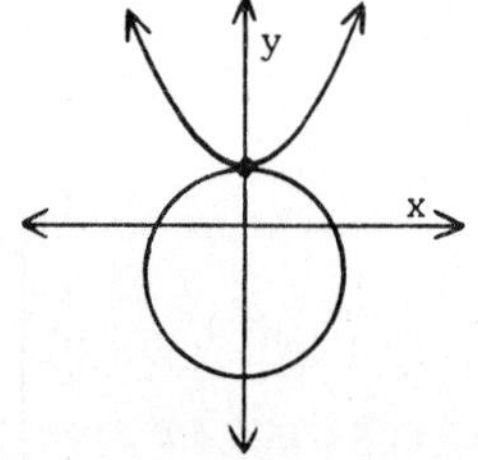

1 real solution

0 real solutions

(2,4) and (1,8)

72. When a system contains two second degree equations, sometimes we use the addition method and sometimes we use the substitution. The addition method is used in the example below.

The system below contains a circle and an ellipse. It is sketched at the right. The four points of intersection are (4,3), (4,-3), (-4,3) and (-4,-3).

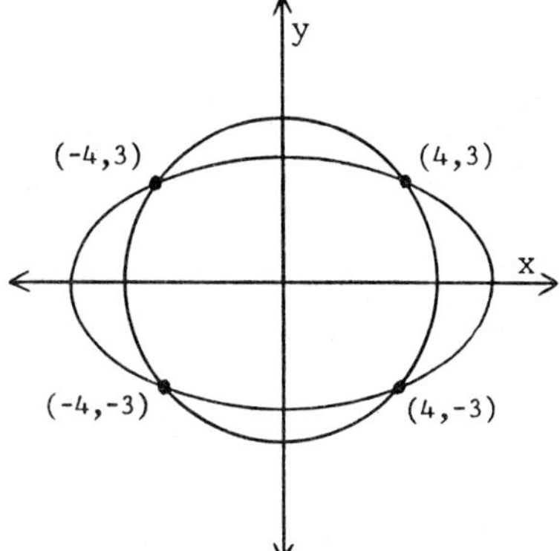

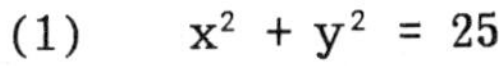

(1) $x^2 + y^2 = 25$

(2) $2x^2 + 3y^2 = 59$

To solve it algebraically, we multiply equation (1) by -2 and then add the two equations to eliminate $\underline{x}$ and solve for $\underline{y}$.

$$\begin{array}{r} -2x^2 - 2y^2 = -50 \\ \underline{2x^2 + 3y^2 = \ \ 59} \\ y^2 = 9 \\ y = \pm 3 \end{array}$$

If $y = 3$ or -3, $y^2 = 9$. Therefore, we can substitute 9 for y^2 in equation (1) and solve for $\underline{x}$.

$$x^2 + 9 = 25$$
$$x^2 = 16$$
$$x = \pm 4$$

Therefore: If $x = 4$, $y = 3$ or $y = -3$.
If $x = -4$, $y = 3$ or $y = -3$.

The four solutions are (4,3), (4,-3), (-4,3), and (-4,-3). The solutions check. They also correspond to the four solutions on the sketch.

---

73. Use the addition method to solve these.

a) $x^2 + y^2 = 16$
$x^2 - y^2 = 16$

b) $3x^2 - 2y^2 = -6$
$5x^2 + y^2 = 29$

74. We used the substitution method to solve the system below. Solve the other system.

(1) $x^2 + y^2 = 10$

(2) $xy = 3$

$x^2 + y^2 = 20$

$xy = 8$

Solving for y in equation (2), we get:

(3) $y = \frac{3}{x}$

Substituting $\frac{3}{x}$ for y in equation (1), we get:

$$x^2 + \left(\frac{3}{x}\right)^2 = 10$$

$$x^2 + \frac{9}{x^2} = 10$$

$$x^4 + 9 = 10x^2$$

$$x^4 - 10x^2 + 9 = 0$$

Substituting a for x^2, we get:

$$a^2 - 10a + 9 = 0$$

$$(a - 1)(a - 9) = 0$$

$$a = 1 \text{ and } a = 9$$

Substituting x^2 for a, we get:

$$x^2 = 1 \text{ and } x^2 = 9$$

$$x = \pm 1 \text{ and } x = \pm 3$$

Substituting those four values for x in equation (3), we get:

If $x = 1$, $y = \frac{3}{1} = 3$

If $x = -1$, $y = \frac{3}{-1} = -3$

If $x = 3$, $y = \frac{3}{3} = 1$

If $x = -3$, $y = \frac{3}{-3} = -1$

Therefore, the four solutions are (1,3), (-1,-3), (3,1) and (-3,-1).

a) (4,0) and (-4,0)

b) (2,3), (2,-3), (-2,3), and (-2,-3)

(2,4), (-2,-4), (4,2), and (-4,-2)

9-11 INVERSES OF FUNCTIONS

In this section we will discuss the inverses of functions.

75. If we interchange the variables in a two-variable function, we get the equation of the inverse relation. The inverse may or may not be a function. If the inverse is a function, we can use the symbol f^{-1} for it. f^{-1} is read "the inverse of f" or "f-inverse". In f^{-1}, the -1 is not an exponent.

To find the inverse of a function, we use these steps.

1. Substitute y for f(x).
2. Interchange x and y.
3. Solve for y.
4. Substitute $f^{-1}(x)$ for y. (if the inverse is a function)

Let's use the steps above to find the inverse of $f(x) = x + 3$.

$f(x) = x + 3$

$y = x + 3$ Substituting y for f(x).

$x = y + 3$ Interchanging x and y

$y = x - 3$ Solving for y

$f^{-1}(x) = x - 3$ Substituting $f^{-1}(x)$ for y because $y = x - 3$ is a function.

Note: In $f(x) = x + 3$, we add 3 to x. In $f^{-1}(x)$ we subtract 3 from x. Therefore, the function and its inverse do opposite things.

Using the same steps, find the inverse of each function.

a) $f(x) = x + 10$ b) $f(x) = x - 1$

a) $f^{-1}(x) = x - 10$

b) $f^{-1}(x) = x + 1$

76. We used the same steps to find the inverse of $f(x) = 2x$ below.

$f(x) = 2x$

$y = 2x$ Substituting y for f(x).

$x = 2y$ Interchanging x and y.

$y = \frac{x}{2}$ Solving for y.

$f^{-1}(x) = \frac{x}{2}$ Substituting $f^{-1}(x)$ for y because $y = \frac{x}{2}$ is a function.

Note: In $f(x) = 2x$, we multiply x by 2. In $f^{-1}(x) = \frac{x}{2}$, we divide x by 2. Therefore, the function and its inverse do opposite things.

Find the inverse of each function.

a) $f(x) = 3x$ b) $f(x) = -5x$

---

77. We found the inverse of $f(x) = 3x - 1$ below.

$f(x) = 3x - 1$

$y = 3x - 1$ Substituting y for f(x).

$x = 3y - 1$ Interchanging x and y.

$y = \frac{x + 1}{3}$ Solving for y.

$f^{-1}(x) = \frac{x + 1}{3}$ Substituting $f^{-1}(x)$ for y because $y = \frac{x + 1}{3}$ is a function.

Notice again that the function and inverse do opposite things. In $f(x) = 3x - 1$, we multiply x by 3 and subtract "1". In $f^{-1}(x) = \frac{x + 1}{3}$, we add "1" and then divide by 3.

Find the inverse of each function.

a) $f(x) = 4x - 5$ b) $f(x) = 2x + 7$

Answers to 76:

a) $f^{-1}(x) = \frac{x}{3}$

b) $f^{-1}(x) = -\frac{x}{5}$

a) $f^{-1}(x) = \frac{x + 5}{4}$

b) $f^{-1}(x) = \frac{x - 7}{2}$

78. In an earlier frame, we found the inverse of $f(x) = 2x$. We got:

$$f(x) = 2x$$

$$f^{-1}(x) = \frac{x}{2} \text{ or } \frac{1}{2}x$$

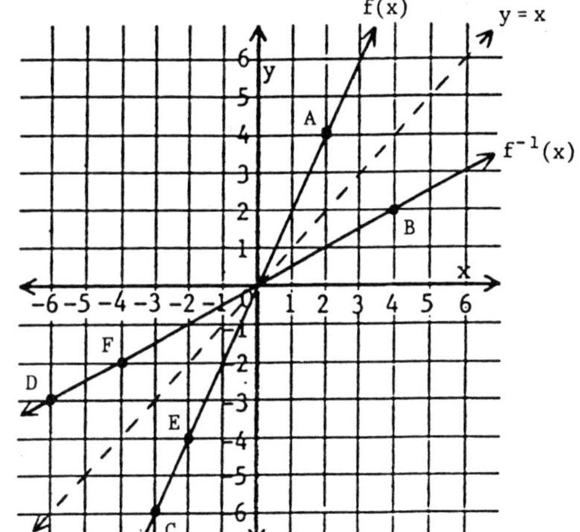

We graphed the function and its inverse at the right. We also drew a dotted line for y = x.

For each point on the function, there is a point on the inverse that is the same distance from y = x. The coordinates of the points in each pair are interchanged. For example:

A and B are the same distance from y = x.

A is (2,4) and B is (4,2).

C and D are the same distance from y = x.

C is (-3,-6) and D is (-6,-3).

Therefore, we say that the inverse is the "mirror image" of the function with respect to y = x.

E and F are the same distance from y = x.

Since E is (-2,-4), F is (,)

(-4,-2)

79. In an earlier frame, we found the inverse of $f(x) = 3x - 1$. We got:

$$f(x) = 3x - 1$$

$$f^{-1}(x) = \frac{x + 1}{3}$$

We graphed the function, its inverse, and the line y = x at the right. Notice again that the inverse is the "mirror image" of the function with respect to y = x. That is:

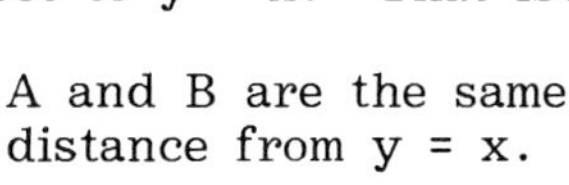

A and B are the same distance from y = x.

C and D are the same distance from y = x.

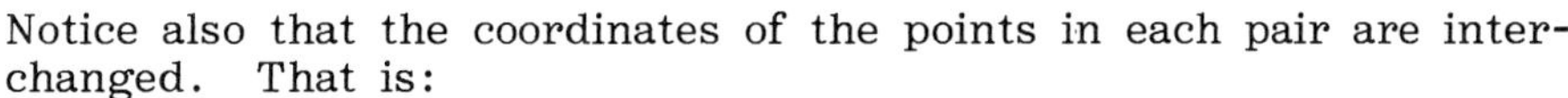

Notice also that the coordinates of the points in each pair are interchanged. That is:

a) A is (2,5) and B is (,).

b) C is (-1,-4) and D is (,).

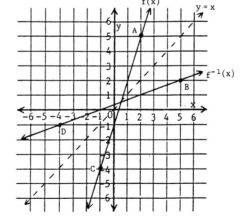

80. We found the inverse of $f(x) = x^2$ below.

$$f(x) = x^2$$
$$y = x^2$$
$$x = y^2$$
$$y = \pm\sqrt{x}$$

Since we get two values of y for each value of x in $y = \pm\sqrt{x}$, the inverse is not a function. We graphed the function and its inverse at the right. Notice that the inverse fails the vertical line test for a function.

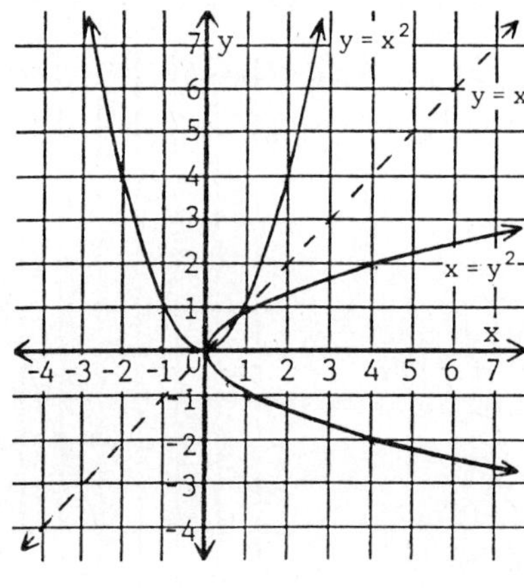

However, if we restrict the domain of $f(x) = x^2$ to nonnegative numbers, then its inverse is a function and we can use $f^{-1}(x)$. We get:

$$f^{-1}(x) = \sqrt{x}$$

We graphed the restricted function and its inverse at the right. Points A and B are the same distance from $y = x$.

Since A is (2,4), B is (,).

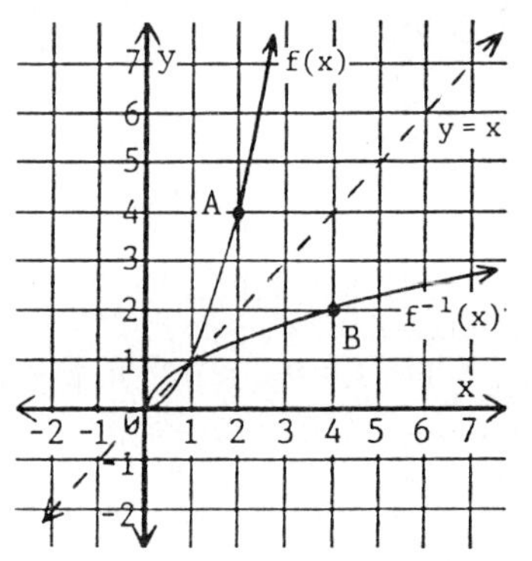

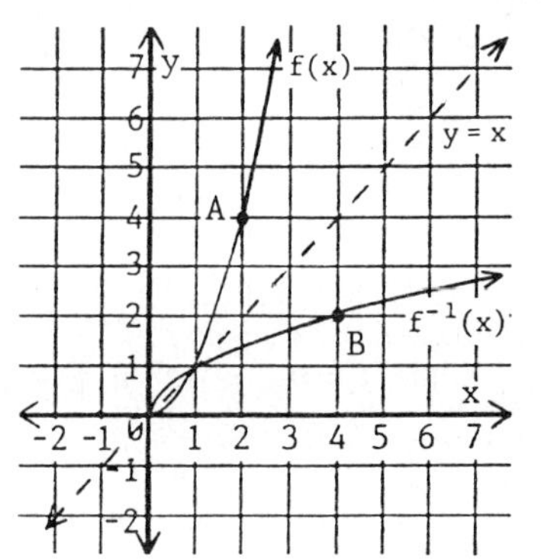

a) (5,2)

b) (-4,-1)

---

81. We found the inverse of $f(x) = x^2 + 1$ below.

$$f(x) = x^2 + 1$$
$$y = x^2 + 1$$
$$x = y^2 + 1$$
$$y^2 = x - 1$$
$$y = \pm\sqrt{x - 1}$$

Since we get two values of y for each value of x in $y = \pm\sqrt{x - 1}$, the inverse is not a function. You can see from the graph at the right that the inverse fails the vertical line test.

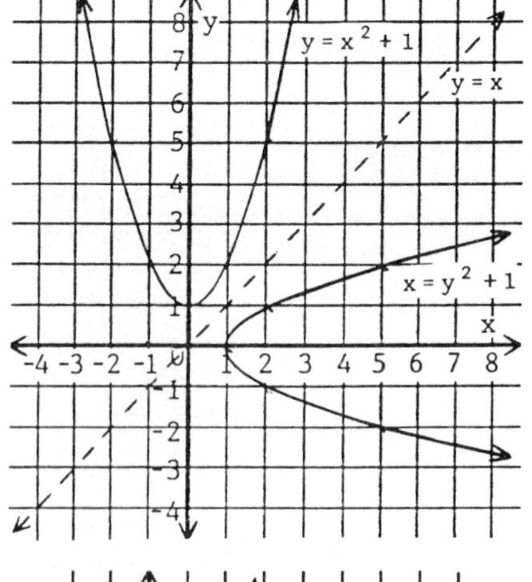

However, if we restrict the domain of $f(x) = x^2 + 1$ to nonnegative values, then its inverse is a function and we can use $f^{-1}(x)$. We get:

$$f^{-1}(x) = \sqrt{x - 1}$$

We graphed the restricted function and its inverse at the right.

Since (3,10) satisfies $f(x)$, we know that (,) must satisfy $f^{-1}(x)$.

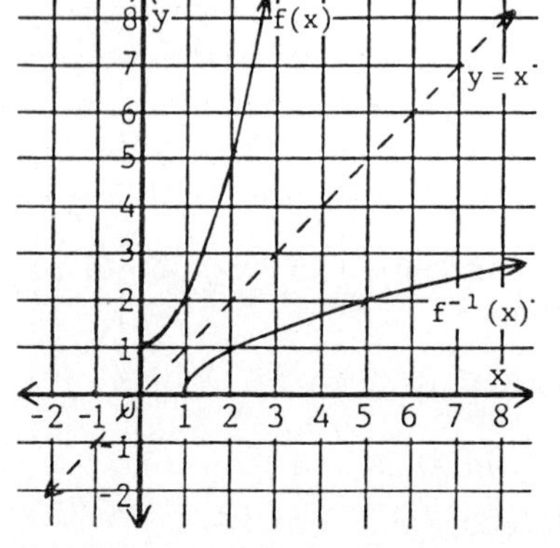

(4,2)

Answer To Frame 81: (10,3)

SELF-TEST 34 (pages 478-493)

Write the intercepts for each hyperbola.

1. $\frac{y^2}{1} - \frac{x^2}{9} = 1$ _______ and _______

2. $\frac{x^2}{25} - \frac{y^2}{4} = 1$ _______ and _______

Write the equations of the asymptotes.

3. $\frac{x^2}{25} - \frac{y^2}{9} = 1$ _______ and _______

4. $\frac{y^2}{36} - \frac{x^2}{9} = 1$ _______ and _______

Graph each hyperbola.

5. $\frac{y^2}{9} - \frac{x^2}{4} = 1$

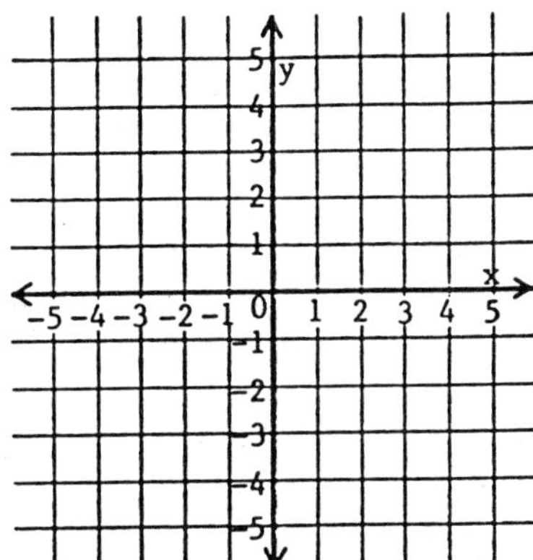

6. $xy = 4$

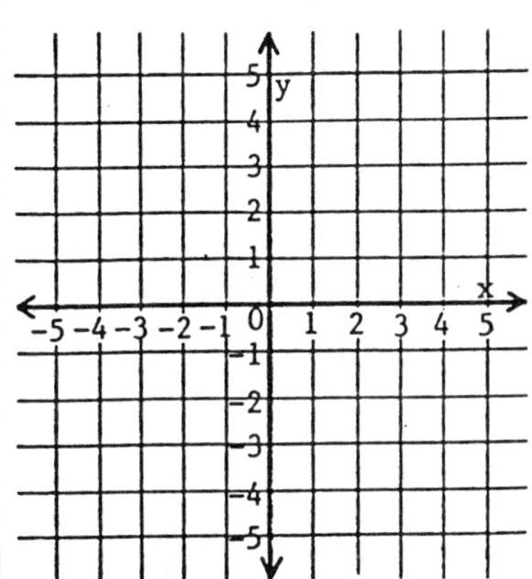

Solve each system.

7. $x = y^2$
   $y = x - 2$

8. $x^2 + y^2 = 29$
   $2x^2 - 3y^2 = 38$

Find the inverse of each function.

9. $f(x) = x - 5$

10. $f(x) = 3x + 4$

Find the inverse of this function if its domain is restricted to nonnegative numbers.

11. $f(x) = x^2 + 7$

ANSWERS:

1. (0,1) and (0,-1)
2. (5,0) and (-5,0)
3. $y = \frac{3}{5}x$ and $y = -\frac{3}{5}x$
4. $y = 2x$ and $y = -2x$

5.

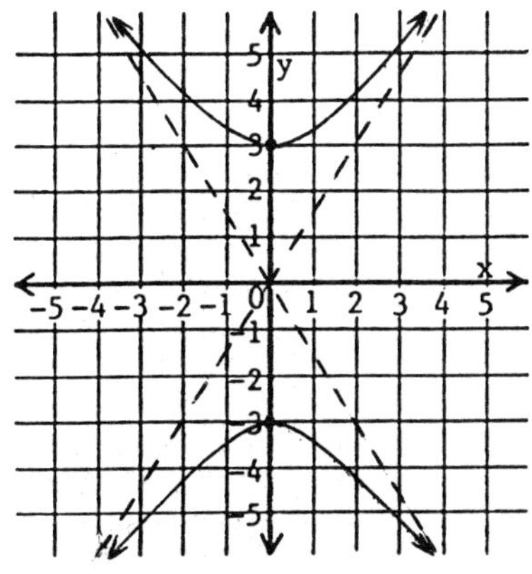

6.

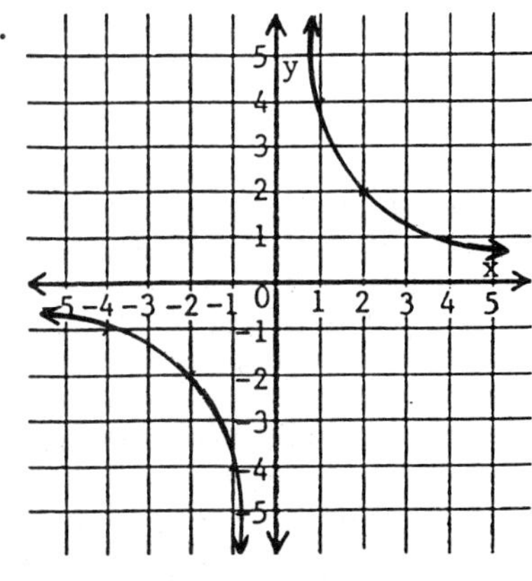

7. (1,-1) and (4,2)
8. (5,2), (5,-2), (-5,2) and (-5,-2)
9. $f^{-1}(x) = x + 5$
10. $f^{-1}(x) = \frac{x - 4}{3}$
11. $f^{-1}(x) = \sqrt{x - 7}$

SUPPLEMENTARY PROBLEMS - CHAPTER 9

Assignment 31

State whether each parabola opens "upward" or "downward".

1. $y = x^2$
2. $y = -3x^2$
3. $y = -x^2 + 5$
4. $y = 2x^2 - 1$
5. $y = (x - 2)^2$
6. $y = -5(x + 3)^2$
7. $y = -\frac{1}{2}(x + 1)^2 - 4$
8. $y = 4(x - 3)^2 + 1$

Give the vertex and axis of symmetry of each parabola.

9. $f(x) = 2x^2$
10. $f(x) = -x^2$
11. $f(x) = (x - 2)^2$
12. $f(x) = x^2 - 2$
13. $f(x) = -2(x + 1)^2$
14. $f(x) = -2x^2 + 1$
15. $f(x) = 3(x - 4)^2$
16. $f(x) = 3x^2 - 4$
17. $f(x) = (x + 5)^2 - 2$
18. $f(x) = -(x - 4)^2 + 3$
19. $f(x) = 5(x - 2)^2 - 1$
20. $f(x) = -\frac{1}{2}(x + 1)^2 - 6$
21. $f(x) = -3(x - 7)^2 - 5$
22. $f(x) = 4(x + 3)^2 + 2$

Graph each function.

23. $f(x) = 3x^2$

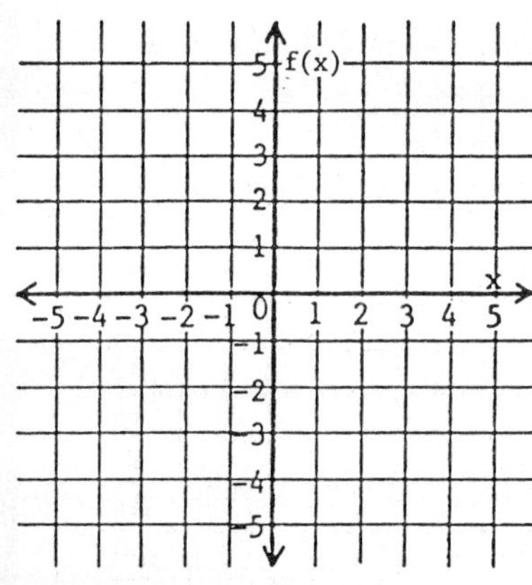

24. $f(x) = -\frac{1}{4}x^2$

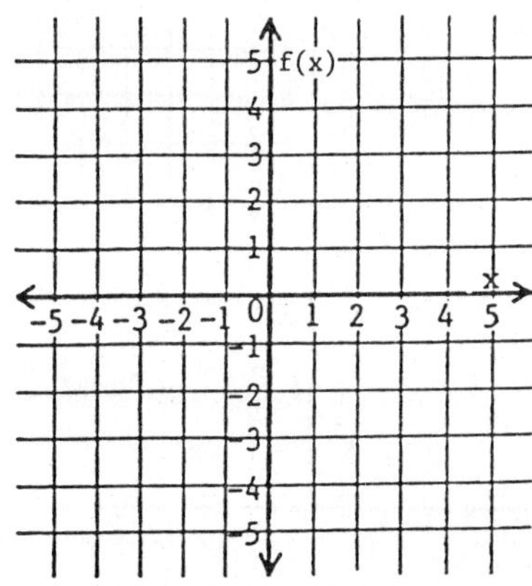

25. $f(x) = (x - 1)^2$

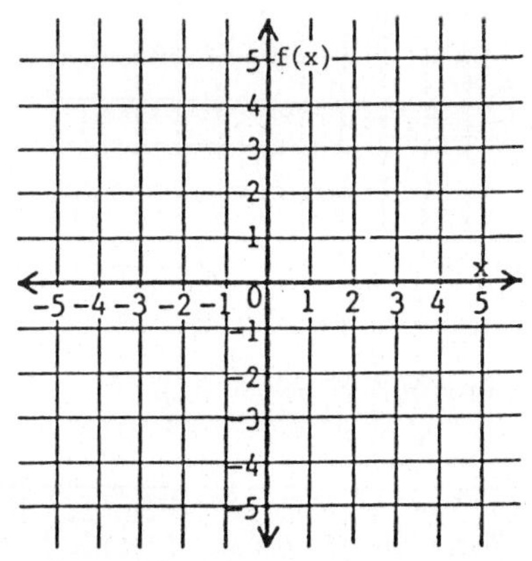

26. $f(x) = -2(x + 3)^2$

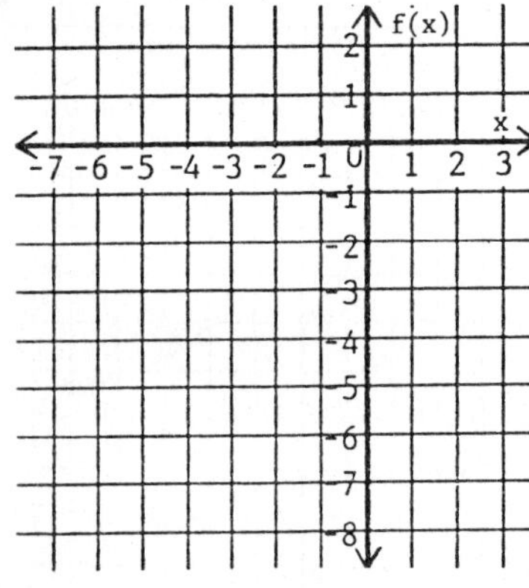

27. $f(x) = \frac{1}{2}(x - 2)^2 - 3$

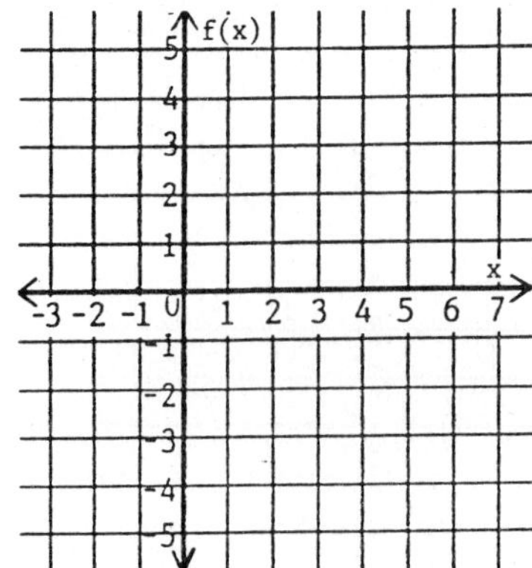

28. $f(x) = -(x + 1)^2 - 2$

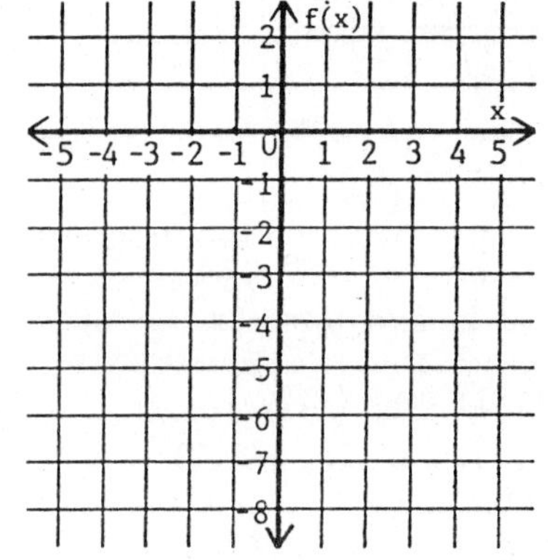

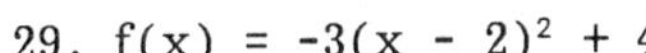

29. $f(x) = -3(x - 2)^2 + 4$

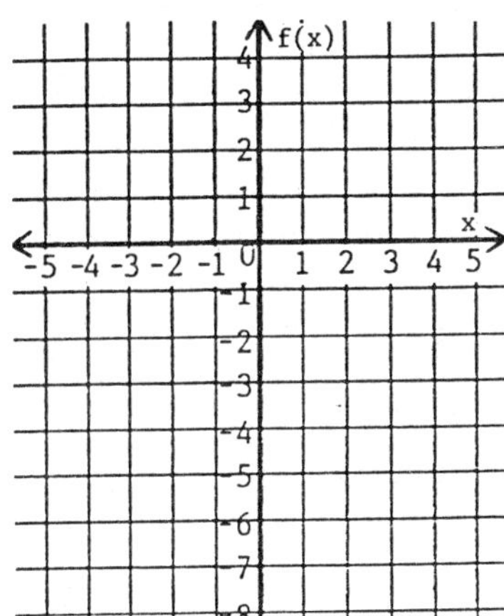

30. $f(x) = x^2 - 4$

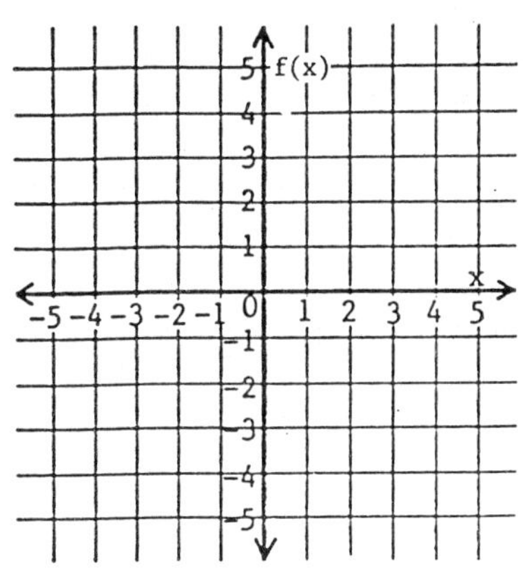

31. $f(x) = -2x^2 + 1$

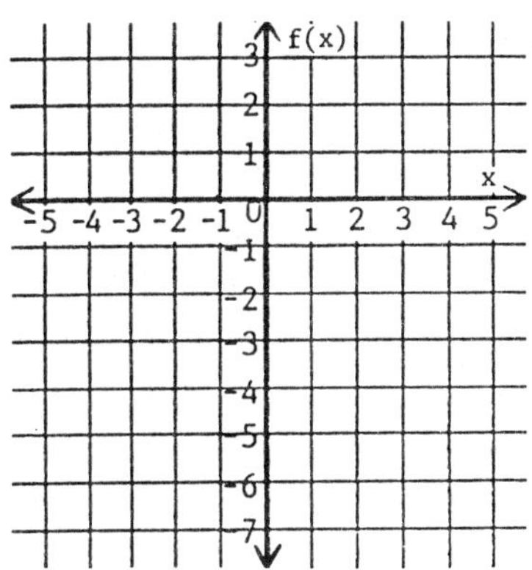

Convert each function to the form $f(x) = a(x - h)^2 + k$.

32. $f(x) = x^2 + 2x - 3$

33. $f(x) = x^2 - 6x + 14$

34. $f(x) = 2x^2 - 20x + 47$

35. $f(x) = 3x^2 + 12x + 22$

36. $f(x) = -2x^2 - 12x - 19$

37. $f(x) = -3x^2 + 12x - 8$

Graph each function.

38. $f(x) = x^2 - 4x + 3$

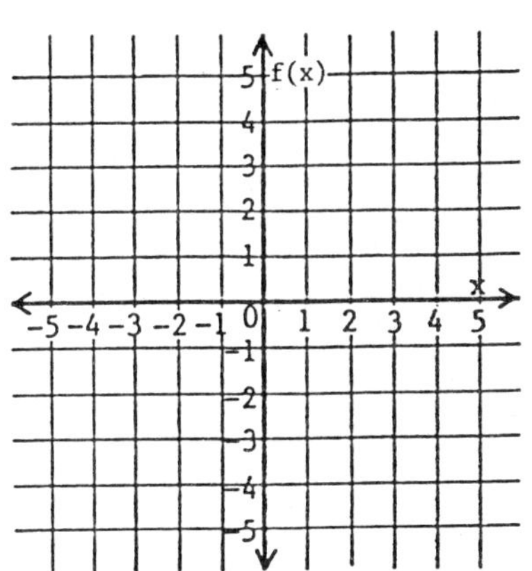

39. $f(x) = 2x^2 + 4x - 1$

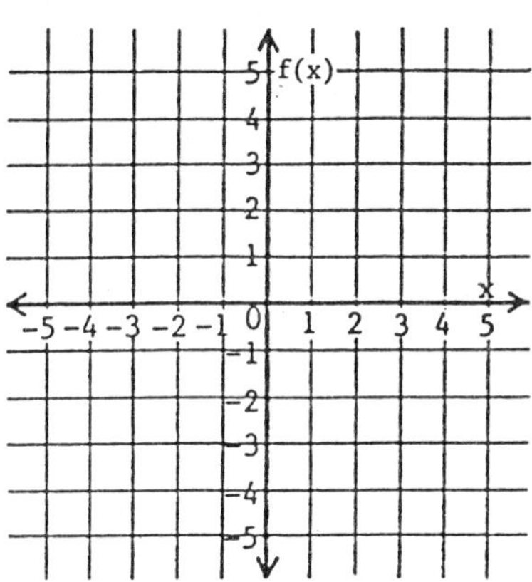

40. $f(x) = -2x^2 - 12x - 14$

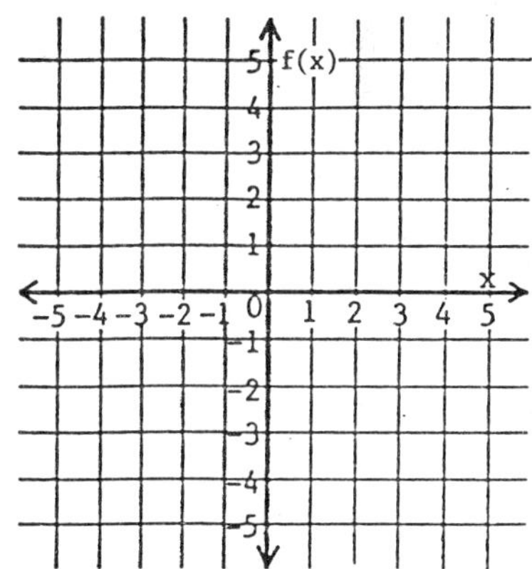

<u>Assignment 32</u>

Find the x-intercepts of each function.

1. $f(x) = x^2 - x - 12$
2. $f(x) = x^2 + 2x - 8$
3. $f(x) = x^2 - 3x$
4. $f(x) = x^2 + x$
5. $f(x) = x^2 - 2x - 5$
6. $f(x) = x^2 + 4x + 2$
7. $f(x) = x^2 + 6x + 9$
8. $f(x) = 9x^2 - 12x + 4$
9. $f(x) = x^2 - 3x + 5$
10. $f(x) = x^2 + 4x + 7$
11. $f(x) = (x - 3)^2 - 1$
12. $f(x) = (x + 4)^2 + 5$
13. $f(x) = (x - 4)^2$
14. $f(x) = 2(x + 1)^2$
15. $f(x) = 4(x - 3)^2$
16. $f(x) = x^2 - 9$
17. $f(x) = x^2 + 9$
18. $f(x) = 3x^2$
19. $f(x) = 2x^2 + 2$

Find the maximum or minimum function value.

20. $f(x) = 3(x - 4)^2 - 2$
21. $f(x) = -2(x + 1)^2 + 5$
22. $f(x) = -\frac{1}{2}(x - 4)^2 - 3$
23. $f(x) = \frac{1}{2}(x + 3)^2 + 6$
24. $f(x) = (x + 7)^2$
25. $f(x) = -3x^2$
26. $f(x) = 2(x - 4)^2 + 1$
27. $f(x) = \frac{1}{2}x^2$
28. $f(x) = -4(x + 2)^2 - 6$
29. $f(x) = x^2 + 4x + 3$
30. $f(x) = -2x^2 - 12x - 11$
31. $f(x) = -3x^2 + 12x - 16$

Solve each problem.

32. Find the maximum area of a rectangle if the sum of the length and width is 24 feet. What length and width give that area?

33. Of all pairs of numbers whose sum is 16, find the pair with the maximum product.

If a model rocket is fired vertically upward at an initial velocity v_o, its distance s (in feet) above the ground at time t (in seconds) is given by the formula $s = v_ot - 16t^2$.

34. Find the maximum height the rocket reaches and the time needed to reach that height if $v_o = 64$ ft/sec.

35. Find the maximum height the rocket reaches and the time needed to reach that height if $v_o = 320$ ft/sec.

Graph each equation.

36. $x = \frac{1}{2}y^2$

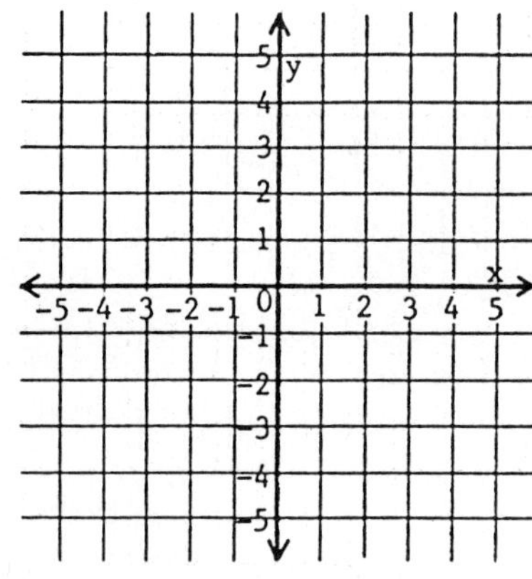

37. $x = -3y^2$

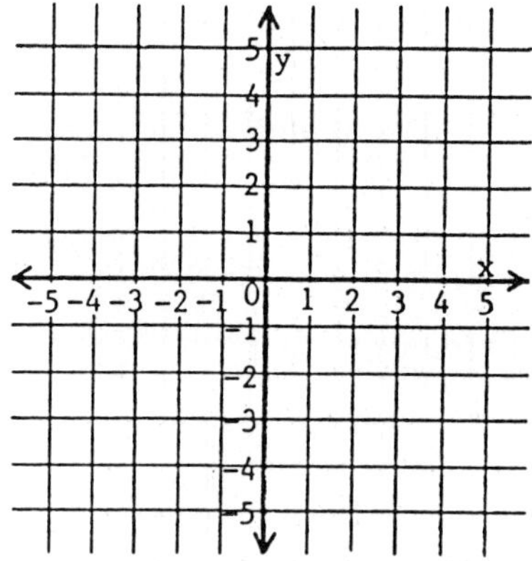

38. $x = y^2 + 1$

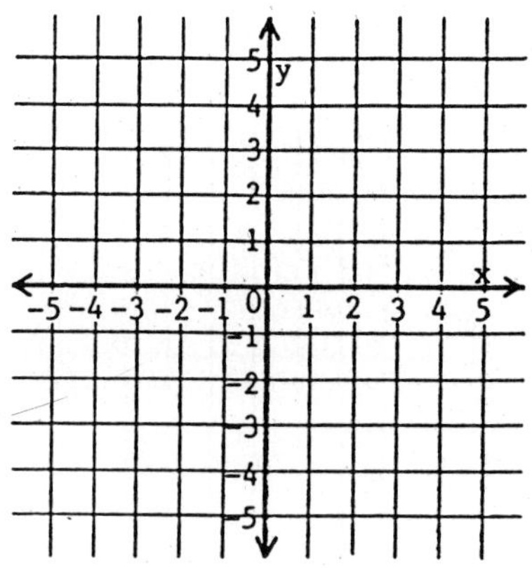

39. $x = -2y^2 + 3$

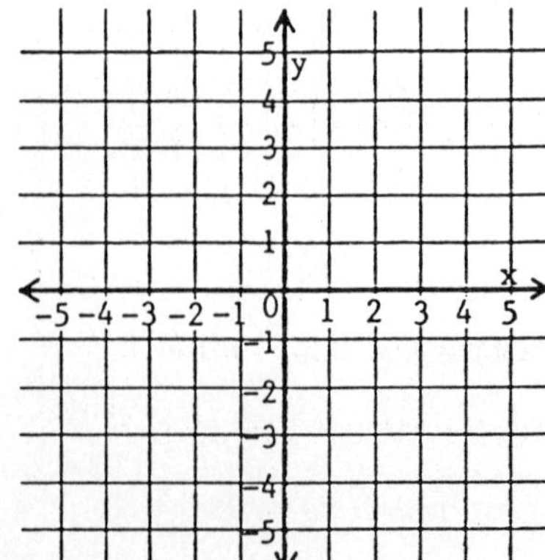

40. $x = 2(y - 1)^2 - 4$

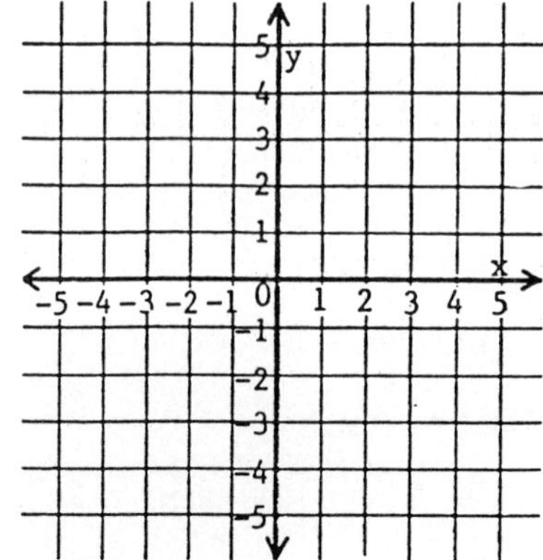

41. $x = -(y + 2)^2 + 1$

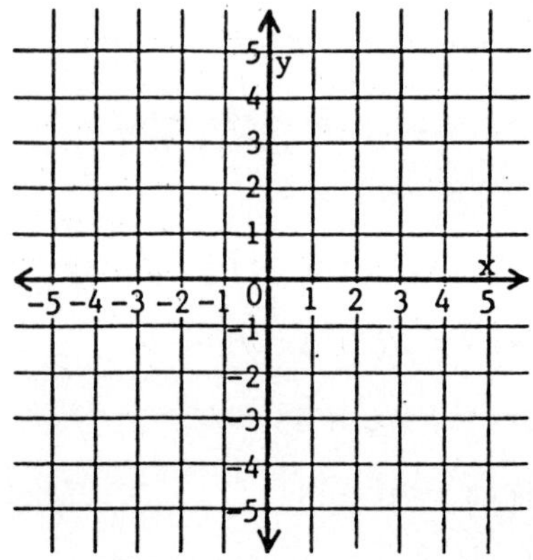

Assignment 33

Find the distance between each pair of points.

1. (2,7) and (2,1) 2. (-6,2) and (3,2) 3. (1,4) and (4,8) 4. (-4,-2) and (4,4)

5. (-3,4) and (1,-3) 6. (-5,-2) and (1,5) 7. (-3,0) and (-3,-4) 8. (8,7) and (6,3)

9. (-1,5) and (-3,2) 10. (0,-6) and (-1,-6) 11. (6,0) and (-3,-1) 12. (0,10) and (5,5)

Identify the center and radius of each circle.

13. $(x - 4)^2 + (y - 2)^2 = 9$ 14. $(x - 1)^2 + (y + 6)^2 = 49$ 15. $(x + 5)^2 + (y - 3)^2 = 64$

16. $x^2 + y^2 = 100$ 17. $(x + 3)^2 + (y + 5)^2 = 81$ 18. $x^2 + y^2 = 1$

Graph each circle.

19. $(x - 1)^2 + (y - 2)^2 = 4$

20. 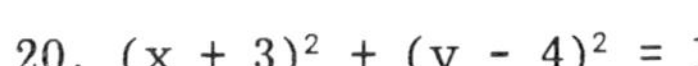
$(x + 3)^2 + (y - 4)^2 = 1$

21. $(x + 2)^2 + (y + 1)^2 = 9$

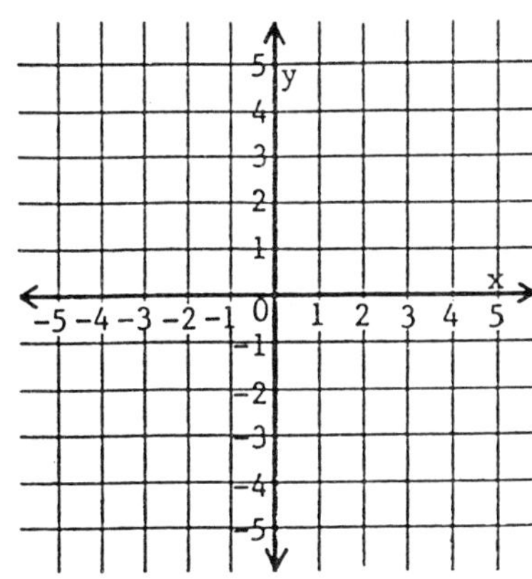

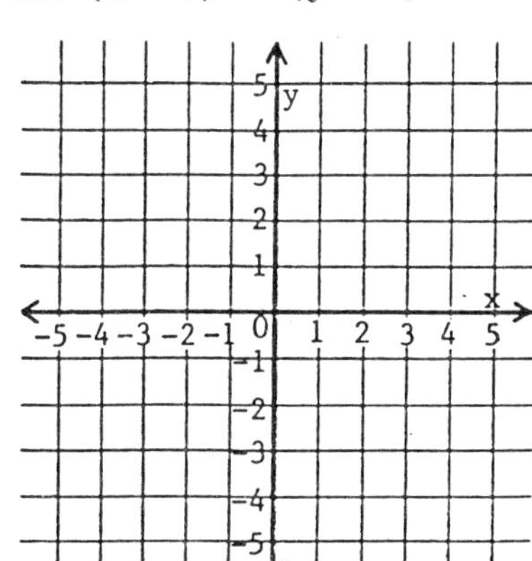

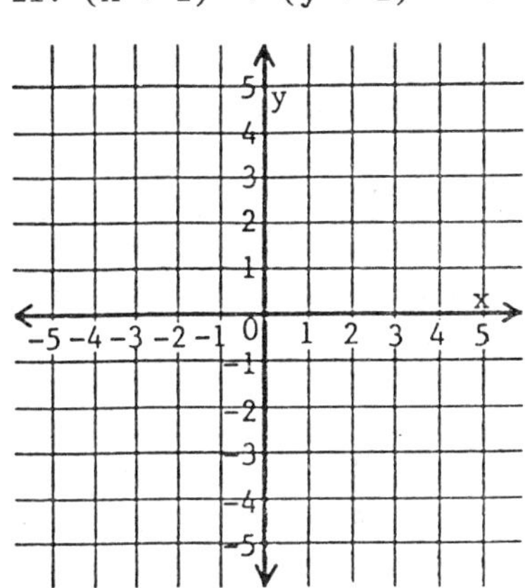

22. $x^2 + y^2 = 16$ 23. $(x - 3)^2 + y^2 = 4$ 24. $(x - 3)^2 + (y + 2)^2 = 5$

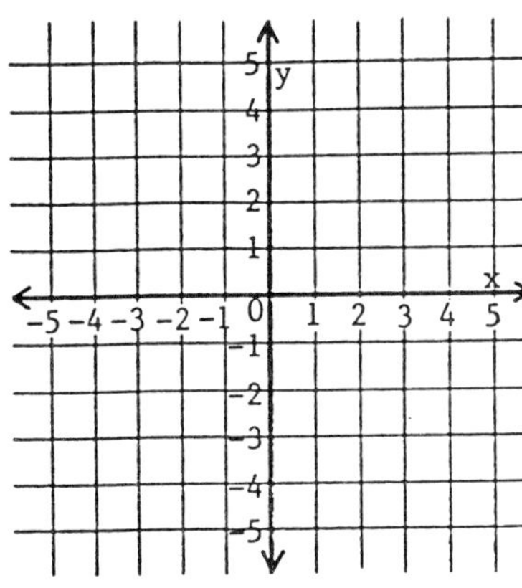

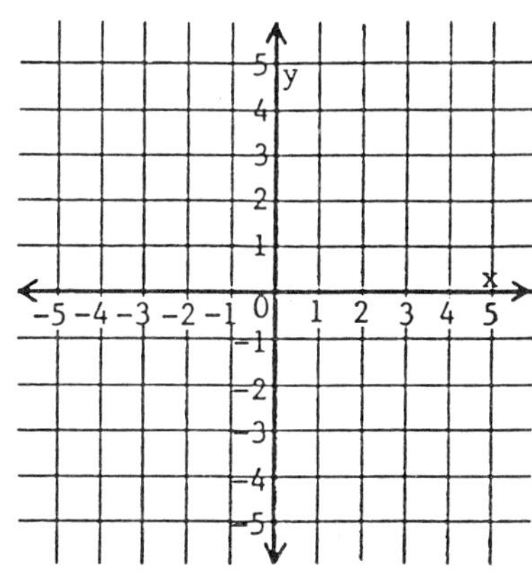

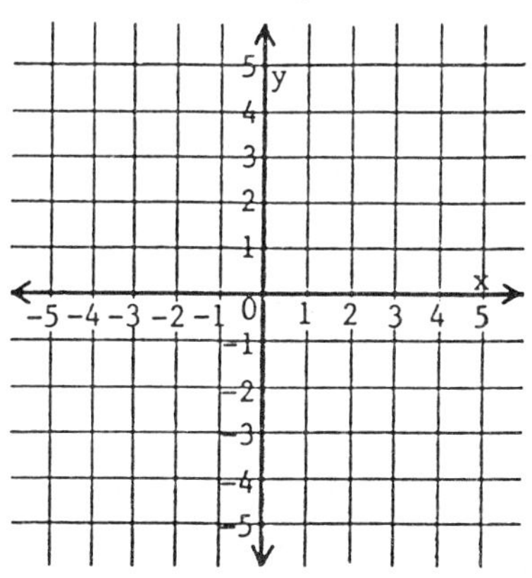

Find the equation of the circle with the given center and radius.

25. center (0,0), radius 6 26. center (4,3), radius 5 27. center (-2,1), radius 3

28. center (3,-5), radius 8 29. center (-3,-1), radius $\sqrt{10}$ 30. center (0,0), radius $\sqrt{17}$

Find the center and radius of each circle.

31. $x^2 + y^2 - 2x - 8y - 19 = 0$ 32. $x^2 + y^2 - 6x - 4y - 12 = 0$ 33. $x^2 + y^2 + 8x - 10y + 40 = 0$

34. $x^2 + y^2 + 6x + 12y + 29 = 0$ 35. $x^2 + y^2 - 10x + 2y + 19 = 0$ 36. $x^2 + y^2 + 16y - 36 = 0$

Find the x-intercepts and y-intercepts of each ellipse.

37. $\frac{x^2}{9} + \frac{y^2}{16} = 1$ 38. $\frac{x^2}{36} + \frac{y^2}{64} = 1$ 39. $\frac{x^2}{25} + \frac{y^2}{100} = 1$

Graph each ellipse.

40. $\frac{x^2}{9} + \frac{y^2}{4} = 1$ 41. $\frac{x^2}{4} + \frac{y^2}{16} = 1$ 42. $\frac{x^2}{16} + \frac{y^2}{9} = 1$

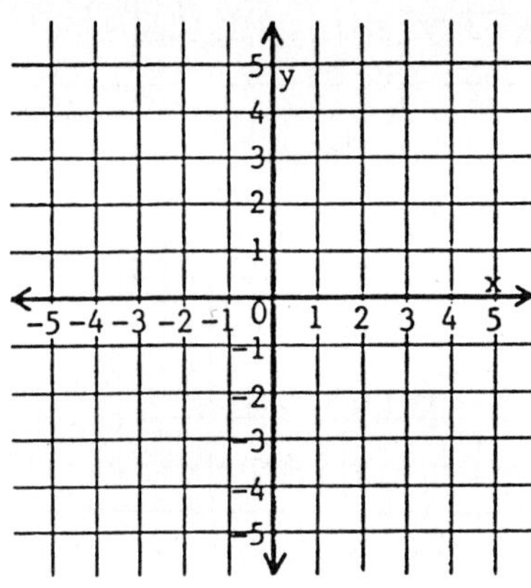

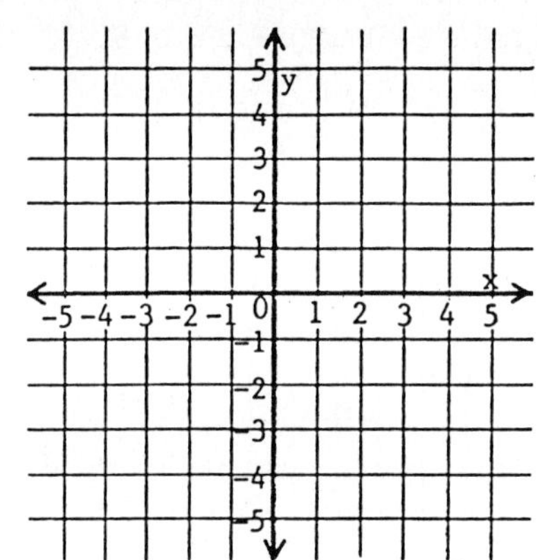

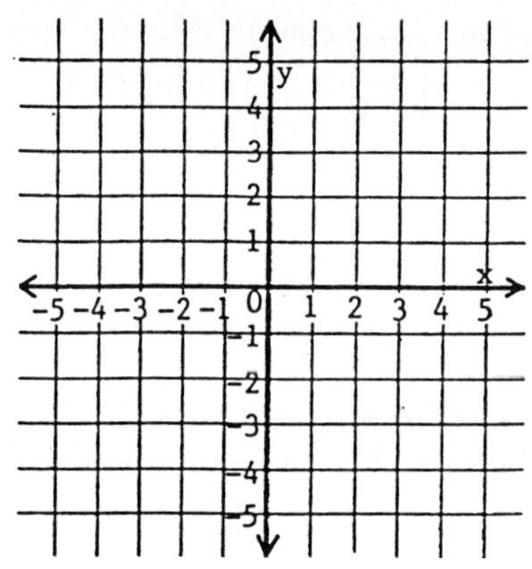

Assignment 34

Find the intercepts of each hyperbola.

1. $\frac{x^2}{25} - \frac{y^2}{9} = 1$ 2. $\frac{y^2}{4} - \frac{x^2}{49} = 1$ 3. $\frac{y^2}{36} - \frac{x^2}{4} = 1$ 4. $\frac{x^2}{1} - \frac{y^2}{81} = 1$

Find the equations of the asymptotes of each hyperbola.

5. $\frac{x^2}{36} - \frac{y^2}{25} = 1$ 6. $\frac{y^2}{4} - \frac{x^2}{36} = 1$ 7. $\frac{x^2}{9} - \frac{y^2}{36} = 1$ 8. $xy = 10$

9. $\frac{y^2}{100} - \frac{x^2}{64} = 1$ 10. $\frac{x^2}{64} - \frac{y^2}{64} = 1$ 11. $xy = 50$ 12. $\frac{y^2}{16} - \frac{x^2}{36} = 1$

Graph each hyperbola.

13. $\frac{x^2}{4} - \frac{y^2}{4} = 1$ 14. $\frac{y^2}{16} - \frac{x^2}{4} = 1$ 15. $\frac{x^2}{9} - \frac{y^2}{16} = 1$

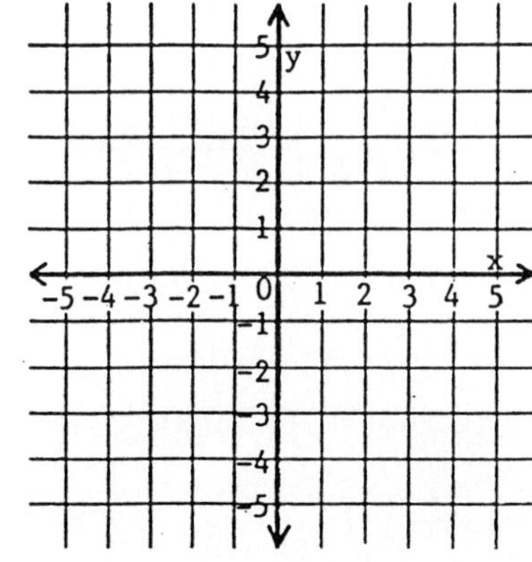

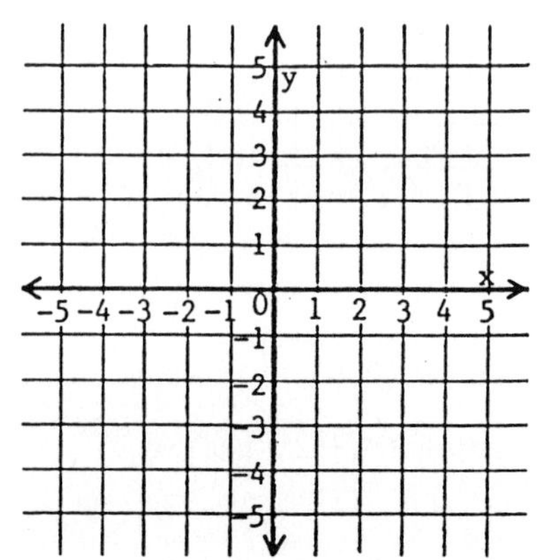

16. $\frac{y^2}{16} - \frac{x^2}{9} = 1$

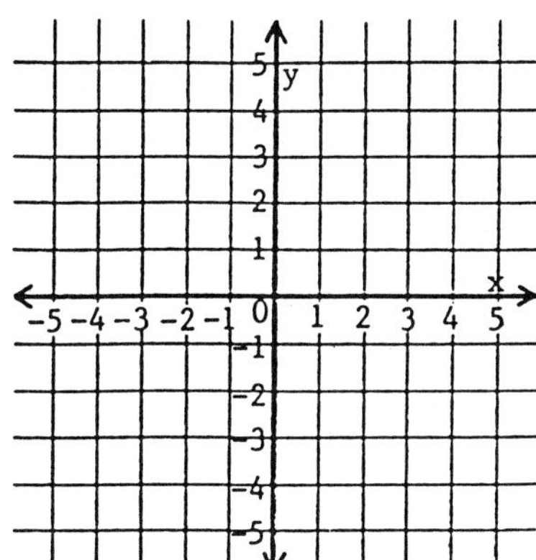

17. $xy = 10$

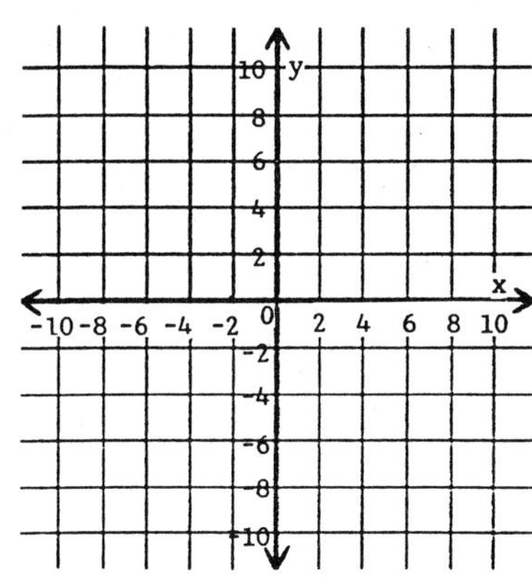

18. $xy = -12$

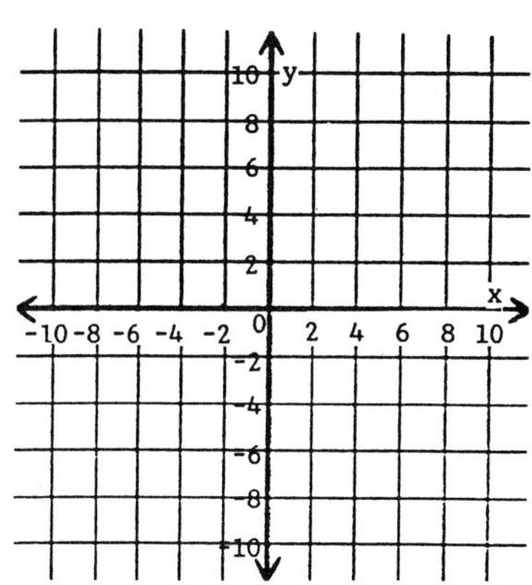

Solve each system of equation.

19. $x^2 + y^2 = 20$
    $x = y - 2$

20. $x^2 + y^2 = 61$
    $y = x - 1$

21. $9x^2 + 4y^2 = 36$
    $3x + 2y = -6$

22. $x^2 + 4y^2 = 16$
    $2y + 4 = x$

23. $y = 2x^2$
    $y + 2 = 4x$

24. $y = 3x^2$
    $y - 3x = 6$

25. $y = -x^2$
    $x + y = -2$

26. $y = -2x^2$
    $y + 8x = 8$

27. $xy = 10$
    $5y + 5 = x$

28. $xy = 16$
    $2y + 4 = x$

29. $xy = -6$
    $x + y = -1$

30. $x^2 + y^2 = 13$
    $3x^2 + 2y^2 = 30$

31. $x^2 + y^2 = 25$
    $x^2 - y^2 = 25$

32. $5x^2 - 2y^2 = -13$
    $3x^2 + 4y^2 = 39$

33. $x^2 + y^2 = 13$
    $xy = 6$

34. $x^2 + y^2 = 26$
    $xy = 5$

Find the inverse of each function.

35. $f(x) = x + 5$

36. $f(x) = x - 2$

37. $f(x) = x + 20$

38. $f(x) = 4x$

39. $f(x) = -7x$

40. $f(x) = 5x + 1$

41. $f(x) = 2x + 3$

42. $f(x) = 9x - 7$

Find the inverse of each function if its domain is restricted to nonnegative numbers.

43. $f(x) = x^2$

44. $f(x) = x^2 + 3$

45. $f(x) = x^2 - 2$

46. $f(x) = x^2 + 10$

10 Exponential and Logarithmic Functions

In this chapter, we will examine exponential and logarithmic functions and their graphs. We will discuss the laws of logarithms, scientific notation, common logarithms, powers of "e", and natural logarithms. We will also discuss methods for solving exponential and logarithmic equations.

10-1 EXPONENTIAL FUNCTIONS

In this section, we will discuss exponential functions and their graphs.

---

1. In an earlier chapter, we saw the meaning of powers with rational exponents.

$5^3 = 125$ $\qquad$ $6^{\frac{1}{2}} = \sqrt{6}$

$3^{-1} = \frac{1}{3}$ $\qquad$ $2^{\frac{5}{4}} = \sqrt[4]{2^5} = \sqrt[4]{32}$

Any power with a fractional exponent can be converted to a power with a decimal exponent. For example:

$6^{\frac{1}{2}} = 6^{0.5}$ $\qquad$ $2^{\frac{5}{4}} = 2^{1.25}$

When the fraction converts to a non-ending decimal number, the decimal number is rounded. In the examples below, we rounded each exponent to hundredths.

$y^{\frac{1}{3}} = y^{0.33}$ $\qquad$ $x^{\frac{7}{6}} = x^{1.17}$

Continued on following page.

1. Continued

Remember the meaning of powers with decimal exponents.

$7^{1.69}$ or $7^{\frac{169}{100}}$ means: Raise 7 to the 169th power and then take the 100th root.

$5^{2.3}$ or $5^{\frac{23}{10}}$ means: ______________________

____

2. Assuming that powers with irrational exponents also have a meaning, we know that any expression of the form b^x has meaning for any real number x. Therefore, we can define exponential functions.

> An exponential function is a function of the form:
>
> $$f(x) = b^x$$
>
> where b is any positive real number.

To graph the exponential function $y = 2^x$ below, we made up a solution table, plotted the points, and drew a smooth curve through them.

$y = 2^x$

x	y
-3	$\frac{1}{8}$
-2	$\frac{1}{4}$
-1	$\frac{1}{2}$
0	1
1	2
2	4
3	8

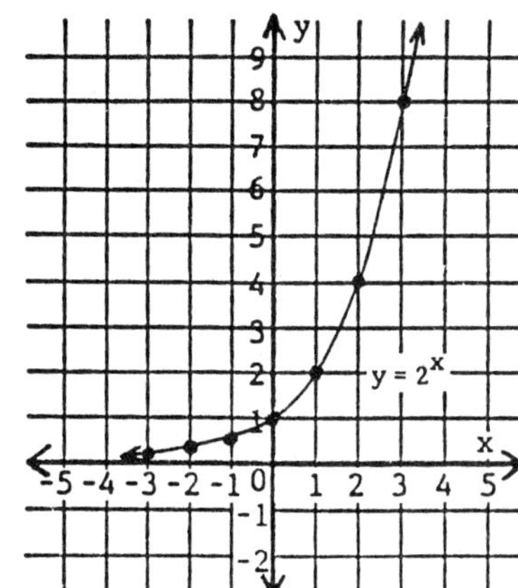

Notice these facts about the function.

1. As x increases, y increases. As x decreases, y decreases.

2. The x-axis is an asymptote. The curve gets closer and closer to it.

Continued on following page.

Raise 5 to the 23rd power and then take the 10th root.

2. Continued

Graph the exponential function $y = 3^x$ below. Use the values in the table.

$y = 3^x$

x	y
-2	
-1	
0	
1	
2	

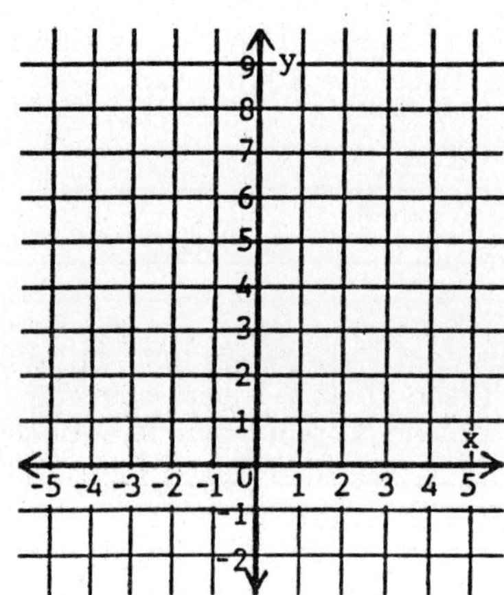

---

3. We evaluated the exponential function $y = \left(\frac{1}{2}\right)^x$ for various x-values below.

If $x = 2$, $y = \left(\frac{1}{2}\right)^2 = \frac{1}{4}$

If $x = 0$, $y = \left(\frac{1}{2}\right)^0 = 1$

If $x = -3$, $y = \left(\frac{1}{2}\right)^{-3} = \frac{1}{\left(\frac{1}{2}\right)^3} = \frac{1}{\frac{1}{8}} = 8$

To graph $y = \left(\frac{1}{2}\right)^x$, we used the values in the table below.

$y = \left(\frac{1}{2}\right)^x$

x	y
-3	8
-2	4
-1	2
0	1
1	$\frac{1}{2}$
2	$\frac{1}{4}$
3	$\frac{1}{8}$

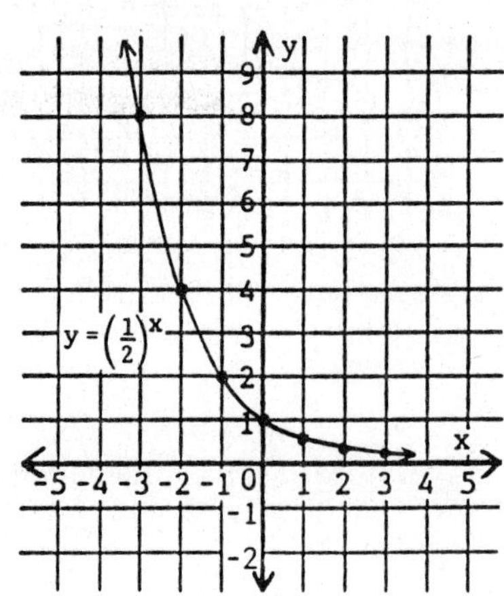

Continued on following page.

Answer To Frame 2:

x	y
-2	$\frac{1}{9}$
-1	$\frac{1}{3}$
0	1
1	3
2	9

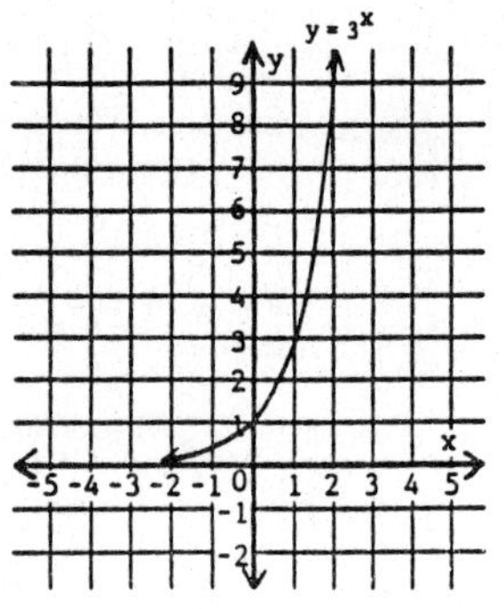

3. Continued

Notice these facts about the function:

1. As x increases, y decreases. As x decreases, y increases.

2. The x-axis is an <u>asymptote</u>. The curve gets closer and closer to it.

Graph the exponential function $y = \left(\frac{1}{3}\right)^x$ below. Use the values in the table.

$y = \left(\frac{1}{3}\right)^x$

x	y
-2	
-1	
0	
1	
2	

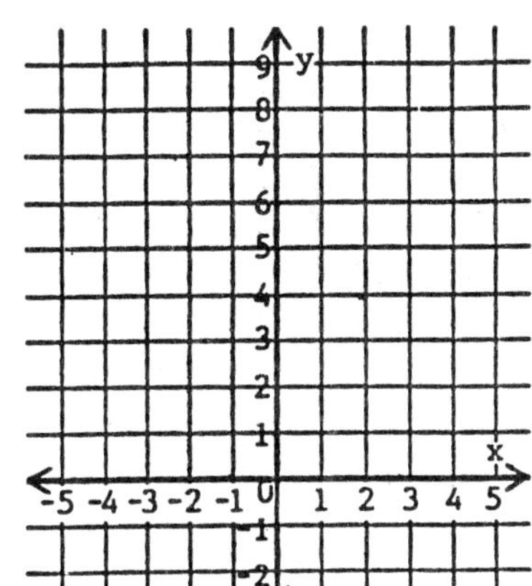

4. The graphs of the four exponential functions from the last two frames are shown together below.

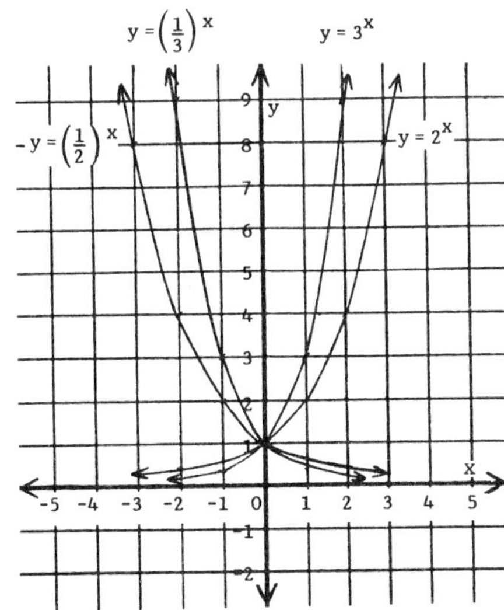

Continued on following page.

<u>Answer To Frame 3:</u>

x	y
-2	9
-1	3
0	1
1	$\frac{1}{3}$
2	$\frac{1}{9}$

4. Continued

All four graphs are called exponential curves. Notice these facts:

1. When $b > 1$, the curve rises from left to right. The larger the value of b, the steeper the rise. For example, the rise of $y = 3^x$ is steeper than the rise of $y = 2^x$.
2. When $b < 1$, the curve falls from left to right. The smaller the value of b, the steeper the fall. For example, the fall of $y = \left(\frac{1}{3}\right)^x$ is steeper than the fall of $y = \left(\frac{1}{2}\right)^x$.
3. All four graphs go through the point (0,1) because any base to the zero power equals "1".
4. All four graphs have the x-axis (or $y = 0$) as an asymptote.

We evaluated the exponential function $y = 1^x$ for various values of x below.

$$\text{If } x = 3, \quad y = 1^3 = 1$$

$$\text{If } x = 0, \quad y = 1^0 = 1$$

$$\text{If } x = -2, \quad y = 1^{-2} = \frac{1}{1^2} = \frac{1}{1} = 1$$

As you can see, for any value of x, $y = 1$. Therefore, the graph of $y = 1^x$ is the line ___________.

---

5. We evaluated the exponential function $y = 2^{x-2}$ for various x-values below. | $y = 1$

$$\text{If } x = 4, \quad y = 2^{4-2} = 2^2 = 4$$

$$\text{If } x = 2, \quad y = 2^{2-2} = 2^0 = 1$$

$$\text{If } x = -1, \quad y = 2^{-1-2} = 2^{-3} = \frac{1}{8}$$

To graph $y = 2^{x-2}$, we used the values in the table below.

$y = 2^{x-2}$

x	y
-1	$\frac{1}{8}$
0	$\frac{1}{4}$
1	$\frac{1}{2}$
2	1
3	2
4	4
5	8

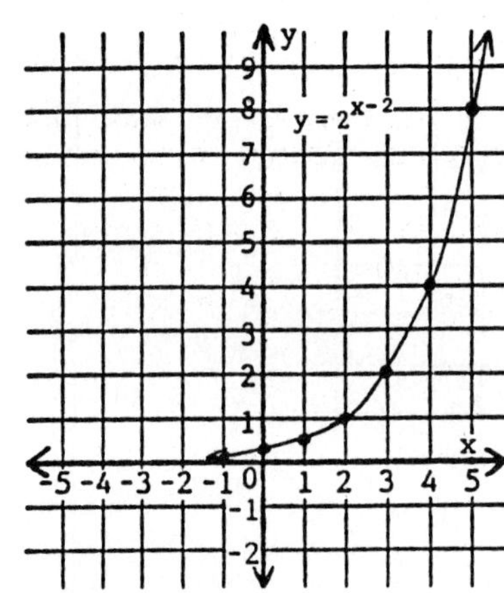

Continued on following page.

5. Graph the exponential function $y = 2^{x + 2}$. Use the values in the table.

$y = 2^{x + 2}$

x	y
-5	
-4	
-3	
-2	
-1	
0	
1	

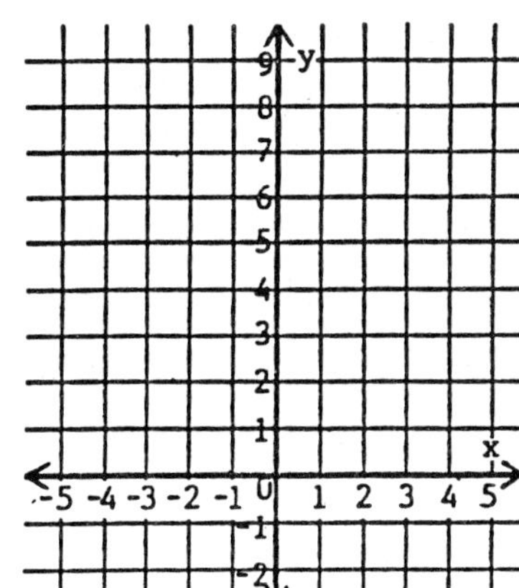

6. The graphs of the two functions from the last frame are shown with the graph of $y = 2^x$ below.

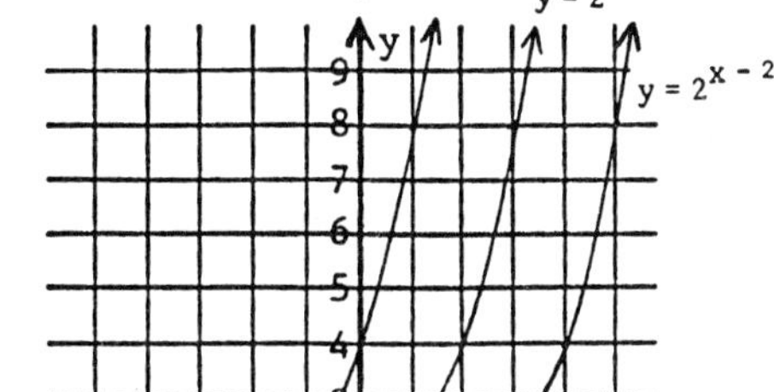

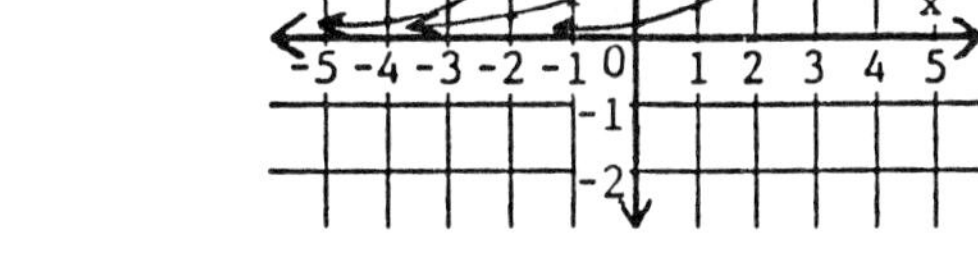

As you can see, the other two graphs have the same shape as the graph of $y = 2^x$ except that they are shifted to the right or left.

$y = 2^{x - 2}$ is shifted two units to the right.

$y = 2^{x + 2}$ is shifted two units to the left.

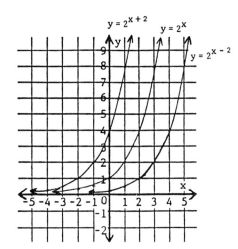

Answer To Frame 5:

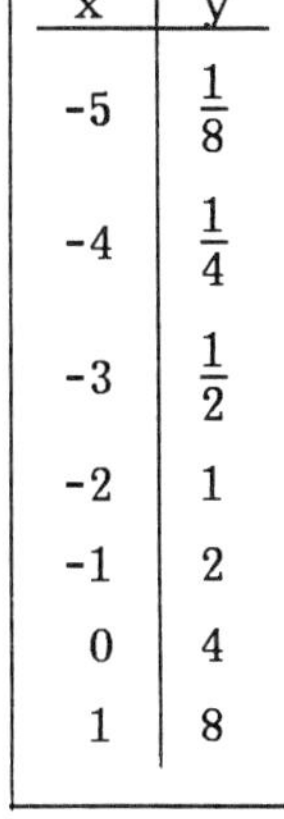

x	y
-5	$\frac{1}{8}$
-4	$\frac{1}{4}$
-3	$\frac{1}{2}$
-2	1
-1	2
0	4
1	8

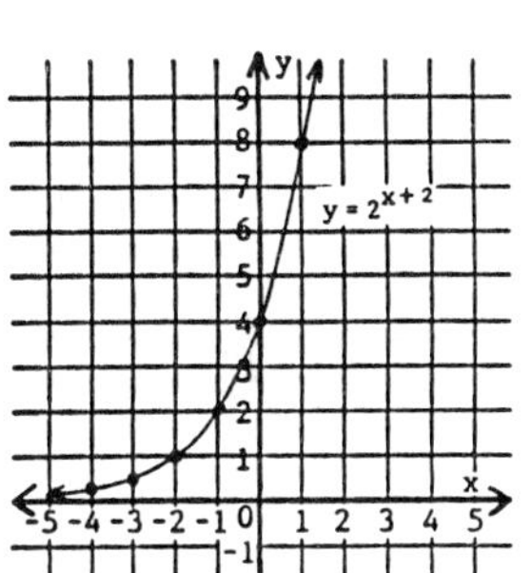

Continued on following page.

6. Continued

Using the facts above, sketch the graphs of $y = 3^{x-1}$ and $y = 3^{x+4}$ on the graph below. The graph of $y = 3^x$ is given.

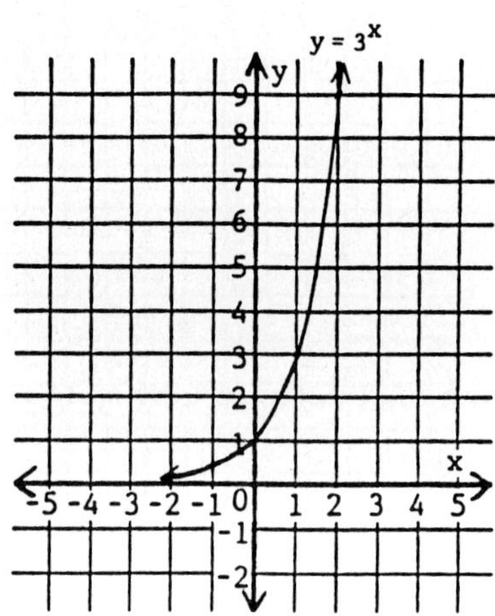

Answer To Frame 6:

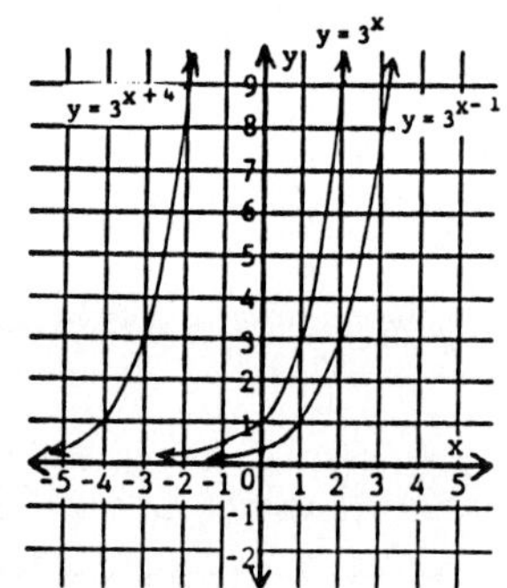

7. To get the inverse of $y = 2^x$, we interchange the $\underline{x}$ and $\underline{y}$ and get $x = 2^y$. We graphed both $y = 2^x$ and $x = 2^y$ below. Notice that the table for $x = 2^y$ is the same as the table for $y = 2^x$ except that the values for $\underline{x}$ and $\underline{y}$ are interchanged.

$y = 2^x$

x	y
-3	$\frac{1}{8}$
-2	$\frac{1}{4}$
-1	$\frac{1}{2}$
0	1
1	2
2	4
3	8

$x = 2^y$

x	y
$\frac{1}{8}$	-3
$\frac{1}{4}$	-2
$\frac{1}{2}$	-1
1	0
2	1
4	2
8	3

Note: 1. The inverse $x = 2^y$ is a function. (Use the vertical line test.)

2. The inverse is the mirror image of $y = 2^x$ with respect to the line $y = x$.

Continued on following page.

7. Continued

Graph $y = 3^x$ and its inverse.

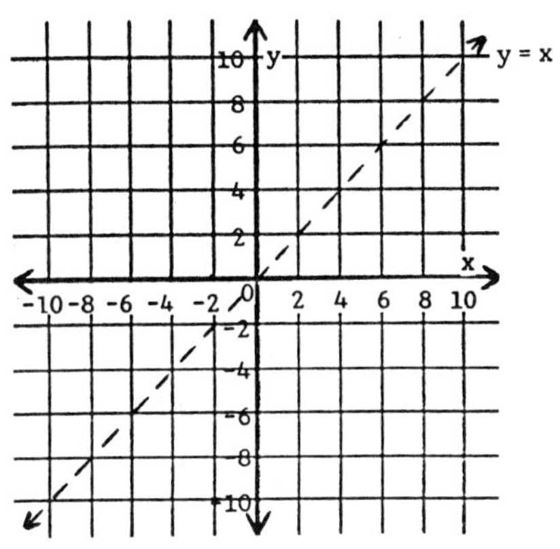

Answer To Frame 7:

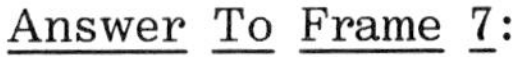

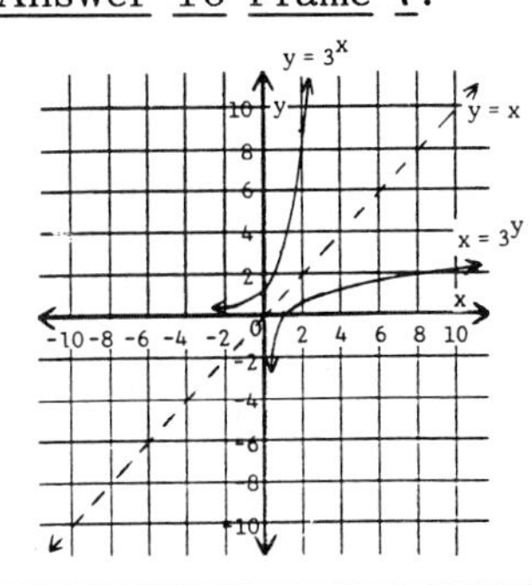

10-2 LOGARITHMIC NOTATION

In this section, we will discuss logarithmic notation, solve some basic logarithmic equations, and find the logarithms of some numbers.

8. We labeled the parts in the exponential statement below.

$$2^3 = 8$$

Exponent: 3; Base: 2; Number: 8

Instead of the word <u>exponent</u>, we can use the word <u>logarithm</u>.

$$2^3 = 8$$

Logarithm: 3; Base: 2; Number: 8

Continued on following page.

8. Continued

If we are given the base and logarithm, we can find the number. For example:

If the base is 5 and the logarithm is 2, the number is $5^2 = 25$.

If the base is 8 and the logarithm is -1, the number is $8^{-1} = \frac{1}{8}$.

Given the following bases and logarithms, find the numbers.

a) base = 3, logarithm = 4, number = ________

b) base = 10, logarithm = 3, number = ________

c) base = 6, logarithm = -2, number = ________

---

9. The statement $2^3 = 8$ is written in logarithmic form below. The abbreviation "log" is used for the word "logarithm".

$2^3 = 8$ can be written $\log_2 8 = 3$

$\log_2 8 = 3$ is read "the logarithm of 8 to the base 2 is 3".

Three more examples of the same statement in exponential form and logarithmic form are given below.

Exponential Form	Logarithmic Form
$7^2 = 49$	$\log_7 49 = 2$
$3^4 = 81$	$\log_3 81 = 4$
$10^{-2} = 0.01$	$\log_{10} 0.01 = -2$

Write each statement below in logarithmic form.

a) $2^5 = 32$ b) $6^3 = 216$ c) $10^{-1} = 0.1$

________ ________ ________

Answers to 8:

a) $3^4 = 81$

b) $10^3 = 1{,}000$

c) $6^{-2} = \frac{1}{36}$

---

10. Any statement in logarithmic form is equivalent to a statement in exponential form. That is:

$\log_b x = y$ is equivalent to $x = b^y$

Therefore, any statement in logarithmic form can be written in exponential form. For example:

$\log_5 125 = 3$ can be written $5^3 = 125$

$\log_2 64 = 6$ can be written $2^6 = 64$

Write each statement below in exponential form.

a) $\log_9 81 = 2$ b) $\log_4 64 = 3$ c) $\log_{10} 0.001 = -3$

________ ________ ________

Answers to 9:

a) $\log_2 32 = 5$

b) $\log_6 216 = 3$

c) $\log_{10} 0.1 = -1$

11. Three more examples of the same statement in logarithmic form and exponential form are shown below. Each statement contains a variable.

Logarithmic Form	Exponential Form
$\log_2 x = 4$	$2^4 = x$
$\log_x 64 = 2$	$x^2 = 64$
$\log_{10} 1000 = x$	$10^x = 1000$

Write each statement below in exponential form.

a) $\log_5 x = 3$ b) $\log_x 32 = 5$ c) $\log_{10} 0.1 = x$

________ ________ ________

a) $9^2 = 81$

b) $4^3 = 64$

c) $10^{-3} = 0.001$

---

12. We solved each logarithmic equation below by converting to exponential form.

$$\log_2 x = 5 \qquad \log_{10} x = -2$$

$$2^5 = x \qquad 10^{-2} = x$$

$$32 = x \qquad 0.01 = x$$

Check: $\log_2 32 = 5$, since $2^5 = 32$

$\log_{10} 0.01 = -2$, since $10^{-2} = 0.01$

Using the same method, solve these equations.

a) $\log_4 x = 2$ b) $\log_{10} x = 4$ c) $\log_2 x = -1$

a) $5^3 = x$

b) $x^5 = 32$

c) $10^x = 0.1$

---

13. We solved each logarithmic equation below by converting to exponential form.

$$\log_x 64 = 3 \qquad \log_x 0.001 = -3$$

$$x^3 = 64 \qquad x^{-3} = 0.001$$

$$x^3 = 4^3 \qquad x^{-3} = 10^{-3}$$

$$x = 4 \qquad x = 10$$

Check: $\log_4 64 = 3$, since $4^3 = 64$

$\log_{10} 0.001 = -3$, since $10^{-3} = 0.001$

Use the same method to solve these equations.

a) $\log_x 32 = 5$ b) $\log_x 125 = 3$ c) $\log_x 0.1 = -1$

a) $x = 16$

b) $x = 10,000$

c) $x = \frac{1}{2}$

14. We solved each equation below by converting to exponential form.

$$\log_2 16 = x \qquad\qquad \log_{10} 0.1 = x$$
$$2^x = 16 \qquad\qquad 10^x = 0.1$$
$$2^x = 2^4 \qquad\qquad 10^x = 10^{-1}$$
$$x = 4 \qquad\qquad x = -1$$

<u>Check</u>: $\log_2 16 = 4$, since $2^4 = 16$

$\log_{10} 0.1 = -1$, since $10^{-1} = 0.1$

Solve these equations.

a) $\log_6 36 = x$ b) $\log_3 27 = x$ c) $\log_{10} 0.001 = x$

a) $x = 2$
b) $x = 5$
c) $x = 10$

---

15. We used equations below to find $\log_2 64$ and $\log_{10} 0.001$.

Find $\log_2 64$.

Let $\log_2 64 = x$
$$2^x = 64$$
$$2^x = 2^6$$
$$x = 6$$

Therefore, $\log_2 64 = 6$

Find $\log_{10} 0.001$.

Let $\log_{10} 0.001 = x$
$$10^x = 0.001$$
$$10^x = 10^{-3}$$
$$x = -3$$

Therefore, $\log_{10} 0.001 = -3$

Find the following:

a) $\log_4 16$ b) $\log_{10} 10{,}000$ c) $\log_{10} 0.1$

a) $x = 2$
b) $x = 3$
c) $x = -3$

---

a) 2
b) 4
c) -1

16. We used an equation to find one logarithm below. Find the other logarithm.

Find $\log_5\left(\frac{1}{25}\right)$.

Fing $\log_2\left(\frac{1}{32}\right)$.

Let $\log_5\left(\frac{1}{25}\right) = x$

$$5^x = \frac{1}{25}$$

$$5^x = \frac{1}{5^2}$$

$$5^x = 5^{-2}$$

$$x = -2$$

Therefore, $\log_5\left(\frac{1}{25}\right) = -2$

-5

17. The logarithm of any number with itself as a base is "1". For example:

Find $\log_5 5$.

Let $\log_5 5 = x$

$$5^x = 5$$

$$5^x = 5^1$$

$$x = 1$$

Therefore, $\log_5 5 = 1$

The logarithm of "1" to any base is 0. For example:

Find $\log_3 1$.

Let $\log_3 1 = x$

$$3^x = 1$$

$$3^x = 3^0$$

$$x = 0$$

Therefore, $\log_3 1 = 0$

Find the following.

a) $\log_9 9$ b) $\log_9 1$ c) $\log_4 1$ d) $\log_2 2$

---

a) 1 c) 0

b) 0 d) 1

10-3 LOGARITHMIC FUNCTIONS

In this section, we will discuss logarithmic functions and their graphs.

18. In an earlier section, we saw that we get the inverse of an exponential function by interchanging x and y. For example:

The inverse of $y = 4^x$ is $x = 4^y$.

The inverse above can be converted to logarithmic form. That is:

$x = 4^y$ is the same as $\log_4 x = y$ or $y = \log_4 x$.

$y = \log_4 x$ is called a logarithmic function. It is equivalent to $x = 4^y$, the inverse of $y = 4^x$.

We graphed $y = 4^x$ and $y = \log_4 x$ below. Notice that the table for $y = \log_4 x$ (or $x = 4^y$) is the same as the table for $y = 4^x$ except that the values for x and y are interchanged.

$y = 4^x$

x	y
-2	$\frac{1}{16}$
-1	$\frac{1}{4}$
0	1
1	4
2	16

$y = \log_4 x$ (or $x = 4^y$)

x	y
$\frac{1}{16}$	-2
$\frac{1}{4}$	-1
1	0
4	1
16	2

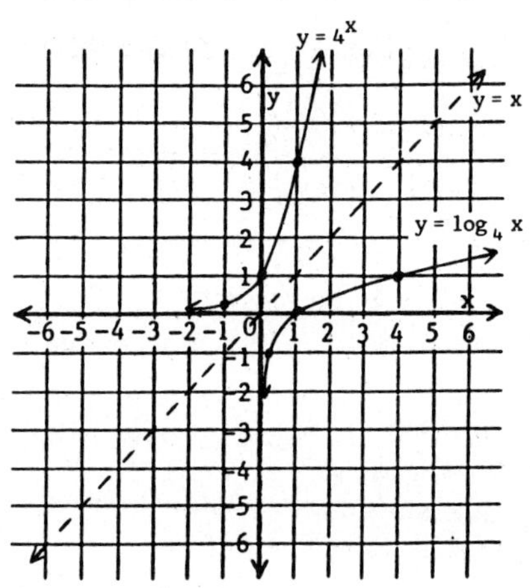

Note: You can see that $y = \log_4 x$ is the mirror image of $y = 4^x$ with respect to the line $y = x$. Therefore, it is the inverse of $y = 4^x$.

19. To graph $y = \log_3 x$ below, we converted to the exponential form $x = 3^y$ and then substituted values for $\underline{y}$. Use the same method to graph $y = \log_5 x$.

$y = \log_3 x$
(or $x = 3^y$)

x	y
$\frac{1}{9}$	-2
$\frac{1}{3}$	-1
1	0
3	1
9	2

$y = \log_5 x$

x	y

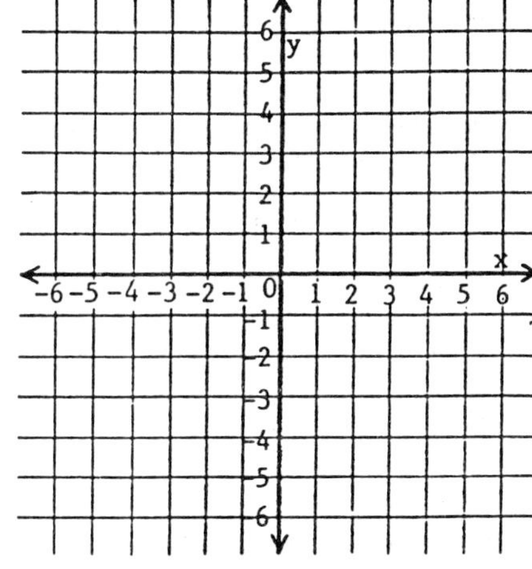

---

20. The inverse of any exponential function $y = b^x$ is a logarithmic function $y = \log_b x$ which is equivalent to $x = b^y$.

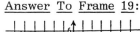

> A $\underline{\text{logarithmic}}$ $\underline{\text{function}}$ is a function of the form:
>
> $$y = \log_b x$$
>
> where $x > 0$ and $\underline{b}$ is a positive number other than 1.

We sketched $y = b^x$ and $y = \log_b x$ at the right. Notice these facts about $y = \log_b x$.

1. The logarithm of the number "1" to any base is 0. That is:

   When $x = 1$, $\log_b x = 0$

2. The logarithm of any number greater than "1" is positive. That is:

   When $x > 1$, $\log_b x$ is positive.

3. The logarithm of any number between 0 and 1 is negative. That is:

   When $0 < x < 1$, $\log_b x$ is negative.

4. There are no logarithms for 0 and negative numbers. That is:

   When $x \leq 0$, $\log_b x$ does not exist.

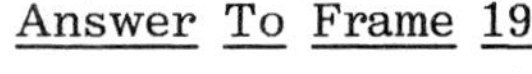
Answer To Frame 19:

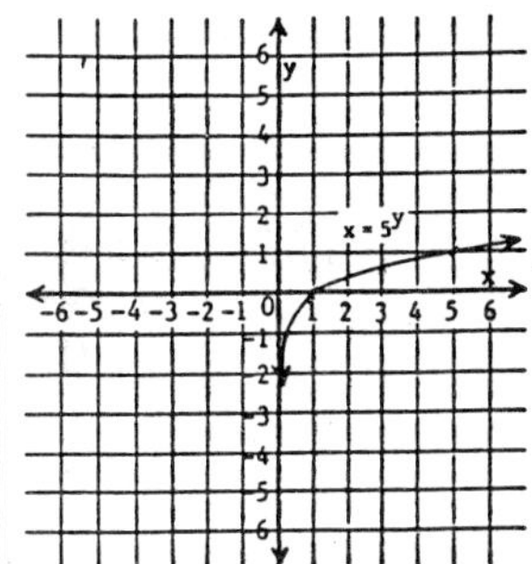

SELF-TEST 35 (pages 500-514)

Graph each function.

1. $y = 3^{x-2}$

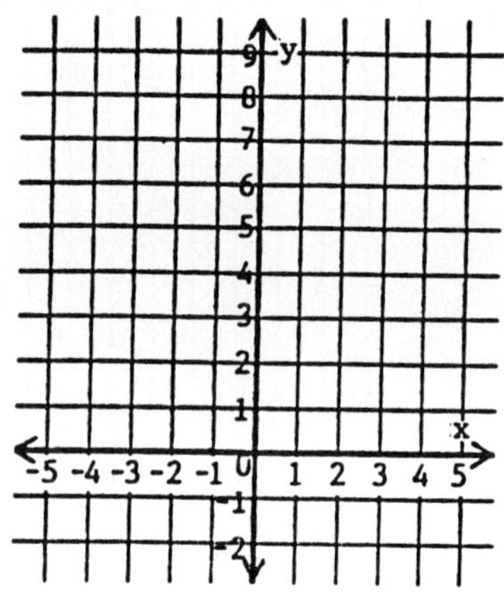

2. $y = \log_6 x$

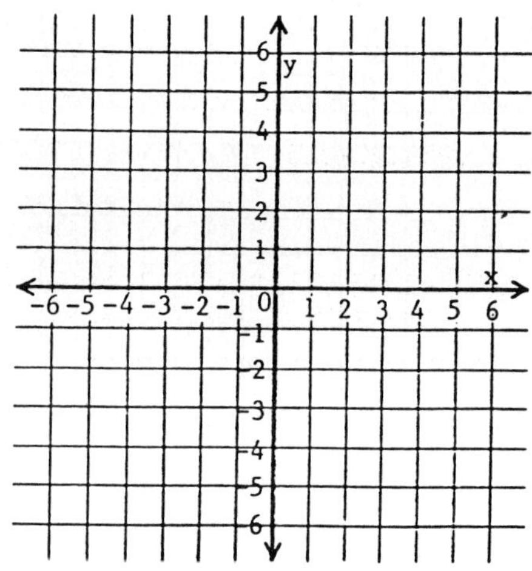

Write in logarithmic form.

3. $5^3 = x$

4. $2^x = 32$

Write in exponential form.

5. $\log_x 27 = 3$

6. $\log_{10} 0.01 = x$

Solve each equation.

7. $\log_3 x = 4$

8. $\log_x 125 = 3$

9. $\log_4 x = -1$

Find the following.

10. $\log_4 64$

11. $\log_{10} 0.01$

12. $\log_7 7$

ANSWERS:

1. 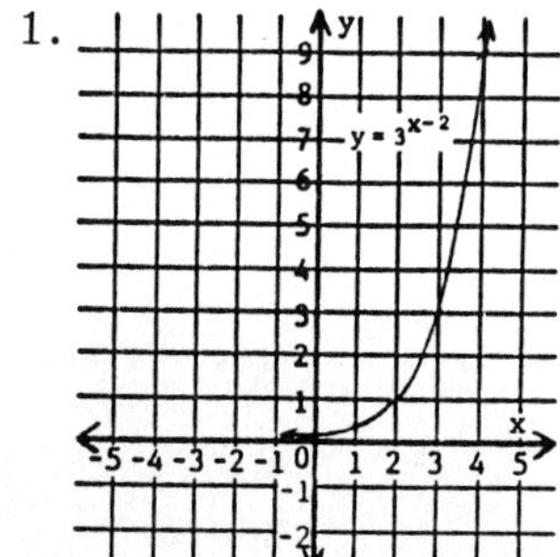

2. 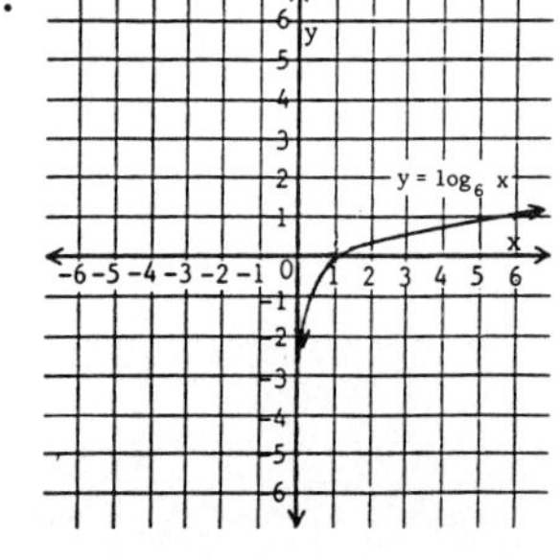

3. $\log_5 x = 3$
4. $\log_2 32 = x$
5. $x^3 = 27$
6. $10^x = 0.01$
7. $x = 81$
8. $x = 5$
9. $x = \frac{1}{4}$
10. 3
11. -2
12. 1

10-4 LAWS OF LOGARITHMS

The laws of logarithms, which are based on the laws of exponents, are used to rewrite expressions in equivalent forms. We will discuss those laws in this section.

---

21. The law of logarithms for products is stated below.

> Product Law. M and N are positive and $b > 1$.
>
> $$\log_b MN = \log_b M + \log_b N$$
>
> (The logarithm of a product is the sum of the logarithms of the factors.)

Following the example, use the product law to complete these:

$\log_3(9)(81) = \log_3 9 + \log_3 81$

a) $\log_4(16)(64) =$ ________ + ________

b) $\log_{10}(100)(1000) =$ ________ + ________

---

a) $\log_4 16 + \log_4 64$

b) $\log_{10}100 + \log_{10}1000$

22. A proof of the product law is given below. It is based on the law of exponents for multiplication.

Letting $\log_b M = x$ and $\log_b N = y$, we can convert to exponential form.

$\log_b M = x \longrightarrow M = b^x$

$\log_b N = y \longrightarrow N = b^y$

Using the law of exponents for multiplication, we get:

$MN = b^x \cdot b^y = b^{x+y}$

Converting the exponential form back to logarithmic form, we get:

$MN = b^{x+y} \longrightarrow \log_b MN = x + y$

Substituting $\log_b M$ for $\underline{x}$ and $\log_b N$ for $\underline{y}$, we get:

$\log_b MN = \log_b M + \log_b N$

Following the example, use the product law to complete these.

$\log_5 3x = \log_5 3 + \log_5 x$

a) $\log_8 10y =$ ________ + ________

b) $\log_b PQ =$ ________ + ________

---

a) $\log_8 10 + \log_8 y$

b) $\log_b P + \log_b Q$

23. Following the example, use the reverse of the product law for these.

$\log_2 4 + \log_2 16 = \log_2 (4)(16)$

a) $\log_7 5 + \log_7 x =$ ____________

b) $\log_b S + \log_b T =$ ____________

---

24. The law of logarithms for quotients is stated below. The proof of the quotient law is similar to the proof of the product law. It is based on the law of exponents for division.

> Quotient Law. M and N are positive and $b > 1$.
>
> $$\log_b\left(\frac{M}{N}\right) = \log_b M - \log_b N$$
>
> (The logarithm of a quotient is the difference of the logarithms of the numerator and denominator.)

Following the example, use the quotient law to complete these.

$\log_2\left(\frac{16}{8}\right) = \log_2 16 - \log_2 8$

a) $\log_4\left(\frac{16}{64}\right) =$ __________ - __________

b) $\log_{10}\left(\frac{1000}{10}\right) =$ __________ - __________

Answers to 23: a) $\log_7 5x$ b) $\log_b ST$

---

25. Following the example, use the quotient law to complete these.

$\log_5\left(\frac{x}{9}\right) = \log_5 x - \log_5 9$

a) $\log_{10}\left(\frac{y}{20}\right) =$ __________ - __________

b) $\log_b\left(\frac{P}{Q}\right) =$ __________ - __________

Answers to 24: a) $\log_4 16 - \log_4 64$ b) $\log_{10} 1000 - \log_{10} 10$

---

26. Following the example, use the reverse of the quotient law for these.

$\log_2 16 - \log_2 32 = \log_2\left(\frac{16}{32}\right)$

a) $\log_7 y - \log_7 3 =$ ____________

b) $\log_b P - \log_b Q =$ ____________

Answers to 25: a) $\log_{10} y - \log_{10} 20$ b) $\log_b P - \log_b Q$

Answers to 26: a) $\log_7\left(\frac{y}{3}\right)$ b) $\log_b\left(\frac{P}{Q}\right)$

27. The law of logarithms for powers is stated below.

> Power Law. M is positive and p is any real number.
>
> $$\log_b M^p = p \log_b M$$
>
> (The logarithm of a power is the product of the exponent and the logarithm of M.)

Following the example, use the power law to complete these.

$$\log_3 6^4 = 4 \log_3 6$$

a) $\log_5 x^3 =$ ________ b) $\log_{10} a^y =$ ________

Answers: a) $3 \log_5 x$ b) $y \log_{10} a$

28. A proof of the power law is given below. It is based on the law of exponents for powers.

Letting $x = \log_b M$, we can convert to exponential form.

$$x = \log_b M \longrightarrow b^x = M$$

Raising both sides of $b^x = M$ to the pth power, we get:

$$(b^x)^p = M^p \quad \text{or} \quad b^{xp} = M^p$$

Converting the exponential form back to logarithmic form, we get:

$$b^{xp} = M^p \longrightarrow \log_b M^p = xp$$

Substituting $\log_b M$ for x, we get:

$$\log_b M^p = (\log_b M)p \quad \text{or} \quad p \log_b M$$

Following the example, use the power law for these.

$$\log_3 \sqrt{5} = \log_3 5^{\frac{1}{2}} = \frac{1}{2}\log_3 5$$

a) $\log_{10} \sqrt{x} =$ ________

b) $\log_2 \sqrt[3]{y} =$ ________

Answers: a) $\frac{1}{2}\log_{10} x$ b) $\frac{1}{3}\log_2 y$

29. Following the examples, use the reverse of the power law for these.

$$3 \log_4 15 = \log_4 15^3$$

$$\frac{1}{2}\log_6 x = \log_6 x^{\frac{1}{2}} = \log_6 \sqrt{x}$$

a) $5 \log_7 y =$ ________ b) $\frac{1}{3}\log_2 13 =$ ________

Answers: a) $\log_7 y^5$ b) $\log_2 \sqrt[3]{13}$

30. There is no law of logarithms for either addition or subtraction. Therefore, $\log_b(x + 2)$ and $\log_b(x - 2)$ cannot be written in an equivalent form.

$$\log_b(x + 2) \neq \log_b x + \log_b 2$$

$$\log_b(x - 2) \neq \log_b x - \log_b 2$$

Which of these statements are true? ________

a) $\log_b MN = \log_b M + \log_b N$

b) $\log_b(M + N) = \log_b M + \log_b N$

c) $\log_b(M - N) = \log_b M - \log_b N$

d) $\log_b\left(\frac{M}{N}\right) = \log_b M - \log_b N$

---

31. We used two laws to write the expression below in an equivalent form. Write the other expression in an equivalent form.

$\log_b x^2y^3$	$= \log_b x^2 + \log_b y^3$	Product Law
	$= 2 \log_b x + 3 \log_b y$	Power Law

$\log_b pq^4 =$

Answer to 30: (a) and (d)

---

32. We used three laws to write the expression below in an equivalent form. Write the other expression in an equivalent form.

$\log_b\left(\frac{c^3d^2}{h^5}\right)$	$= \log_b c^3d^2 - \log_b h^5$	Quotient Law
	$= \log_b c^3 + \log_b d^2 - \log_b h^5$	Product Law
	$= 3 \log_b c + 2 \log_b d - 5 \log_b h$	Power Law

$\log_b\left(\frac{x^4y^5}{z^2}\right) =$

Answer to 31: $\log_b p + 4 \log_b q$

Answer to 32: $4 \log_b x + 5 \log_b y - 2 \log_b z$

33. We also used three laws to write the expression below in an equivalent form. Write the other expression in an equivalent form.

$$\log_b\left(\frac{\sqrt{3}}{dh^6}\right) = \log_b\sqrt{3} - \log_b dh^6 \quad \text{Quotient Law}$$

$$= \log_b\sqrt{3} - (\log_b d + \log_b h^6) \quad \text{Product Law}$$

$$= \log_b 3^{\frac{1}{2}} - \log_b d - \log_b h^6$$

$$= \frac{1}{2}\log_b 3 - \log_b d - 6\log_b h \quad \text{Power Law}$$

$$\log_b\left(\frac{x^3}{y\sqrt{z}}\right) =$$

---

34. Following the example, write the other expression in an equivalent form.

$3\log_b x - \log_b y - \frac{1}{2}\log_b z$

$$\log_b\left(\sqrt[3]{\frac{cd}{h^2}}\right) = \log_b\left(\frac{cd}{h^2}\right)^{\frac{1}{3}}$$

$$= \frac{1}{3}\log_b\left(\frac{cd}{h^2}\right)$$

$$= \frac{1}{3}(\log_b cd - \log_b h^2)$$

$$= \frac{1}{3}(\log_b c + \log_b d - 2\log_b h)$$

$$\log_b\left(\sqrt{\frac{x^3}{yx}}\right) =$$

---

35. Following the example, write the other expression as a single logarithm.

$\frac{1}{2}(3\log_b x - \log_b y - \log_b z)$

$$2\log_b x + \log_b y = \log_b x^2 + \log_b y \quad \text{Power Law}$$

$$= \log_b x^2 y \quad \text{Product Law}$$

$$\log_b 3 + 4\log_b t =$$

36. Following the example, write the other expression as a single logarithm.

$$5\log_b x - 3\log_b y = \log_b x^5 - \log_b y^3 \quad \text{Power Law}$$
$$= \log_b\left(\frac{x^5}{y^3}\right) \quad \text{Quotient Law}$$

$2\log_b c - 4\log_b d =$

$\log_b 3t^4$

---

37. Following the example, write the other expression as a single logarithm.

$$3\log_b c + \log_b d - 2\log_b h = \log_b c^3 + \log_b d - \log_b h^2 \quad \text{Power Law}$$
$$= \log_b c^3 d - \log_b h^2 \quad \text{Product Law}$$
$$= \log_b\left(\frac{c^3 d}{h^2}\right) \quad \text{Quotient Law}$$

$\log_b x + 5\log_b y - 3\log_b z =$

$\log_b\left(\frac{c^2}{d^4}\right)$

---

38. Following the example, write the other expression as a single logarithm.

$$\log_b p - 2\log_b q - \log_b t = \log_b p - \log_b q^2 - \log_b t \quad \text{Power Law}$$
$$= \log_b p - (\log_b q^2 + \log_b t)$$
$$= \log_b p - \log_b q^2 t \quad \text{Product Law}$$
$$= \log_b\left(\frac{p}{q^2 t}\right) \quad \text{Quotient Law}$$

$3\log_b x - \log_b y - 4\log_b z =$

$\log_b\left(\frac{xy^5}{z^3}\right)$

$\log_b\left(\frac{x^3}{yz^4}\right)$

39. Following the example, write the other expression as a single logarithm.

$$4 \log_b x + \frac{1}{2} \log_b y - \log_b z = \log_b x^4 + \log_b \sqrt{y} - \log_b z$$
$$= \log_b x^4\sqrt{y} - \log_b z$$
$$= \log_b\left(\frac{x^4\sqrt{y}}{z}\right)$$

$$2 \log_b c - \log_b d - \frac{1}{2} \log_b h =$$

$\log_b\left(\frac{c^2}{d\sqrt{h}}\right)$

10-5 SCIENTIFIC NOTATION

The very large and very small numbers that occur in science are frequently written in scientific notation. We will discuss scientific notation in this section.

---

40. Some powers of ten are shown in the table at the right. You can see this fact:

The exponent equals the number of 0's in decimal notation.

10^6	=	1,000,000
10^5	=	100,000
10^4	=	10,000
10^3	=	1,000
10^2	=	100
10^1	=	10
10^0	=	1

Using the above fact, convert each of these to a power of ten.

a) 10,000,000 = ______

b) 1,000,000,000 = ______

---

a) 10^7

b) 10^9

41. We can use the decimal-point-shift method to multiply by powers of ten like 10^1, 10^2, 10^3, etc. The exponent tells us how many places to shift the decimal point to the right. For example:

$2.56 \times 10^3 = 2.560_\wedge = 2{,}560$ (Shifted three places)

$.0084 \times 10^7 = .0084000_\wedge = 84{,}000$ (Shifted seven places)

Do these multiplications.

a) 9.33×10^4 = __________ b) $.000125 \times 10^8$ = __________

---

a) 93,300 b) 12,500

42. Some powers of ten with negative exponents are shown in the table at the right. You can see this fact:

The absolute value of the exponent equals the number of decimal places in decimal notation.

$10^{-1} = 0.1$
$10^{-2} = 0.01$
$10^{-3} = 0.001$
$10^{-4} = 0.0001$
$10^{-5} = 0.00001$
$10^{-6} = 0.000001$

Using the fact above, convert each of these to a power of ten.

a) .00000001 = ________

b) .0000000001 = ________

---

43. We can use the decimal-point-shift method to multiply by powers of ten like 10^{-1}, 10^{-2}, 10^{-3}, etc. The absolute value of the exponent tells us how many places to shift the decimal point to the left. For example:

3.6×10^{-4} = .0003.6 = .00036 (Shifted four places)

$90{,}000{,}000 \times 10^{-7}$ = 9.0,000,000. = 9 (Shifted seven places)

Do these multiplications.

a) 6.4×10^{-3} = ________ b) $45{,}000{,}000 \times 10^{-8}$ = ________

a) 10^{-8}

b) 10^{-10}

---

44. A number is written in scientific notation when the first factor is a number between 1 and 10 and the second factor is a power of ten. Some numbers are written in scientific notation below.

$6{,}200{,}000 = 6.2 \times 10^{6}$
$17{,}000 = 1.7 \times 10^{4}$
$.0084 = 8.4 \times 10^{-3}$
$.000039 = 3.9 \times 10^{-5}$

To convert a number written in scientific notation to decimal notation, we perform the multiplication. For example:

5.1×10^{7} = 5.1000000. = 51,000,000

3.9×10^{-9} = .000000003.9 = .0000000039

Convert to decimal notation.

a) 7.13×10^{9} = ________ b) 5×10^{-10} = ________

a) .0064 b) .45

---

a) 7,130,000,000

b) .0000000005

45. To convert a number larger than "1" to scientific notation, we write a caret (^) after the first digit to get a number between 1 and 10. Then we find the exponent of the power of ten by counting the number of places from the caret to the decimal point. The exponent is positive. For example:

7^3,000. = 7.3×10^4 (Four places to the decimal point)

8^00,000. = 8×10^5 (Five places to the decimal point)

9^1.7 = 9.17×10^1 (One place to the decimal point)

Write each number in scientific notation.

a) 47,000 = ______________ c) 88.55 = ______________

b) 200 = ______________ d) 9,160,000 = ______________

---

46. To convert a number smaller than "1" to scientific notation, we write a caret after the first non-zero digit in the number to get a number between 1 and 10. Then we find the exponent of the power of ten by counting the number of places from the caret to the decimal point. The exponent is negative. For example:

.06^8 = 6.8×10^{-2} (Two places to the decimal point)

.1^95 = 1.95×10^{-1} (One place to the decimal point)

.00007^ = 7×10^{-5} (Five places to the decimal point)

Write each number in scientific notation.

a) .00025 = ______________ c) .06718 = ______________

b) .4 = ______________ d) .0000019 = ______________

Answers to 45:

a) 4.7×10^4

b) 2×10^2

c) 8.855×10^1

d) 9.16×10^6

---

47. Scientific notation is used to express the very large and very small numbers that occur in science. For example:

The speed of light is 2.998×10^8 meters per second.

The diameter of a large molecule is 1.7×10^{-7} centimeter.

Convert each measurement above to decimal notation by multiplying.

a) 2.998×10^8 meters per second = ______________ meters per second.

b) 1.7×10^{-7} centimeter = ______________ centimeter

Answers to 46:

a) 2.5×10^{-4}

b) 4×10^{-1}

c) 6.718×10^{-2}

d) 1.9×10^{-6}

Answers to 47:

a) 299,800,000

b) .00000017

48. Two more measurements are given below.

A light-year (the distance light travels in one year) is 5,870,000,000,000 miles.

One cycle of a television broadcast signal takes .00000000481 second.

Ordinarily measurements of that size would be expressed in scientific notation. Convert them to scientific notation below.

a) 5,870,000,000,000 miles = ________________ miles

b) .00000000481 second = ________________ second

---

49. To multiply the two numbers in scientific notation below, we multiplied the two whole number factors and the two powers of ten. Complete the other multiplication.

$(3 \times 10^4) \times (2 \times 10^5) = (3 \times 2) \times (10^4 \times 10^5) = 6 \times 10^9$

$(2 \times 10^{-3}) \times (4 \times 10^{-7}) =$ ________________

Answer (48): a) 5.87×10^{12} miles b) 4.81×10^{-9} second

---

50. In the multiplication below, the product is not in scientific notation because 142 is not a number between 1 and 10. We converted the product to scientific notation by writing 142 in scientific notation and then multiplying the two powers of ten.

$20 \times (7.1 \times 10^6) = 142 \times 10^6 = (1.42 \times 10^2) \times 10^6 = 1.42 \times 10^8$

Use the same method to convert the product below to scientific notation.

$.04 \times (5.3 \times 10^{-5}) = .212 \times 10^{-5} =$ ________________

Answer (49): 8×10^{-10}

---

51. To divide the two numbers in scientific notation below, we divided the two whole numbers and the two powers of ten. Complete the other division.

$$\frac{8 \times 10^9}{2 \times 10^4} = \frac{8}{2} \times \frac{10^9}{10^4} = 4 \times 10^5$$

$$\frac{9 \times 10^{-8}}{3 \times 10^{-6}} = \frac{9}{3} \times \frac{10^{-8}}{10^{-6}} = \text{\_\_\_\_\_\_\_\_}$$

Answer (50): 2.12×10^{-6}

---

52. Following the example, convert the other quotient below to scientific notation.

$$\frac{3.01 \times 10^7}{7} = .43 \times 10^7 = (4.3 \times 10^{-1}) \times 10^7 = 4.3 \times 10^6$$

$$\frac{6.66 \times 10^{10}}{8.88 \times 10^{20}} = .75 \times 10^{-10} = \text{\_\_\_\_\_\_\_\_}$$

Answer (51): 3×10^{-2}

Answer (52): 7.5×10^{-11}

10-6 COMMON LOGARITHMS

There are two types of logarithms that are more frequently used:

1. Common logarithms (base 10)
2. Natural logarithms (base e = 2.71828...)

We will discuss common logarithms in this section. Natural logarithms are discussed in a later section.

Note: The instruction in this section assumes that you have a calculator with a [log] key. If you do not, Table 1 in the back of the text is a table of common logarithms. Your instructor can tell you how to use it.

---

53. Base 10 logarithms are called common logarithms. By converting to exponential form, we can see that the common logarithm of a number is the exponent of the power of ten.

$$\log_{10} N = x \longrightarrow N = 10^x$$

When writing common logarithms, we ordinarily omit the base 10. That is:

Instead of $\log_{10} N$, we write $\log N$.

By converting to powers of ten, we found the logarithms of the numbers below.

$$\log 1000 = \log 10^3 = 3$$
$$\log 100 = \log 10^2 = 2$$
$$\log 10 = \log 10^1 = 1$$
$$\log 1 = \log 10^0 = 0$$
$$\log .1 = \log 10^{-1} = -1$$
$$\log .01 = \log 10^{-2} = -2$$

Find these.

a) log 10,000 = ______ b) log 1,000,000 = ______ c) log .001 = ______

---

54. Since log 1 = 0 and log 10 = 1, the logarithm of any number between 1 and 10 is a number between 0 and 1. To show that fact, use a calculator to find the logarithms below. To do so, enter the number and press [log]. Round to four decimal places.

a) log 6 = ____________ b) log 2.76 = ____________

Answers to 53: a) 4 b) 6 c) -3

---

55. Since log 10 = 1 and log 100 = 2, the logarithm of any number between 10 and 100 is a number between 1 and 2. Find these. Round to four decimal places.

a) log 27 = ____________ b) log 78.3 = ____________

Answers to 54: a) 0.7782 b) 0.4409

Answers to 55: a) 1.4314 b) 1.8938

56. a) 375 lies between 100 and 1,000. Therefore, its logarithm lies between 2 and 3 since log 100 = 2 and log 1,000 = 3.

Rounding to four decimal places, log 375 = ______________

b) 836,000 lies between 100,000 and 1,000,000. Therefore, its logarithm lies between 5 and 6 since log 100,000 = 5 and log 1,000,000 = 6.

Rounding to four decimal places, log 836,000 = ______________

---

57. The logarithm of any number between 0 and 1 is negative. To show that fact, do these. Round to four decimal places.

a) log 0.25 = ______________ b) log 0.0829 = ______________

a) 2.5740

b) 5.9222

---

58. A common logarithm is <u>the exponent of the power-of-ten form</u> of a number. Therefore, when given the logarithm of a number, we can write its power-of-ten form. For example:

If log N = 2.4099, N = $10^{2.4099}$

If log N = -1.6378, N = $10^{-1.6378}$

We can use the "power" key [y^x] to convert the powers above to decimal notation. To do so for log N = 2.4099, we enter 10, press [y^x], enter 2.4099, and press [=]. To get a negative exponent, we press the [+/-] key.

a) Round to the nearest whole number.

If log N = 2.4099, N = $10^{2.4099}$ = __________

b) Round to thousandths.

If log N = -1.6378, N = $10^{-1.6378}$ = __________

a) -0.6021

b) -1.0814

---

59. If log N = 3.1056:

a) In power-of-ten form, N = ___________

b) Rounded to the nearest whole number, N = ___________

a) 257

b) 0.023

---

60. If log N = -0.7622:

a) In power-of-ten form, N = ___________

b) Rounded to thousandths, N = ___________

a) $10^{3.1056}$

b) 1,275

---

61. a) If log N = 0.8549, N = ___________

(Round to hundredths.)

b) If log N = -2.0633, N = ___________

(Round to five decimal places.)

a) $10^{-0.7622}$

b) 0.173

---

a) 7.16

b) 0.00864

62. To find N in log N = 1.9238, we must find the number whose logarithm is 1.9238. That process is called "finding the antilogarithm" or "finding the inverse logarithm". To find the inverse logarithm, we can also use [INV] [log] on a calculator. Use [INV] [log] to complete these.

a) log N = 1.9238 Rounded to tenths, N = ____________

b) log N = -0.56 Rounded to thousandths, N = ____________

---

63. To find the inverse logarithm, we can use either [INV] [log] or the "power" key [y^x]. Use either method for these.

a) 83.9

b) 0.275

a) log N = 5.7699 Rounded to thousands, N = ____________

b) log N = -1.29 Rounded to thousandths, N = ____________

---

64. When converting from a logarithm to a number, the information in the following table is helpful.

a) 589,000

b) 0.051

LOGARITHM	NUMBER
NEGATIVE	BETWEEN 0 AND 1
0 TO 1	BETWEEN 1 AND 10
1 TO 2	BETWEEN 10 AND 100
2 TO 3	BETWEEN 100 AND 1,000
3 TO 4	BETWEEN 1,000 AND 10,000
4 TO 5	BETWEEN 10,000 AND 100,000
5 TO 6	BETWEEN 100,000 AND 1,000,000

Using the facts in the table, complete these.

If log N = 2.6033, N is a number between 100 and 1,000.

a) If log N = 0.9655, N is a number between _____ and _____.

b) If log N = 5.2471, N is a number between _____ and _____.

c) If log N = -1.9127, N is a number between _____ and _____.

---

65. Negative numbers and "0" do not have logarithms. Therefore, finding the logarithm of a negative number or "0" is an IMPOSSIBLE operation. A calculator shows that fact with a flashing display, by printing out "Error", or in some other way. Try these on a calculator.

a) 1 and 10

b) 100,000 and 1,000,000

c) 0 and 1

log(-155) log 0

Note: When you are using a calculator for a "log" problem and the calculator display shows an IMPOSSIBLE operation, you know that you have made a mistake.

66. Common logarithms appear in some formulas. In chemistry, for example, the number pH is a measure of the acidity or alkalinity of a solution. If (H+) is the hydronium ion concentration measured in moles per liter, then pH = -log(H+) Find the pH of a solution with a hydronium ion concentration of .004. Round to tenths. pH = -log(.004) = ________	
67. The formula below gives the gain in an amplifier measured in decibels (D) in terms of the ratio (R) of output voltage to input voltage. D = 20 log(R) Find D when R = 600. Round to tenths. D = 20 log(600) = ________	2.4
	55.6

SELF-TEST 36 (pages 515-528)

Use the laws of logarithms to express each of these in terms of log x, log y, and log z.

1. $\log_b x^3y^2$	2. $\log_b\left(\frac{xy^4}{z^5}\right)$	3. $\log_b\left(\frac{\sqrt{x}}{y^3z}\right)$

Express each of these as a single logarithm.

4. $6 \log_b x - 7 \log_b y$	5. $\log_b c + \frac{1}{2} \log_b d - 2 \log_b t$	6. $3 \log_b x - \log_b y - 4 \log_b z$

Convert to decimal notation.

7. 4.5×10^6 =	8. 2.1×10^{-3} =	9. 7.12×10^{-7} =

Convert to scientific notation.

10. 51,000 =	11. .0608 =	12. .00000085 =

Continued on following page.

SELF-TEST 36 - Continued

Find the following. Round to four decimal places.

13. log 93,600	14. log 5.18	15. log 0.0072

Find N in each equation.

16. Round to tenths.	17. Round to thousands.	18. Round to thousandths.
$\log N = 1.8069$	$\log N = 5.2354$	$\log N = -0.3916$

ANSWERS:

1. $3 \log_b x + 2 \log_b y$	7. 4,500,000	13. 4.9713
2. $\log_b x + 4 \log_b y - 5 \log_b z$	8. 0.0021	14. 0.7143
3. $\frac{1}{2} \log_b x - 3 \log_b y - \log_b z$	9. 0.000000712	15. -2.1427
4. $\log_b\left(\frac{x^6}{y^7}\right)$	10. 5.1×10^4	16. N = 64.1
5. $\log_b\left(\frac{c\sqrt{d}}{t^2}\right)$	11. 6.08×10^{-2}	17. N = 172,000
6. $\log_b\left(\frac{x^3}{yz^4}\right)$	12. 8.5×10^{-7}	18. N = 0.406

10-7 EXPONENTIAL EQUATIONS

In this section, we will discuss two methods for solving exponential equations.

68. An exponential equation is an equation in which the variable appears in an exponent. Some examples are:

$$5^x = 11 \qquad\qquad 3^{2x-1} = 17$$

When both sides are powers with the same base, we can equate the exponents. That is:

Property Of Powers

If $b^x = b^y$

then $x = y$

Continued on following page.

68. Continued

We used the preceding principle to solve each equation below.

$$4^{2x} = 4^{10} \qquad 2^{3x-1} = 2^8$$
$$2x = 10 \qquad 3x - 1 = 8$$
$$x = 5 \qquad 3x = 9$$
$$x = 3$$

Following the examples, solve these:

a) $7^x = 7^2$ b) $6^{3y} = 6^2$ c) $10^{2x+3} = 10^{15}$

---

69. We solved the equations below by expressing both sides as powers with the same base and then equating the exponents.

$9^x = 27$	$16^x = \frac{1}{4}$
$(3^2)^x = 3^3$	$(2^4)^x = 2^{-2}$
$3^{2x} = 3^3$	$2^{4x} = 2^{-2}$
$2x = 3$	$4x = -2$
$x = \frac{3}{2}$	$x = -\frac{2}{4} = -\frac{1}{2}$

Use the same method to solve these.

a) $81^x = 9$ b) $16^y = 64$ c) $125^x = \frac{1}{25}$

Answers to 68: a) $x = 2$ b) $y = \frac{2}{3}$ c) $x = 6$

---

70. To solve exponential equations when both sides cannot be expressed as powers with the same base, we use the <u>log</u> principle for equations.

<u>Log Principle For Equations</u>

If $M = N$ $M > 0$ and $N > 0$

then $\log M = \log N$

Answers to 69: a) $x = \frac{1}{2}$ b) $y = \frac{3}{2}$ c) $x = -\frac{2}{3}$

Continued on following page.

70. Continued

We used the log principle for equations to solve one equation below. Use it to solve the other equation. Round to hundredths.

$5^x = 12$ | $3^y = 49$

$\log 5^x = \log 12$

$x \log 5 = \log 12$

$$x = \frac{\log 12}{\log 5}$$

$$x = \frac{1.0792}{0.6990}$$

$x = 1.54$

---

71. In the last frame, we got these expressions:

$$\frac{\log 12}{\log 5} \qquad \frac{\log 49}{\log 3}$$

Don't confuse the expressions above with those below.

$$\log\left(\frac{12}{5}\right) \qquad \log\left(\frac{49}{3}\right)$$

The law of logarithms for division applies only to the logarithm of a quotient, not to a quotient of two logarithms. That is:

$$\log\left(\frac{12}{5}\right) = \log 12 - \log 5 \qquad \frac{\log 12}{\log 5} \neq \log 12 - \log 5$$

Which statement is true? ________

a) $\frac{\log 49}{\log 3} = \log 49 - \log 3$ b) $\log\left(\frac{49}{3}\right) = \log 49 - \log 3$

Answer to 70: $y = 3.54$, from: $y = \frac{\log 49}{\log 3}$

---

72. We solved one equation below. Solve the other equation. Round to thousandths.

$69.7 = 1.4^y$ | $456 = 559^t$

$\log 69.7 = \log 1.4^y$

$\log 69.7 = y \log 1.4$

$$y = \frac{\log 69.7}{\log 1.4}$$

$$y = \frac{1.8432}{0.1461}$$

$y = 12.6$

Answer to 71: (b)

Answer to 72: $t = 0.968$

73. We solved one equation below. Solve the other equation. Round to hundredths.

$$7^{2x - 1} = 3 \qquad\qquad 5^{y + 1} = 23$$

$$\log 7^{2x - 1} = \log 3$$

$$(2x - 1)\log 7 = \log 3$$

$$2x - 1 = \frac{\log 3}{\log 7}$$

$$2x - 1 = \frac{0.4771}{0.8451}$$

$$2x - 1 = 0.56$$

$$2x = 1.56$$

$$x = \frac{1.56}{2}$$

$$x = 0.78$$

$y = 0.95$

74. One of the applications of exponential equations is compound interest (interest paid on interest). The amount A that principal P will be worth after t years at interest rate r, compounded annually, is given by the formula.

$$A = P(1 + r)^t$$

We can use the formula above to solve this problem.

If the average annual rate of inflation is 8%, how long would it take for the average price level to double?

To find when \$1 will double to \$2 at an inflation rate of 8%, we let P = \$1, A = \$2, and r = 8% or .08. We get:

$$2 = 1(1 + .08)^t$$

$$2 = 1(1.08)^t$$

$$2 = 1.08^t$$

$$\log 2 = t \log 1.08$$

$$t = \frac{\log 2}{\log 1.08}$$

$$= \frac{0.3010}{0.0334}$$

$$= 9.0 \text{ years}$$

Therefore, it would take 9.0 years for prices to double at an 8% inflation rate.

How long (rounded to tenths) would it take prices to double:

a) at a 5% inflation rate? b) at a 12% inflation rate?

---

a) 14.2 years

b) 6.1 years

10-8 LOGARITHMIC EQUATIONS

In this section, we will discuss methods for solving some types of logarithmic equations.

---

75. A logarithmic equation is an equation that contains one or more logarithmic expressions. Some examples are:

$$\log(x + 1) = \log 9 \qquad \log x - \log(x - 3) = 1$$

When both sides contain a single logarithm, we use the log principle for equations.

Log Principle For Equations

If $\log M = \log N$
then $M = N$

$M > 0$ and $N > 0$

We used the log principle to solve one equation below. Use it to solve the other equation. Check your solution.

$$\log(x + 1) = \log 9 \qquad \log x = \log(6 - x)$$

$$x + 1 = 9$$

$$x = 8$$

Check

$$\log(x + 1) = \log 9$$

$$\log(8 + 1) = \log 9$$

$$\log 9 = \log 9$$

Answer to 75: $x = 3$

Check:

$$\log x = \log(6 - x)$$

$$\log 3 = \log(6 - 3)$$

$$\log 3 = \log 3$$

76. To solve the equation below, we began by using the product law of logarithms to get a single logarithm on the left side. Solve and check the other equation.

$$\log(2x - 1) + \log 3 = \log(4x + 7) \qquad \log(x - 3) + \log 5 = \log(2x + 3)$$

$$\log 3(2x - 1) = \log(4x + 7)$$

$$3(2x - 1) = 4x + 7$$

$$6x - 3 = 4x + 7$$

$$2x = 10$$

$$x = 5$$

Check

$$\log(2x - 1) + \log 3 = \log(4x + 7)$$

$$\log[2(5) - 1] + \log 3 = \log[4(5) + 7]$$

$$\log 9 + \log 3 = \log 27$$

$$\log(9)(3) = \log 27$$

$$\log 27 = \log 27$$

77. To solve the equation below, we began by using the quotient law of logarithms to get a single logarithm on the left side. Solve the other equation.

$$\log(x - 1) - \log 2 = \log(x - 4)$$

$$\log\left(\frac{x - 1}{2}\right) = \log(x - 4)$$

$$\frac{x - 1}{2} = x - 4$$

$$2\left(\frac{x - 1}{2}\right) = 2(x - 4)$$

$$x - 1 = 2x - 8$$

$$7 = x$$

$$\log(x + 4) - \log(x - 2) = \log 4$$

$x = 6$

Check:

$$\log(x - 3) + \log 5 = \log(2x + 3)$$

$$\log(6 - 3) + \log 5 = \log[2(6) + 3]$$

$$\log 3 + \log 5 = \log 15$$

$$\log(3)(5) = \log 15$$

$$\log 15 = \log 15$$

---

78. We got two solutions for the equation below.

$$\log x + \log(x - 2) = \log 8$$

$$\log x(x - 2) = \log 8$$

$$x(x - 2) = 8$$

$$x^2 - 2x = 8$$

$$x^2 - 2x - 8 = 0$$

$$(x - 4)(x + 2) = 0$$

$$x = 4 \text{ and } -2$$

Whenever we get two solutions, we must check for an extraneous root. We checked both solutions below.

Checking $x = 4$

$$\log x + \log(x - 2) = \log 8$$

$$\log 4 + \log(4 - 2) = \log 8$$

$$\log 4 + \log 2 = \log 8$$

$$\log(4)(2) = \log 8$$

$$\log 8 = \log 8$$

Checking $x = -2$

$$\log x + \log(x - 2) = \log 8$$

$$\log(-2) + \log(-2 - 2) = \log 8$$

$$\log(-2) + \log(-4) = \log 8$$

As you can see, $x = -2$ is not a solution because it leads to logarithms of negative numbers and negative numbers do not have logarithms. Therefore, the only solution is $x = 4$.

Continued on following page.

$x = 4$

78. Continued

Solve this equation. Be sure to check for an extraneous root.

$$\log x + \log(x + 1) = \log 12$$

---

79. When an equation contains a logarithm on only one side, we solve it by converting to exponential form. We use the definition below which we saw earlier.

If $\log M = x$, then $M = 10^x$

We solved the equation below by converting to exponential form. Solve and check the other equation.

$$\log(x + 3) = 1 \qquad\qquad \log(x - 20) = 2$$

$$x + 3 = 10^1$$

$$x + 3 = 10$$

$$x = 7$$

<u>Check</u>

$$\log(x + 3) = 1$$

$$\log(7 + 3) = 1$$

$$\log 10 = 1$$

$$1 = 1$$

Answer to 78: $x = 3$ (not $x = -4$)

80. Notice how we got a single logarithm on the left side below before converting to exponential form. Solve the other equation.

$\log 15x - \log(x + 2) = 1$ $\qquad$ $\log(x + 6) - \log x = 1$

$$\log\left(\frac{15x}{x+2}\right) = 1$$

$$\frac{15x}{x+2} = 10$$

$$\cancel{x+2}\left(\frac{15x}{\cancel{x+2}}\right) = 10(x+2)$$

$$15x = 10x + 20$$

$$5x = 20$$

$$x = 4$$

$x = 120$

Check:

$$\log(x - 20) = 2$$

$$\log(120 - 20) = 2$$

$$\log 100 = 2$$

$$2 = 2$$

81. Solve the equation below. Check for an extraneous root.

$$\log x + \log(x - 9) = 1$$

$x = \frac{2}{3}$

82. The equation below contains base 2 logarithms. Notice how we converted to exponential form. Solve the other equation. It contains base 3 logarithms.

$\log_2(x + 2) - \log_2(x - 1) = 2$ $\qquad$ $\log_3(x + 13) - \log_3(x + 1) = 2$

$$\log_2\left(\frac{x+2}{x-1}\right) = 2$$

$$\frac{x+2}{x-1} = 2^2$$

$$\frac{x+2}{x-1} = 4$$

$$(\cancel{x-1})\left(\frac{x+2}{\cancel{x-1}}\right) = 4(x-1)$$

$$x + 2 = 4x - 4$$

$$6 = 3x$$

$$x = 2$$

$x = 10$

(not $x = -1$)

$x = \frac{1}{2}$

10-9 POWERS OF "e" AND NATURAL LOGARITHMS

In this section, we will discuss powers of the number "e" and natural logarithms. Powers of "e" and natural logarithms occur in many applications, especially those involving growth and decay.

Note: The instructions in this section assume that you have a calculator that gives powers of "e" and natural logarithms. If you do not have such a calculator, Tables 2 and 3 at the back of the text give some values of e^x and e^{-x} and the natural logarithms of some numbers.

---

83. Many useful exponential functions contain powers of an irrational number called "e". The numerical value of "e" is 2.7182818... To convert powers of "e" to decimal notation with a calculator, we enter the exponent and then press [INV] [ln x].

Convert to decimal notation. Round to the indicated place.

a) $e^{2.1}$ = ________ (Round to hundredths.)

b) $e^{-3.8}$ = ________ (Round to four decimal places.)

---

84. Let's confirm two facts on a calculator.

1. Since $e = e^1$, confirm the fact that e = 2.7182818... by entering "1" and pressing [INV] [ln x].

2. Just as $10^0 = 1$, $e^0 = 1$. Confirm that fact by entering 0 and pressing [INV] [ln x].

Any power of "e" with a positive exponent equals a number larger than "1". To confirm that fact, convert these powers to decimal notation.

a) $e^{0.5}$ = ________ (Round to hundredths.)

b) $e^{3.6}$ = ________ (Round to tenths.)

c) $e^{6.7}$ = ________ (Round to hundreds.)

Answers to 83: a) 8.17 b) 0.0224

---

85. Any power of "e" with a negative exponent equals a number between 0 and "1". To confirm that fact, convert these powers to decimal notation.

a) $e^{-0.5}$ = ________ (Round to thousandths.)

b) $e^{-3.4}$ = ________ (Round to four decimal places.)

c) $e^{-6.8}$ = ________ (Round to five decimal places.)

Answers to 84: a) 1.65 b) 36.6 c) 812

Answers to 85: a) 0.607 b) 0.0334 c) 0.00111

86. $y = e^x$ and $y = e^{-x}$ are exponential functions. In $y = e^{-x}$, the "-x" means "the additive inverse of x". Therefore:

If $x = 2$, $y = e^{-x}$ equals $y = e^{-2}$.

If $x = -2$, $y = e^{-x}$ equals $y = e^{2}$.

Using the tables below, we graphed $y = e^x$ and $y = e^{-x}$ at the right.

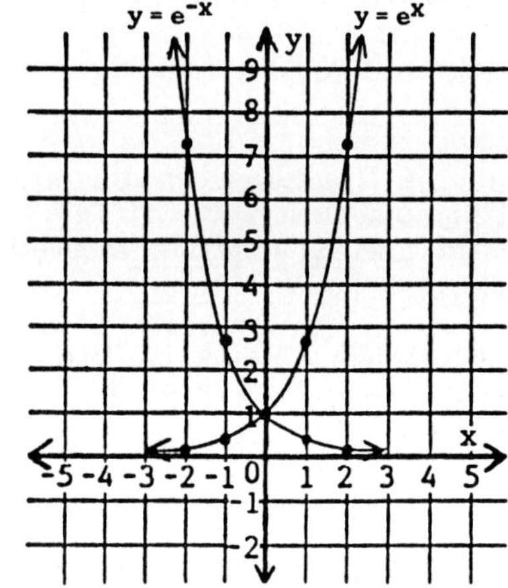

$y = e^x$

x	y
-2	0.14
-1	0.37
0	1.00
1	2.72
2	7.39

$y = e^{-x}$

x	y
-2	7.39
-1	2.72
0	1.00
1	0.37
2	0.14

Notice these facts:

1. The graph of $y = e^x$ is a rising exponential curve. The graph of $y = e^{-x}$ is a falling exponential curve.

2. The asymptote for each curve is the x-axis or $y = 0$.

3. Since $y = 1$ when $x = 0$ because $e^0 = 1$, the y-intercept for each graph is (0,1).

---

87. In many growth or decay situations, the amount of growth or decay at time t can be represented by an exponential function of the form:

$$y = y_o e^{kt}$$

where y_o is the number or amount present at time $t = 0$ and k is a constant. When k is positive, the function represents a growth process. When k is negative, the function represents a decay process. In the example below, the function represents a growth process.

Suppose the population P of a city is given by the growth function below where t represents time measured in years.

$$P = 100{,}000e^{.06t}$$

To show that 100,000 is the population at $t = 0$ (the beginning of the measured period of growth), we substitute 0 for t.

$$P = 100{,}000e^{.06(0)}$$

$$= 100{,}000e^{0}$$

$$= 100{,}000(1) \qquad \text{(Since } e^0 = 1\text{)}$$

$$= 100{,}000$$

Continued on following page.

87. Continued

We found the population at t = 5 (after 5 years) below. Find the population at t = 10 (after 10 years).

$$P = 100{,}000e^{.06(5)}$$
$$= 100{,}000e^{.3}$$
$$= 100{,}000(1.34986)$$
$$= 134{,}986$$

---

88. Here is an example of an exponential function that represents a decay process.

Suppose the amount A (measured in grams) of a radioactive substance present at time t (measured in days) is given by the decay function below.

$$A = 1{,}000e^{-.04t}$$

To show that 1,000 grams is the amount initially present, we substitute 0 for t.

$$A = 1{,}000e^{-.04(0)}$$
$$= 1{,}000e^{0}$$
$$= 1{,}000(1)$$
$$= 1{,}000 \text{ grams}$$

We found the amount left after 5 days (t = 5) below. Find the amount left after 10 days (t = 10).

$$A = 1{,}000e^{-.04(5)}$$
$$= 1{,}000e^{-.2}$$
$$= 1{,}000(0.8187)$$
$$= 818.7 \text{ grams}$$

Answer: P = 182,212

---

89. When a number is written in power-of-"e" form, the exponent of the "e" is called the natural logarithm of the number. The abbreviation ln is used for "natural logarithm". Any power-of-"e" statement can be converted to logarithmic form. That is:

Definition Of Natural Logarithm
If $e^x = N$, then $\ln N = x$

We converted the statement below to logarithmic form.

Since $e^{2.5} = 12.2$, $\ln 12.2 = 2.5$

Continued on following page.

Answer: A = 670.3 grams

89. Continued

Convert each statement to logarithmic form.

a) $e^{0.5} = 1.65$ b) $e^{6.1} = 446$ c) $e^{-1.9} = 0.15$

__________ __________ __________

---

90. To find the natural logarithm of a number on a calculator, we enter the number and press [ln x]. Find these. Round to four decimal places.

a) ln 34 = __________ b) ln 500 = __________

a) ln 1.65 = 0.5

b) ln 446 = 6.1

c) ln 0.15 = -1.9

---

91. Since $1 = e^0$, the natural logarithm of "1" is 0. That is:

$$\ln 1 = 0$$

The natural logarithm of any number greater than "1" is a <u>positive</u> number. To confirm that fact, find these.

a) ln 1.66 = __________ (Round to thousandths.)

b) ln 250 = __________ (Round to hundredths.)

The natural logarithm of any number between 0 and 1 is a <u>negative</u> number. To confirm that fact, find these.

c) ln 0.8 = __________ (Round to thousandths.)

d) ln 0.006 = __________ (Round to hundredths.)

a) 3.5264

b) 6.2146

---

92. We used the tables below to graph $y = e^x$ and $y = \ln x$ at the right. You can see that $y = \ln x$ is the mirror image of $y = e^x$ with respect to the line $y = x$.

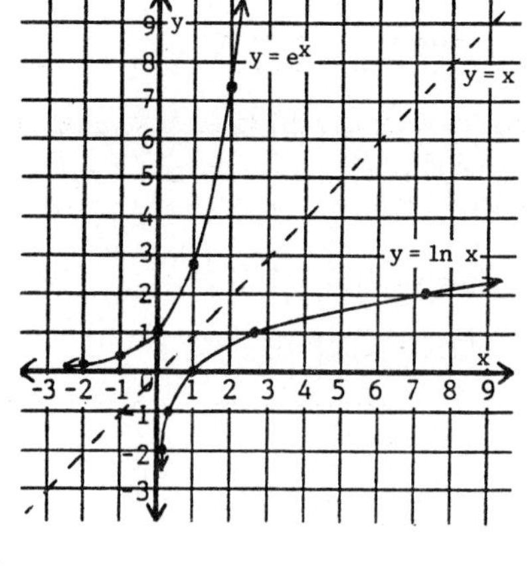

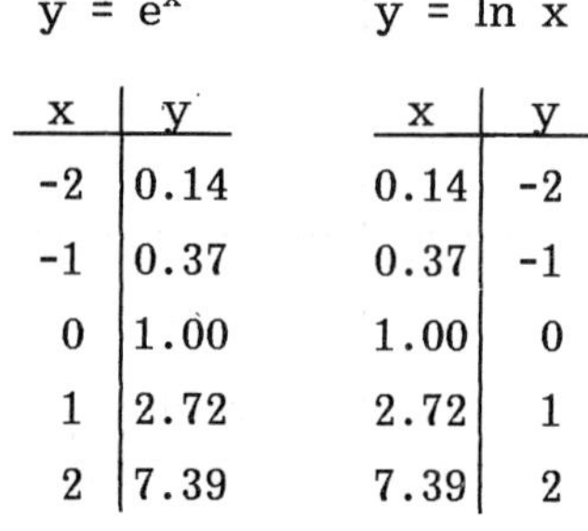

$y = e^x$

x	y
-2	0.14
-1	0.37
0	1.00
1	2.72
2	7.39

$y = \ln x$

x	y
0.14	-2
0.37	-1
1.00	0
2.72	1
7.39	2

The point (1.5,4.48) lies on the graph of $y = e^x$. Therefore, what point must lie on the graph of $y = \ln x$? __________

a) 0.507
b) 5.52

c) -0.223
d) -5.12

(4.48,1.5)

93. If we know the natural logarithm of a number, we can find the number on a calculator. To do so, we enter the logarithm and press [INV] [ln x]. Do these.

a) If ln N = 4.382, N = ________ (Round to a whole number.)

b) If ln N = 2.0412, N = ________ (Round to tenths.)

c) If ln N = -1.204, N = ________ (Round to tenths.)

---

94. The natural logarithm of a number is the exponent of its power-of-"e" form. Therefore, if we know the natural logarithm of a number, we can write it in power-of-"e" form. For example:

Since ln 8.8 = 2.1748, $8.8 = e^{2.1748}$

Since ln 76 = 4.3307, $76 = e^{4.3307}$

Write each number in power-of-"e" form by finding its natural logarithm. Round to four decimal places.

a) 1.9 = ________ b) 17 = ________ c) 400 = ________

Answers to 93: a) N = 80 b) N = 7.7 c) N = 0.3

---

95. The formula D = k ln P contains an "ln" expression. We did one evaluation with the formula below. Do the other evaluation. Round to tenths.

If k = 10 and P = 70:

D = k ln P = 10(ln 70) = 10(4.2485) = 42.5

If k = 10 and P = 95:

D = k ln P = ____________

Answers to 94: a) $e^{0.6419}$ b) $e^{2.8332}$ c) $e^{5.9915}$

---

96. In Q = -ln B, -ln B means "the additive inverse of ln B". Therefore, after finding ln B on a calculator, we press [+/-] to get its additive inverse. Complete these. Round to hundredths.

In Q = -ln B: a) when B = 4.2, Q = ________

b) when B = 150, Q = ________

Answer to 95: D = 45.5

Answers to 96: a) -1.44 b) -5.01

SELF-TEST 37 (pages 529-542)

Solve each equation.

1. $8^x = 32$

2. $12^x = 11$ (Round to hundredths.)

3. $7^{y - 2} = 13$ (Round to hundredths.)

4. $\log x = \log(2x - 9)$

5. $\log(x + 5) - \log(x - 3) = \log 2$

6. $\log x + \log(x + 3) = 1$

Find these. Round to the indicated places.

7. $e^{2.9}$ = ________ (tenths)

8. $e^{-0.8}$ = ________ (thousandths)

Find these. Round to four decimal places.

9. ln 66 = ________

10. ln 0.4 = ________

Round to tenths.

11. If ln N = 2.0412, N = ________

12. $28 = e^{\square}$

13. Find D when k = 100 and P = 8. (Round to a whole number.)

$$D = k \ln P$$

ANSWERS:

1. $x = \frac{5}{3}$
2. x = 0.96
3. y = 3.32
4. x = 9
5. x = 11
6. x = 2 (not x = -5)
7. 18.2
8. 0.449
9. 4.1897
10. -0.9163
11. N = 7.7
12. $e^{3.3}$
13. D = 208

SUPPLEMENTARY PROBLEMS - CHAPTER 10

Assignment 35

Graph each function.

1. $y = 2^x$

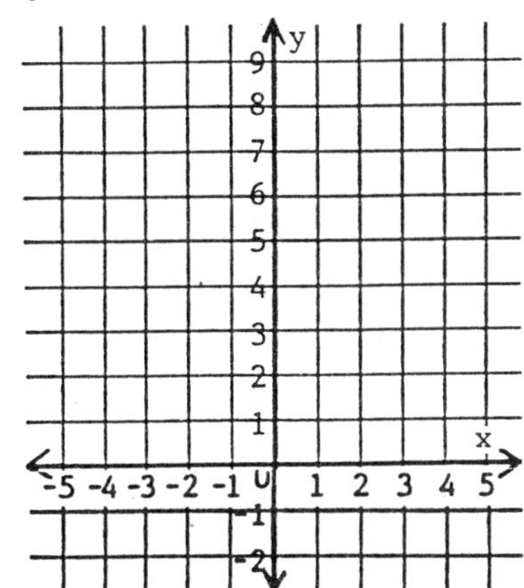

2. $y = 4^x$

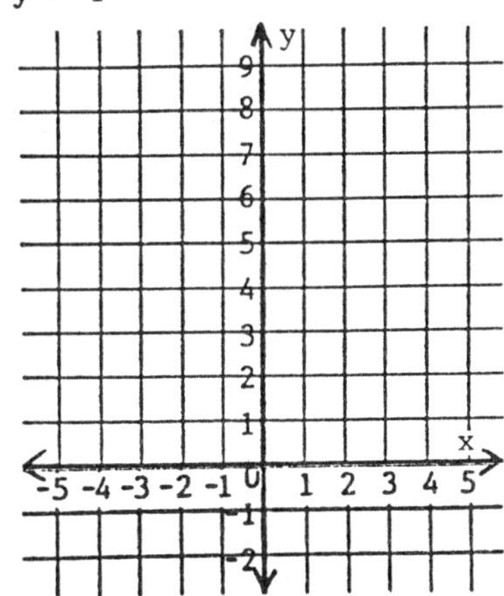

3. $y = \left(\frac{1}{2}\right)^x$

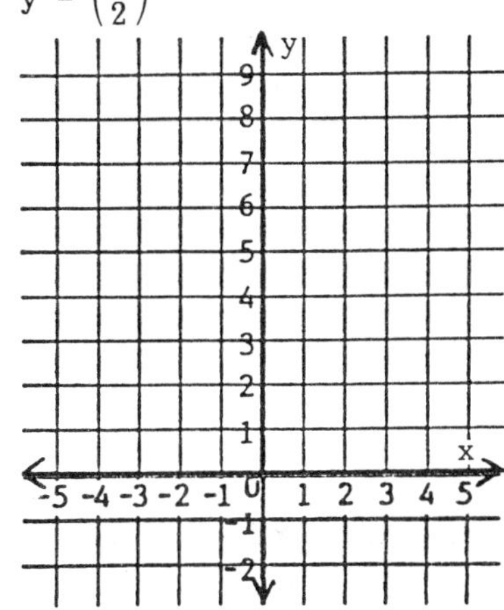

4. $y = \left(\frac{1}{4}\right)^x$

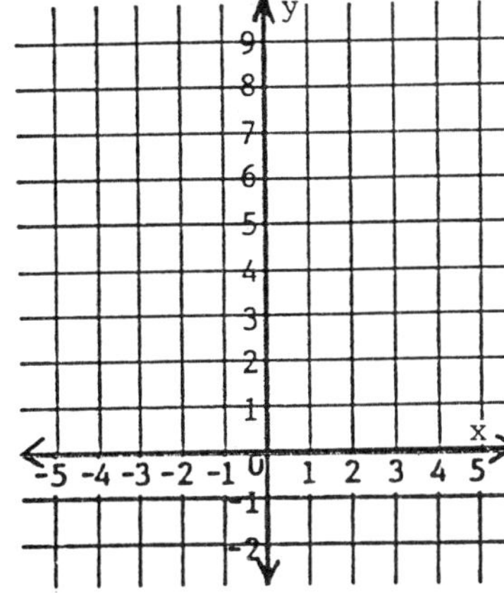

5. $y = 2^{x - 1}$

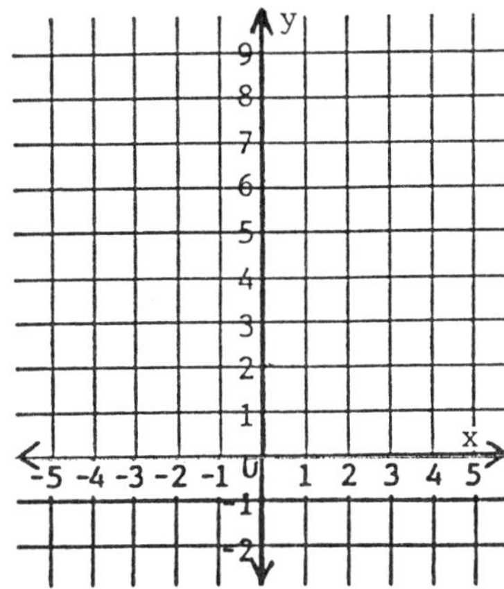

6. $y = 2^{x + 1}$

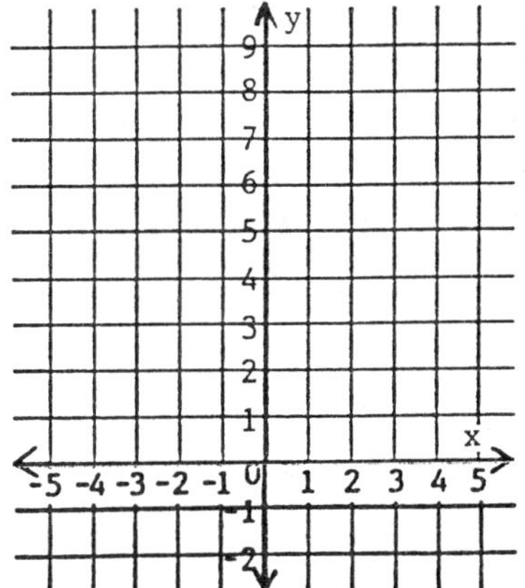

7. $y = 3^{x + 3}$

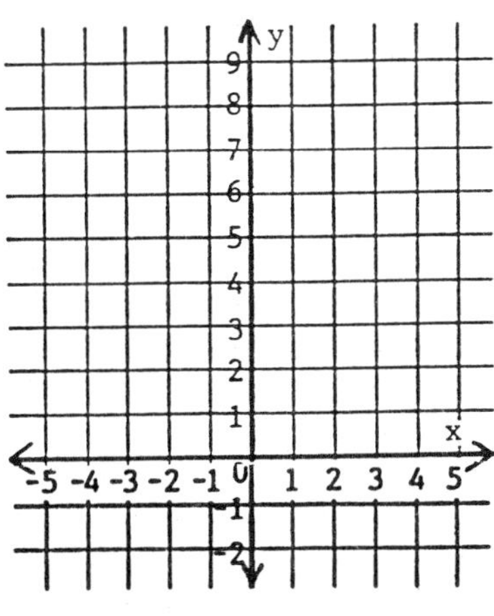

8. $x = 3^y$

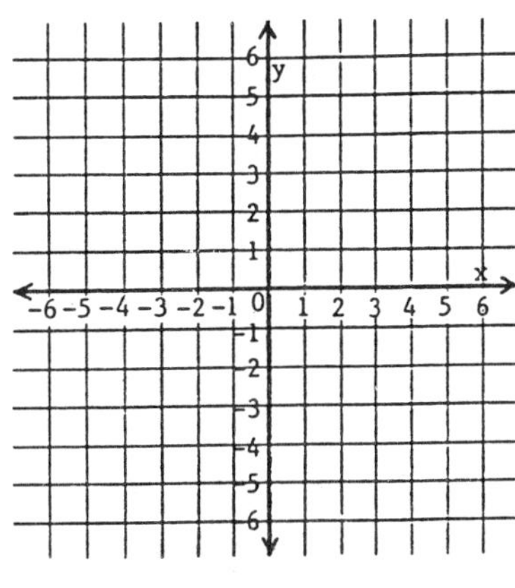

9. $x = 5^y$

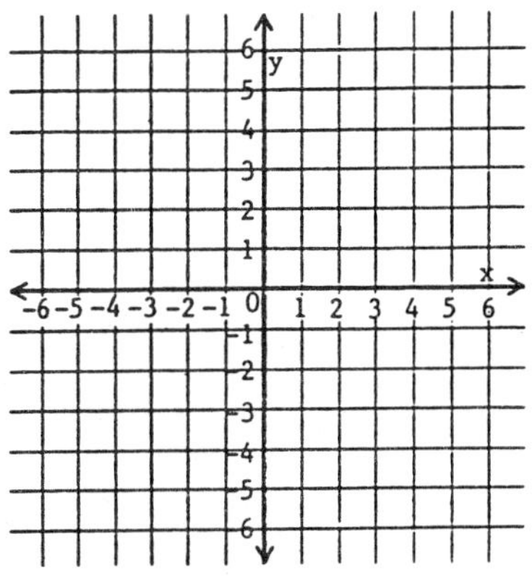

Convert to logarithmic form.

10. $3^2 = 9$ 11. $5^3 = 125$ 12. $2^5 = 32$ 13. $10^{-3} = 0.001$

14. $x^3 = 64$ 15. $6^x = 36$ 16. $3^5 = x$ 17. $10^x = 0.1$

Convert to exponential form.

18. $\log_7 49 = 2$ 19. $\log_3 81 = 4$ 20. $\log_6 216 = 3$ 21. $\log_{10} 0.01 = -2$

22. $\log_5 x = 4$ 23. $\log_x 27 = 3$ 24. $\log_2 64 = x$ 25. $\log_x 100 = 2$

Solve each equation.

26. $\log_3 x = 4$ 27. $\log_9 x = 2$ 28. $\log_4 x = -1$ 29. $\log_x 32 = 5$

30. $\log_x 125 = 3$ 31. $\log_x 0.01 = -2$ 32. $\log_2 8 = x$ 33. $\log_{10} 0.1 = x$

Find these:

34. $\log_2 16$ 35. $\log_3 27$ 36. $\log_5 25$ 37. $\log_{10} 1000$

38. $\log_6\left(\frac{1}{6}\right)$ 39. $\log_{10} 0.01$ 40. $\log_2\left(\frac{1}{4}\right)$ 41. $\log_3\left(\frac{1}{27}\right)$

42. $\log_7 7$ 43. $\log_7 1$ 44. $\log_3 3$ 45. $\log_5 1$

Graph each function.

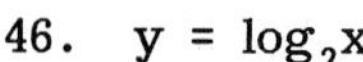
46. $y = \log_2 x$

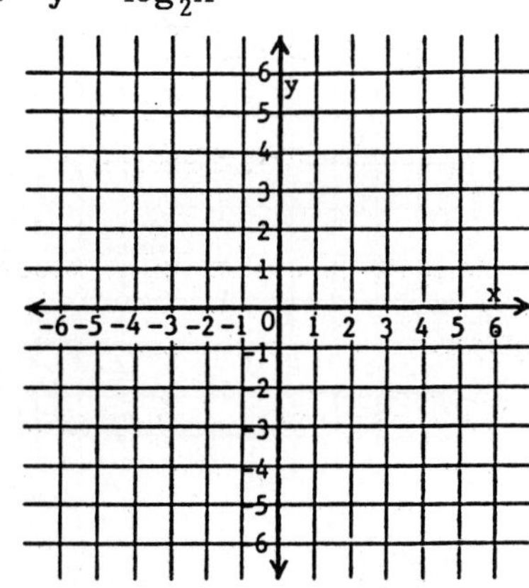

47. $y = \log_3 x$

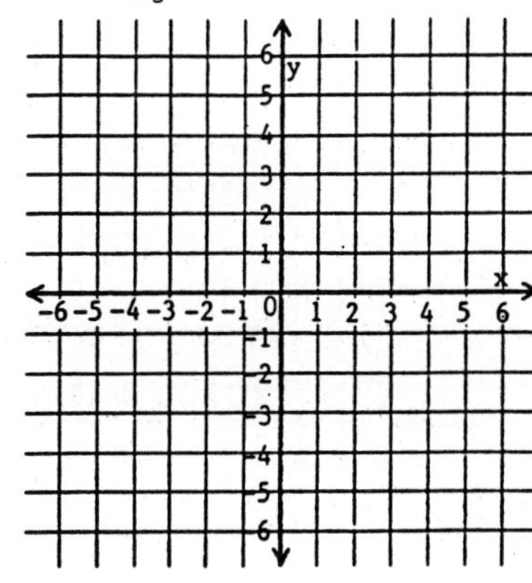

48. $y = \log_5 x$

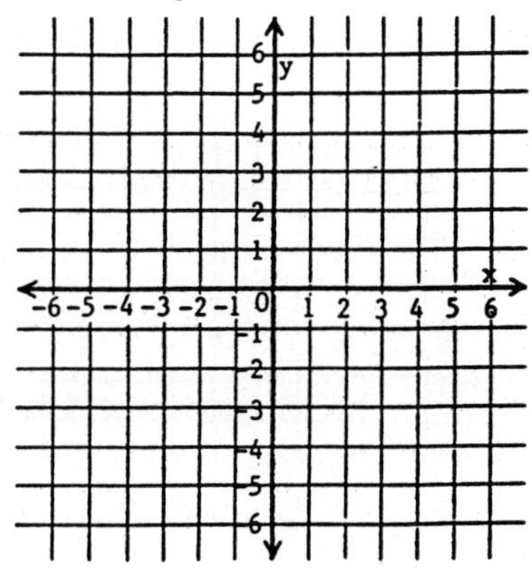

Assignment 36

Use the laws of logarithms to write in an equivalent form.

1. $\log_2 (4)(8)$ 2. $\log_5 10x$ 3. $\log_3 (y + 1)$ 4. $\log_3\left(\frac{27}{9}\right)$

5. $\log_7\left(\frac{a}{b}\right)$ 6. $\log_{10} (B - 1)$ 7. $\log_6 x^2$ 8. $\log_b 2^t$

Express as a single logarithm.

9. $\log_3 9 + \log_3 81$ 10. $\log_6 9 + \log_6 x$ 11. $\log_2 32 - \log_2 8$

12. $\log_b R - \log_b S$ 13. $4 \log_3 5$ 14. $x \log_b y$

Express in terms of log x, log y, and log z.

15. $\log_b xy^2$ 16. $\log_b 3y^3z^4$ 17. $\log_b 5x^5z$ 18. $\log_b\left(\frac{x^2y}{z}\right)$

19. $\log_b\left(\frac{2y^4}{z^6}\right)$ 20. $\log_b\left(\frac{\sqrt{x}}{y^2z}\right)$ 21. $\log_b\left(\frac{x^2}{y^3\sqrt{z}}\right)$ 22. $\log_b\sqrt{\frac{x^4}{yz}}$

Express as a single logarithm.

23. $2\log_b x + \log_b y + 3\log_b z$ 24. $\log_b a + 3\log_b c + 5\log_b d$

25. $4\log_b x + \log_b z - 2\log_b y$ 26. $\log_b p - \log_b q - 3\log_b t$

27. $3\log_b x + \frac{1}{2}\log_b y - \log_b z$ 28. $5\log_b a - \log_b c - \frac{1}{2}\log_b d$

Convert to decimal notation.

29. 2.8×10^1 30. 6.56×10^4 31. 7×10^7 32. 9.4×10^9

33. 4.1×10^{-1} 34. 1.14×10^{-3} 35. 2×10^{-6} 36. 8.8×10^{-8}

Convert to scientific notation.

37. 27.7 38. 519 39. 49,600 40. 6,000,000 41. 425,000,000

42. .314 43. .075 44. .000004 45. .0000639 46. .00000011

Write each answer in scientific notation.

47. $(4 \times 10^2) \times (2 \times 10^7)$ 48. $51 \times (7 \times 10^5)$ 49. $\frac{8 \times 10^{-7}}{2 \times 10^{-2}}$ 50. $\frac{2.46 \times 10^{-5}}{6}$

51. In chemistry, the quantity 602,000,000,000,000,000,000,000 molecules is called Avogadro's number. Write that number in scientific notation.

52. The wavelength of an x-ray is 7.82×10^{-7} centimeter. Write that length as a decimal number.

53. One second equals .000011574 day. Write that time in scientific notation.

54. The "half life" of a radioactive uranium 238 is 4.5×10^{10} years. Write that time as a whole number.

Find each logarithm. Round to four decimal places.

55. log 7 56. log 1.99 57. log 68.4 58. log 504

59. log 37,200 60. log 45,000,000 61. log 0.7 62. log 0.55

63. log 0.0129 64. log 0.0008 65. log 0.00616 66. log 0.0000071

Find N. Round to the indicated place.

67. log N = 0.8752 (hundredths) 68. log N = 1.6034 (tenths)

69. log N = 2.0195 (whole number) 70. log N = 5.2361 (thousands)

71. log N = -0.2356 (thousandths) 72. log N = -1.4252 (four decimal places)

73. log N = -2.1086 (five decimal places) 74. log N = -3.8547 (six decimal places)

If pH = -log(H+):

75. Find pH when H+ = .005. (Round to hundredths.)

76. Find pH when H+ = .007. (Round to hundredths.)

If D = 20 log R:

77. Find D when R = 500. (Round to tenths.)

78. Find D when R = 800. (Round to tenths.)

Assignment 37

Solve each equation.

1. $5^x = 5^7$ 2. $3^{4x} = 3^{12}$ 3. $8^{x-1} = 8^3$ 4. $2^x = 16$

5. $27^x = 9$ 6. $8^x = 64$ 7. $9^x = \frac{1}{3}$ 8. $16^x = \frac{1}{64}$

Solve each equation. Round to hundredths.

9. $3^x = 5$ 10. $7^x = 9$ 11. $27^x = 33$ 12. $6.5 = 7.5^x$

13. $8^{x+1} = 11$ 14. $10^{x-3} = 7$ 15. $2^{2x-1} = 7$ 16. $12^{4x-5} = 8$

Use the compound interest formula $A = P(1 + r)^t$ for these. Round to tenths.

17. How many years would it take an investment of $1000 to double itself when interest is compounded annually at 8%?

18. How many years would it take an investment of $1000 to triple itself when interest is compounded annually at 8%?

Solve each equation.

19. $\log(x - 7) = \log 4$ 20. $\log x = \log(10 - x)$ 21. $\log(2x + 7) = \log 3x$

22. $\log(x + 2) + \log 2 = \log(x + 7)$ 23. $\log 5 + \log(x - 2) = \log(x + 10)$

24. $\log(x + 1) - \log 3 = \log(x - 1)$ 25. $\log(x + 5) - \log(x - 3) = \log 5$

26. $\log x + \log(x - 1) = \log 12$ 27. $\log(x) + \log(x + 2) = \log 15$

28. $\log(x + 6) = 1$ 29. $\log(x - 50) = 2$ 30. $\log 9x - \log(x - 2) = 1$

31. $\log(x + 3) - \log x = 1$ 32. $\log x + \log(x - 3) = 1$ 33. $\log_2(2x + 5) - \log_2(x + 1) = 2$

Find each power. Round to the indicated place.

34. $e^{0.7}$ (hundredths) 35. $e^{3.1}$ (tenths) 36. $e^{5.8}$ (whole number)

37. $e^{-1.1}$ (thousandths) 38. $e^{-2.5}$ (four decimal places) 39. $e^{-6.2}$ (five decimal places)

40. The number of ants in an ant hill grows according to the formula $y = 300e^{.2t}$, where t is measured in days. How many ants would there be after 10 days?

41. When a bactericide is introduced into a given culture, the number of bacteria goes down according to the formula $N = 40{,}000e^{-.01t}$, where t is measured in hours. How many bacteria are left after 30 hours?

Convert to logarithmic form.

42. $e^{0.6} = 1.82$ 43. $e^{4.9} = 134$ 44. $e^{-1.2} = 0.301$ 45. $e^{-2.9} = 0.055$

Convert to exponential form.

46. $\ln 11 = 2.4$ 47. $\ln 545 = 6.3$ 48. $\ln 0.549 = -0.6$ 49. $\ln 0.0166 = -4.1$

Find each logarithm. Round to four decimal places.

50. $\ln 0.7$ 51. $\ln 2.6$ 52. $\ln 71$ 53. $\ln 400$

If $D = 100 \ln P$:

54. Find D when $P = 2.1$ (Round to tenths.)

55. Find D when $P = 36$ (Round to a whole number.)

If $Q = -\ln B$:

56. Find Q when $B = 5.8$ (Round to hundredths.)

57. Find Q when $B = 35$ (Round to hundredths.)

11 Sequences and Series

In this section, we will define both sequences and series and discuss arithmetic and geometric progressions. We will also discuss the expansion of binomials.

11-1 SEQUENCES

In this section, we will define sequences and discuss sequence functions.

1. A sequence of numbers is a set of numbers arranged in a definite order. The set of numbers below is a sequence. Each number is obtained by adding 5 to the preceding number.

   10, 15, 20, 25, 30, ...

   Write the next three numbers in each sequence below.

   a) 4, 7, 10, 13, 16, _______, _______, _______, ...

   b) 20, 18, 16, 14, 12, _____, _____, _____, ...

2. The numbers in a sequence are called the terms of the sequence. The general form of a sequence is shown below. Notice that the subscript of each term represents the term number.

   $a_1, a_2, a_3, \ldots, a_n, \ldots$

   Answers to frame 1:

   a) 19, 22, 25, ...

   b) 10, 8, 6, ...

Continued on following page.

2. Continued

where a_1 = first term

a_2 = second term

a_3 = third term

⋮

a_n = $\underline{n}^{th}$ term (called the general term of the sequence)

⋮

In the sequence 2, 4, 6, 8, 10, 12, 14, 16, 18, 20, ...

a) a_1 = ______ b) a_3 = ______ c) a_6 = ______ d) a_8 = ______

---

3. A sequence is called a <u>finite</u> sequence if, when counting the terms of the sequence, the counting comes to an end. The last term of a finite sequence is represented by the symbol a_n.

The sequence of odd numbers less than 20 is a <u>finite</u> sequence. Each term (except the first term) is found by adding 2 to the preceding term. The period at the end indicates that the sequence ends at that point.

1, 3, 5, 7, 9, 11, 13, 15, 17, 19.

The sequence of natural numbers is an <u>infinite</u> sequence. Each term (except the first term) is found by adding "1" to the preceding term. The three dots mean "and so on". They indicate that the sequence never ends.

1, 2, 3, 4, 5, 6, 7, 8, 9, 10, ...

State whether each sequence is "finite" or "infinite".

a) -10, -5, 0, 5, 10, 15, 20, 25, ... ____________

b) 0, 1, 2, 3, 4, 5, 6, 7, 8, 9. ____________

Answers to 2:

a) $a_1 = 2$

b) $a_3 = 6$

c) $a_6 = 12$

d) $a_8 = 16$

---

4. Sometimes each term of a sequence is a function of $\underline{n}$, where $\underline{n}$ is the term number. Therefore, the domain of the function is the set of natural numbers. That is:

$$a_n = f(n)$$

If we are given the function, we can generate the term in the sequence.

If $a_n = n^2$:
$$a_1 = (1)^2 = 1$$
$$a_2 = (2)^2 = 4$$
$$a_3 = (3)^2 = 9$$
$$a_4 = (4)^2 = 16$$
⋮

Continued on following page.

Answers to 3:

a) infinite

b) finite

4. Continued

Write the first four terms of each sequence.

a) $a_n = n + 5$ b) $a_n = n^3 + 1$

---

5. We generated some terms in the sequence below.

$$\text{If } a_n = \frac{n+1}{3}: \quad a_1 = \frac{1+1}{3} = \frac{2}{3}$$

$$a_2 = \frac{2+1}{3} = \frac{3}{3} = 1$$

$$a_3 = \frac{3+1}{3} = \frac{4}{3}$$

$$a_4 = \frac{4+1}{3} = \frac{5}{3}$$

$$\vdots$$

Write the first four terms in each sequence.

a) $a_n = \frac{3n}{4}$ b) $a_n = \frac{n(n-1)}{2}$

Answers to 4: a) 6, 7, 8, 9 b) 2, 9, 28, 65

---

6. Find the 10th term of each sequence.

a) $a_n = 5n - 1$ ________ c) $a_n = n^2 + 7$ ________

b) $a_n = \frac{n+5}{2n}$ ________ d) $a_n = \frac{n^2 - 5n}{5}$ ________

Answers to 5: a) $\frac{3}{4}$, $\frac{3}{2}$, $\frac{9}{4}$, 3 b) 0, 1, 3, 6

Answers to 6:

a) 49

b) $\frac{3}{4}$

c) 107

d) 10

11-2 SERIES

In this section, we will define series and discuss partial sums.

---

7. A series is the indicated sum of a sequence of terms. It is a finite series if the sequence is finite. It is an infinite series if the sequence is infinite.

   The general form of a finite series is:

   $a_1 + a_2 + a_3 + \ldots + a_n$

   The general form of an infinite series is:

   $a_1 + a_2 + a_3 + \ldots + a_n + \ldots$

   Given the sequence function $a_n = 2n - 1$:

   The sequence is: 1, 3, 5, 7, 9, ...

   The series is: 1 + 3 + 5 + 7 + 9 + ...

   Given the sequence function $a_n = \frac{n}{2}$:

   a) The sequence is: ______________________

   b) The series is: ______________________

Answers to frame 7:

a) $\frac{1}{2}, 1, \frac{3}{2}, 2, \frac{5}{2}, \ldots$

b) $\frac{1}{2} + 1 + \frac{3}{2} + 2 + \frac{5}{2} + \ldots$

---

8. A partial sum of a series is the sum of a finite number of consecutive terms, beginning with the first term. For example:

$S_1 = a_1$	First partial sum
$S_2 = a_1 + a_2$	Second partial sum
$S_3 = a_1 + a_2 + a_3$	Third partial sum
$\vdots$	
$S_n = a_1 + a_2 + a_3 + \ldots + a_n$	n^{th} partial sum
$\vdots$	

   Given the sequence function $a_n = n^2$:

   The sequence is 1, 4, 9, 16, ...

   The series is 1 + 4 + 9 + 16 + ...

Continued on following page.

8. Continued

The partial sums are:

$S_1 = 1 \qquad\qquad\quad = 1$

$S_2 = 1 + 4 \qquad\quad\;\; = 5$

$S_3 = 1 + 4 + 9 \qquad = 14$

$S_4 = 1 + 4 + 9 + 16 = 30$

Find S_4 for each of the following:

a) $a_n = 3n + 2$ b) $a_n = 2^n$

---

9. Given the sequence function $a_n = \frac{1}{n}$:

The sequence is: $\frac{1}{1}, \frac{1}{2}, \frac{1}{3}, \frac{1}{4}, \ldots$

The series is: $1 + \frac{1}{2} + \frac{1}{3} + \frac{1}{4} + \ldots$

The partial sums are:

$S_1 = 1 \qquad\qquad\qquad = 1$

$S_2 = 1 + \frac{1}{2} \qquad\qquad = \frac{3}{2}$

$S_3 = 1 + \frac{1}{2} + \frac{1}{3} \qquad = \frac{11}{6}$

$S_4 = 1 + \frac{1}{2} + \frac{1}{3} + \frac{1}{4} = \frac{25}{12}$

Find S_3 for each of the following.

a) $a_n = \frac{1}{2^n}$ b) $a_n = \frac{n - 1}{n + 1}$

Answers to 8:

a) $5 + 8 + 11 + 14 = 38$

b) $2 + 4 + 8 + 16 = 30$

Answers to 9:

a) $\frac{1}{2} + \frac{1}{4} + \frac{1}{8} = \frac{7}{8}$

b) $0 + \frac{1}{3} + \frac{1}{2} = \frac{5}{6}$

11-3 ARITHMETIC PROGRESSIONS

In this section, we will discuss arithmetic progressions (or arithmetic sequences). We will use the formulas for finding the $\underline{n}^{th}$ term and the sum of $\underline{n}$ terms of an arithmetic progression.

---

10. An arithmetic progression or arithmetic sequence is a sequence in which each term after the first differs from the previous term by a fixed amount. The fixed amount is called the common difference (d). AP is the abbreviation for arithmetic progression.

Let's find the first five terms of the arithmetic progression having a first term $a_1 = 2$ and a common difference $d = 4$.

$a_1 = 2$

$a_2 = 2 + 4 = 6$ Adding 4 to a_1

$a_3 = 6 + 4 = 10$ Adding 4 to a_2

$a_4 = 10 + 4 = 14$ Adding 4 to a_3

$a_5 = 14 + 4 = 18$ Adding 4 to a_4

The AP is 2, 6, 10, 14, 18, ...

Find the first five terms of the arithmetic progression having a first term $a_1 = 5$ and a common difference $d = 10$.

---

11. Let's find the first five terms of the arithmetic progression having a first term $a_1 = 12$ and a common difference $d = -5$.

Answer: 5, 15, 25, 35, 45

$a_1 = 12$

$a_2 = 12 + (-5) = 7$

$a_3 = 7 + (-5) = 2$

$a_4 = 2 + (-5) = -3$

$a_5 = -3 + (-5) = -8$

The AP is 12, 7, 2, -3, -8, ...

Find the first five terms of the arithmetic progression having a first term $a_1 = 4$ and a common difference $d = -3$.

12. To determine whether a sequence is an arithmetic progression, we subtract each term from the following term to see if there is a common difference. Two examples are discussed.

2, 5, 8, 11, 14, ...	3, 0, -5, -9, -13, ...
14 - 11 = 3	-13 - (-9) = -4
11 - 8 = 3	-9 - (-5) = -4
8 - 5 = 3	-5 - 0 = -5
5 - 2 = 3	0 - 3 = -3
It is an AP. d = 3	It is not an AP. There is no common difference.

Determine whether each sequence is an arithmetic progression.

a) 5, 1, -3, -7, -11, ...   b) 2, $3\frac{1}{2}$, 5, 6, $7\frac{1}{2}$, ...

4, 1, -2, -5, -8

---

13. An arithmetic progression with first term a_1 and common difference d has the following terms.

$a_1 = a_1$

$a_2 = a_1 + d$

$a_3 = (a_1 + d) + d = a_1 + 2d$

$a_4 = (a_1 + 2d) + d = a_1 + 3d$ ... and so on.

We get the following pattern for terms.

$a_1 = a_1$

$a_2 = a_1 + 1d$

$a_3 = a_1 + 2d$

$a_4 = a_1 + 3d$ ... and so on.

As you can see, the coefficient of d is always one less than the term number. Therefore, the general term of an AP has the following form.

The nth Term Of An AP

$$a_n = a_1 + (n - 1)d$$

Continued on following page.

a) It is an AP.
d = -4

b) It is not an AP.

13. Continued

We can use the formula above to find the 11th term of the AP with $a_1 = 15$ and $d = 7$.

$$a_n = a_1 + (n - 1)d$$

$$a_{11} = 15 + (11 - 1)(7)$$

$$a_{11} = 15 + (10)(7)$$

$$a_{11} = 15 + 70 = 85$$

a) Find the 9th term of the AP with $a_1 = 12$ and $d = 5$.

b) Find the 21st term of the AP with $a_1 = 17$ and $d = 3$.

---

14. We used the same formula to find the 31st term of the AP with $a_1 = 50$ and $d = -4$.

$$a_n = a_1 + (n - 1)d$$

$$a_{31} = 50 + (31 - 1)(-4)$$

$$a_{31} = 50 + 30(-4)$$

$$a_{31} = 50 + (-120) = -70$$

a) Find the 8th term of the AP with $a_1 = 17$ and $d = -5$.

b) Find the 51st term of the AP with $a_1 = -7$ and $d = -2$.

Answers to 13:

a) $a_9 = 52$

b) $a_{21} = 77$

---

15. a) Find the 50th term of the AP with $a_1 = 10$ and $d = 2$.

b) Find the 100th term of the AP with $a_1 = 200$ and $d = -5$.

Answers to 14:

a) $a_8 = -18$

b) $a_{51} = -107$

Answers to 15:

a) $a_{50} = 108$

b) $a_{100} = -295$

16. An arithmetic series is the sum of the terms of an arithmetic progression. For the arithmetic progression 1, 5, 9, 13, 17, ..., the arithmetic series is:

$$1 + 5 + 9 + 13 + 17 + \ldots$$

The natural numbers are an arithmetic progression in which $a_1 = 1$ and $d = 1$. There is a short way to find the sum of the first 50 natural numbers. To do so, we can write the terms of S_{50} in opposite directions and add. We get:

$$\begin{aligned} S_{50} &= 1 + 2 + 3 + 4 + \ldots + 49 + 50 \\ S_{50} &= 50 + 49 + 48 + 47 + \ldots + 2 + 1 \\ \hline 2S_{50} &= 51 + 51 + 51 + 51 + \ldots + 51 + 51 \end{aligned}$$

Since we added 50 pairs, we can find $2S_{50}$ by multiplying 51 by 50. Then we can solve for S_{50}. We get:

$$2S_{50} = 50(51)$$

$$2S_{50} = 2,550$$

$$S_{50} = 1,275$$

Let's use the same method to find the sum of the first 100 natural numbers. Solve for S_{100} below.

$$\begin{aligned} S_{100} &= 1 + 2 + 3 + 4 + \ldots + 99 + 100 \\ S_{100} &= 100 + 99 + 98 + 97 + \ldots + 2 + 1 \\ \hline 2S_{100} &= 101 + 101 + 101 + 101 + \ldots + 101 + 101 \end{aligned}$$

$2S_{100} = 100(101)$

$2S_{100} = 10,100$

$S_{100} = 5,050$

17. We wrote the sum of any finite arithmetic series below. Notice that we got the second-last term by subtracting d from a_n and the third last term by subtracting 2d from a_n.

$$S_n = a_1 + (a_1 + d) + (a_1 + 2d) + \ldots + (a_n - 2d) + (a_n - d) + a_n$$

Using the method from the last frame, we can find a formula for the sum of any arithmetic series. To do so, we write the terms of S_n in opposite directions and add.

$$\begin{aligned} S_n &= a_1 + (a_1 + d) + (a_1 + 2d) + \ldots + (a_n - 2d) + (a_n - d) + a_n \\ S_n &= a_n + (a_n - d) + (a_n - 2d) + \ldots + (a_1 + 2d) + (a_1 + d) + a_1 \\ \hline 2S_n &= (a_1 + a_n) + (a_1 + a_n) + (a_1 + a_n) + \ldots + (a_1 + a_n) + (a_1 + a_n) + (a_1 + a_n) \end{aligned}$$

Since we added n pairs, we can find $2S_n$ by multiplying $(a_1 + a_n)$ by n. We get:

$$2S_n = n(a_1 + a_n)$$

Continued on following page.

17. Continued

Therefore, the formula for the sum of n terms of an arithmetic progression is:

The Sum Of n Terms Of An AP

$$S_n = \frac{n(a_1 + a_n)}{2}$$

Let's use the formula to find the sum of the first 30 natural numbers. Since $n = 30$, $a_1 = 1$, $a_n = 30$, we get:

$$S_n = \frac{n(a_1 + a_n)}{2}$$

$$S_{30} = \frac{30(1 + 30)}{2} = \frac{30(31)}{2} = \frac{930}{2} = 465$$

Use the formula to find the sum of:

a) the first 20 natural numbers. b) the first 60 natural numbers.

---

18. To find the sum of the first 10 terms of the AP with $a_1 = 5$ and $d = 4$, we must begin by finding a_{10}.

$$a_n = a_1 + (n - 1)d$$

$$a_{10} = 5 + (10 - 1)(4)$$

$$a_{10} = 5 + 36 = 41$$

Now we can use the formula to find the sum of the first 10 terms.

$$S_n = \frac{n(a_1 + a_n)}{2}$$

$$S_{10} = \frac{10(5 + 41)}{2} = \frac{10(46)}{2} = \frac{460}{2} = 230$$

Following the example, find the sum of the first 10 terms of:

a) the AP with $a_1 = 9$, and $d = 8$ b) the AP with $a_1 = 12$ and $d = -6$

a) $S_{20} = 210$

b) $S_{60} = 1{,}830$

19. Even natural numbers are an AP with $a_1 = 2$ and $d = 2$. We used that fact to solve one problem below. Solve the other problem.

Find the sum of the first 20 even natural numbers.	Find the sum of the first 50 even natural numbers.

$$a_{20} = 2 + (20 - 1)(2)$$
$$= 2 + 19(2)$$
$$= 2 + 38 = 40$$

$$S_{20} = \frac{20(2 + 40)}{2}$$
$$= \frac{20(42)}{2}$$
$$= \frac{840}{2} = 420$$

a) $S_{10} = 450$

b) $S_{10} = -150$

---

20. Odd natural numbers are an AP with $a_1 = 1$ and $d = 2$. We used that fact to solve one problem below. Solve the other problem.

Find the sum of the first 40 odd natural numbers.	Find the sum of the first 100 odd natural numbers.

$$a_{40} = 1 + (40 - 1)(2)$$
$$= 1 + 39(2)$$
$$= 1 + 78 = 79$$

$$S_{40} = \frac{40(1 + 79)}{2}$$
$$= \frac{40(80)}{2}$$
$$= \frac{3,200}{2} = 1,600$$

$S_{50} = 2,550$

$S_{100} = 10,000$

SELF-TEST 38 (pages 548-559)

1. Write the next three terms in this sequence.

   $5, 3\frac{1}{2}, 2, \frac{1}{2}$, ____, ____, ____, ...

2. Write the first four terms of this sequence.

   $a_n = \frac{3n}{2}$

Find the 10th term of each sequence.

3. $a_n = 2n - 7$

4. $a_n = \frac{n-1}{n^2}$

Find S_4 for each sequence.

5. $a_n = n^2 - 1$

6. $a_n = \frac{n-1}{n+2}$

Find the first five terms of each arithmetic progression.

7. $a_1 = 4$ and $d = 3$

8. $a_1 = 10$ and $d = -4$

Decide whether each sequence is an arithmetic progression.

9. 7, 11, 16, 20, 25, ...

10. 8, 5, 2, -1, -4, ...

11. Find the 21st term of the AP with $a_1 = 50$ and $d = -3$.

12. Find the sum of the first 90 natural numbers.

13. Find the sum of the first 10 terms of the AP with $a_1 = 10$ and $d = 5$.

14. Find the sum of the first 100 even natural numbers.

ANSWERS:

1. $-1, -2\frac{1}{2}, -4, \ldots$
2. $\frac{3}{2}, 3, \frac{9}{2}, 6$
3. $a_{10} = 13$
4. $a_{10} = \frac{9}{100}$
5. $S_4 = 26$
6. $S_4 = \frac{23}{20}$
7. 4, 7, 10, 13, 16
8. 10, 6, 2, -2, -6
9. Not an AP
10. An AP with $d = -3$
11. $a_{11} = -10$
12. $S_{90} = 4{,}095$
13. $S_{10} = 325$
14. $S_{100} = 10{,}100$

11-4 GEOMETRIC PROGRESSIONS

In this section, we will discuss geometric progressions (or geometric sequences). We will use the formulas for finding the nth term and the sum of n terms of a geometric progression. We will also use the formula for finding the sum of an infinite geometric series when $|r| < 1$.

---

21. A geometric progression (or geometric sequence) is a sequence in which each term after the first is found by multiplying the previous term by a fixed number called the common ratio (r). GP is the abbreviation for geometric progression.

Let's find the first five terms of the geometric progression having a first term $a_1 = 10$ and a common ratio $r = 2$.

$a_1 = 10$

$a_2 = 10(2) = 20$ Multiplying a_1 by 2

$a_3 = 20(2) = 40$ Multiplying a_2 by 2

$a_4 = 40(2) = 80$ Multiplying a_3 by 2

$a_5 = 80(2) = 160$ Multiplying a_4 by 2

The GP is 10, 20, 40, 80, 160, ...

Find the first five terms of the geometric progression having a first term $a_1 = 4$ and a common ratio $r = -3$.

---

22. Let's find the first five terms of the geometric progression having a first term $a_1 = 48$ and a common ratio $r = -\frac{1}{2}$.

Answer to 21: 4, -12, 36, -108, 324, ...

$a_1 = 48$

$a_2 = 48\left(-\frac{1}{2}\right) = -24$

$a_3 = -24\left(-\frac{1}{2}\right) = 12$

$a_4 = 12\left(-\frac{1}{2}\right) = -6$

$a_5 = -6\left(-\frac{1}{2}\right) = 3$

The GP is 48, -24, 12, -6, 3, ...

Find the first five terms of the geometric progression having a first term $a_1 = 18$ and a common ratio $r = \frac{1}{3}$.

Answer to 22: 18, 6, 2, $\frac{2}{3}$, $\frac{2}{9}$, ...

23. To determine whether a sequence is a geometric progression, we divide each term by the previous term to see if there is a common ratio. Two examples are discussed.

4, -8, 16, -32, 64, ...

$$\frac{64}{-32} = -2$$

$$\frac{-32}{16} = -2$$

$$\frac{16}{-8} = -2$$

$$\frac{-8}{4} = -2$$

It is a GP. r = -2.

72, 18, 9, 3, 1, ...

$$\frac{1}{3} = \frac{1}{3}$$

$$\frac{3}{9} = \frac{1}{3}$$

$$\frac{9}{18} = \frac{1}{2}$$

$$\frac{18}{72} = \frac{1}{4}$$

It is not a GP. There is no common ratio.

Determine whether each sequence is a geometric progression.

a) 2, 8, 32, 96, 192, ...

b) 81, -27, 9, -3, 1, ...

---

a) It is not a GP.

b) It is a GP.

$r = -\frac{1}{3}$

24. A geometric progression with first term a_1 and common ratio r has the following terms.

$$a_1 = a_1$$

$$a_2 = a_1r$$

$$a_3 = (a_1r)r = a_1r^2$$

$$a_4 = (a_1r^2)r = a_1r^3$$

$$a_5 = (a_1r^3)r = a_1r^4$$ ... and so on

We get the following pattern for terms:

$$a_1 = a_1$$

$$a_2 = a_1r$$

$$a_3 = a_1r^2$$

$$a_4 = a_1r^3$$

$$a_5 = a_1r^4$$

Continued on following page.

24. Continued

As you can see, the exponent of r is always one less than the term number. Therefore, the general term of a GP has the following form.

The nth Term Of A GP:

$$a_n = a_1 r^{n-1}$$

We can use the formula above to find the 7th term of the GP with $a_1 = 5$ and $r = 2$.

$$a_n = a_1 r^{n-1}$$

$$a_7 = 5(2)^{7-1}$$

$$a_7 = 5(2)^6$$

$$a_7 = 5(64) = 320$$

a) Find the 5th term of the GP with $a_1 = 2$ and $r = 3$.

b) Find the 4th term of the GP with $a_1 = 10$ and $r = -4$.

---

25. We used the same formula to find the 5th term of the GP with $a_1 = 36$ and $r = \frac{1}{3}$.

a) $a_5 = 162$

b) $a_4 = -640$

$$a_n = a_1 r^{n-1}$$

$$a_5 = 36\left(\frac{1}{3}\right)^4$$

$$a_5 = 36\left(\frac{1}{81}\right) = \frac{36}{81} = \frac{4}{9}$$

a) Find the 3rd term of the GP with $a_1 = 32$ and $r = \frac{3}{4}$.

b) Find the 6th term of the GP with $a_1 = 8$ and $r = -\frac{1}{2}$.

---

a) $a_3 = 18$

b) $a_6 = -\frac{1}{4}$

26. A geometric series is the sum of the terms of a geometric progression. We wrote the sum of any finite geometric series below.

$$S_n = a_1 + a_1r + a_1r^2 + \ldots + a_1r^{n-1}$$

Multiplying both sides of the equation above by -r, we get:

$$-rS_n = -a_1r - a_1r^2 - a_1r^3 - \ldots - a_1r^{n-1} - a_1r^n$$

By adding the two equations above and then solving for S_n, we can find a formula for the sum of any geometric series.

$$\begin{array}{rcl} S_n & = & a_1 + a_1r + a_1r^2 + \ldots + a_1r^{n-1} \\ -rS_n & = & \quad - a_1r - a_1r^2 - \ldots - a_1r^{n-1} - a_1r^n \\ \hline S_n - rS_n & = & a_1 + 0 + 0 + \ldots + 0 - a_1r^n \end{array}$$

$$S_n - rS_n = a_1 - a_1r^n$$

$$S_n(1 - r) = a_1(1 - r^n)$$

$$S_n = \frac{a_1(1 - r^n)}{1 - r}$$

Therefore, the formula for the sum of n terms of a geometric progression is:

The Sum Of n Terms Of A GP

$$S_n = \frac{a_1(1 - r^n)}{1 - r}, \text{ where } r \neq 1$$

Let's use the formula to find the sum of the first 6 terms of the GP with $a_1 = 4$ and $r = 2$.

$$S_n = \frac{a_1(1 - r^n)}{1 - r}$$

$$S_6 = \frac{4(1 - 2^6)}{(1 - 2)} = \frac{4(1 - 64)}{-1} = -4(-63) = 252$$

Use the formula to find the sum of the first 5 terms of the GP:

a) with $a_1 = 8$ and $r = 2$

b) with $a_1 = 10$ and $r = 3$

27. We used the formula to find the sum of the first 6 terms of the GP with $a_1 = 16$ and $r = \frac{1}{2}$.

$$S_n = \frac{a_1(1 - r^n)}{1 - r}$$

$$S_6 = \frac{16\left[1 - \left(\frac{1}{2}\right)^6\right]}{1 - \frac{1}{2}} = \frac{16\left(1 - \frac{1}{64}\right)}{\frac{1}{2}} = \frac{16\left(\frac{63}{64}\right)}{\frac{1}{2}} = 16\left(\frac{63}{64}\right)(2) = \frac{63}{2}$$

Use the formula for these:

a) Find the sum of the first 5 terms of the GP with $a_1 = 27$ and $r = \frac{1}{3}$.

b) Find the sum of the first 4 terms of the GP with $a_1 = 9$ and $r = \frac{2}{3}$.

a) $S_5 = 248$

b) $S_5 = 1,210$

---

28. The symbol S_∞ represents the sum of a geometric series when n, the number of terms, becomes infinitely large. If $|r| > 1$, the absolute value of the terms becomes larger and larger. Therefore, S_∞ is infinitely large. Two of the possible cases are discussed below.

For the series below, $a_1 = 5$ and $r = 4$. You can see that S_∞ would be an infinitely larger positive number.

5, 20, 80, 320, 1280, 5120, ...

For the series below, $a_1 = -10$ and $r = 3$. You can see that S_∞ would be an infinitely large negative number.

-10, -30, -90, -270, -810, -2430, ...

However, when $|r| < 1$, the absolute value of the terms becomes smaller and smaller. The partial sums of such a sequence approach a limiting value. An example is discussed.

$$2 + 1 + \frac{1}{2} + \frac{1}{4} + \frac{1}{8} + \ldots \qquad \left(r = \frac{1}{2}\right)$$

Some partial sums are shown below. They approach 4 as a limiting value.

$S_3 = 3.5000$

$S_6 = 3.9375$

$S_{10} = 3.9961$

$S_{20} = 3.9999$

Continued on following page.

a) $S_5 = \frac{121}{3}$

b) $S_4 = \frac{65}{3}$

28. Continued

The following formula can be used to find the limiting value of the sum of an infinite geometric series when $|r| < 1$.

Sum Of An Infinite Geometric Series

$$S_\infty = \frac{a_1}{1 - r}, \text{ when } |r| < 1$$

For the series on the previous page, $a_1 = 2$ and $r = \frac{1}{2}$. Using the formula, we get:

$$S_\infty = \frac{2}{1 - \frac{1}{2}} = \frac{2}{\frac{1}{2}} = 2(2) = 4$$

Find the sum of each series. (You will have to find r to use the formula.)

a) $4 + 1 + \frac{1}{4} + \frac{1}{16} + \ldots$ b) $-9 - 6 - 4 - \frac{8}{3} - \ldots$

---

a) $S_\infty = 5\frac{1}{3}$ $\left(r = \frac{1}{4}\right)$

b) $S_\infty = -27$ $\left(r = \frac{2}{3}\right)$

29. We used the formula to find the sum of the geometric series below.

$8 - 6 + \frac{9}{2} - \frac{27}{8} + \ldots$ $\left(r = \frac{-6}{8} = -\frac{3}{4}\right)$

$$S_\infty = \frac{a_1}{1 - r} = \frac{8}{1 - \left(-\frac{3}{4}\right)} = \frac{8}{\frac{7}{4}} = 8\left(\frac{4}{7}\right) = \frac{32}{7} = 4\frac{4}{7}$$

Find the sum of each geometric series.

a) $3 - 1 + \frac{1}{3} - \frac{1}{9} + \ldots$ b) $10 - 4 + \frac{8}{5} - \frac{16}{25} + \ldots$

---

a) $S_\infty = 2\frac{1}{4}$ $\left(r = -\frac{1}{3}\right)$

b) $S_\infty = 7\frac{1}{7}$ $\left(r = -\frac{2}{5}\right)$

11-5 BINOMIAL EXPANSION

In this section, we will discuss a method for expanding binomials of the form $(a + b)^n$.

30. Binomial expressions of the form $(a + b)^n$, where n is a whole number, can be expanded to produce a family of series. For example:

$$\begin{aligned}
(a + b)^0 &= 1 \\
(a + b)^1 &= a + b \\
(a + b)^2 &= a^2 + 2ab + b^2 \\
(a + b)^3 &= a^3 + 3a^2b + 3ab^2 + b^3 \\
(a + b)^4 &= a^4 + 4a^3b + 6a^2b^2 + 4ab^3 + b^4 \\
(a + b)^5 &= a^5 + 5a^4b + 10a^3b^2 + 10a^2b^3 + 5ab^4 + b^5 \\
&\vdots
\end{aligned}$$

Notice the following patterns about the expansions above:

1. Each expansion of $(a + b)^n$ has $n + 1$ terms.

   Example: The expansion of $(a + b)^4$ has 5 terms.

2. The first term is always a^n and the last term is b^n.

   Example: For $(a + b)^5$, the first term is a^5 and the last term is b^5.

3. The coefficients of the first term and last term are "1".

   Example: For $(a + b)^3$, the coefficients of a^3 and b^3 are "1".

4. The coefficient of the second term and second-last term are n.

   Example: For $(a + b)^4$, the coefficients of $4a^3b$ and $4ab^3$ are 4.

5. The exponent of a is n in the first term and decreases by "1" in each succeeding term. The exponent of b is 0 in the first term and increases by "1" in each succeeding term.

   Example: $(a + b)^4 = a^4 + 4a^3b + 6a^2b^2 + 4ab^3 + b^4$

	a^4	$4a^3b$	$6a^2b^2$	$4ab^3$	b^4
Exponents of a:	4	3	2	1	0
Exponents of b:	0	1	2	3	4

6. The sum of the exponents in each term is n.

   Example: $(a + b)^4 = a^4 + 4a^3b + 6a^2b^2 + 4ab^3 + b^4$

   (The sum of the exponents in each term is 4.)

31. The coefficients of the terms in the expansions in the last frame form the symmetrical, triangular pattern below. This pattern is called Pascal's triangle.

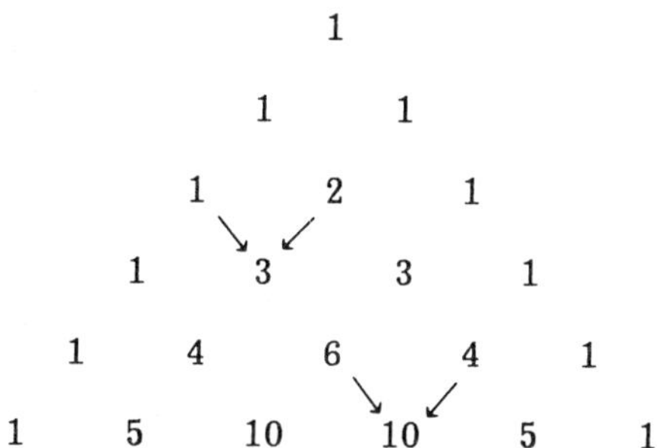

Notice these facts about the numbers in Pascal's triangle.

1. The first and last number in each row are "1".
2. Every other number in the triangle is the sum of the two closest numbers in the row above it. For example, the arrows above show that:

   3 is the sum of 1 and 2

   10 is the sum of 6 and 4

Using the two facts above, find the next row of coefficients for $(a + b)^6$. Use the row below to do so.

1 5 10 10 5 1

---

32. Though Pascal's triangle is easy to use, it is time consuming to construct it for larger values of n. Therefore, we use a quicker method to find the coefficients for the expansion of $(a + b)^n$. The quicker method involves the factorial function n! which is:

> **Factorial Function n!**
>
> $n! = n(n - 1)(n - 2)(n - 3) \ldots 3 \cdot 2 \cdot 1$

The symbol n! is read "n factorial". Some examples of the factorial function are:

$1! = 1$

$2! = 2 \cdot 1 = 2$

$3! = 3 \cdot 2 \cdot 1 = 6$

$4! = 4 \cdot 3 \cdot 2 \cdot 1 = 24$

Using the definition, compute these:

a) $5! =$ b) $6! =$

1 5 10 10 5 1

1 6 15 20 15 6 1

33. Factorial functions can also be partially expanded. For example:

$$5! = 5 \cdot 4!$$

$$8! = 8 \cdot 7 \cdot 6!$$

We used the fact above to evaluate each expression below.

$$\frac{5!}{4!} = \frac{5 \cdot 4!}{4!} = 5 \qquad \frac{8!}{6!} = \frac{8 \cdot 7 \cdot 6!}{6!} = 8 \cdot 7 = 56$$

Evaluate each expression.

a) $\frac{9!}{8!} =$ b) $\frac{10!}{8!} =$

Answers: a) $5! = 120$ b) $6! = 720$

---

34. We also used partial expansions to evaluate each expression below.

$$\frac{7!}{5!3!} = \frac{7 \cdot 6 \cdot 5!}{5!3!} = \frac{7 \cdot 6}{3!} = \frac{7 \cdot 6}{3 \cdot 2 \cdot 1} = 7$$

$$\frac{6!}{3!3!} = \frac{6 \cdot 5 \cdot 4 \cdot 3!}{3!3!} = \frac{6 \cdot 5 \cdot 4}{3!} = \frac{6 \cdot 5 \cdot 4}{3 \cdot 2 \cdot 1} = 20$$

Evaluate each expression.

a) $\frac{10!}{8!2!} =$

b) $\frac{8!}{4!4!} =$

Answers: a) 9 b) 90

---

35. The expansion of $(a + b)^n$ can be written in factorial notation. We get:

Answers: a) 45 b) 70

$$(a + b)^n = a^n + \frac{n!}{(n-1)!1!}a^{n-1}b + \frac{n!}{(n-2)!2!}a^{n-2}b^2 + \dots + \frac{n!}{1!(n-1)!}ab^{n-1} + b^n$$

We used the formula above to expand $(a + b)^7$ below. Notice in each term that the factors in the denominator are the same numbers as the exponents of <u>a</u> and <u>b</u>.

$$(a + b)^7 = a^7 + \frac{7!}{6!1!}a^6b + \frac{7!}{5!2!}a^5b^2 + \frac{7!}{4!3!}a^4b^3 + \frac{7!}{3!4!}a^3b^4 + \frac{7!}{2!5!}a^2b^5 + \frac{7!}{1!6!}ab^6 + b^7$$

$$= a^7 + 7a^6b + 21a^5b^2 + 35a^4b^3 + 35a^3b^4 + 21a^2b^5 + 7ab^6 + b^7$$

Use the formula to expand $(x + y)^6$ below.

$(x + y)^6$

$$= x^6 + \frac{6!}{5!1!}x^5y + \frac{6!}{4!2!}x^4y^2 + \frac{6!}{3!3!}x^3y^3 + \frac{6!}{2!4!}x^2y^4 + \frac{6!}{1!5!}xy^5 + y^6$$

$$= x^6 + 6x^5y + 15x^4y^2 + 20x^3y^3 + 15x^2y^4 + 6xy^5 + y^6$$

36. We used the formula to expand $(x + 2)^6$ below. Notice how we multiplied the coefficients and the powers of 2.

$$(x + 2)^6 = x^6 + \frac{6!}{5!1!}x^5 \cdot 2 + \frac{6!}{4!2!}x^4 \cdot 2^2 + \frac{6!}{3!3!}x^3 \cdot 2^3 + \frac{6!}{2!4!}x^2 \cdot 2^4 + \frac{6!}{1!5!}x \cdot 2^5 + 2^6$$

$$= x^6 + 6 \cdot 2 \cdot x^5 + 15 \cdot 2^2 \cdot x^4 + 20 \cdot 2^3 \cdot x^3 + 15 \cdot 2^4 \cdot x^2 + 6 \cdot 2^5 \cdot x + 2^6$$

$$= x^6 + 12x^5 + 60x^4 + 160x^3 + 240x^2 + 192x + 64$$

Use the formula to expand $(y + 3)^5$ below.

$(y + 3)^5 =$

$$= y^5 + \frac{5!}{4!1!}y^4 \cdot 3 + \frac{5!}{3!2!}y^3 \cdot 3^2 + \frac{5!}{2!3!}y^2 \cdot 3^3 + \frac{5!}{1!4!}y \cdot 3^4 + 3^5$$

$$= y^5 + 15y^4 + 90y^3 + 270y^2 + 405y + 243$$

---

37. We used the formula to expand $(a - b)^4$ or $[a + (-b)]^4$ below. Notice that the term is <u>negative</u> when the exponent of $-b$ is <u>odd</u>, and <u>positive</u> when the exponent of $-b$ is <u>even</u>.

$$(a - b)^4 = a^4 + \frac{4!}{3!1!}a^3 \cdot (-b)^1 + \frac{4!}{2!2!}a^2 \cdot (-b)^2 + \frac{4!}{1!3!}a \cdot (-b)^3 + (-b)^4$$

$$= a^4 - 4a^3b + 6a^2b^2 - 4ab^3 + b^4$$

Use the formula to expand $(x - y)^5$ below.

$(x - y)^5 =$

$$= x^5 + \frac{5!}{4!1!}x^4 \cdot (-y)^1 + \frac{5!}{3!2!}x^3 \cdot (-y)^2 + \frac{5!}{2!3!}x^2 \cdot (-y)^3 + \frac{5!}{1!4!}x \cdot (-y)^4 + (-y)^5$$

$$= x^5 - 5x^4y + 10x^3y^2 - 10x^2y^3 + 5xy^4 - y^5$$

38. We used the formula to expand $(x - 2)^5$ below. Notice how we multiplied the coefficients and the powers of -2.

$$(x - 2)^5 = x^5 + \frac{5!}{4!1!}x^4 \cdot (-2)^1 + \frac{5!}{3!2!}x^3 \cdot (-2)^2 + \frac{5!}{2!3!}x^2 \cdot (-2)^3 + \frac{5!}{1!4!}x \cdot (-2)^4 + (-2)^5$$

$$= x^5 + 5 \cdot (-2) \cdot x^4 + 10 \cdot 4 \cdot x^3 + 10 \cdot (-8) \cdot x^2 + 5 \cdot 16 \cdot x + (-32)$$

$$= x^5 - 10x^4 + 40x^3 - 80x^2 + 80x - 32$$

Use the formula to expand $(1 - x)^4$ below.

$(1 - x)^4 =$

$$= 1^4 + \frac{4!}{3!1!} \cdot 1^3 \cdot (-x) + \frac{4!}{2!2!} \cdot 1^2 \cdot (-x)^2 + \frac{4!}{1!3!} \cdot 1 \cdot (-x)^3 + (-x)^4$$

$$= 1 - 4x + 6x^2 - 4x^3 + x^4$$

SELF-TEST 39 (pages 560-571)

Find the first five terms of each geometric progression.	Decide whether each sequence is a geometric progression.
1. $a_1 = 5$, $r = -2$	3. 64, 16, 4, 1, $\frac{1}{4}$, ...
2. $a_1 = 48$, $r = \frac{1}{4}$	4. 3, 9, 27, 54, 108, ...
5. Find the 5th term of the GP with $a_1 = 10$ and $r = 3$.	6. Find the fourth term of the GP with $a_1 = 3$ and $r = -\frac{3}{4}$. 4

Continued on following page.

SELF-TEST 39 - Continued

7. Find the sum of the first five terms of the GP with $a_1 = 8$ and $r = 3$.

8. Find the sum of the first four terms of the GP with $a_1 = 3$ and $r = \frac{1}{4}$.

Find the sum of each infinite geometric series.

9. $4 - 2 + 1 - \frac{1}{2} + \ldots$

10. $25 + 20 + 16 + \frac{64}{5} + \ldots$

Evaluate each expression.

11. $\frac{11!}{9!}$

12. $\frac{9!}{6!3!}$

Expand each binomial.

13. $(p + q)^6$

14. $(y - 1)^5$

ANSWERS:

1. 5, -10, 20, -40, 80
2. 48, 12, 3, $\frac{3}{4}$, $\frac{3}{16}$
3. A GP with $r = \frac{1}{4}$
4. Not a GP
5. $a_5 = 810$
6. $a_4 = -\frac{27}{16}$
7. $S_5 = 968$
8. $S_4 = \frac{255}{64}$
9. $S_\infty = 2\frac{2}{3}$
10. $S_\infty = 125$
11. 110
12. 84
13. $p^6 + 6p^5q + 15p^4q^2 + 20p^3q^3 + 15p^2q^4 + 6pq^5 + q^6$
14. $y^5 - 5y^4 + 10y^3 - 10y^2 + 5y - 1$

SUPPLEMENTARY PROBLEMS - CHAPTER 11

Assignment 38

Write the next three terms in each sequence.

1. 5, 9, 13, 17, ____, ____, ____, ...

2. 3, 10, 17, 24, ____, ____, ____, ...

3. 14, 11, 8, 5, ____, ____, ____, ...

4. $\frac{7}{2}$, 3, $\frac{5}{2}$, 2, ____, ____, ____, ...

Write the first four terms of each sequence.

5. $a_n = n + 3$

6. $a_n = n^2 - 1$

7. $a_n = n^3 + 2$

8. $a_n = \frac{n + 2}{4}$

9. $a_n = \frac{2n}{3}$

10. $a_n = \frac{n(n + 1)}{2}$

Find the 10th term of each sequence.

11. $a_n = 3n - 1$

12. $a_n = \frac{n + 4}{n - 3}$

13. $a_n = \frac{n^2 + 1}{100}$

Find S_4 for each sequence:

14. $a_n = 2n - 1$

15. $a_n = 4n + 3$

16. $a_n = n^3$

Find S_3 for each sequence:

17. $a_n = \frac{1}{2n}$

18. $a_n = \frac{1}{n^2}$

19. $a_n = \frac{n + 1}{n + 2}$

Find the first five terms of each arithmetic progression.

20. $a_1 = 3$ and $d = 5$

21. $a_1 = 15$ and $d = -4$

22. $a_1 = 0$ and $d = -5$

Determine whether each sequence is an arithmetic progression.

23. 3, 9, 15, 20, 27, ...

24. 10, 7, 4, 1, -2, ...

25. 6, 1, -4, -8, -13, ...

26. Find the 7th term of the AP with $a_1 = 10$ and $d = 4$.

27. Find the 11th term of the AP with $a_1 = 5$ and $d = 8$.

28. Find the 41st term of the AP with $a_1 = 100$ and $d = -3$.

29. Find the 61st term of the AP with $a_1 = 160$ and $d = -5$.

30. Find the 80th term of the AP with $a_1 = 25$ and $d = 2$.

31. Find the 100th term of the AP with $a_1 = 800$ and $d = -7$.

32. Find the sum of the first 40 natural numbers.

33. Find the sum of the first 70 natural numbers.

34. Find the sum of the first 10 terms of the AP with $a_1 = 5$ and $d = 3$.

35. Find the sum of the first 10 terms of the AP with $a_1 = 25$ and $d = -10$.

36. Find the sum of the first 30 even natural numbers.

37. Find the sum of the first 80 even natural numbers.

38. Find the sum of the first 50 odd natural numbers.

39. Find the sum of the first 90 odd natural numbers.

Assignment 39

Find the first five terms of each geometric progression.

1. $a_1 = 5,\ r = 2$ 2. $a_1 = 10,\ r = -3$ 3. $a_1 = 24,\ r = \frac{1}{2}$ 4. $a_1 = 36,\ r = -\frac{2}{3}$

Determine whether each sequence is a geometric progression.

5. 2, 8, 32, 128, 512, ... 6. 3, 6, 12, 18, 36, ... 7. 64, -32, 16, -8, 4, ...

8. Find the 6th term of the GP with $a_1 = 10$ and $r = 2$.

9. Find the 4th term of the GP with $a_1 = 4$ and $r = 3$.

10. Find the 5th term of the GP with $a_1 = 6$ and $r = -2$.

11. Find the 4th term of the GP with $a_1 = 2$ and $r = -4$.

12. Find the 5th term of the GP with $a_1 = 9$ and $r = \frac{2}{3}$.

13. Find the 4th term of the GP with $a_1 = 8$ and $r = -\frac{1}{4}$.

14. Find the sum of the first five terms of the GP with $a_1 = 5$ and $r = 2$.

15. Find the sum of the first five terms of the GP with $a_1 = 8$ and $r = 3$.

16. Find the sum of the first four terms of the GP with $a_1 = 3$ and $r = 4$.

17. Find the sum of the first three terms of the GP with $a_1 = 10$ and $r = 6$.

18. Find the sum of the first six terms of the GP with $a_1 = 32$ and $r = \frac{1}{2}$.

19. Find the sum of the first three terms of the GP with $a_1 = 2$ and $r = \frac{3}{4}$.

Find the sum of each infinite geometric series.

20. $4 + 2 + 1 + \frac{1}{2} + \ldots$ 21. $16 + 12 + 9 + \frac{27}{4} + \ldots$ 22. $-25 - 5 - 1 - \frac{1}{5} + \ldots$

23. $9 - 3 + 1 - \frac{1}{3} + \ldots$ 24. $27 - 18 + 12 - 8 + \ldots$ 25. $5 - 3 + \frac{9}{5} - \frac{27}{25} + \ldots$

Evaluate each expression.

26. $3!$ 27. $6!$ 28. $\frac{10!}{9!}$ 29. $\frac{5!}{4!}$ 30. $\frac{9!}{7!}$

31. $\frac{7!}{4!}$ 32. $\frac{8!}{6!2!}$ 33. $\frac{10!}{7!3!}$ 34. $\frac{4!}{2!2!}$ 35. $\frac{10!}{5!5!}$

Expand each binomial.

36. $(c + d)^6$ 37. $(x + y)^5$ 38. $(x + 1)^7$ 39. $(y + 2)^5$

40. $(a - b)^6$ 41. $(c - d)^5$ 42. $(x - 3)^5$ 43. $(1 - x)^6$

TABLE 1

COMMON LOGARITHMS

x	0	1	2	3	4	5	6	7	8	9
1.0	.0000	.0043	.0086	.0128	.0170	.0212	.0253	.0294	.0334	.0374
1.1	.0414	.0453	.0492	.0531	.0569	.0607	.0645	.0682	.0719	.0755
1.2	.0792	.0828	.0864	.0899	.0934	.0969	.1004	.1038	.1072	.1106
1.3	.1139	.1173	.1206	.1239	.1271	.1303	.1335	.1367	.1399	.1430
1.4	.1461	.1492	.1523	.1553	.1584	.1614	.1644	.1673	.1703	.1732
1.5	.1761	.1790	.1818	.1847	.1875	.1903	.1931	.1959	.1987	.2014
1.6	.2041	.2068	.2095	.2122	.2148	.2175	.2201	.2227	.2253	.2279
1.7	.2304	.2330	.2355	.2380	.2405	.2430	.2455	.2480	.2504	.2529
1.8	.2553	.2577	.2601	.2625	.2648	.2672	.2695	.2718	.2742	.2765
1.9	.2788	.2810	.2833	.2856	.2878	.2900	.2923	.2945	.2967	.2989
2.0	.3010	.3032	.3054	.3075	.3096	.3118	.3139	.3160	.3181	.3201
2.1	.3222	.3243	.3263	.3284	.3304	.3324	.3345	.3365	.3385	.3404
2.2	.3424	.3444	.3464	.3483	.3502	.3522	.3541	.3560	.3579	.3598
2.3	.3617	.3636	.3655	.3674	.3692	.3711	.3729	.3747	.3766	.3784
2.4	.3802	.3820	.3838	.3856	.3874	.3892	.3909	.3927	.3945	.3962
2.5	.3979	.3997	.4014	.4031	.4048	.4065	.4082	.4099	.4116	.4133
2.6	.4150	.4166	.4183	.4200	.4216	.4232	.4249	.4265	.4281	.4298
2.7	.4314	.4330	.4346	.4362	.4378	.4393	.4409	.4425	.4440	.4456
2.8	.4472	.4487	.4502	.4518	.4533	.4548	.4564	.4579	.4594	.4609
2.9	.4624	.4639	.4654	.4669	.4683	.4698	.4713	.4728	.4742	.4757
3.0	.4771	.4786	.4800	.4814	.4829	.4843	.4857	.4871	.4886	.4900
3.1	.4914	.4928	.4942	.4955	.4969	.4983	.4997	.5011	.5024	.5038
3.2	.5051	.5065	.5079	.5092	.5105	.5119	.5132	.5145	.5159	.5172
3.3	.5185	.5198	.5211	.5224	.5237	.5250	.5263	.5276	.5289	.5302
3.4	.5315	.5328	.5340	.5353	.5366	.5378	.5391	.5403	.5416	.5428
3.5	.5441	.5453	.5465	.5478	.5490	.5502	.5514	.5527	.5539	.5551
3.6	.5563	.5575	.5587	.5599	.5611	.5623	.5635	.5647	.5658	.5670
3.7	.5682	.5694	.5705	.5717	.5729	.5740	.5752	.5763	.5775	.5786
3.8	.5798	.5809	.5821	.5832	.5843	.5855	.5866	.5877	.5888	.5899
3.9	.5911	.5922	.5933	.5944	.5955	.5966	.5977	.5988	.5999	.6010
4.0	.6021	.6031	.6042	.6053	.6064	.6075	.6085	.6096	.6107	.6117
4.1	.6128	.6138	.6149	.6160	.6170	.6180	.6191	.6201	.6212	.6222
4.2	.6232	.6243	.6253	.6263	.6274	.6284	.6294	.6304	.6314	.6325
4.3	.6335	.6345	.6355	.6365	.6375	.6385	.6395	.6405	.6415	.6425
4.4	.6435	6444	.6454	.6464	.6474	.6484	.6493	.6503	.6513	.6522
4.5	.6532	.6542	.6551	.6561	.6571	.6580	.6590	.6599	.6609	.6618
4.6	.6628	.6637	.6646	.6656	.6665	.6675	.6684	.6693	.6702	.6712
4.7	.6721	.6730	.6739	.6749	.6758	.6767	.6776	.6785	.6794	.6803
4.8	.6812	.6821	.6830	.6839	.6848	.6857	.6866	.6875	.6884	.6893
4.9	.6902	.6911	.6920	.6928	.6937	.6946	.6955	.6964	.6972	.6981
5.0	.6990	.6998	.7007	.7016	.7024	.7033	.7042	.7050	.7059	.7067
5.1	.7076	.7084	.7093	.7101	.7110	.7118	.7126	.7135	.7143	.7152
5.2	.7160	.7168	.7177	.7185	.7193	.7202	.7210	.7218	.7226	.7235
5.3	.7243	.7251	.7259	.7267	.7275	.7284	.7292	.7300	.7308	.7316
5.4	.7324	.7332	.7340	.7348	.7356	.7364	.7372	.7380	.7388	.7396
x	**0**	**1**	**2**	**3**	**4**	**5**	**6**	**7**	**8**	**9**

TABLE 1 *(cont.)*

x	0	1	2	3	4	5	6	7	8	9
5.5	.7404	.7412	.7419	.7427	.7435	.7443	.7451	.7459	.7466	.7474
5.6	.7482	.7490	.7497	.7505	.7513	.7520	.7528	.7536	.7543	.7551
5.7	.7559	.7566	.7574	.7582	.7589	.7597	.7604	.7612	.7619	.7627
5.8	.7634	.7642	.7649	.7657	.7664	.7672	.7679	.7686	.7694	.7701
5.9	.7709	.7716	.7723	.7731	.7738	.7745	.7752	.7760	.7767	.7774
6.0	.7782	.7789	.7796	.7803	.7810	.7818	.7825	.7832	.7839	.7846
6.1	.7853	.7860	.7868	.7875	.7882	.7889	.7896	.7903	.7910	.7917
6.2	.7924	.7931	.7938	.7945	.7952	.7959	.7966	.7973	.7980	.7987
6.3	.7993	.8000	.8007	.8014	.8021	.8028	.8035	.8041	.8048	.8055
6.4	.8062	.8069	.8075	.8082	.8089	.8096	.8102	.8109	.8116	.8122
6.5	.8129	.8136	.8142	.8149	.8156	.8162	.8169	.8176	.8182	.8189
6.6	.8195	.8202	.8209	.8215	.8222	.8228	.8235	.8241	.8248	.8254
6.7	.8261	.8267	.8274	.8280	.8287	.8293	.8299	.8306	.8312	.8319
6.8	.8325	.8331	.8338	.8344	.8351	.8357	.8363	.8370	.8376	.8382
6.9	.8388	.8395	.8401	.8407	.8414	.8420	.8426	.8432	.8439	.8445
7.0	.8451	.8457	.8463	.8470	.8476	.8482	.8488	.8494	.8500	.8506
7.1	.8513	.8519	.8525	.8531	.8537	.8543	.8549	.8555	.8561	.8567
7.2	.8573	.8579	.8585	.8591	.8597	.8603	.8609	.8615	.8621	.8627
7.3	.8633	.8639	.8645	.8651	.8657	.8663	.8669	.8675	.8681	.8686
7.4	.8692	.8698	.8704	.8710	.8716	.8722	.8727	.8733	.8739	.8745
7.5	.8751	.8756	.8762	.8768	.8774	.8779	.8785	.8791	.8797	.8802
7.6	.8808	.8814	.8820	.8825	.8831	.8837	.8842	.8848	.8854	.8859
7.7	.8865	.8871	.8876	.8882	.8887	.8893	.8899	.8904	.8910	.8915
7.8	.8921	.8927	.8932	.8938	.8943	.8949	.8954	.8960	.8965	.8971
7.9	.8976	.8982	.8987	.8993	.8998	.9004	.9009	.9015	.9020	.9025
8.0	.9031	.9036	.9042	.9047	.9053	.9058	.9063	.9069	.9074	.9079
8.1	.9085	.9090	.9096	.9101	.9106	.9112	.9117	.9122	.9128	.9133
8.2	.9138	.9143	.9149	.9154	.9159	.9165	.9170	.9175	.9180	.9186
8.3	.9191	.9196	.9201	.9206	.9212	.9217	.9222	.9227	.9232	.9238
8.4	.9243	.9248	.9253	.9258	.9263	.9269	.9274	.9279	.9284	.9289
8.5	.9294	.9299	.9304	.9309	.9315	.9320	.9325	.9330	.9335	.9340
8.6	.9345	.9350	.9555	.9360	.9365	.9370	.9375	.9380	.9385	.9390
8.7	.9395	.9400	.9405	.9410	.9415	.9420	.9425	.9430	.9435	.9440
8.8	.9445	.9450	.9455	.9460	.9465	.9469	.9474	.9479	.9484	.9489
8.9	.9494	.9499	.9504	.9509	.9513	.9518	.9523	.9528	.9533	.9538
9.0	.9542	.9547	.9552	.9557	.9562	.9566	.9571	.9576	.9581	.9586
9.1	.9590	.9595	.9600	.9605	.9609	.9614	.9619	.9624	.9628	.9633
9.2	.9638	.9643	.9647	.9652	.9657	.9661	.9666	.9671	.9675	.9680
9.3	.9685	.9689	.9694	.9699	.9703	.9708	.9713	.9717	.9722	.9727
9.4	.9731	.9736	.9741	.9745	.9750	.9754	.9759	.9763	.9768	.9773
9.5	.9777	.9782	.9786	.9791	.9795	.9800	.9805	.9809	.9814	.9818
9.6	.9823	.9827	.9832	.9836	.9841	.9845	.9850	.9854	.9859	.9863
9.7	.9868	.9872	.9877	.9881	.9886	.9890	.9894	.9899	.9903	.9908
9.8	.9912	.9917	.9921	.9926	.9930	.9934	.9939	.9943	.9948	.9952
9.9	.9956	.9961	.9965	.9969	.9974	.9978	.9983	.9987	.9991	.9996
x	**0**	**1**	**2**	**3**	**4**	**5**	**6**	**7**	**8**	**9**

TABLE 2

e^x and e^{-x}

x	e^x	e^{-x}	x	e^x	e^{-x}
0.0	1.00	1.00	3.5	33.1	0.0302
0.1	1.11	0.905	3.6	36.6	0.0273
0.2	1.22	0.819	3.7	40.4	0.0247
0.3	1.35	0.741	3.8	44.7	0.0224
0.4	1.49	0.670	3.9	49.4	0.0202
0.5	1.65	0.607	4.0	54.6	0.0183
0.6	1.82	0.549	4.1	60.3	0.0166
0.7	2.01	0.497	4.2	66.7	0.0150
0.8	2.23	0.449	4.3	73.7	0.0136
0.9	2.46	0.407	4.4	81.5	0.0123
1.0	2.72	0.368	4.5	90.0	0.0111
1.1	3.00	0.333	4.6	99.5	0.0101
1.2	3.32	0.301	4.7	110.	0.00910
1.3	3.67	0.273	4.8	122.	0.00823
1.4	4.06	0.247	4.9	134.	0.00745
1.5	4.48	0.223	5.0	148.	0.00674
1.6	4.95	0.202	5.1	164.	0.00610
1.7	5.47	0.183	5.2	181.	0.00552
1.8	6.05	0.165	5.3	200.	0.00499
1.9	6.69	0.150	5.4	221.	0.00452
2.0	7.39	0.135	5.5	245.	0.00409
2.1	8.17	0.122	5.6	270.	0.00370
2.2	9.02	0.111	5.7	299.	0.00335
2.3	9.97	0.100	5.8	330.	0.00303
2.4	11.0	0.0907	5.9	365.	0.00274
2.5	12.2	0.0821	6.0	403.	0.00248
2.6	13.5	0.0743	6.1	446.	0.00224
2.7	14.9	0.0672	6.2	493.	0.00203
2.8	16.4	0.0608	6.3	545.	0.00184
2.9	18.2	0.0550	6.4	602.	0.00166
3.0	20.1	0.0498	6.5	665.	0.00150
3.1	22.2	0.0450	6.6	735.	0.00136
3.2	24.5	0.0408	6.7	812.	0.00123
3.3	27.1	0.0369	6.8	898.	0.00111
3.4	30.0	0.0334	6.9	992.	0.00101

TABLE 3

NATURAL LOGARITHMS

N	ln N	N	ln N	N	ln N	N	ln N
0.00	-------	5.00	1.6094	10.0	2.3026	60.0	4.0943
0.10	-2.3026	5.10	1.6292	11.0	2.3979	61.0	4.1109
0.20	-1.6094	5.20	1.6487	12.0	2.4849	62.0	4.1271
0.30	-1.2040	5.30	1.6677	13.0	2.5650	63.0	4.1431
0.40	-0.9163	5.40	1.6864	14.0	2.6391	64.0	4.1589
0.50	-0.6932	5.50	1.7048	15.0	2.7080	65.0	4.1744
0.60	-0.5108	5.60	1.7228	16.0	2.7726	66.0	4.1896
0.70	-0.3567	5.70	1.7405	17.0	2.8332	67.0	4.2047
0.80	-0.2231	5.80	1.7579	18.0	2.8904	68.0	4.2195
0.90	-0.1054	5.90	1.7750	19.0	2.9444	69.0	4.2341
1.00	0.0000	6.00	1.7918	20.0	2.9957	70.0	4.2485
1.10	0.0953	6.10	1.8083	21.0	3.0445	71.0	4.2627
1.20	0.1823	6.20	1.8246	22.0	3.0910	72.0	4.2767
1.30	0.2624	6.30	1.8406	23.0	3.1355	73.0	4.2905
1.40	0.3365	6.40	1.8563	24.0	3.1780	74.0	4.3041
1.50	0.4055	6.50	1.8718	25.0	3.2189	75.0	4.3175
1.60	0.4700	6.60	1.8871	26.0	3.2581	76.0	4.3307
1.70	0.5306	6.70	1.9021	27.0	3.2958	77.0	4.3438
1.80	0.5878	6.80	1.9169	28.0	3.3322	78.0	4.3567
1.90	0.6418	6.90	1.9315	29.0	3.3673	79.0	4.3694
2.00	0.6932	7.00	1.9459	30.0	3.4012	80.0	4.3820
2.10	0.7419	7.10	1.9601	31.0	3.4340	81.0	4.3944
2.20	0.7885	7.20	1.9741	32.0	3.4657	82.0	4.4067
2.30	0.8329	7.30	1.9879	33.0	3.4965	83.0	4.4188
2.40	0.8755	7.40	2.0015	34.0	3.5264	84.0	4.4308
2.50	0.9163	7.50	2.0149	35.0	3.5554	85.0	4.4426
2.60	0.9555	7.60	2.0282	36.0	3.5835	86.0	4.4544
2.70	0.9932	7.70	2.0412	37.0	3.6109	87.0	4.4659
2.80	1.0296	7.80	2.0541	38.0	3.6376	88.0	4.4773
2.90	1.0647	7.90	2.0669	39.0	3.6636	89.0	4.4886
3.00	1.0986	8.00	2.0794	40.0	3.6889	90.0	4.4998
3.10	1.1314	8.10	2.0919	41.0	3.7136	91.0	4.5109
3.20	1.1632	8.20	2.1041	42.0	3.7377	92.0	4.5218
3.30	1.1939	8.30	2.1163	43.0	3.7612	93.0	4.5326
3.40	1.2238	8.40	2.1282	44.0	3.7842	94.0	4.5433
3.50	1.2528	8.50	2.1401	45.0	3.8067	95.0	4.5539
3.60	1.2809	8.60	2.1518	46.0	3.8286	96.0	4.5644
3.70	1.3083	8.70	2.1633	47.0	3.8502	97.0	4.5747
3.80	1.3350	8.80	2.1748	48.0	3.8712	98.0	4.5850
3.90	1.3610	8.90	2.1860	49.0	3.8918	99.0	4.5951
4.00	1.3863	9.00	2.1972	50.0	3.9120	100.	4.6052
4.10	1.4110	9.10	2.2083	51.0	3.9318	150.	5.0106
4.20	1.4351	9.20	2.2192	52.0	3.9512	200.	5.2983
4.30	1.4586	9.30	2.2300	53.0	3.9703	250.	5.5215
4.40	1.4816	9.40	2.2407	54.0	3.9890	300.	5.7038
4.50	1.5041	9.50	2.2513	55.0	4.0073	350.	5.8579
4.60	1.5261	9.60	2.2618	56.0	4.0254	400.	5.9915
4.70	1.5476	9.70	2.2721	57.0	4.0430	450.	6.1092
4.80	1.5686	9.80	2.2824	58.0	4.0604	500.	6.2146
4.90	1.5892	9.90	2.2925	59.0	4.0775		

ANSWERS FOR SUPPLEMENTARY EXERCISES

CHAPTER 1 - REAL NUMBERS

Assignment 1

1. 2, 7   2. 0, 2, 7   3. -6, 0, 3, 13   4. $-\sqrt{5}$, $\sqrt{11}$ 5. true   6. false   7. false   8. true
9. >   10. >   11. >   12. <   13. true   14. false   15. true   16. false
17.   18.   19.   20.

21. 7   22. $\frac{1}{2}$ 23. 0   24. 15.7   25. -4   26. -3   27. 6   28. -30   29. $-\frac{1}{3}$ 30. $-\frac{4}{5}$

31. $\frac{1}{4}$ 32. $-\frac{1}{8}$ 33. 0   34. -7.7   35. .14   36. -.7967   37. 3   38. -4   39. -8

40. 5   41. $-\frac{3}{4}$ 42. 9.4   43. -8   44. 10   45. -6   46. 1   47. 0   48. -5.2   49. -7.4

50. .44   51. $-\frac{2}{3}$ 52. $-\frac{3}{4}$ 53. $-\frac{3}{8}$ 54. $\frac{5}{4}$

Assignment 2

1. -56   2. -20   3. 27   4. 0   5. 10   6. -3.9   7. -6.3   8. 1.44   9. $-\frac{3}{8}$ 10. $\frac{1}{2}$
11. $-\frac{5}{2}$ 12. 6   13. 28   14. -30   15. 0   16. 126   17. -6   18. -5   19. 8   20. 0
21. -59   22. -17.6   23. impossible   24. 8   25. $-\frac{1}{6}$ 26. 6   27. $-\frac{4}{7}$ 28. 25
29. identity   30. identity   31. associative   32. commutative   33. associative
34. commutative   35. identity   36. commutative   37. identity   38. 8   39. 256   40. 1
41. -9   42. 0   43. 1   44. $\frac{1}{8}$ 45. $\frac{1}{49}$ 46. $\frac{1}{216}$ 47. $-\frac{1}{32}$ 48. -.125   49. 2.25
50. 1   51. $-\frac{9}{5}$ 52. $\frac{1}{64}$

Assignment 3

1. 4^2 2. x^{-6} 3. y^{-3} 4. b^3 5. $12a^2$ 6. -12   7. $7a^6b^{-5}$ 8. $c^4d^{-3}t^{-3}$ 9. 3^5
10. x^{-4} 11. y^6 12. ab^2 13. 8   14. $-cd^2$ 15. $5x^{-3}y^{-1}$ 16. $4t^4$ 17. 4^{-6} 18. x^{15}
19. b^2c^{-3} 20. x^4y^{-12} 21. $64p^{-12}q^3$ 22. $\frac{x^{-4}}{y^8}$ 23. $\frac{c^{-2}d^{-2}}{t^{-2}}$ 24. $\frac{4p^{-4}q^{10}}{9t^{-2}}$ 25. 1   26. 23
27. -6   28. -9   29. 5   30. -77   31. 9   32. $\frac{5}{3}$ 33. $-\frac{1}{4}$ 34. -4   35. -7   36. -2
37. 0   38. 5   39. 32   40. $\frac{7}{8}$ 41. $\frac{9}{2}$ 42. 4   43. A = 96   44. P = 80   45. B = 70
46. V = 400   47. F = 68   48. s = 64   49. m = -2   50. A = 1,180

Assignment 4

1. 3x + 21   2. 50y + 10   3. 7x - 14   4. 5 - 20d   5. 20a - 24b   6. bcR - bcP

7. -4m - 12   8. -7 + 6P   9. 6x - 12y + 9   10. 10b - 20c - 10d   11. -2ap - 6aq + 2a

12. 6x   13. 7y   14. -7m   15. 5d   16. -6r   17. -2t   18. 0   19. 5p - 1   20. 9a - 3b

21. -3x - 6   22. 2p + q + 2   23. 5x - 4   24. 2y + 5   25. -2t - 1   26. 3d - 5

27. 2 - 2b   28. 3a - 10   29. 4x + 2y   30. 18 + 3t   31. 4x - 15   32. 30 + 15m

33. 3y - 12   34. -11b + 16c   35. -6x - 6   36. 2a - 6b   37. 5y + 24   38. -4p + 15q

39. -3x + 9   40. -10y - 12   41. 15t - 5   42. y + 8   43. -4x + 27   44. 10p - 13

CHAPTER 2 - LINEAR EQUATIONS AND INEQUALITIES

Assignment 5

1. $x = -4$ 2. $y = 41$ 3. $d = -7.5$ 4. $t = \frac{5}{6}$ 5. $p = \frac{4}{9}$ 6. $h = 0$ 7. $y = -3$ 8. $p = \frac{4}{3}$

9. $x = 4$ 10. $H = -\frac{3}{2}$ 11. $s = \frac{1}{4}$ 12. $x = -1$ 13. $p = 0$ 14. $t = -4$ 15. $m = 4.5$

16. $A = 10$ 17. $x = \frac{3}{16}$ 18. $r = \frac{5}{4}$ 19. $x = -\frac{2}{3}$ 20. $a = \frac{1}{2}$ 21. $t = 3$ 22. $N = \frac{1}{3}$

23. $E = 1$ 24. $y = 2$ 25. $w = -\frac{5}{2}$ 26. $k = -1$ 27. $y = 3$ 28. $s = -\frac{1}{2}$ 29. $x = \frac{1}{6}$

30. $r = \frac{2}{5}$ 31. $t = \frac{1}{5}$ 32. $y = 10$ 33. $V = -5$ 34. $d = -\frac{1}{4}$ 35. $x = -\frac{15}{4}$ 36. $E = -\frac{1}{2}$

37. $r = 3$ 38. $y = 15$ 39. $b = -\frac{4}{7}$ 40. $x = \frac{1}{5}$ 41. $y = \frac{4}{3}$ 42. $m = 16$ 43. $p = -\frac{1}{7}$

44. $y = \frac{7}{6}$ 45. $d = \frac{8}{5}$ 46. $y = \frac{3}{2}$ 47. $t = 5$ 48. $x = \frac{15}{7}$ 49. $y = \frac{12}{5}$ 50. $x = 6$

51. $m = 5$ 52. $d = \frac{8}{3}$

Assignment 6

1. $W = 8$ 2. $K = 323$ 3. $W = 8$ 4. $h = 10$ 5. $I = 10$ 6. $L = 62$ 7. $F = 86$

8. $S = 80$ 9. $s = \frac{P}{4}$ 10. $L = \frac{A}{W}$ 11. $r = \frac{C}{2\pi}$ 12. $H = \frac{V}{LW}$ 13. $h = \frac{V}{\pi r^2}$ 14. $r = \frac{I}{Pt}$

15. $p = \frac{m}{q + r}$ 16. $D = \frac{2C}{a - b}$ 17. $b = \frac{2A}{h}$ 18. $h = \frac{3V}{B}$ 19. $T = \frac{2S}{V + W}$ 20. $t^2 = \frac{2A}{B + b}$

21. $K = D + L$ 22. $a^2 = c^2 - b^2$ 23. $a = 180 - b - c$ 24. $V = \frac{H - E}{P}$ 25. $W = \frac{P - 2L}{2}$

26. $N = \frac{4M}{3} + 23$ 27. $t = \frac{2R - ps}{p}$ 28. $c = \frac{3P - Rd}{R}$ 29. 16 30. 35 31. 116 and 118

32. 26, 27, and 28 33. 45 cm and 80 cm 34. 799 adults 35. 220 36. 60% 37. 700

38. \$473.50 39. 76% 40. \$12,500 41. L = 70m, W = 50m 42. W = 40 ft, L = 65 ft

43. 24°, 72°, and 84° 44. 32°, 64°, and 84° 45. 100 pounds 46. 40 liters 47. $\frac{2}{3}$ hour

48. 4 hours

Assignment 7

1. $x > -1$

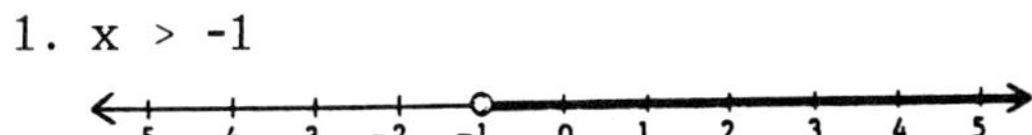

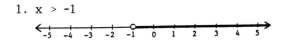

2. $x < 0$

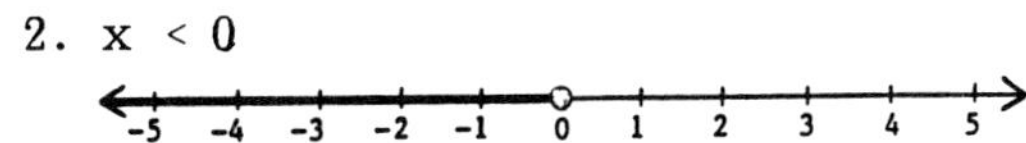

3. $x \geq 2$

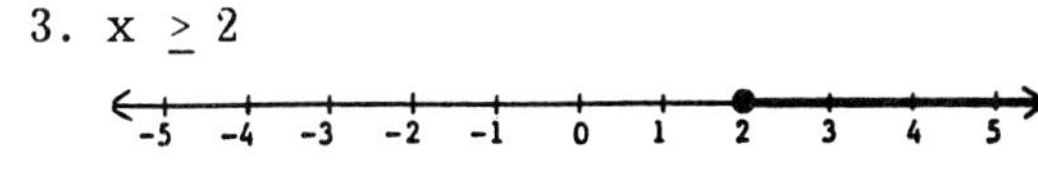

4. $x \leq 1$

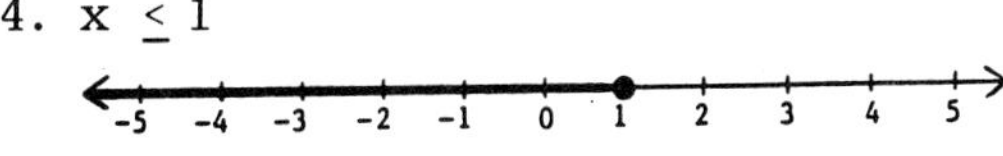

5. $y \geq 3.6$ 6. $x < \frac{5}{6}$ 7. $y \geq -\frac{1}{4}$ 8. $x < -2$ 9. $x \geq 5$ 10. $y > -5$ 11. $x \leq 0$

12. $y < \frac{2}{3}$ 13. $x < -\frac{1}{12}$ 14. $y < -\frac{2}{5}$ 15. $x \leq 6$ 16. $y < \frac{5}{4}$ 17. $x < \frac{1}{2}$ 18. $y \leq 1$

19. $x < 4$ 20. $x \leq -\frac{3}{2}$ 21. $x > \frac{8}{9}$ 22. $y \leq 3$ 23. $t \leq 9$ 24. $b > 2$ 25. $x \leq -2$

26. $x > 0$ 27. $x < 4$ 28. $x \geq -3$ 29. $x + 5 < 7$ 30. $x - 4 > -1$ 31. $3x \geq -6$

32. $5 \leq 2x - 3$ 33. $x \geq 60$ 34. $x < 3$ 35. $L \geq 15$ cm 36. $W \leq 15$m 37. 75 or more

38. 76 or more 39. more than 100 miles 40. less than 100 miles 41. gross sales under \$20,000

42. gross sales over \$15,000 43. \$10,000 at 8% 44. \$20,000 at 6%

Assignment 8

1. 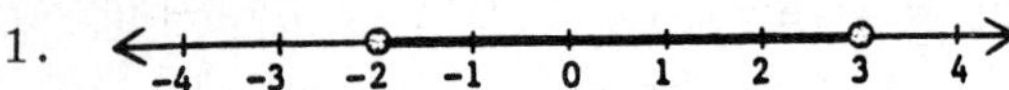

2. 

3.

4.

5. $-4 < x < 4$ 6. $-3 \le x < 5$ 7. $-1 < x \le 4$ 8. $\frac{3}{4} \le x \le 3$ 9. $-2 < x < \frac{1}{3}$
10. $\frac{3}{5} < x \le \frac{11}{5}$ 11. $-5 \le x \le 0$ 12. $-15 < x < 33$

13.

14.

15. $x < -4$ or $x > 5$ 16. $x < -2$ or $x \ge 4$ 17. $x \le 1$ or $x > 5$ 18. $x = 5$ and -5
19. $x = 0$ 20. $x = \frac{1}{2}$ and $-\frac{1}{2}$ 21. No solution 22. $x = 1$ and -11 23. $y = 12$ and 6
24. No solution 25. $x = 1$ and $-\frac{1}{2}$ 26. $x = 3$ and -10 27. $m = 6$ and -2 28. $x = 18$ and -6
29. $x = 4$ and $-\frac{7}{2}$

30. $-6 < x < 6$

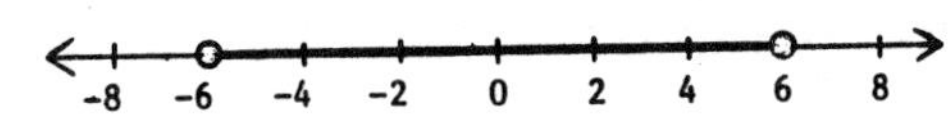

31. $-5 \le x \le 5$

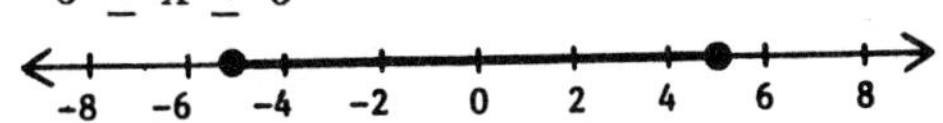

32. $x < -2.5$ or $x > 2.5$

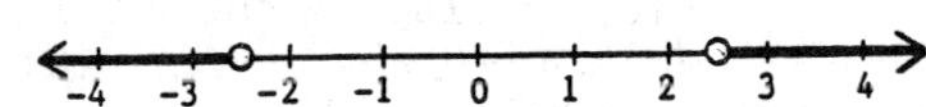

33. $x \le -6$ or $x \ge 6$

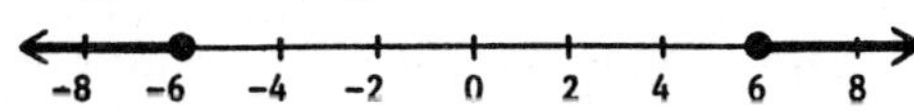

34. $-17 < x < 3$ 35. $4 \le y \le 6$ 36. $-\frac{7}{3} < x < 1$ 37. $-5 \le x \le 10$ 38. $a < -8$ or $a > 2$
39. $b \le 5$ or $b \ge 9$ 40. $x < -2$ or $x > \frac{4}{5}$ 41. $x \le -1$ or $x \ge \frac{7}{3}$

CHAPTER 3 - POLYNOMIALS

Assignment 9

1. binomial 2. monomial 3. trinomial 4. binomial 5. 1 6. 2 7. 4 8. 5 9. 10
10. 2 11. 7 12. 9 13. $-2x^2 + 6x$ 14. $ay^4 - by^2$ 15. $5t^3 - 5$ 16. $3x^2 - x - 1$
17. $-3x^3y^2 + 4xy + 5y^2$ 18. $3bx^2 + ax - 6$ 19. $2x^2y + 1$ 20. $3x^3 - x$ 21. $4x^4 + x^2 - 2$
22. $4x^3 + x^2 - 2$ 23. $kx^4 + ax^2 - b$ 24. $2y^2 + 4$ 25. $-y^3 + 3y + 2$ 26. $y^2 - y + 3$
27. $xy^4 - 3$ 28. $2by^2 + dy$ 29. $ty + 2w$ 30. $-5t^2$ 31. $18x^3$ 32. $-12r^4t^3$ 33. abx^3y^6
34. $10d^4k^6$ 35. $-18a^3p^4s^2$ 36. $25x^6$ 37. $16a^2b^8c^{12}$ 38. $3y^2 + 7y$ 39. $-5d^2 - 5d$
40. $12x^5 - 6x^2$ 41. $3my^4 - 2m^2y$ 42. $x^3y + x^2y^2 + xy^3$ 43. $15h^3s^3 - 5h^4s^2 - 10h^3s^2$
44. $2x^4 - x^2 - 6$ 45. $a^2b^2 - ab - 30$ 46. $x^3 - y^3$ 47. $2d^2 + 3d - 5$ 48. $x^3 - 3x^2 - 4x + 12$
49. $a^4 + 2a^2b^2 + ab^3 + 2b^4$ 50. $y^4 - 2y^3 + 2y^2 + 2y - 3$

Assignment 10

1. $6x^2 + 19x + 10$ 2. $2p^2 + 7pq + 3q^2$ 3. $5y^2 + 19y - 4$ 4. $2a^2 - 7ab - 4b^2$
5. $15x^2 + 11x - 12$ 6. $4x^2 - 7xy - 2y^2$ 7. $a^2b^2 + 6ab - 7$ 8. $12a^2 - 25a + 12$
9. $20p^2q^2 - 13pq + 2$ 10. $3x^4 - 7x^2y^2 + 2y^4$ 11. $3ap - 6aq - bq + 2bq$
12. $3b^2d - 12a^2d - 2ab^3 + 8a^3b$ 13. $x^3 - 25$ 14. $81 - 4y^2$ 15. $49a^2 - 1$ 16. $9a^2 - b^2$
17. $16p^2 - 9q^2$ 18. $x^2y^2 - 9$ 19. $x^4 - y^6$ 20. $16b^8 - 25d^{10}$ 21. $x^2 + 18x + 81$

22. $25y^2 - 60y + 36$ 23. $81 + 36t + 4t^2$ 24. $100 - 20m + m^2$ 25. $9x^2 + 24xy + 16y^2$
26. $a^2b^2 - 2abc + c^2$ 27. $4t^4 - 4t^2v^3 + v^6$ 28. $x^4y^2 + 8x^3y + 16x^2$ 29. $3x^4y$ 30. $-5q^2$

31. $\frac{3b}{ac^2}$ 32. $\frac{6}{t}$ 33. $\frac{1}{c^2d}$ 34. $\frac{2x^4}{5}$ 35. $\frac{4}{3}$ 36. $\frac{b}{3a^2}$ 37. $2x^2 + 3x + 1$

38. $\frac{4y^4}{3} - y^2 - \frac{1}{3}$ 39. $a^2b^2 - a + b^3$ 40. $\frac{2}{3} + \frac{y}{x} - \frac{2y^2}{x^2}$ 41. $\frac{3}{2} - \frac{2}{pq} + \frac{5q}{2p^2}$

42. $2y + \frac{1}{3} - \frac{1}{y}$ 43. $x + 6$ 44. $a - 6$ 45. $3y - 4$ 46. $5t - 2$ 47. $m + 3 + \frac{1}{m + 2}$

48. $4x^2 + 6x + 9$ 49. $2y + 3 + \frac{6y + 5}{y^2 - 3}$ 50. $5b + 1 + \frac{5b - 3}{b^2 - 2b + 1}$

Assignment 11

1. $5(x + 7)$ 2. $4(1 - 2y)$ 3. $t(6t - 1)$ 4. $3m^2(m^2 + 3)$ 5. $4d^3(3-2d^4)$ 6. $xy(xy + 7x^2)$
7. $a^3b(8a^2 - 1)$ 8. $2(3x + 4y + 2z)$ 9. $y^3(y^4 + 3y^2 + 1)$ 10. $5t^2(2t^2 + 3t - 6)$
11. $p^2q^2(3p^2 - pq + 5q^2)$ 12. $3a^2b^2(a^2b^2 + 2ab - 1)$ 13. $(4t + 5)(t + 2)$
14. $(4p + 5q)(3p - 2q)$ 15. $(x - 2)(x + 5)$ 16. $(d - 4)(d - 3)$ 17. $(4a^2 + b)(2a - b^2)$
18. $(x - y)(p + 3q)$ 19. $(t - 3)(t - 1)$ 20. $(w - 4)(w + 3)$ 21. $(x + 7)(x - 3)$
22. $(p + 3q)(p + 5q)$ 23. $(cd - 1)(cd - 5)$ 24. $(xy + 7)(xy - 2)$ 25. $(t^4 - 3)(t^4 - 5)$
26. $(p^3q^3 - 6)(p^3q^3 + 4)$ 27. $x(x + 5)(x + 7)$ 28. $y^2(y - 6)(y + 5)$ 29. $(y - 2)(y + 6)$
30. $(a - b - 1)(a - b - 4)$ 31. $(3x - 4)(2x - 1)$ 32. $(2y - 1)(2y + 3)$ 33. $(1 - 6t)(1 + 2t)$
34. $(2b + t)(b + 3t)$ 35. $(3tw + 2)(tw - 3)$ 36. $(4ab - 3)(2ab + 3)$ 37. $(2x^4 - 1)(3x^4 - 4)$
38. $(5p^3q^3 + 2)(2p^3q^3 + 3)$ 39. $3(y + 3)(y + 7)$ 40. $5x(2x + 1)(x - 2)$
41. $cd(cd + 8)(cd - 2)$ 42. $(6y + 19)(y + 3)$ 43. $(7x + 1)(7x - 1)$ 44. $(3a + 4b)(3a - 4b)$
45. $(5m^3 + 8t^2)(5m^3 - 8t^2)$ 46. $(2 + 3pq^4)(2 - 3pq^4)$ 47. $(y^2 + 9)(y + 3)(y - 3)$
48. $2(t + 3w)(t - 3w)$ 49. $a(8x + 7)(8x - 7)$ 50. $(b + 7)(b - 3)$

Assignment 12

1. $(x + 4)^2$ 2. $(y - 8)^2$ 3. $(3w + 2)^2$ 4. $(2a - 5b)^2$ 5. $(4dh - 1)^2$ 6. $(t^2 + 1)^2$
7. $(7m^5 - 3)^2$ 8. $5(x + 2)^2$ 9. $3y(y - 1)^2$ 10. $(x + 1)(x^2 - x + 1)$
11. $(y - 3)(y^2 + 3y + 9)$ 12. $(2b + 5d)(4b^2 - 10bd + 25d^2)$ 13. $(4p - t)(16p^2 + 4pt + t^2)$
14. $(x^2 - 3y)(x^4 + 3x^2y + 9y^2)$ 15. $b(a + 2b)(a^2 - 2ab + 4b^2)$ 16. $2(m - 4)(m^2 + 4m + 16)$
17. $(2y + 5)(y^2 + 5y + 25)$ 18. $5(x + y)(x - y)$ 19. $2(a - 2b^2)(a^2 + 2ab^2 + 4b^4)$
20. $3(c - 4)(c - 1)$ 21. $(p^2 + q^2)(p + q)(p - q)$ 22. $(y^2 - 2)(y + 2)(y - 2)$
23. $(x + 10)(x + 5)$ 24. $a(4p + 1)(4p - 1)$ 25. $2b(2t - 3)(4t^2 + 6t + 9)$ 26. $(y - 5)(y + 3)$
27. $(4x^2 + 1)(2x + 1)(2x - 1)$ 28. $3a(x - 4y)^2$ 29. $(x^2 + y^2)(x - y)(x^2 + xy + y^2)$

30. $x = 2$ and 3 31. $y = 1$ and $-\frac{1}{3}$ 32. $p = 3$ and -3 33. $t = \frac{4}{5}$ and $-\frac{4}{5}$ 34. $x = 0$ and -7

35. $y = 0$ and $\frac{2}{3}$ 36. $a = -\frac{4}{3}$ and $\frac{3}{2}$ 37. $b = 0$ and $\frac{3}{5}$ 38. $h = \frac{1}{7}$ and $-\frac{1}{7}$ 39. $x = \frac{3}{2}$ and -2

40. $x = 8$ and -10 41. $x = 5$ and -5 42. -2 and 5 43. $-\frac{5}{2}$ and 3 44. 6 and -6

45. 8 and 9 or -9 and -8 46. 7 and 9 or -9 and -7 47. 10 and 12 or -12 and -10

48. L = 10 ft, W = 4 ft 49. h = 7m, b = 10m 50. 3 seconds

CHAPTER 4 - RATIONAL EXPRESSIONS

Assignment 13

1. 3 2. $\frac{1}{5m}$ 3. $\frac{4x^2}{7y^3}$ 4. $\frac{3d^2}{2c}$ 5. $\frac{4}{3}$ 6. $\frac{y - 1}{2(y + 1)}$ 7. $\frac{2x + 1}{3}$ 8. $\frac{2t}{t - 4}$ 9. $\frac{t - 2}{t + 3}$

10. $\frac{y + 5}{y - 2}$ 11. -1 12. $-(c + d)$ or $-c - d$ 13. $\frac{2x}{3y}$ 14. $\frac{6}{5}$ 15. $\frac{5x}{4}$ 16. $\frac{5}{6y}$ 17. $x - 1$

18. $\frac{y - 1}{y - 2}$ 19. $\frac{1}{a}$ 20. $\frac{x(x + 5)}{2(2x + 3)}$ 21. $\frac{t - 3}{t + 1}$ 22. $\frac{p - 5}{3}$ 23. $\frac{1}{x - 4}$ 24. $3p - q$

25. $\frac{x}{3}$ 26. $\frac{1}{c^2t^2}$ 27. $\frac{a}{b}$ 28. $\frac{2(x - 6)}{x + 3}$ 29. $\frac{3y}{4(x + y)}$ 30. $\frac{x + 1}{x + 3}$ 31. $\frac{a + 3}{a - 1}$

32. $\frac{2c - d}{2(2c + d)}$ 33. $\frac{d^2 + 4d + 16}{d^2 - 2d + 1}$ 34. $\frac{1}{x - y}$ 35. p(p - q) 36. $\frac{k - 3}{5}$

Assignment 14

1. $\frac{2x + 1}{3}$ 2. $\frac{4t - 5}{t - 1}$ 3. $\frac{3x}{5}$ 4. $\frac{2y - 5}{y - 1}$ 5. $\frac{2p - 7}{p + 3}$ 6. $\frac{2}{3x}$ 7. $\frac{d + 2}{d - 3}$ 8. $\frac{3 + 2x}{x}$

9. $\frac{1}{y + 2}$ 10. $\frac{1}{t - 3}$ 11. $\frac{m - 3}{m - 4}$ 12. $\frac{7p - 3}{2p - q}$ 13. 72 14. 60 15. 24x 16. 4xy

17. $30t^2$ 18. 3y(y - 3) 19. 2x(2x + 1) 20. (y + 1)(y - 1)(y + 4) 21. $(m + 2)^2(m - 5)$

22. $12x^2$ 23. (y + 3)(y - 1) 24. 6(2m + 1)(2m - 1) 25. $\frac{17}{20x}$ 26. $\frac{2m^2 + 3m - 3}{2m(m - 1)}$

27. $\frac{2y^2 - 5y + 17}{(y - 2)(y + 3)}$ 28. $\frac{x^2 + 15}{3x(x - 2)}$ 29. $\frac{2y^2 + 7y + 2}{(y + 3)(y - 3)(y + 2)}$ 30. $\frac{5}{6x}$ 31. $\frac{p - 7}{(p - 2)(p - 3)}$

32. $\frac{3x - 8}{(x + 4)(x - 4)}$ 33. $\frac{x^2 + y^2}{(x - y)(x - y)(x + y)}$ 34. $\frac{x^2 + 6x - 8}{4x^2}$ 35. $\frac{5y + 5}{(y - 2)(y + 2)}$

36. $\frac{5a^2 - 2a + 2}{(3a - 2)(a + 1)}$ 37. $\frac{5(x + 3)}{2x}$ 38. $\frac{2(y + 12)}{7}$ 39. $\frac{15}{14}$ 40. $\frac{t - 1}{t + 1}$ 41. a - b

42. $\frac{2(3x - 1)}{2x + 1}$ 43. $\frac{b - 1}{b + 1}$ 44. $\frac{x^2}{y^2}$ 45. $\frac{a + 5}{a - 5}$

Assignment 15

1. $x = \frac{6}{5}$ 2. $x = \frac{20}{7}$ 3. $x = \frac{7}{12}$ 4. $y = \frac{9}{8}$ 5. y = -1 6. $y = \frac{5}{3}$ 7. x = 1 8. $t = \frac{6}{5}$

9. $x = \frac{10}{7}$ 10. $x = -\frac{1}{4}$ 11. t = 9 12. m = 11 13. $y = -\frac{8}{3}$ 14. x = -6 15. $p = \frac{1}{2}$

16. No solution 17. $x = \frac{24}{7}$ 18. y = 6 19. y = 3 20. m = 3 21. p = -3

22. m = 5 and -4 23. $x = \frac{4}{3}$ and 1 24. x = -3 (not 3) 25. x = -8 26. $x = \frac{3}{2}$

27. $F_1 = 36$ 28. $V_1 = 36$ 29. $R_2 = 30$ 30. D = 180 31. R = 5 32. p = 3

33. $d_2 = \frac{d_1F_2}{F_1}$ 34. $H = \frac{M}{P(t_2 - t_1)}$ 35. $C_o = vt - C_i$ 36. $P_1 = \frac{P_2V_2T_1}{V_1T_2}$ 37. $F = \frac{Df}{d}$

38. $P_2 = \frac{P_1V_1}{V_2}$ 39. $T_2 = \frac{V_2T_1}{V_1}$ 40. $A = \frac{B}{C + D}$ 41. $K = \frac{AB}{T - R}$ 42. $t = \frac{mp}{m + p}$ 43. $R = \frac{FS}{F - S}$

44. $D = \frac{C}{1 - C}$ 45. $T = \frac{HP}{P - H}$ 46. $t = \frac{bD}{B - b}$ 47. $a = \frac{a_od - a_fD}{D - d}$

Assignment 16

1. 2 and 12 2. 16 and 20 3. $\frac{8}{11}$ 4. $\frac{10}{30}$ 5. 40 6. 36 7. 15 and 16 8. 20 and 21

9. 5 hours 10. $16.50 11. 99 inches 12. 4.5 ounces 13. $2\frac{2}{5}$ hours 14. $2\frac{2}{9}$ hours

15. $3\frac{3}{7}$ hours 16. 9 hours for Joe, 18 hours for Mike 17. 46 mph and 54 mph

18. 400 mph and 450 mph 19. 140 mph 20. $1\frac{1}{2}$ mph

CHAPTER 5 - EXPONENTS AND RADICALS

Assignment 17

1. 7 2. -4 3. No real number root 4. 6.083 5. -9.434 6. 4 7. -5 8. -2 9. 6
10. -1 11. $5\sqrt{2}$ 12. $2\sqrt[3]{3x}$ 13. y^3 14. $(x + y)^4$ 15. 2t 16. $y^4\sqrt[3]{4}$ 17. $2xy^2$
18. $3b^3\sqrt{2a}$ 19. $m^3\sqrt{m}$ 20. $y^2\sqrt[5]{y^3}$ 21. $6x\sqrt{2x}$ 22. $3y\sqrt[3]{y^2}$ 23. $9xy^2\sqrt{y}$ 24. $2b\sqrt[3]{2a^2b}$
25. $x - 3$ 26. $b\sqrt[3]{x + y}$ 27. $3\sqrt{2}$ 28. $2xy\sqrt[3]{2y^2}$ 29. $(x + 1)\sqrt[4]{x + 1}$ 30. $6x\sqrt[2]{x}$
31. $5y^2\sqrt[3]{y}$ 32. $6x\sqrt{3x}$ 33. $2\sqrt{2}$ 34. 3 35. $4xy\sqrt{y}$ 36. $m\sqrt[4]{2m}$ 37. $2\sqrt[3]{2x}$ 38. $\frac{2y\sqrt{3x}}{5}$
39. $15\sqrt{d}$ 40. $\sqrt[3]{ab}$ 41. $\frac{2}{3}$ 42. $\frac{6x}{5}$ 43. $\frac{1}{2}$ 44. $\frac{2\sqrt[3]{2y}}{3}$ 45. $\frac{8}{3\sqrt{2t}}$ 46. $\frac{3x\sqrt[3]{x^2}}{4y\sqrt[3]{y}}$ 47. $\frac{x\sqrt{7x}}{4y}$
48. $\frac{1}{d^3\sqrt[3]{b}}$

Assignment 18

1. $9\sqrt{3}$ 2. $-5\sqrt[3]{7}$ 3. $4\sqrt{x}$ 4. $5\sqrt[4]{9}$ 5. 0 6. $13\sqrt{a - 3}$ 7. $(a - b)\sqrt{t}$ 8. $(x - 3)\sqrt{y}$
9. $3\sqrt[3]{x^2} + \sqrt[4]{x^2}$ 10. $11\sqrt{2}$ 11. $4\sqrt[3]{3x}$ 12. $(a + 3)\sqrt[3]{2a}$ 13. $(5x - 1)\sqrt{x} + 3x$ 14. $\sqrt{2y - 1}$
15. $(x^2 + 5)\sqrt{x + 1}$ 16. $x\sqrt{5} - \sqrt{10}$ 17. $\sqrt{6} + \sqrt{15}$ 18. $2 - 5\sqrt{2}$ 19. $\sqrt{21} - 7$ 20. $3\sqrt{10} + 2\sqrt{35}$
21. $5\sqrt[3]{14} - \sqrt[3]{21}$ 22. 5 23. -4 24. a - b 25. 5 26. 30 27. -30
28. $\sqrt{10} - \sqrt{6} + \sqrt{35} - \sqrt{21}$ 29. $7 + 3\sqrt{3}$ 30. $6 + \sqrt{10}$ 31. $14 - 31\sqrt{5}$ 32. $19 + 8\sqrt{3}$
33. $41 - 12\sqrt{5}$ 34. $\frac{2\sqrt{5x}}{5x}$ 35. $\frac{7\sqrt{2}}{6}$ 36. $\frac{\sqrt{3y}}{15y}$ 37. $\sqrt{7}$ 38. $\frac{\sqrt{x}}{9}$ 39. $\frac{4\sqrt{3}}{9}$ 40. $\frac{x\sqrt{5}}{5y^2}$
41. $\frac{\sqrt{15}}{5}$ 42. $\frac{\sqrt{35x}}{7x}$ 43. $\frac{4 - \sqrt{7}}{9}$ 44. $\sqrt{5} + 1$ 45. $4(\sqrt{5} + \sqrt{3})$ 46. $\frac{x + 2\sqrt{xy} + y}{x - y}$ 47. $\frac{4\sqrt[3]{3}}{3}$
48. $\frac{\sqrt[3]{20}}{2}$ 49. $\frac{x\sqrt[3]{15y^2}}{5y^2}$

Assignment 19

1. x = 4 2. m = 50 3. $x = \frac{5}{7}$ 4. No solution 5. V = 13 6. x = 10 7. t = 3
8. w = 1 and 2 9. x = 2 (not x = -2) 10. y = 4 11. x = 16 12. x = 26 13. x = 17
14. x = 5 and 13 15. y = 2 and 6 16. m = 14 17. t = 2 18. x = 7 19. t = 3
20. a = 4 21. N = 10 22. V = 64 23. a = 8 24. R = 4 25. s = 64 26. a = 48
27. $A = s^2$ 28. $V = s^3$ 29. $T = \frac{P^2}{S}$ 30. $p = m^2 - q$ 31. $V = \frac{T}{M^2}$ 32. $a = \frac{bh^2}{3}$
33. $E = \frac{FH^2}{P^2}$ 34. $d = \frac{b^2c}{a^2}$

Assignment 20

1. $\sqrt{5}$ 2. $\sqrt[5]{x}$ 3. $\sqrt[3]{7^2}$ 4. $\sqrt{10^3}$ 5. $\sqrt[8]{x^7}$ 6. $3^{\frac{1}{2}}$ 7. $8^{\frac{1}{4}}$ 8. $y^{\frac{5}{2}}$ 9. $2^{\frac{3}{4}}$ 10. $m^{\frac{5}{6}}$
11. 7 12. 5 13. $\frac{1}{2}$ 14. 8 15. $\frac{1}{4}$ 16. $5^{\frac{5}{6}}$ 17. $x^{\frac{13}{12}}$ 18. $3^{\frac{4}{5}}$ 19. $y^{\frac{7}{6}}$ 20. $m^{-\frac{3}{2}}$
21. $2^{\frac{1}{2}}$ 22. $7^{-\frac{1}{3}}$ 23. $m^{\frac{1}{4}}$ 24. $\sqrt[3]{x^2}$ 25. $\sqrt[4]{m}$ 26. $\sqrt{5y}$ 27. $\sqrt[3]{3a^2b}$ 28. $\sqrt[6]{288}$ 29. $\sqrt[4]{100}$
30. $\sqrt[6]{x^5}$ 31. $y\sqrt[10]{y^3}$ 32. $m^2\sqrt[8]{m}$ 33. $\sqrt{t}$ 34. $\sqrt[8]{5^3}$ or $\sqrt[8]{125}$ 35. $\frac{1}{\sqrt[4]{x}}$ 36. 2i 37. 9i
38. 10i 39. $\sqrt{7}i$ 40. $\sqrt{37}i$ 41. 7 + 9i 42. 7 - 3i 43. 11 - 2i 44. 4i 45. 4 + i
46. 2 + 4i 47. -1 - 10i 48. -4 49. -2 + 2i 50. 10 + 6i 51. -4 - 4i 52. -42 + 12i
53. 2 + 23i 54. 18 - 38i 55. -47 + 14i 56. -11 + 7i 57. 35 + 12i 58. 16 - 30i

59. 58 60. 82 61. 20 62. 16 63. -64 64. -125i 65. 256 66. $2 - \frac{2}{3}i$ 67. $-\frac{1}{2} - \frac{5}{2}i$
68. $\frac{5}{17} + \frac{14}{17}i$ 69. 3 + i 70. $\frac{15}{13} + \frac{16}{13}i$ 71. -4i

CHAPTER 6 - QUADRATIC EQUATIONS

Assignment 21

1. x = -5 and 8 2. $x = \frac{5}{3}$ and 1 3. $t = \frac{3}{5}$ and -2 4. m = 9 and -9 5. x = 0 and 12
6. $d = 0$ and $-\frac{6}{5}$ 7. $t = \frac{1}{3}$ and $\frac{3}{2}$ 8. $m = \frac{1}{5}$ and $-\frac{1}{5}$ 9. b = 0 and 5 10. x = 6 and -6
11. m = 1 and -1 12. $y = \frac{6}{7}$ and $-\frac{6}{7}$ 13. $b = \frac{3}{2}$ and $-\frac{3}{2}$ 14. $h = \pm\sqrt{5} = \pm 2.236$ 15. $x = \pm\frac{\sqrt{30}}{5}$
16. y = ±3i 17. $p = \pm\frac{1}{4}i$ 18. x = 2 and -10 19. $y = \frac{5}{3}$ and -1 20. y = 7 ± 5i
21. $b = \frac{-5 \pm 7i}{2}$ 22. $x^2 + 6x + 9 = (x + 3)^2$ 23. $y - 14y + 49 = (y - 7)^2$
24. $d^2 + d + \frac{1}{4} = \left(d + \frac{1}{2}\right)^2$ 25. $m - 5m + \frac{25}{4} = \left(m - \frac{5}{2}\right)^2$ 26. x = 1 and 7 27. $x = 2 \pm\sqrt{15}$
28. $y = -5 \pm\sqrt{5}$ 29. $t = \frac{3 \pm \sqrt{5}}{2}$ 30. $y = \frac{2}{3}$ and -2 31. $b = \frac{5 \pm \sqrt{33}}{4}$ 32. x = 4 and -2
33. $t = \frac{1}{2}$ and $-\frac{2}{3}$ 34. w = 1 and $-\frac{3}{4}$ 35. $d = \frac{5 \pm \sqrt{37}}{6}$ or d = 1.847 and -0.181
36. $p = \frac{4 \pm \sqrt{6}}{5}$ or p = 1.29 and 0.31 37. $F = \frac{-1 \pm \sqrt{57}}{4}$ or F = 1.638 and -2.138
38. $h = -1 \pm\sqrt{2}$ or h = 0.414 and -2.414 39. $x = \frac{3 \pm \sqrt{11}i}{2}$ 40. $y = \frac{-1 \pm \sqrt{19}i}{5}$

Assignment 22

1. $x^2 - 5x + 7 = 0$ 2. $y^2 + 3y - 9 = 0$ 3. $t^2 - 3t + 4 = 0$ 4. $x^2 - 10x - 2 = 0$
5. $3y^2 - 8y - 7 = 0$ 6. $2x^2 + x - 9 = 0$ 7. $m^2 - m - 25 = 0$ 8. $x^2 + 4x + 1 = 0$
9. $x^2 - 4x - 9 = 0$ 10. x = 1 and $\frac{2}{3}$ 11. t = 3 and $-\frac{1}{2}$ 12. x = 3 and 11
13. $x = -1 \pm 2\sqrt{2}$ or x = 1.828 and -3.828 14. $x = 2 \pm\sqrt{7}$ or x = 4.646 and -0.646
15. $y = 3 \pm\sqrt{13}$ or y = 6.606 and -0.606 16. 12 hours for Joan, 6 hours for Judy
17. 24 for the smaller pipe, 12 hours for the larger pipe 18. L = 4.5 ft, W = 1.5 ft
19. W = 2.5m, L = 4.5m 20. 3 in 21. 1 in 22. 6 mph 23. 7 mph 24. A = 600
25. s = 270 26. t = 2 27. $r = \sqrt{2} = 1.414$ 28. v = 3 29. d = 5 30. R = 12 31. $t = \frac{5}{2}$
32. $P = \sqrt{Q^2 + R^2}$ 33. $a = \sqrt{\frac{b}{c}}$ 34. $r = \sqrt{\frac{A}{\pi}}$ 35. $v = \sqrt{\frac{2E}{m}}$ 36. $d = \sqrt{\frac{m_1 m_2}{Fr}}$
37. $G = \sqrt{B^2 - F^2}$ 38. $s = \frac{-a \pm \sqrt{a^2 + 4b}}{2}$ 39. $r = \frac{-\pi s \pm \sqrt{\pi^2 s^2 + 4\pi A}}{2\pi}$

Assignment 23

1. h = 9.434 cm 2. d = 4.243 ft 3. leg = 8 in 4. L = 5.657m 5. h = 12 ft
6. 40 miles (due south), 30 miles (due west) 7. two real, rational solutions ($b^2 - 4ac = 100$)
8. one real, rational solution ($b^2 - 4ac = 0$) 9. two real, irrational solutions ($b^2 - 4ac = 72$)

10. two complex solutions (b^2 - 4ac = -23) 11. two real, irrational solutions (b^2 - 4ac = 52)

12. two complex solutions (b^2 - 4ac = -15) 13. two real, rational solutions (b^2 - 4ac = 1)

14. one real, rational solution (b^2 - 4ac = 0) 15. two real, irrational solutions (b^2 - 4ac = 85)

16. $x = \pm\sqrt{2}$ and ± 3 17. $x = \pm\sqrt{3}$ and 0 18. x = -3, 1, 2, and -4 19. $y = -\frac{2}{3}$, 1, $\frac{4}{3}$, and -1

20. $t = \pm\sqrt{2}$ and $\pm\sqrt{3}$ 21. $m = \pm 1$ and $\pm\sqrt{7}$ 22. $x = \pm\sqrt{\frac{3}{2}}$ and ± 2 23. $y = \pm\sqrt{\frac{1}{3}}$ and ± 3

24. $x = \pm\sqrt{5}$

25. $x < -1$ or $x > 3$

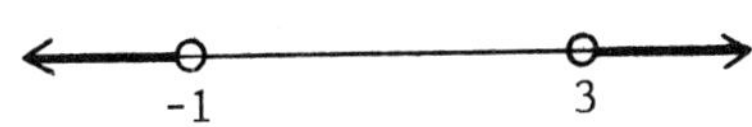

26. $-2 < y < 1$

27. $m \le -3$ or $m \ge 2$

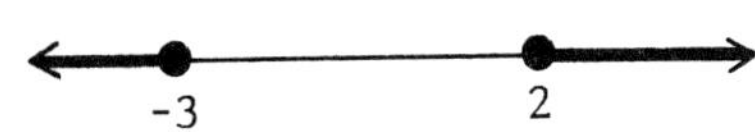

28. $-1 \le b \le 4$

29. $x < -\frac{1}{2}$ or $x > 3$

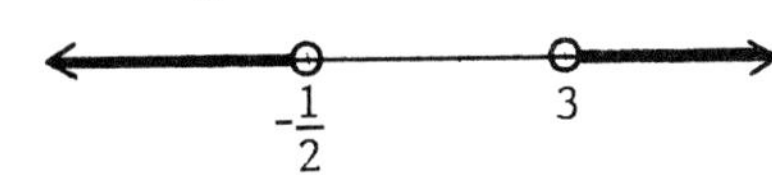

30. $-2 \le t \le \frac{3}{2}$

31. $\frac{1}{2} < x < 3$

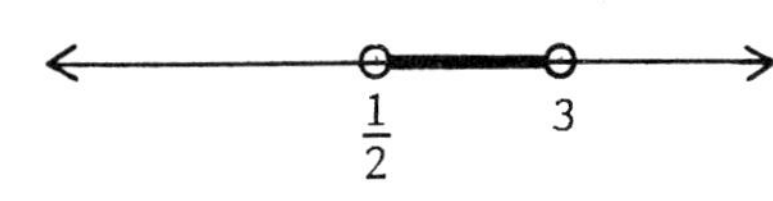

32. $d \le -4$ or $d \ge -1$

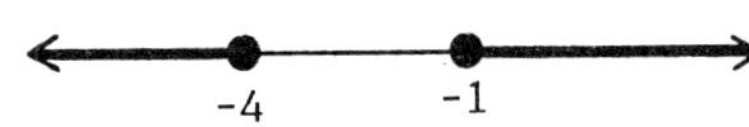

33. $x < -2$ or $x > 2$

34. $-1 \le x \le 4$

Assignment 24

1. Yes 2. Yes 3. No 4. Yes 5. No 6. No 7. (-1,3) 8. (-4,-6) 9. (0,6) 10. (-2,0) 11. (3,-4) 12. (1,4) 13. (0,8) 14. (2,0) 15. A(3,3), B(3,-2), C(-2,-1), D (-4,2), E(2,0) F(0,-3) 16. G(1,25), H($-2\frac{1}{2}$,40), I(-4,-23), J($1\frac{1}{2}$,-32), K($-3\frac{1}{2}$,0), L(0,15) 17. a) 2 b) 4 c) 1 d) 3 e) 2 18. b, c, d 19. a, d, e 20. the origin

21-22.

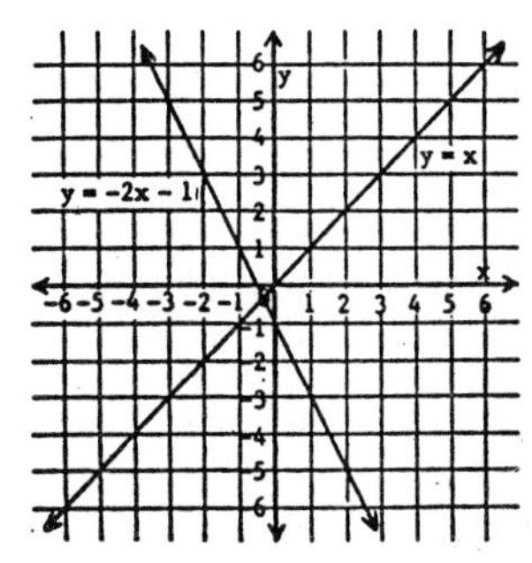

23-24.

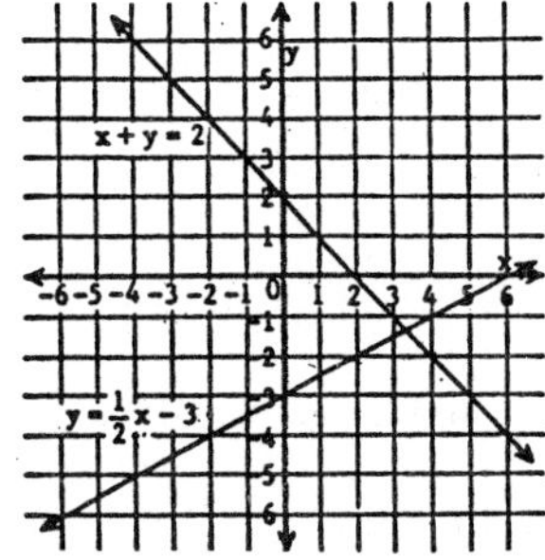

25. (-3,0)   26. (5,0)   27. $(\frac{1}{2},0)$ 28. $(-\frac{8}{5},0)$ 29. (0,-9)   30. (0,2)   31. $(0,\frac{1}{3})$ 32. $(0,-\frac{5}{2})$

33.

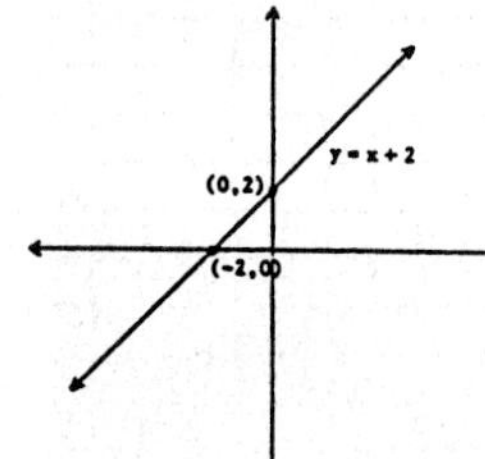

34.

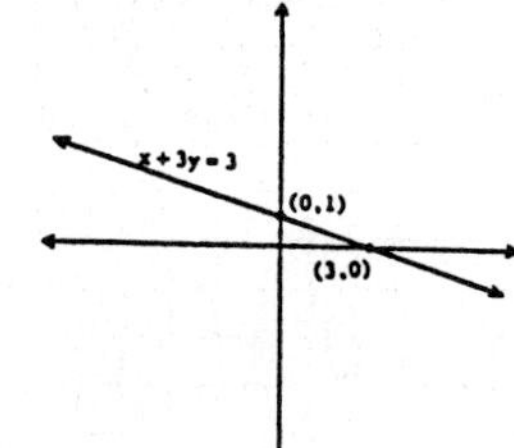

35.

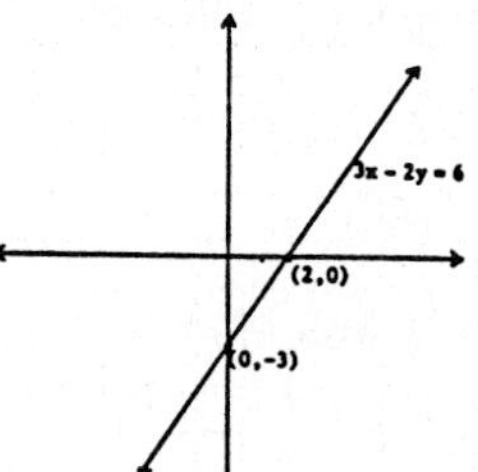

36. y = 3   37. x = -2   38. x = 1   39. y = -4   40. y = 0   41. x = 0

Assignment 25

1. 1   2. $-\frac{1}{2}$ 3. -4   4. $\frac{1}{6}$ 5. A, D   6. A   7. B, C, E   8. C   9. 2   10. $-\frac{3}{2}$ 11. 1   12. -1   13. $\frac{1}{4}$ 14. $-\frac{3}{5}$ 15. $\frac{1}{2}$ 16. 2   17. -4   18. 0   19. undefined   20. m = 5, (0,2)   21. m = 1, (0,-3)   22. $m = -\frac{2}{5}$, $(0,\frac{3}{2})$ 23. $m = \frac{1}{2}$, (0,0)   24. y = -x + 2   25. $y = \frac{1}{2}x - 5$ 26. $y = \frac{1}{3}x - 1$ 27. $y = \frac{8}{3}x$ 28. y = -4x + 6   29. y = 3x + 2   30. y = x - 5   31. $y = \frac{1}{4}x - 2$ 32. $y = -\frac{1}{2}x + \frac{5}{3}$ 33. $y = \frac{1}{5}x$ 34-35.

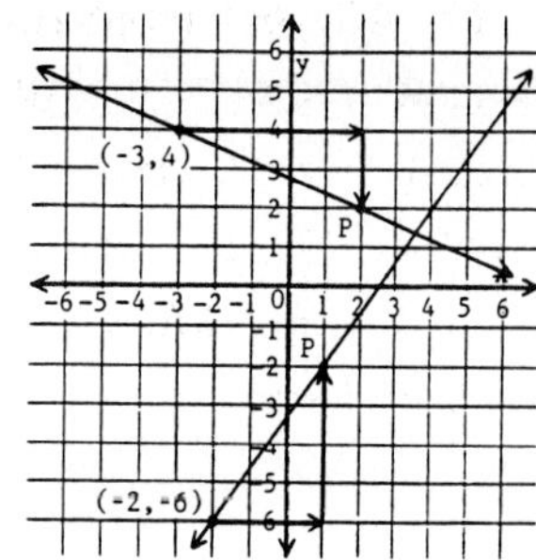

36-37.

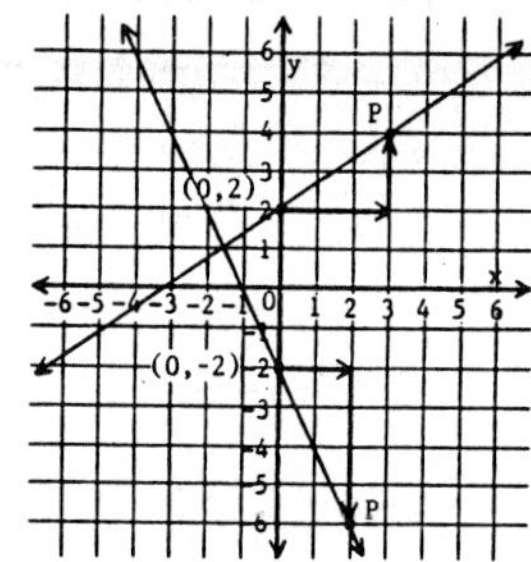

38. y = 3x - 6   39. y = -x + 2   40. $y = \frac{1}{3}x + 5$ 41. y = -4x + 6   42. $y = -\frac{1}{2}x + 3$ 43. y = 5x   44. $y = \frac{3}{4}x - \frac{5}{4}$ 45. y = -3x + 1   46. $y = -\frac{1}{2}x + 5$ 47. parallel   48. perpendicular   49. perpendicular   50. parallel   51. parallel   52. perpendicular   53. y = 3x + 2   54. y = -x - 4   55. $y = \frac{1}{2}x + 4$ 56. $y = \frac{1}{2}x - 1$ 57. y = 4x - 3   58. $y = -\frac{4}{3}x + 6$

Assignment 26

1.

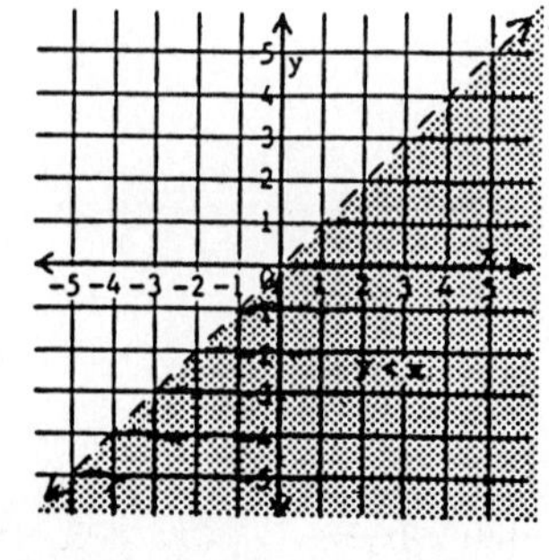

2.

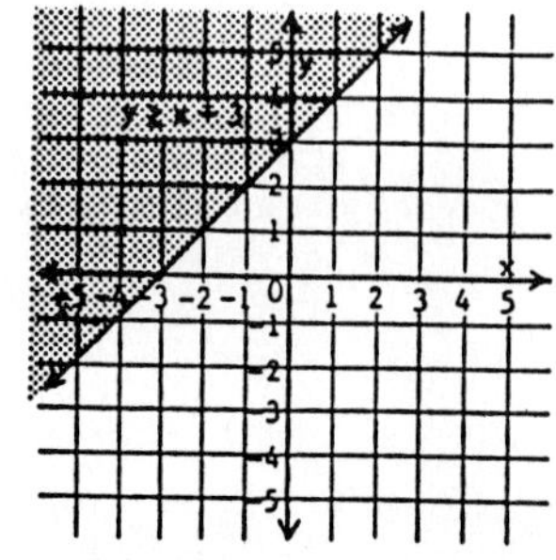

3.

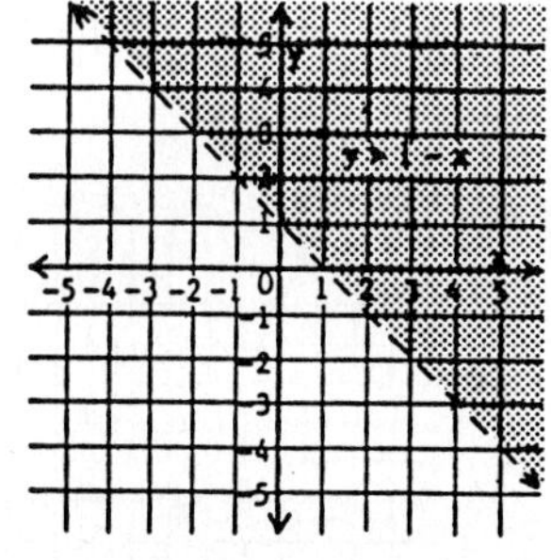

4.

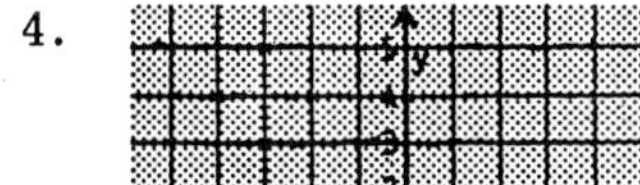

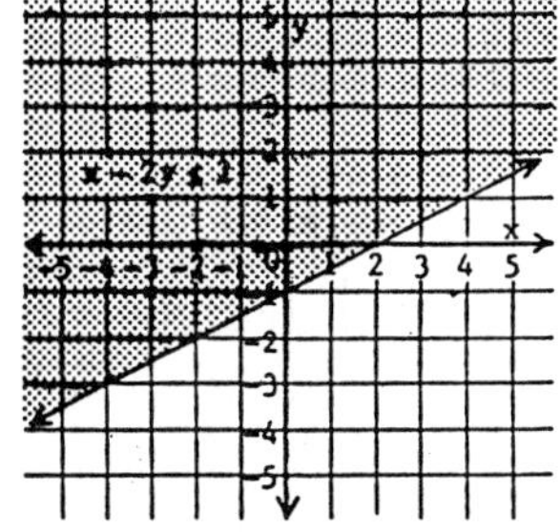

5.

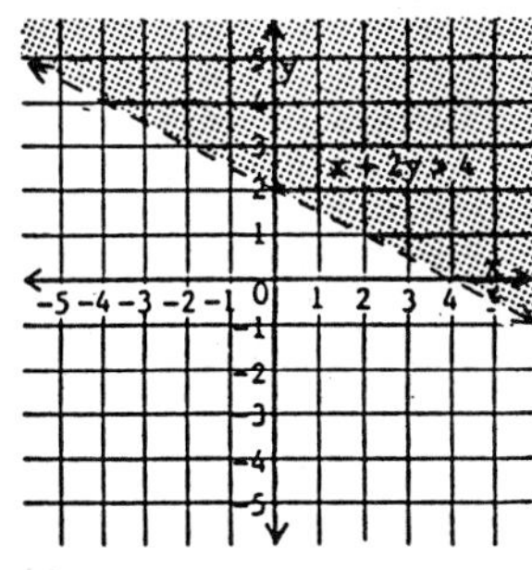

6.

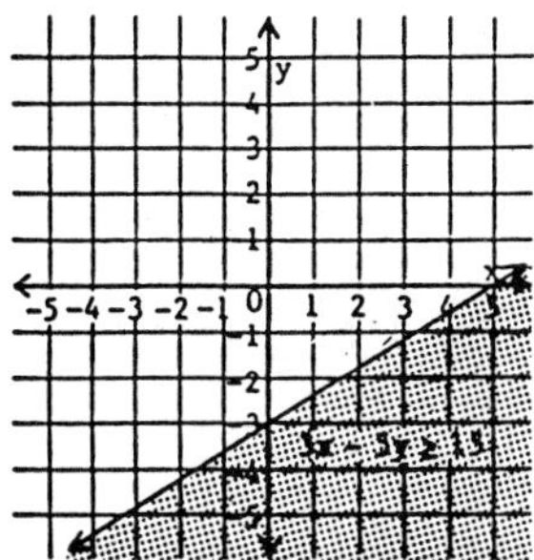

7.

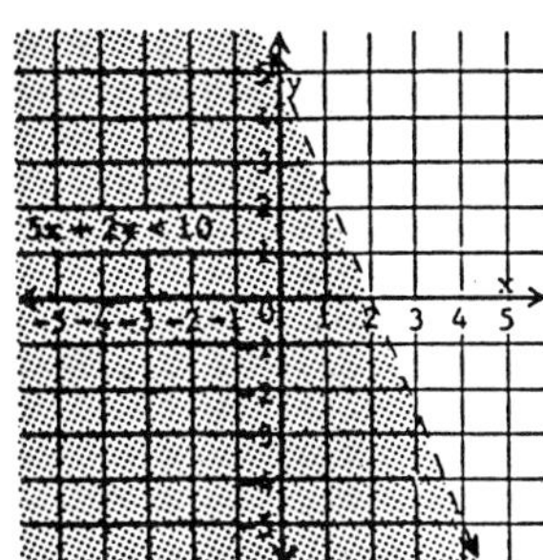

8.

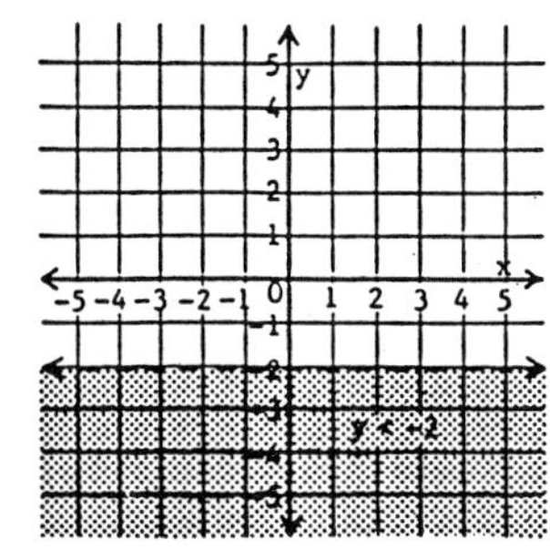

9.

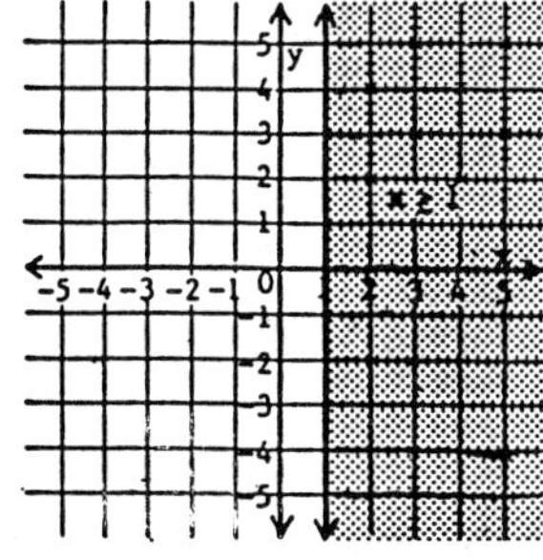

10. All real numbers except x = 0 11. All real numbers 12. All real numbers except x = 5
13. All real numbers except x = 2 14. All real numbers except x < 0 15. All real numbers
16. All real numbers except x < 9 17. All real numbers except x < 3 18. (a) and (d) 19. 13
20. 7 21. 4 22. 5 23. 3 24. 24 25. 5 26. 5 27. 21 28. 4 29. 2b + 7 30. 2a + 1
31. 6a - 16 32. $a^2 + 4a - 5$ 33. $2a + 2b + a^2 + 2ab + b^2$ 34. $a^2 - 2ab + b^2 - 3a + 3b$
35. 6x + 3h 36. 2x + h - 2 37. 400 38. 7π 39. 144 40. 25π

Assignment 27

1. y = 105 2. P = 750 3. d = 595 kilometers 4. I = 9 amperes 5. y = 10 6. p = 5
7. V = 20 liters 8. t = 32.5 min 9. y = 135 10. G = 600 11. d = 400 ft 12. d = 245 ft
13. y = 16 14. S = 200 15. I = 3 microwatts 16. F = 10 dynes 17. y = 500 18. b = 48
19. I = $465 20. V = 220 cm^3

CHAPTER 8 - SYSTEMS OF EQUATIONS

Assignment 28

1. (3,1)

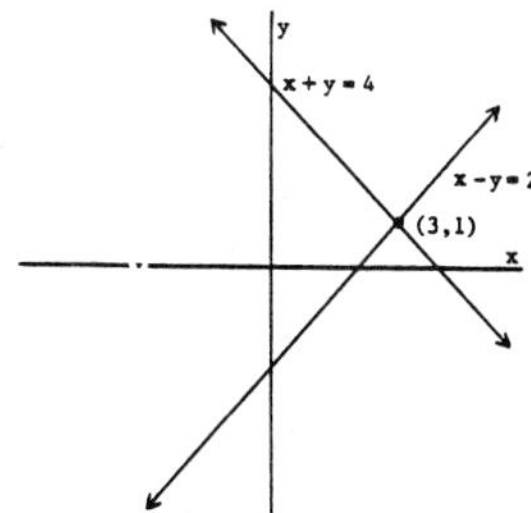

2. (-1,2)

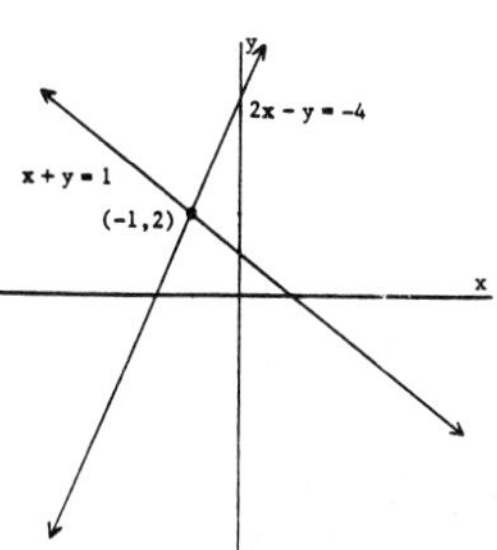

3. (3,-2) 4. (20,5) 5. (6,5) 6. (-1,4) 7. (3,-1) 8. (2,4) 9. (5,1) 10. (-2,-5)
11. (5,3) 12. (8,-2) 13. (4,1) 14. $(2,\frac{1}{2})$ 15. (6,2) 16. (12,3) 17. (1,-3)
18. (-1,-5) 19. (10,4) 20. (6,8) 21. (20,10) 22. $(2,-\frac{2}{3})$ 23. (4,-2) 24. (-3,2)
25. $(10,\frac{5}{2})$ 26. (1,-3) 27. (4,6) 28. (7,30) 29. (20,5) 30. $(\frac{3}{4},5)$

Assignment 29

1. 73 and 141 2. 56 and 14 3. L = 18 ft, W = 12 ft 4. L = 14 cm, W = 7 cm 5. \$4,000 at 7%, \$3,500 at 8% 6. \$2,500 at 6%, \$4,500 at 8% 7. speed of wind is 34 mph, speed in still air is 450 mph 8. speed of current is $1\frac{1}{2}$ mph, speed in still water is $4\frac{1}{2}$ mph 9. 22 white, 28 blue 10. flat rate is \$3, amount per mile is \$1.25 11. 75 milliliters of 10% solution, 125 milliliters of 50% solution 12. 15 lbs of the \$4.80 candy, 25 lbs of the \$2.40 candy 13. (3,-1,2) 14. (4,-2,-3) 15. (-2,5,1) 16. (3,4,-2) 17. (1,-1,2) 18. (5,7,3) 19. first is -1, second is 3, third is 5 20. first is 6, second is -2, third is -1 21. A = 60°, B = 20°, C = 100° 22. A = 45°, B = 80°, C = 55° 23. first is 90, second is 85, third is 93 24. A = 205 blocks, B = 220 blocks, C = 225 blocks

Assignment 30

1. 18 2. -5 3. -6 4. 22 5. -5 6. 20 7. 54 8. -3 9. -18 10. -22 11. 93 12. -7 13. -2 14. 8 15. (5,-2) from: D = -7, D_x = -35, D_y = 14 16. (-2,-6) from: D = -1, D_x = 2, D_y = 6 17. (6,-3) from: D = -31, D_x = -186, D_y = 93 18. (1.5,2.5) from: D = 94, D_a = 141, D_b = 235 19. (-1,3,1) from: D = 3, D_x = -3, D_y = 9, D_z = 3 20. (4,2,0) from: D = -12, D_x = -48, D_y = -24, D_z = 0 21. (1,-2,5) from: D = 12, D_x = 12, D_y = -24, D_z = 60 22. (3,-5,2) from: D = 9, D_x = 27, D_y = -45, D_z = 18 23. (9,2,1) from: D = -3, D_x = -27, D_y = -6, D_z = -3 24. (-2,1,-1) from: D = 1, D_p = -2, D_q = 1, D_r = -1

CHAPTER 9 - SECOND-DEGREE EQUATIONS AND THEIR GRAPHS

Assignment 31

1. upwards 2. downwards 3. downwards 4. upwards 5. upwards 6. downwards 7. downwards 8. upwards 9. (0,0), x = 0 10. (0,0), x = 0 11. (2,0), x = 2 12. (0,-2), x = 0 13. (-1,0), x = -1 14. (0,1), x = 0 15. (4,0), x = 4 16. (0,-4), x = 0 17. (-5,-2), x = -5 18. (4,3), x = 4 19. (2,-1), x = 2 20. (-1,-6), x = -1 21. (7,-5), x = 7 22. (-3,2), x = -3

23.
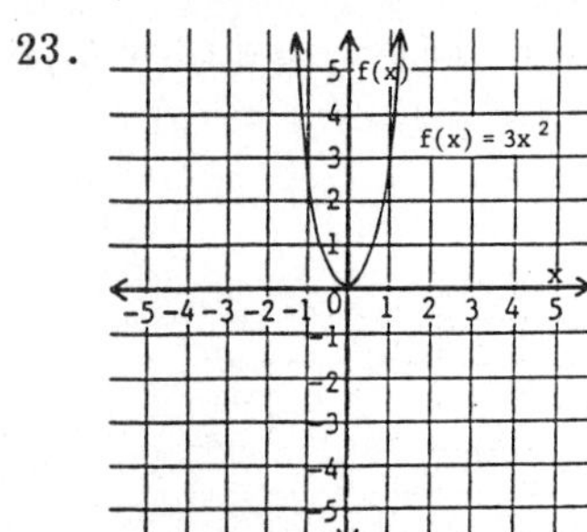

24.
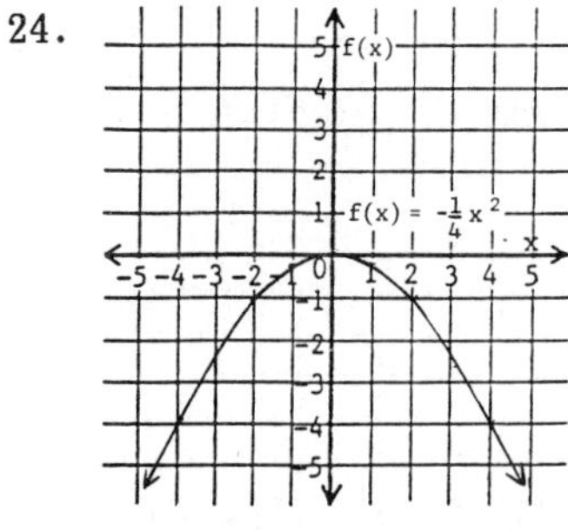

25.
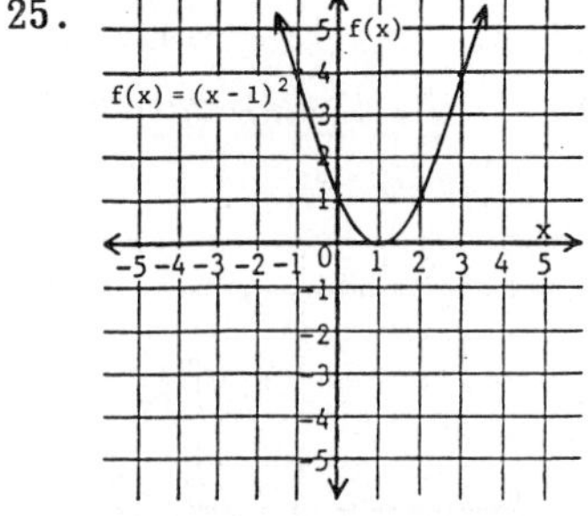

26.
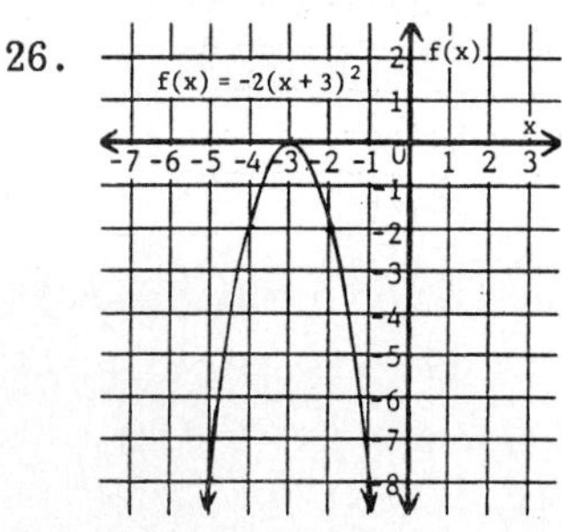

27.
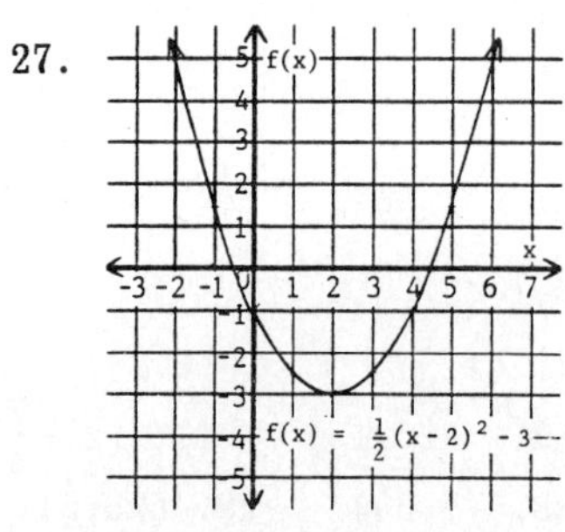

28.
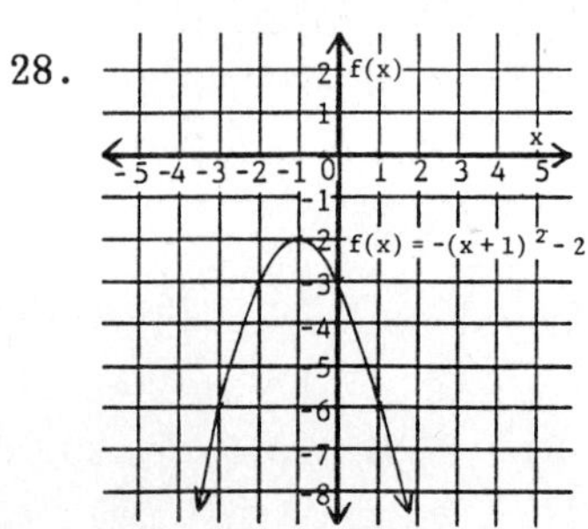

29. 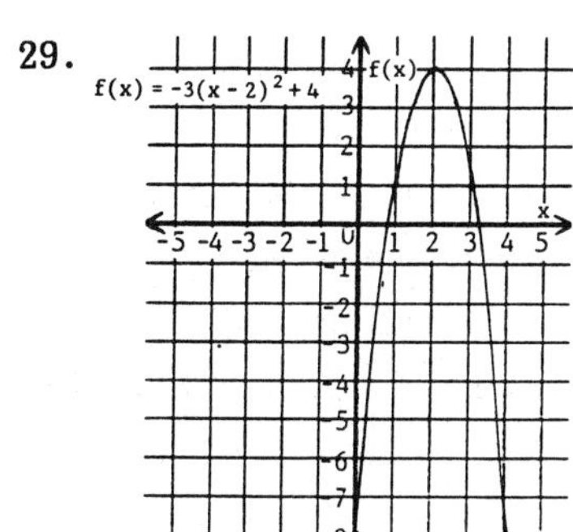

30. 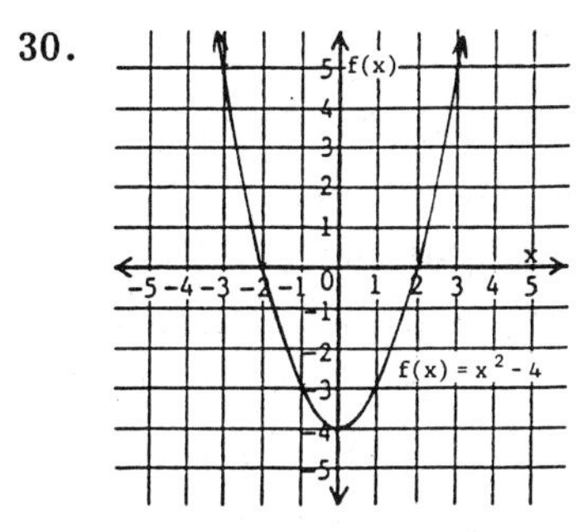

31. 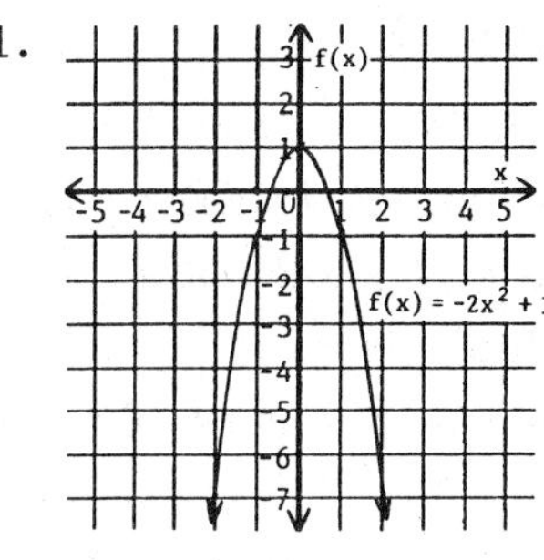

32. $f(x) = (x + 1)^2 - 4$ 33. $f(x) = (x - 3)^2 + 5$ 34. $f(x) = 2(x - 5)^2 - 3$

35. $f(x) = 3(x + 2)^2 + 10$ 36. $f(x) = -2(x + 3)^2 - 1$ 37. $f(x) = -3(x - 2)^2 + 4$

38. 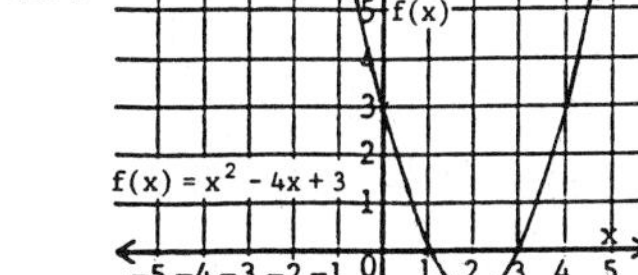

39. 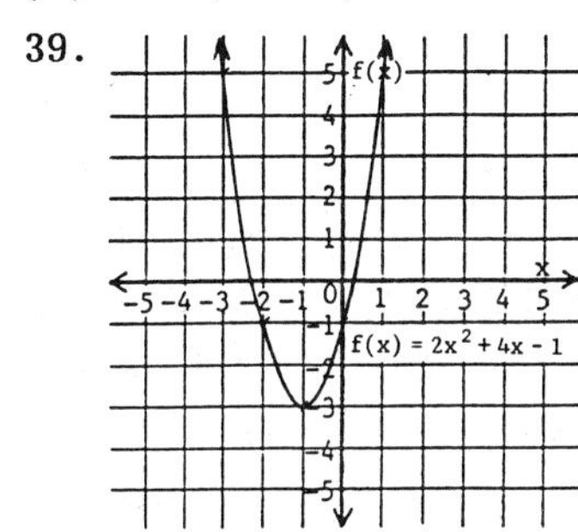

40. 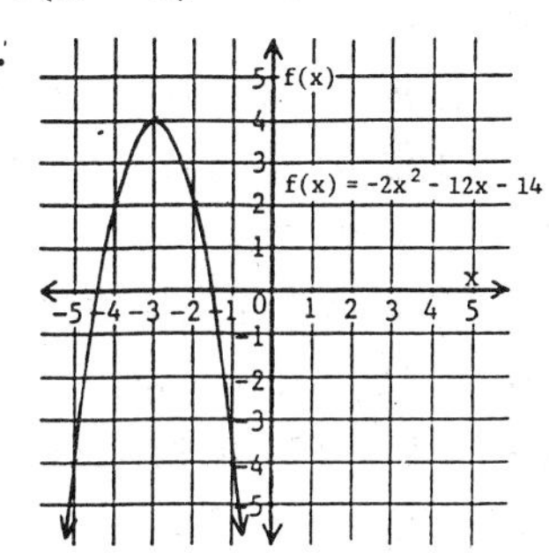

Assignment 32

1. (4,0) and (-3,0) 2. (2,0) and (-4,0) 3. (0,0) and (3,0) 4. (-1,0) and (0,0) 5. $(1 - \sqrt{6},0)$ and $(1 + \sqrt{6},0)$ or (3.449,0) and (-1.449,0) 6. $(-2 - \sqrt{2},0)$ and $(-2 + \sqrt{2},0)$ or (-3.414,0) and (-0.586,0) 7. (-3,0) 8. $(\frac{2}{3},0)$ 9. None 10. None 11. (2,0) and (4,0) 12. None 13. (4,0) 14. (-1,0) 15. (3,0) 16. (-3,0) and (3,0) 17. None 18. (0,0) 19. None 20. Minimum: -2 21. Maximum: 5 22. Maximum: -3 23. Minimum: 6 24. Minimum: 0 25. Maximum: 0 26. Minimum: 1 27. Minimum: 0 28. Maximum: -6 29. Minimum: -1 30. Maximum: 7 31. Maximum: -4 32. Maximum area = 144 ft², L = 12 ft and W = 12 ft 33. Numbers are 8 and 8; maximum product is 64 34. 64 ft in 2 sec 35. 1600 ft in 10 sec

36. 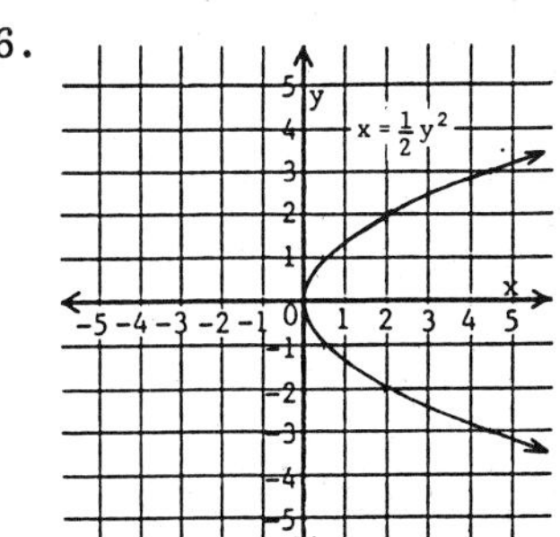

37. 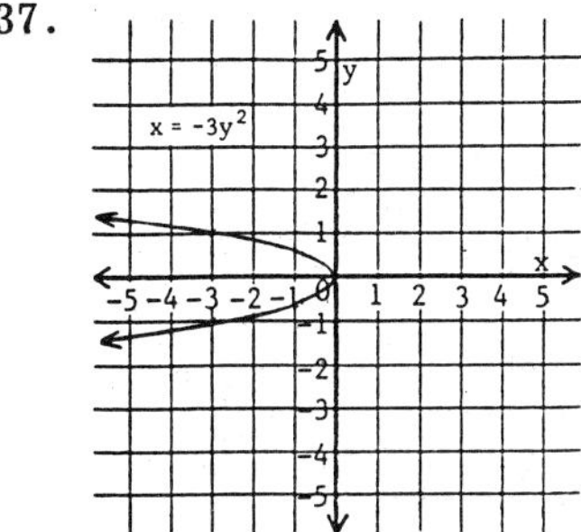

38. 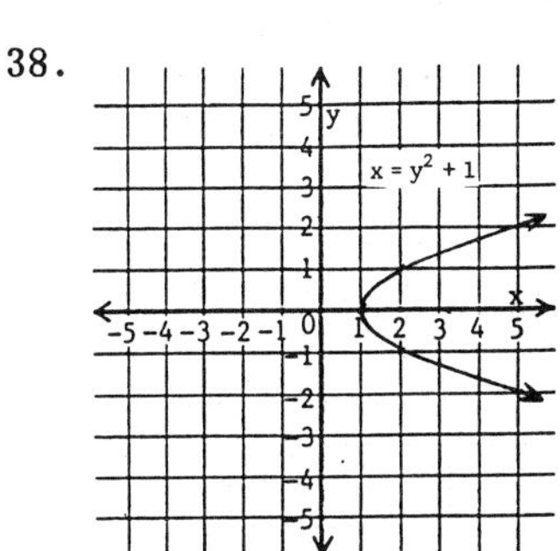

39.

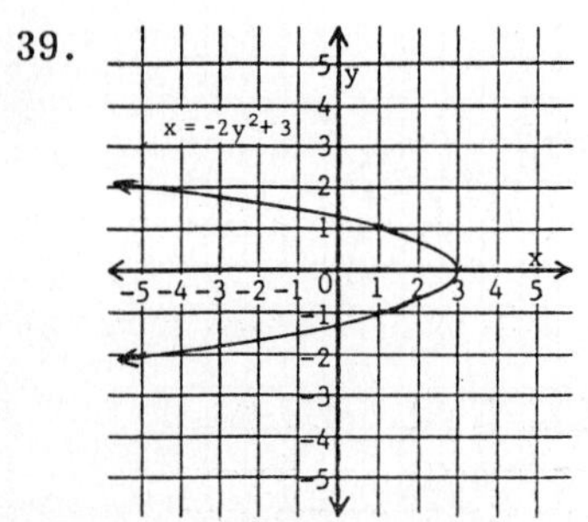

40.

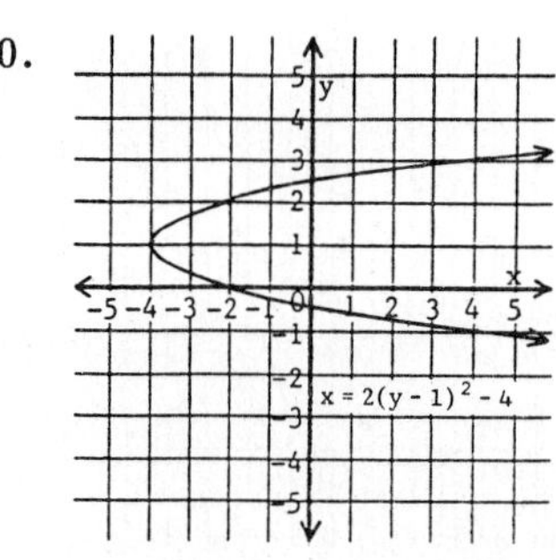

41.

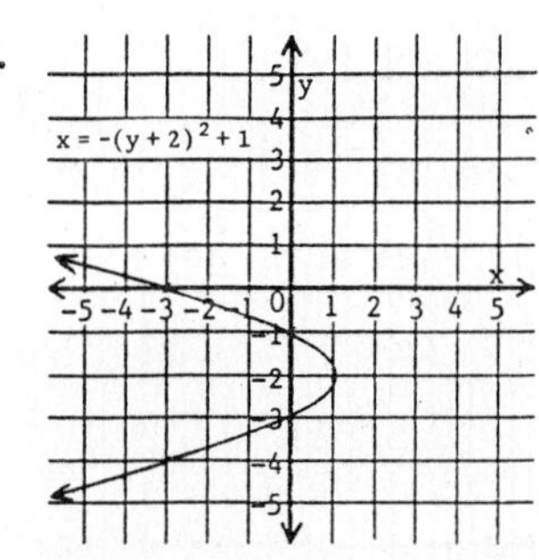

Assignment 33

1. 6 2. 9 3. 5 4. 10 5. $\sqrt{65} = 8.062$ 6. $\sqrt{85} = 9.22$ 7. 4 8. $\sqrt{20} = 4.472$
9. $\sqrt{13} = 3.606$ 10. 1 11. $\sqrt{82} = 9.055$ 12. $\sqrt{50} = 7.071$ 13. (4,2), r = 3 14. (1,-6), r = 7
15. (-5,3), r = 8 16. (0,0), r = 10 17. (-3,-5), r = 9 18. (0,0), r = 1

19.

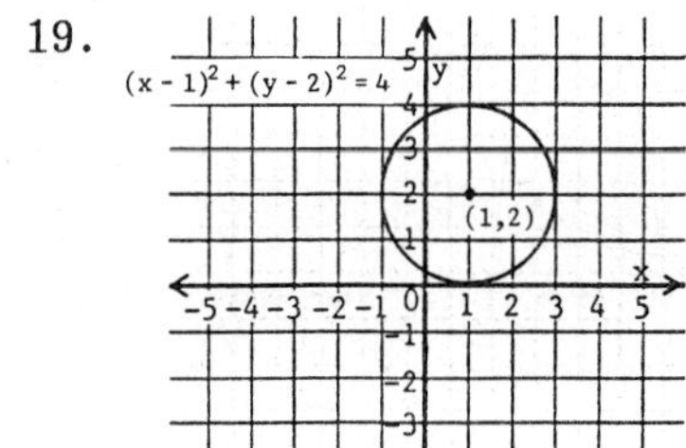

20.

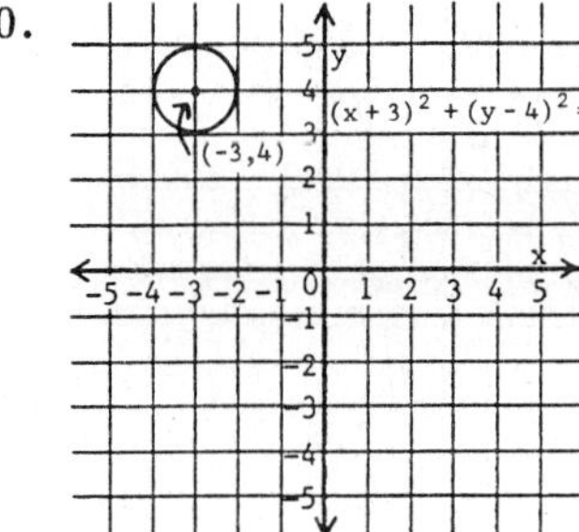

21.

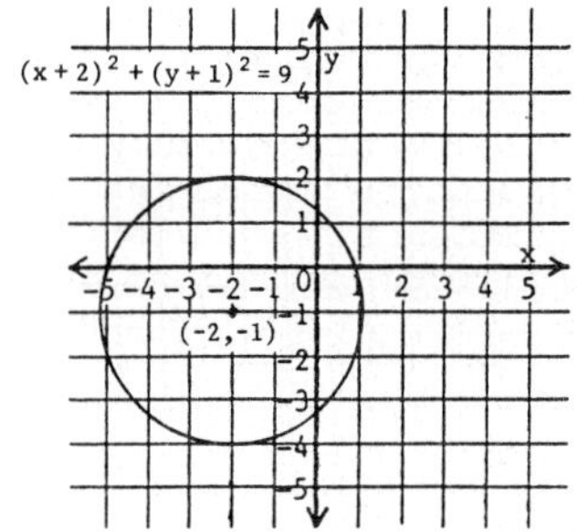

22.

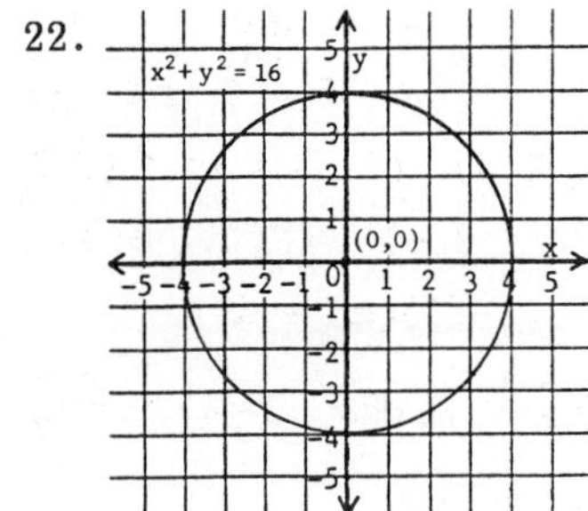

23.

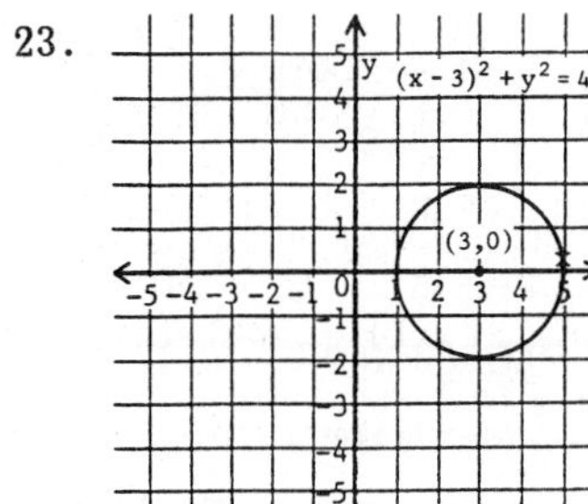

24.

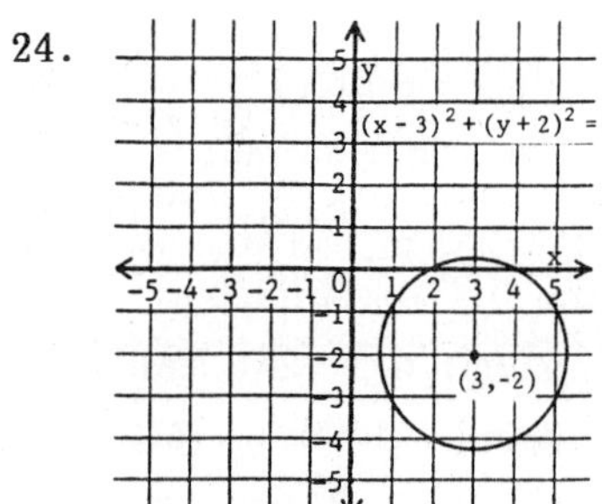

25. $x^2 + y^2 = 36$ 26. $(x - 4)^2 + (y - 3)^2 = 25$ 27. $(x + 2)^2 + (y - 1)^2 = 9$
28. $(x - 3)^2 + (y + 5)^2 = 64$ 29. $(x + 3)^2 + (y + 1)^2 = 10$ 30. $x^2 + y^2 = 17$ 31. (1,4), r = 6
32. (3,2), r = 5 33. (-4,5), r = 1 34. (-3,-6), r = 4 35. (5,-1), r = $\sqrt{7}$ 36. (0,-8), r = 10
37. (3,0) and (-3,0), (0,4) and (0,-4) 38. (6,0) and (-6,0), (0,8) and (0,-8)
39. (5,0) and (-5,0), (0,10) and (0,-10)

40.

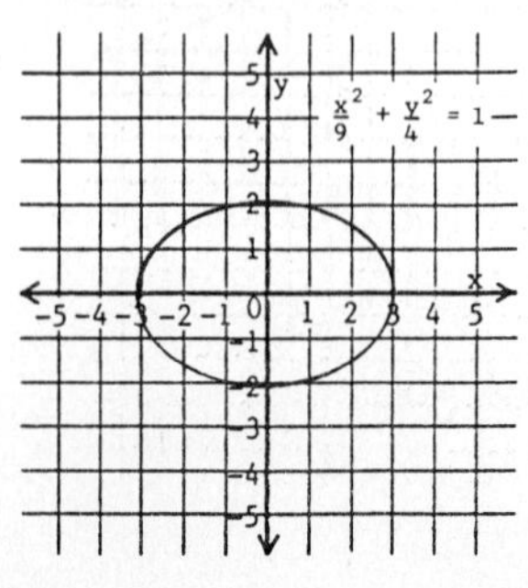

41.

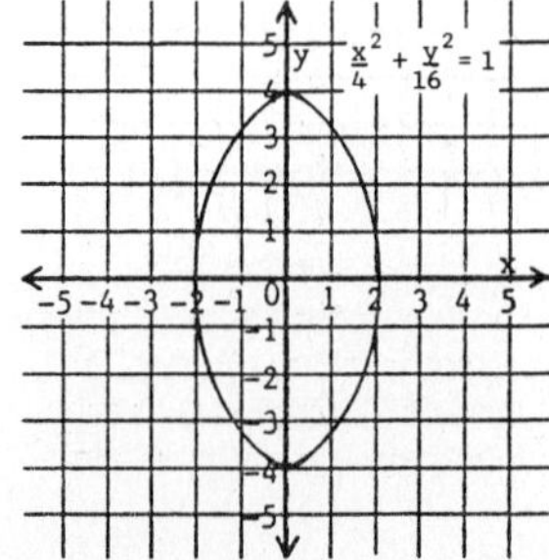

42.

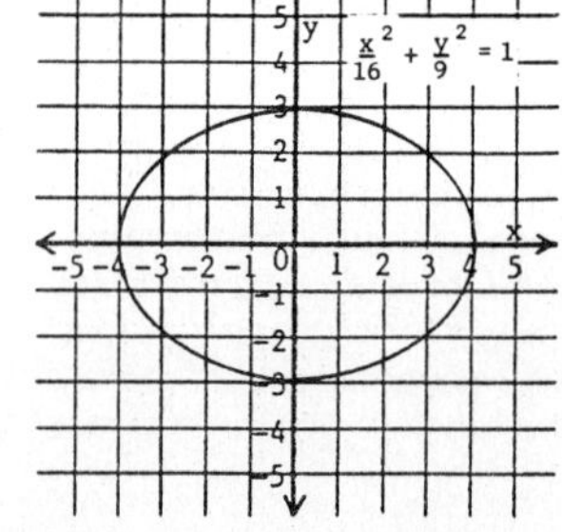

Assignment 34

1. (5,0) and (-5,0) 2. (0,2) and (0,-2) 3. (0,6) and (0,-6) 4. (1,0) and (-1,0)

5. $y = \frac{5}{6}x$ and $y = -\frac{5}{6}x$ 6. $y = \frac{1}{3}x$ and $y = -\frac{1}{3}x$ 7. $y = 2x$ and $y = -2x$ 8. $x = 0$ and $y = 0$

9. $y = \frac{5}{4}x$ and $y = -\frac{5}{4}x$ 10. $y = x$ and $y = -x$ 11. $x = 0$ and $y = 0$ 12. $y = \frac{2}{3}x$ and $y = -\frac{2}{3}x$

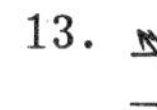

13. 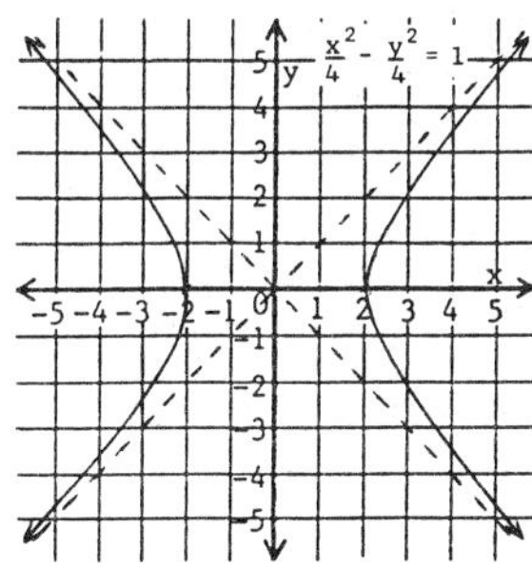

14. 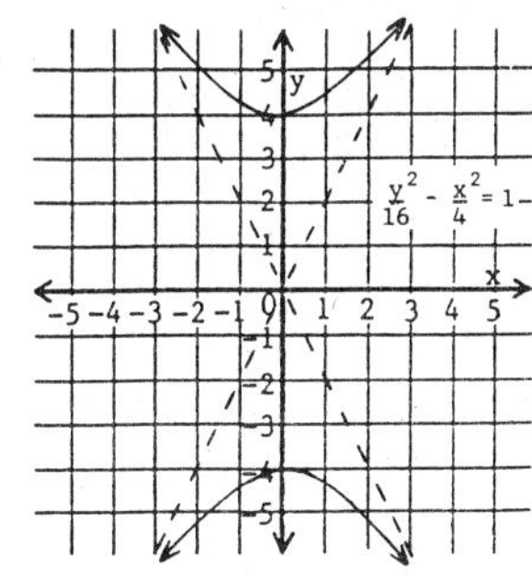

15. 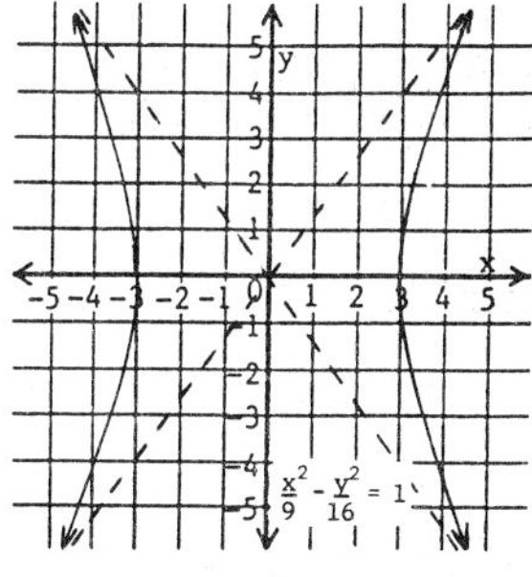

16. 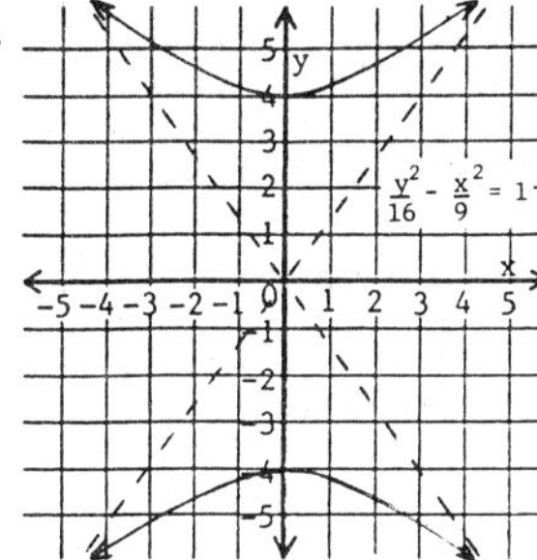

17. 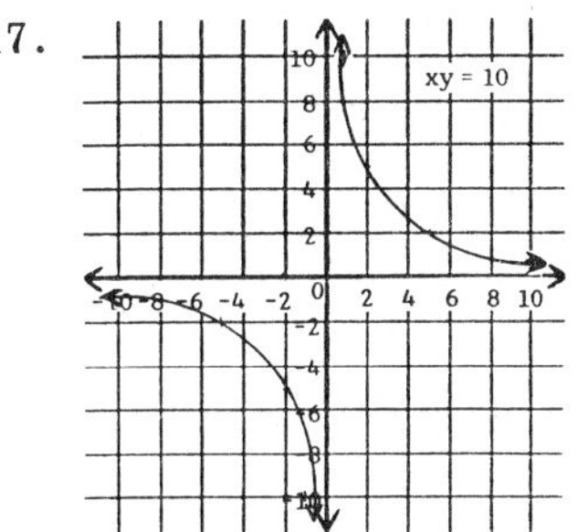

18. 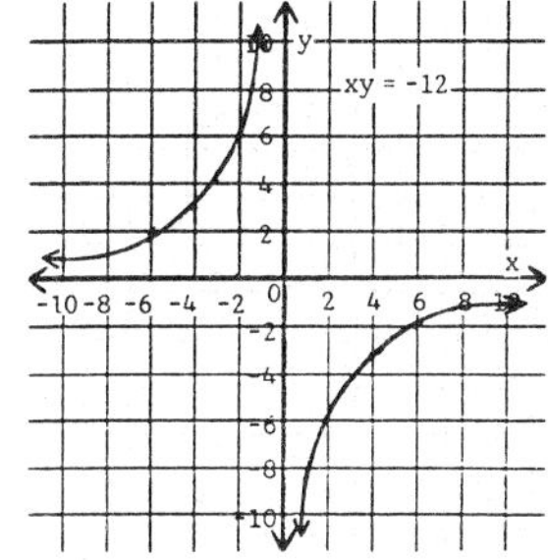

19. (2,4) and (-4,-2) 20. (6,5) and (-5,-6) 21. (-2,0) and (0,-3) 22. (4,0) and (0,-2) 23. (1,2) 24. (-1,3) and (2,12) 25. (-1,-1) and (2,-4) 26. (2,-8) 27. (-5,-2) and (10,1) 28. (-4,-4) and (8,2) 29. (2,-3) and (-3,2) 30. (-2,-3), (-2,3), (2,-3), and (2,3) 31. (-5,0) and (5,0) 32. (-1,-3), (-1,3), (1,-3) and (1,3) 33. (3,2), (-3,-2), (2,3), and (-2,-3) 34. (1,5), (-1,-5), (5,1), and (-5,-1) 35. $f^{-1}(x) = x - 5$ 36. $f^{-1}(x) = x + 2$

37. $f^{-1}(x) = x - 20$ 38. $f^{-1}(x) = \frac{x}{4}$ 39. $f^{-1}(x) = -\frac{x}{7}$ 40. $f^{-1}(x) = \frac{x - 1}{5}$ 41. $f^{-1}(x) = \frac{x - 3}{2}$

42. $f^{-1}(x) = \frac{x + 7}{9}$ 43. $f^{-1}(x) = \sqrt{x}$ 44. $f^{-1}(x) = \sqrt{x - 3}$ 45. $f^{-1}(x) = \sqrt{x + 2}$

46. $f^{-1}(x) = \sqrt{x - 10}$

CHAPTER 10 - EXPONENTIAL AND LOGARITHMIC FUNCTIONS

Assignment 35

1. 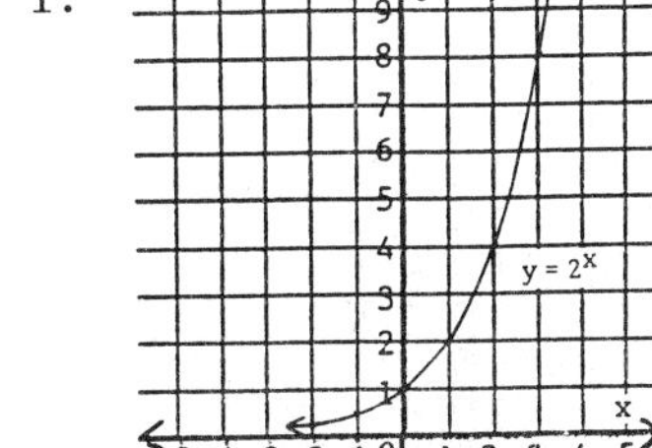

2. 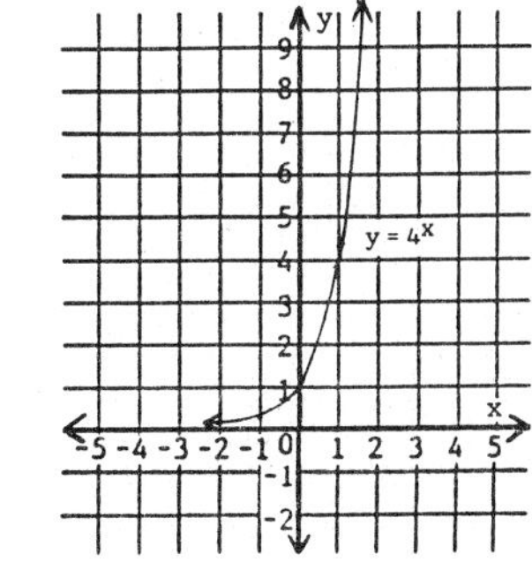

3. 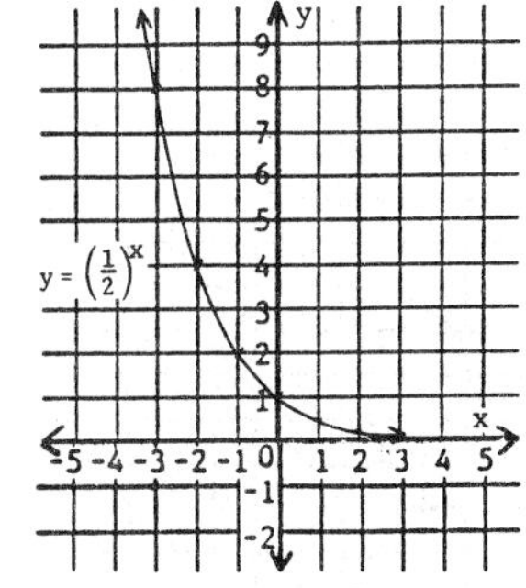

4.
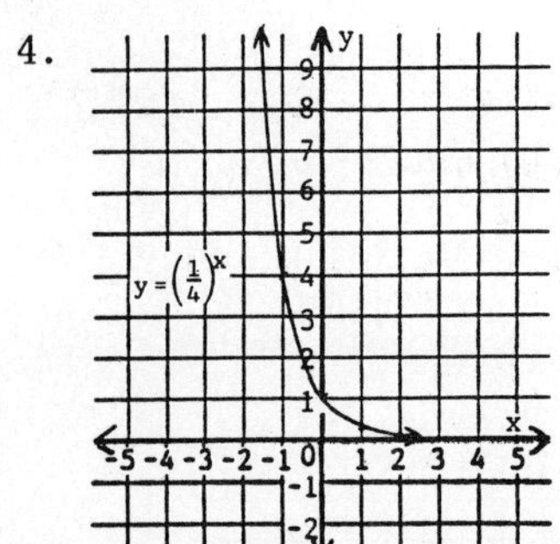

5.
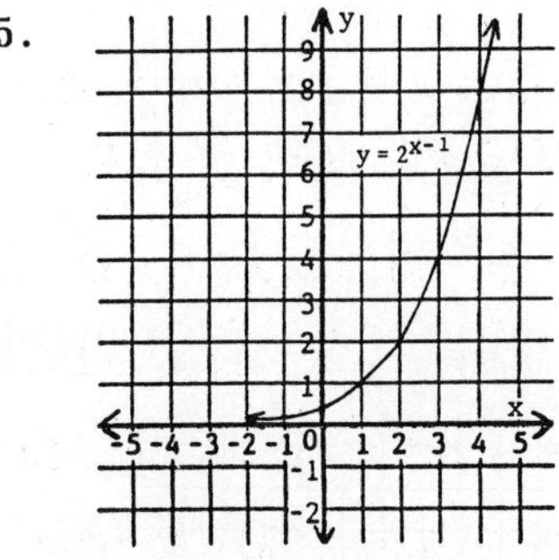

6.
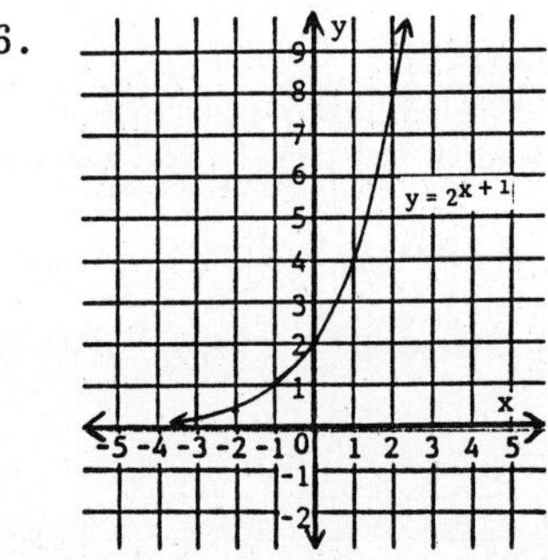

7.
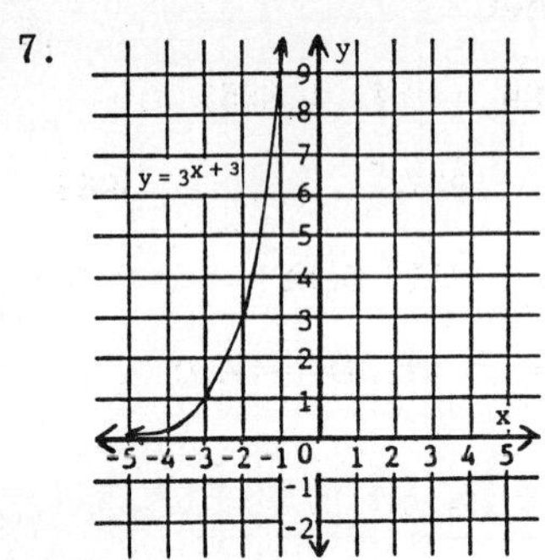

8.
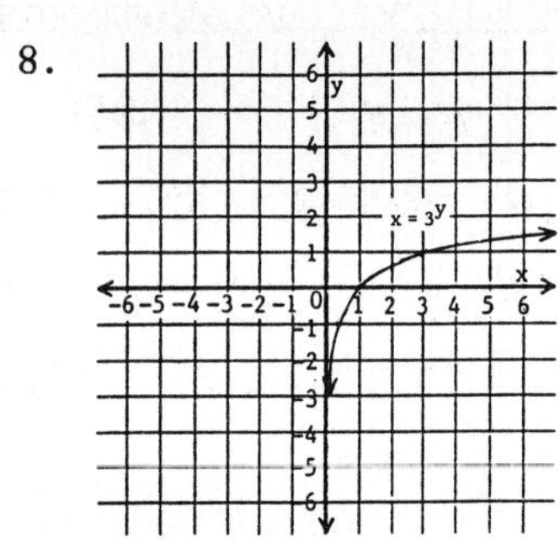

9.
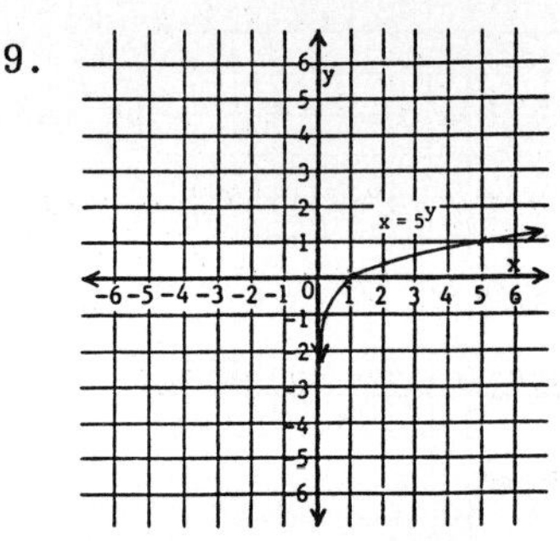

10. $\log_3 9 = 2$ 11. $\log_5 125 = 3$ 12. $\log_2 32 = 5$ 13. $\log_{10} 0.001 = -3$ 14. $\log_x 64 = 3$

15. $\log_6 36 = x$ 16. $\log_3 x = 5$ 17. $\log_{10} 0.1 = x$ 18. $7^2 = 49$ 19. $3^4 = 81$

20. $6^3 = 216$ 21. $10^{-2} = 0.01$ 22. $5^4 = x$ 23. $x^3 = 27$ 24. $2^x = 64$ 25. $x^2 = 100$

26. $x = 81$ 27. $x = 81$ 28. $x = \frac{1}{4}$ 29. $x = 2$ 30. $x = 5$ 31. $x = 10$ 32. $x = 3$

33. $x = -1$ 34. 4 35. 3 36. 2 37. 3 38. -1 39. -2 40. -2 41. -3

42. 1 43. 0 44. 1 45. 0

46.
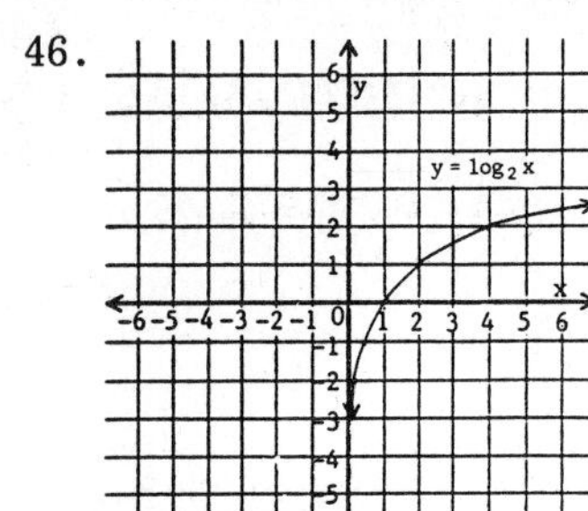

47.
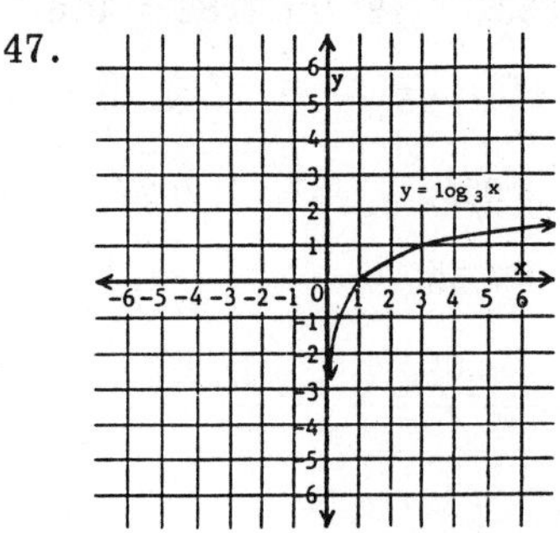

48.
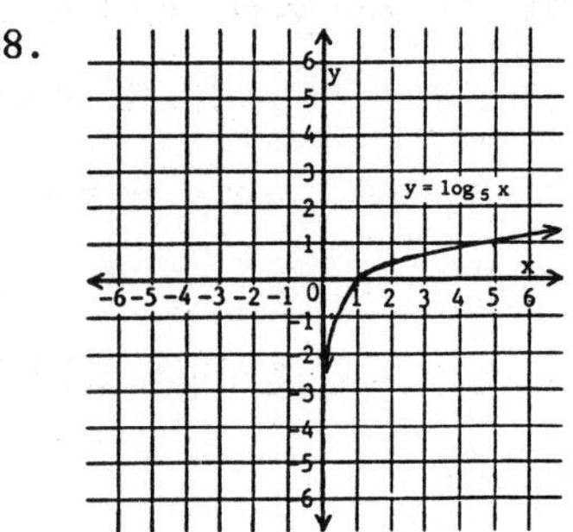

Assignment 36

1. $\log_2 4 + \log_2 8$ 2. $\log_5 10 + \log_5 x$ 3. Not possible 4. $\log_3 27 - \log_3 9$ 5. $\log_7 a - \log_7 b$

6. Not possible 7. $2 \log_6 x$ 8. $t \log_b 2$ 9. $\log_3 (9)(81)$ 10. $\log_6 9x$ 11. $\log_2\left(\frac{32}{8}\right)$

12. $\log_b\left(\frac{R}{S}\right)$ 13. $\log_3 5^4$ 14. $\log_b y^x$ 15. $\log_b x + 2 \log_b y$ 16. $\log_b 3 + 3 \log_b y + 4 \log_b z$

17. $\log_b 5 + 5 \log_b x + \log_b z$ 18. $2 \log_b x + \log_b y - \log_b z$ 19. $\log_b 2 + 4 \log_b y - 6 \log_b z$

20. $\frac{1}{2}\log_b x - 2\log_b y - \log_b z$ 21. $2\log_b x - 3\log_b y - \frac{1}{2}\log_b z$ 22. $\frac{1}{2}(4\log_b x - \log_b y - \log_b z)$

23. $\log_b x^2yz^3$ 24. $\log_b ac^3d^5$ 25. $\log_b\left(\frac{x^4z}{y^2}\right)$ 26. $\log_b\left(\frac{p}{qt^3}\right)$ 27. $\log_b\left(\frac{x^3\sqrt{y}}{z}\right)$

28. $\log_b\left(\frac{a^5}{c\sqrt{d}}\right)$ 29. 28 30. 65,600 31. 70,000,000 32. 9,400,000,000 33. 0.41

34. 0.00114 35. 0.000002 36. 0.000000088 37. 2.77×10^1 38. 5.19×10^2

39. 4.96×10^4 40. 6×10^6 41. 4.25×10^8 42. 3.14×10^{-1} 43. 7.5×10^{-2}

44. 4×10^{-6} 45. 6.39×10^{-5} 46. 1.1×10^{-7} 47. 8×10^9 48. 3.57×10^7

49. 4×10^{-5} 50. 4.1×10^{-6} 51. 6.02×10^{23} 52. 0.000000782 centimeter

53. 1.1574×10^{-5} day 54. 45,000,000,000 years 55. 0.8451 56. 0.2989 57. 1.8351

58. 2.7024 59. 4.5705 60. 7.6532 61. -0.1549 62. -0.2596 63. -1.8894 64. -3.0969

65. -2.2104 66. -5.1487 67. 7.50 68. 40.1 69. 105 70. 172,000 71. 0.581

72. 0.0376 73. 0.00779 74. 0.000140 75. pH = 2.30 76. pH = 2.15 77. D = 54.0

78. D = 58.1

Assignment 37

1. x = 7 2. x = 3 3. x = 4 4. x = 4 5. $x = \frac{2}{3}$ 6. x = 2 7. $x = -\frac{1}{2}$ 8. $x = -\frac{3}{2}$

9. x = 1.46 10. x = 1.13 11. x = 1.06 12. x = 0.93 13. x = 0.15 14. x = 3.85

15. x = 1.90 16. x = 1.46 17. 9.0 years 18. 14.3 years 19. x = 11 20. x = 5

21. x = 7 22. x = 3 23. x = 5 24. x = 2 25. x = 5 26. x = 4 (not x = -3)

27. x = 3 (not x = -5) 28. x = 4 29. x = 150 30. x = 20 31. $x = \frac{1}{3}$ 32. x = 5 (not x = -2) 33. $x = \frac{1}{2}$ 34. 2.01 35. 22.2 36. 330 37. 0.333 38. 0.0821

39. 0.00203 40. 2,217 ants 41. 29,633 42. ln 1.82 = 0.6 43. ln 134 = 4.9

44. ln 0.301 = -1.2 45. ln 0.055 = -2.9 46. $e^{2.4} = 11$ 47. $e^{6.3} = 545$ 48. $e^{-0.6} = 0.549$

49. $e^{-4.1} = 0.0166$ 50. -0.3567 51. 0.9555 52. 4.2627 53. 5.9915 54. D = 74.2

55. D = 358 56. Q = -1.76 57. Q = -3.56

CHAPTER 11 - SEQUENCES AND SERIES

Assignment 38

1. 21, 25, 29, ... 2. 31, 38, 45, ... 3. 2, -1, -4, ... 4. $\frac{3}{2}, 1, \frac{1}{2}, \ldots$ 5. 4, 5, 6, 7

6. 0, 3, 8, 15 7. 3, 10, 29, 66 8. $\frac{3}{4}, 1, \frac{5}{4}, \frac{3}{2}$ 9. $\frac{2}{3}, \frac{4}{3}, 2, \frac{8}{3}$ 10. 1, 3, 6, 10

11. $a_{10} = 29$ 12. $a_{10} = 2$ 13. $a_{10} = \frac{101}{100}$ 14. $S_4 = 16$ 15. $S_4 = 52$ 16. $S_4 = 100$

17. $S_3 = \frac{11}{12}$ 18. $S_3 = \frac{49}{36}$ 19. $S_3 = \frac{133}{60}$ 20. 3, 8, 13, 18, 23 21. 15, 11, 7, 3, -1

22. 0, -5, -10, -15, -20 23. Not an AP 24. An AP with d = -3 25. Not an AP

26. $a_7 = 34$ 27. $a_{11} = 85$ 28. $a_{41} = -20$ 29. $a_{61} = -140$ 30. $a_{80} = 183$ 31. $a_{100} = 107$

32. $S_{40} = 820$ 33. $S_{70} = 2{,}485$ 34. $S_{10} = 185$ 35. $S_{10} = -200$ 36. $S_{30} = 930$

37. $S_{80} = 3{,}280$ 38. $S_{50} = 2{,}500$ 39. $S_{90} = 8{,}100$

Assignment 39

1. 5, 10, 20, 40, 80 2. 10, -30, 90, -270, 810 3. 24, 12, 6, 3, $\frac{3}{2}$ 4. 36, -24, 16, $-\frac{32}{3}$, $\frac{64}{9}$

5. A GP with $r = 4$ 6. Not a GP 7. A GP with $r = -\frac{1}{2}$ 8. $a_6 = 320$ 9. $a_4 = 108$

10. $a_5 = 96$ 11. $a_4 = -128$ 12. $a_5 = \frac{16}{9}$ 13. $a_4 = -\frac{1}{8}$ 14. $S_5 = 155$ 15. $S_5 = 968$

16. $S_4 = 255$ 17. $S_3 = 430$ 18. $S_6 = 63$ 19. $S_3 = \frac{37}{8}$ 20. $S_\infty = 8$ 21. $S_\infty = 64$

22. $S_\infty = -31\frac{1}{4}$ 23. $S_\infty = 6\frac{3}{4}$ 24. $S_\infty = 16\frac{1}{5}$ 25. $S_\infty = 3\frac{1}{8}$ 26. 6 27. 720 28. 10

29. 5 30. 72 31. 210 32. 28 33. 120 34. 6 35. 252

36. $c^6 + 6c^5d + 15c^4d^2 + 20c^3d^3 + 15c^2d^4 + 6cd^5 + d^6$

37. $x^5 + 5x^4y + 10x^3y^2 + 10x^2y^3 + 5xy^4 + y^5$

38. $x^7 + 7x^6 + 21x^5 + 35x^4 + 35x^3 + 21x^2 + 7x + 1$

39. $y^5 + 10y^4 + 40y^3 + 80y^2 + 80y + 32$

40. $a^6 - 6a^5b + 15a^4b^2 - 20a^3b^3 + 15a^2b^4 - 6ab^5 + b^6$

41. $c^5 - 5c^4d + 10c^3d^2 - 10c^2d^3 + 5cd^4 - d^5$

42. $x^5 - 15x^4 + 90x^3 - 270x^2 + 405x - 243$

43. $1 - 6x + 15x^2 - 20x^3 + 15x^4 - 6x^5 + x^6$

Index